AF358788

Artificial Intelligence in Cancer, Biology and Oncology

Artificial Intelligence in Cancer, Biology and Oncology

Editors

Hamid Khayyam
Ali Hekmatnia
Rahele Kafieh

Basel • Beijing • Wuhan • Barcelona • Belgrade • Novi Sad • Cluj • Manchester

Editors
Hamid Khayyam
School of Engineering,
RMIT University
Melbourne
Australia

Ali Hekmatnia
Isfahan University of
Medical Sciences
Isfahan
Iran

Rahele Kafieh
Durham University
Durham
UK

Editorial Office
MDPI
St. Alban-Anlage 66
4052 Basel, Switzerland

This is a reprint of articles from the Topic published online in the open access journals *Cancers* (ISSN 2072-6694), *Onco* (ISSN 2673-7523), and *Current Oncology* (ISSN 1718-7729) (available at: https://www.mdpi.com/topics/R92PU1325L).

For citation purposes, cite each article independently as indicated on the article page online and as indicated below:

Lastname, A.A.; Lastname, B.B. Article Title. *Journal Name* **Year**, *Volume Number*, Page Range.

ISBN 978-3-7258-1111-3 (Hbk)
ISBN 978-3-7258-1112-0 (PDF)
doi.org/10.3390/books978-3-7258-1112-0

Contents

About the Editors

Hamid Khayyam

Hamid Khayyam obtained his B.Sc. degree (Hons.) from the University of Isfahan, followed by an M.Sc. degree from the Iran University of Science and Technology, and ultimately a Ph.D. in mechanical engineering, specializing in intelligent systems, from Deakin University in Australia. With over a decade of experience in automation and energy productivity across various industrial sectors, Dr. Khayyam has notably led initiatives in the modeling, control, and optimization of energy systems at the Carbon Nexus production line within Deakin University, leveraging artificial intelligence and machine learning technologies. Currently, Dr. Khayyam serves as a Senior Lecturer in the Department of Mechanical Engineering at RMIT University in Australia. His prolific contributions to the professional literature include over 600 journal articles as both editor and reviewer, over 100 published journal articles, two books (one as sole editor), and nine book chapters. Additionally, he holds positions on several Editorial Boards of ISI journals. Dr. Khayyam's research endeavors are dedicated to pioneering innovative approaches by integrating artificial intelligence and machine learning to address complex systems. He frequently serves as a keynote and invited speaker, academic committee member, and participant in international conferences, webinars, and workshops focusing on technology development, artificial intelligence, and machine learning. As a senior academic member, Dr. Khayyam is affiliated with the Intelligent Automation Research Group (IARG) at RMIT University in Australia and The Materials and Manufacturing Research Institute (MMRI) at The University of British Columbia in Canada. He also holds the esteemed title of Senior Member of IEEE.

Ali Hekmatnia

Ali Hekmatnia is a professor in radiology at Isfahan University of Medical Sciences, Iran. He received his medical degree from Shahid Beheshti University, Tehran, in 1989, and held an internship and radiology residency at Isfahan University of Medical Sciences, Iran, in 1995. He has held two visiting fellowships in pediatric radiology, one at Great Ormond Street Hospital for Sick Children, London, UK, in 2001, and a visiting fellowship in neuroradiology at the National Hospital for Neurology and Neurosurgery, London, UK, in 2001 too. He received an Msc degree in medical education from Isfahan University of Medical Sciences, Iran, in 2007. Dr Hekmatnia has contributed more than 120 peer-reviewed articles to professional journals. His research interests include CT scan, MRI, pediatric radiology, and neuroradiology.

Rahele Kafieh

Rahele Kafieh received a BSc (2005) followed by an MSc (2008) and a Ph.D. (2014) in Biomedical Engineering from Sahand University of Technology and Isfahan University of Medical Sciences in Iran. She obtained her first tenure post as an assistant professor in the Department of Advanced Technologies in Medicine at Isfahan University of Medical Sciences in 2014, and remained there for 7 years, during which she obtained valuable experience in teaching and supervision of projects at the undergraduate and graduate levels. She has led many successful previously unexplored projects on medical images, including but not limited to the classification and segmentation of retinal images with AI; applications of AI in CT-scan and X-ray images; and multi-modality MRI data analysis. Her two stays at Charité university hospital, Germany, and one stay at Sabanci University in Turkey were funded by competitive research scholarships from the Einstein forum and TUBITAK, respectively. Then she moved to Newcastle University, where she worked as a Research Associate in an interdisciplinary team, a position that provided her with the great opportunity to conduct in-depth research on the role of artificial intelligence (AI) in the detection of neurological diseases of the eye. Since July 2022, she has been working as an Assistant Professor in the Department of Engineering at Durham University. Her current research is on medical image processing in different organs and parts of the body (eye, chest, teeth, brain, heart, and breast), from different modalities (OCT, Fundoscopy, CT-scan, X-ray, cone-beam CT, MRI, fMRI, and infrared), facing challenges like high dimensionality, noisiness, and imbalance.

Preface

Cancer stands as the second leading cause of death globally. In 2020, approximately 10 million individuals succumbed to this disease, as reported by the World Health Organization (WHO). Early cancer identification remains paramount for successful treatment and halting metastasis. However, the complexity within and between tumors complicates this task and the implementation of effective therapies. Unveiling the detection, diagnosis, and treatment of cancer reveals a concealed structure amidst seemingly chaotic medical occurrences, necessitating methodologies capable of grasping cancer's complexity to devise optimal diagnostic systems and therapies. In biology and oncology, numerous complex problems arise, including genomic analysis, drug discovery and development, disease diagnosis and prognosis, personalized medicine, understanding tumor heterogeneity, clinical trial optimization, and healthcare resource allocation.

Artificial intelligence (AI) and machine learning have sparked a revolution in the realms of discovery, diagnosis, and treatment design, particularly in the field of oncology. They offer invaluable support not only in cancer detection but also in crafting tailored therapies, pinpointing novel therapeutic targets to expedite drug discovery, and enhancing cancer surveillance through the analysis of patient and cancer statistics. AI-driven cancer care holds promise for improving clinical screening and management, thus yielding better health outcomes. Machine learning (ML) algorithms, rooted in both biological and computer sciences, play a pivotal role in elucidating the intricate biological systems underlying cancer initiation, growth, and metastasis. They empower scientists to expedite the discovery process and assist physicians and surgeons in devising effective diagnostic and treatment strategies across various cancer types. Moreover, AI and machine learning stand to revolutionize biotechnology and pharmaceutical industries by streamlining drug discovery processes. In essence, AI embodies the intelligent behavior of computer science, with its techniques capable of learning from data and generalizing insights garnered from them. The potential of AI extends beyond innovation; it holds the power to revolutionize biology and oncology by accelerating research endeavors, enhancing diagnostic capabilities, and personalizing treatment modalities, ultimately fostering improved outcomes for cancer patients. This transformative potential spans across biology and oncology research, clinical practice, and patient care, promising a brighter future for those affected by cancer.

This book covers some significant impacts in recent research AI and machine learning in both the private and public sectors of cancer, biology, and oncology. The book is divided in 19 chapters in Cancer, Biology, and Oncology.

Hamid Khayyam, Ali Hekmatnia, and Rahele Kafieh

Editors

Article

Deep Learning-Based Classification and Targeted Gene Alteration Prediction from Pleural Effusion Cell Block Whole-Slide Images

Wenhao Ren, Yanli Zhu, Qian Wang, Haizhu Jin, Yiyi Guo and Dongmei Lin *

Key Laboratory of Carcinogenesis and Translational Research (Ministry of Education), Department of Pathology, Peking University Cancer Hospital and Institute, Beijing 100142, China
* Correspondence: lindm3@163.com

Simple Summary: For many patients with advanced cancer, pleural effusion is the only accessible specimen for establishing a pathological diagnosis. Some pleural effusion cell blocks have not undergone adequate morphological, immunohistochemical, or genetic analysis due to problems with the specimen itself or cost. Deep learning is a potential way to solve the above problems. In this study, on the basis of scanning whole slide images of pleural effusion cell blocks, we investigated the identification of benign and malignant pleural effusion, the determination of the primary site of pleural effusion common metastatic carcinoma, and the alteration of common targeted genes using a weakly supervised deep learning model. We achieved good results in these tasks. Although deep learning cannot be the gold standard for diagnosis, it can be a useful tool to aid in cytology diagnosis.

Abstract: Cytopathological examination is one of the main examinations for pleural effusion, and especially for many patients with advanced cancer, pleural effusion is the only accessible specimen for establishing a pathological diagnosis. The lack of cytopathologists and the high cost of gene detection present opportunities for the application of deep learning. In this retrospective analysis, data representing 1321 consecutive cases of pleural effusion were collected. We trained and evaluated our deep learning model based on several tasks, including the diagnosis of benign and malignant pleural effusion, the identification of the primary location of common metastatic cancer from pleural effusion, and the prediction of genetic alterations associated with targeted therapy. We achieved good results in identifying benign and malignant pleural effusions (0.932 AUC (area under the ROC curve)) and the primary location of common metastatic cancer (0.910 AUC). In addition, we analyzed ten genes related to targeted therapy in specimens and used them to train the model regarding four alteration statuses, which also yielded reasonable results (0.869 AUC for ALK fusion, 0.804 AUC for KRAS mutation, 0.644 AUC for EGFR mutation and 0.774 AUC for NONE alteration). Our research shows the feasibility and benefits of deep learning to assist in cytopathological diagnosis in clinical settings.

Keywords: deep learning; pleural effusion; cell blocks; classification; gene alteration prediction

Citation: Ren, W.; Zhu, Y.; Wang, Q.; Jin, H.; Guo, Y.; Lin, D. Deep Learning-Based Classification and Targeted Gene Alteration Prediction from Pleural Effusion Cell Block Whole-Slide Images. *Cancers* 2023, 15, 752. https://doi.org/10.3390/cancers15030752

Academic Editors: Hamid Khayyam, Ali Madani, Rahele Kafieh and Ali Hekmatnia

Received: 29 December 2022
Revised: 23 January 2023
Accepted: 23 January 2023
Published: 25 January 2023

1. Introduction

Serous effusion cytology is a common clinical method used to differentiate benign from malignant serous effusions due to its minimal discomfort and risk to patients [1–3]. With the gradual increase in treatment methods and the emergence of cell block technology, clinical cytologists are required not only to distinguish benign and malignant pleural effusions but also to identify the primary location of metastatic carcinomas or mutant genes using auxiliary methods such as immunohistochemistry or molecular detection [4,5]. However, due to the subjective and regionally dependent diagnostic level of cytopathologists, there is a problem of low consistency in the diagnosis of benign and malignant pleural effusions [3,6]. In addition, malignant pleural effusions have not yet been recognized as routine substrates for the immunohistochemical or molecular testing pipeline due to their occasionally low tumor fraction and sparse cellularity [7–9]. Low tumor cellularity means it is not always possible to perform sufficient immunohistochemical and molecular analyses to accurately diagnose gene mutation and the primary site of metastatic cancer. In cases with sufficient cellularity, the cost burden is another reason some patients fail to undergo immunohistochemistry or genetic testing.

In recent years, artificial intelligence in the form of deep learning has been extensively utilized in the field of pathology and has the potential to solve several clinical pathology problems [10–13], but fewer studies have focused on clinical cytopathology [14,15]. In this study, we used a weakly supervised deep learning approach to investigate the determination of benign and malignant pleural effusion, the identification of the primary site of metastatic cancer, and the prediction of genetic alterations associated with targeted therapy using whole-slide images (WSIs) of pleural effusion cell blocks in an effort to solve some urgent clinical issues with deep learning.

2. Materials and Methods

2.1. Materials

From January 2018 to September 2022, 1321 consecutive pleural effusion specimens from Peking University Cancer Hospital were embedded to the greatest extent possible and the successfully embedded cases were then scanned as whole-slide images (WSIs) with Pannoramic 250 Flash III scanner (3DHISTECH, Hungary). Several WSIs with unclear scanning were rescanned. Patient demographics, clinical presentation, cytology and histology reports, auxiliary tests, and patient management information were extracted from pathology databases and electronic medical records. All sections were assessed blindly by two senior cytopathologists (W.R. and Y.Z.), and in the case of inconsistent diagnosis, a unified diagnosis was negotiated with the participation of a third cytopathologist (Q.W.). Ultimately, 1307 digitized WSIs were included in the subsequent analysis.

For all malignant tumors, cases with a clear and unique tumor history did not confirm their primary site through immunohistochemistry, whereas the remaining cases (those with multiple prior malignancies or ambiguous primary locations) were confirmed with immunohistochemistry.

The AmoyDx® Essential next-generation sequencing (NGS) Panel (Amoy Diagnostics, Xiamen, China) was used to detect genetic abnormalities in FFPE cell block tumor tissues (http://www.amoydiagnostics.com/productDetail_9.html, accessed on 1 January 2023). The AmoyDx® Essential NGS Panel is an NGS-based in vitro diagnostic assay intended for qualitative detection of single nucleotide variants (SNVs), insertions and deletions (InDels), gene fusions, and copy number variations (CNVs) in driver genes. An amplification refractory mutation system polymerase chain reaction (ARMS-PCR) and a mutation detection kit (Amoy Diagnostics) were used to identify the gene alteration in driver genes. This kit is designed for the detection of common mutations in 10 genes in two categories: (1) mutation gene detection (EGFR gene (exons 18, 19, 20, 21), BRAF gene (V600E mutation), KRAS gene (codons 12 and 13 of exon 2), NRAS gene (codon 61 of exon 3), HER2 gene (exon 20), PIK3CA gene (exons 9 and 20), MET gene (exon 14 skipping mutation)); and (2) fusion gene detection (ALK, ROS1, RET fusion gene detection). In this study, we labeled a gene's alteration status as "NONE" if none of the 10 genes listed above contained abnormalities.

2.2. Datasets

All the HE-stained slides were digitized with a 40× magnification objective and a resolution of 0.25 μm/pixel and saved in MRXS format according to the manufacturer's protocol (Pannoramic 250 Flash III scanner, 3DHISTECH, Budapest, Hungary).

The dataset was partitioned into three parts as shown in Figure 1: (1) the benign vs. malignant dataset which was used to differentiate between benign and malignant pleural effusion and contained 1307 WSIs (representing 533 benign lesions and 774 malignant tumors); (2) the primary site dataset which was used to identify the primary site of common metastatic cancers and contained 560 WSIs (representing 94 breast invasive ductal carcinomas, 56 gastric adenocarcinomas, and 410 lung adenocarcinomas); and (3) the gene alteration dataset, which was further divided into the ALK dataset (23 ALK fusions and 335 ALK wild-types), the EGFR dataset (215 EGFR mutations and 143 EGFR wild-types), the KRAS dataset (31 KRAS mutations and 327 KRAS wild-types) and the NONE dataset (53 NONE alterations and 305 gene mutations).

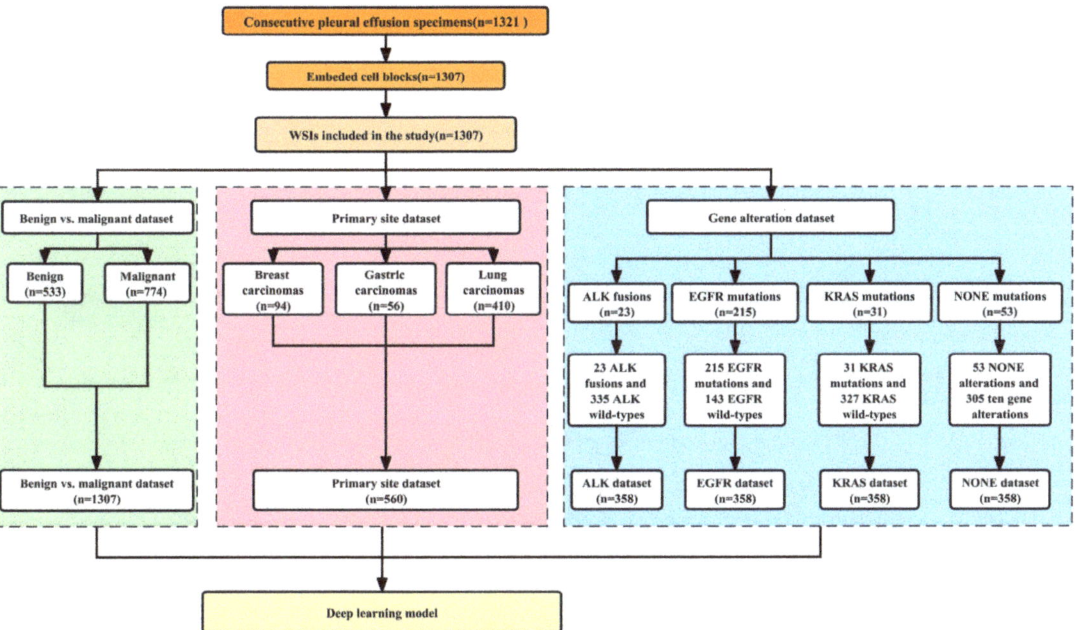

Figure 1. Case screening and establishment of datasets for three tasks.

According to the distribution of the dataset, whole-slide image (WSI) cases were randomly assigned to the training set, validation set, and test set in a ratio of 6:2:2. In order to solve the problem of unbalanced data, we performed different levels of data augmentation (such as rotating, random flipping, Gaussian blurring, and clipping) on the training set in different datasets. To analyze the predictive performance of the model more accurately, we employed the 10-fold cross-validation method.

2.3. Image Preparation

2.3.1. Preprocessing

Our pipeline (as shown in Figure 2) began with automatic segmentation of the tissue regions based on the operations associated with the Python Open-CV application programming interface. Except for the manual markings that we removed from the slides before scanning, we did not undertake any extra work, such as stain normalization or removing artifacts, on our images and used the entire tissue region of each slide during evaluation. Because the sizes of the tissue regions were still too large (~1.62 GB per image) for direct input into a neural network, all the tissue regions were cropped from the original microscopic images without overlapping and then resized to 299 × 299 pixels as input for model training with Qupath (Version 0.3.0) [16]. The number of patches per slide depended on the specimen size, and mean slide patches were 7989 ± 4568.

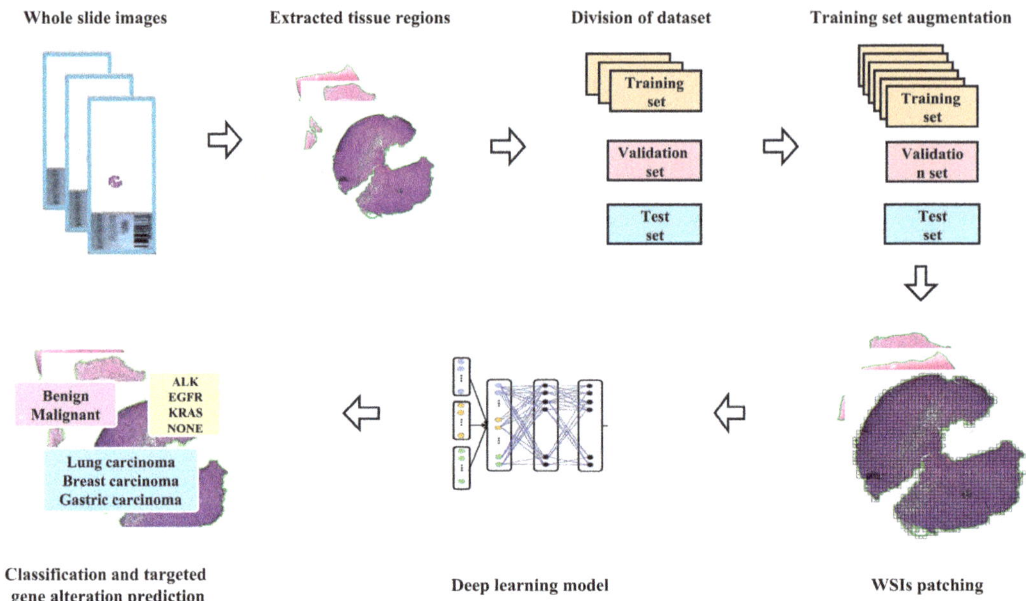

Figure 2. Flowchart of the deep learning framework presented in this study.

2.3.2. Deep Learning Model Training

In our approach, inception-ResNet-v2 serves as the backbone [17], and the weakly supervised WSI classification model is based on multiple instance learning (MIL), which treats each WSI as a collection of many smaller regions or patches. According to MIL, a slide should be labeled as positive if at least one of its patches is in the positive class and as negative if all of its patches are in the negative class [18]. In order to improve the interpretability of the model, we added an attention-based pooling function to the model [19]. Using attention-based learning, our model can generate interpretable heatmaps that enable clinicians to visualize the regions that the model focuses on when making a prediction. Without pixel-level annotation during training, we can determine for every tissue region the relative contribution to and correlation with the model prediction.

During training, the slides were randomly sampled using a batch size of 299. The pretrained classifier inferred all of the patches in the training dataset using the weights learned from the ImageNet dataset [20]. The attention module's weights and bias parameters were initialized at random, and the module was trained in conjunction with the rest of the model using the slide-level labels. To generate the slide-level prediction for inference during validation and testing, we utilized the model to create predictions for each patch on the slide and then averaged their probability scores, according to the method of a previous study [21].

If the validation loss did not reduce during a span of 20 validation epochs, the model was terminated early. This process was repeated every 100,000 patches. The model with the lowest validation loss at each checkpoint was chosen for evaluation on the test set. We used a cross-entropy loss function and optimized the models' parameters with stochastic gradient descent and the Adam optimizer at a learning rate of 0.0002 and a weight decay of 0.00001. To illustrate the relative contribution and significance of each tissue location, heatmaps were utilized.

2.4. Hardware and Software for Data Processing

We used multiple hard drives to store the raw files of the WSIs. Segmentation and patching of WSIs were performed on a computer with 40 Intel(R) Xeon(R) CPU E5-2640 v4 @ 2.40 GHz and Qupath (Version 0.3.0). Model training, validation, and testing were accelerated through data batch parallelization across three NVIDIA A5000 on local workstations. Our whole-slide processing pipeline is written in Python (3.6.2) and utilizes image-processing packages such as OpenSlide (3.4.1), OpenCV (4.1.1), and NumPy (version 1.18.1). The TensorFlow (version 2.9.0) deep learning library was used to load data and train deep learning models.

2.5. Evaluation Metrics and Graph

We measured model performance using AUC (area under the ROC curve), accuracy, sensitivity, specificity, positive predictive value (PPV), negative predictive value (NPV), and F1-score at the slide level. All plots were generated in R version 4.1.0.

3. Results

3.1. Patient Demographics and Clinicopathological Characteristics

The patient population consisted of 655 males and 652 females, with a median age of 62 years (range 14–91 years). Figure 3A,B depicts the distribution of benign and malignant pleural effusions by age and sex. One hundred and eight instances (14.0%) among 774 malignancies had pleural effusion as the initial symptom, while 10 cases (1.3%) had multiple prior tumors. Malignant WSI that contained fewer than 10 tumor cells (contained very few tumor cells) were present in 7.0% (54/774), 10–100 tumor cells were present in 29.8% (231/774), and more than 100 tumor cells were present in 63.2% (489/774). Figure 4 depicts the specific pathological categories of 774 cases of malignant pleural effusion, the most prevalent of which were lung adenocarcinomas (410, 53.0%), breast invasive ductal carcinomas (94, 12.1%), and gastric adenocarcinomas (56, 7.2%). In 360 of 410 cases of metastatic lung cancers, genetic testing was successfully performed. Two cases containing two types of genetic mutations were excluded from subsequent analysis, and a total of 358 cases with a single gene alteration were included in subsequent predictive analysis of genetic alterations. The specific types of gene alterations are shown in Figure 5A,B. The most prevalent mutations were EGFR mutations (215/358, 60.1%), ALK fusions (23/358, 6.4%), KRAS mutations (31/358, 8.7%), and NONE alterations (53/358, 14.8%). 21 L858R (101/215, 47.0%) and 19 del (67/215, 31.2%) were the most common EGFR mutant subtypes.

3.2. Deep Learning Models for Differentiating between Benign and Malignant Pleural Effusions

According to our research aim, we initially trained our model to distinguish between "benign" and "malignant" WSIs, with the terms simply representing the pathological characterization of the sampled cells. This task was trained on 320 benign and 464 malignant WSIs, validated on 107 benign and 155 malignant WSIs, and tested on 106 benign and 155 malignant WSIs. In terms of data set division, the benign and malignant tumors were randomly allocated to the training set, validation set, and test set in proportion as a whole. We did not deliberately allocate equal proportions to different types of malignant tumors. Since we adopted the 10-fold cross-validation method, the bias caused by the imbalance in the proportions of different types of malignant tumors in the training set, verification set, and test set can be reduced to a certain extent. The learning curve, evaluation metrics and confusion matrix of this task are shown in Figure 6A–C. In the test set, the results yielded an average AUC of 0.932 (range: 0.899 to 0.993), an average accuracy of 0.891 (range: 0.847 to 0.962), an average sensitivity of 0.911 (range: 0.821 to 0.987), an average specificity of 0.870 (range: 0.793 to 0.960), an average PPV of 0.910 (range: 0.845 to 0.974), an average NPV of 0.864 (range: 0.698 to 0.981), and an average F1-score of 0.909 (range: 0.873 to 0.967). The heatmaps (as shown in Figure 6D-E) demonstrate that our models are generally capable of identifying the boundary between malignant and benign tissue and are able to differentiate between tumor and nearby normal tissue without the usage of normal slides or region-of-interests during training. The largest proportion of false-negative cases was WSI with very few tumor cells (49.3%), and the largest proportion of false-positive cases was WSI with hyperplastic mesothelial cells or hyperplastic lymphocytes (48.6%).

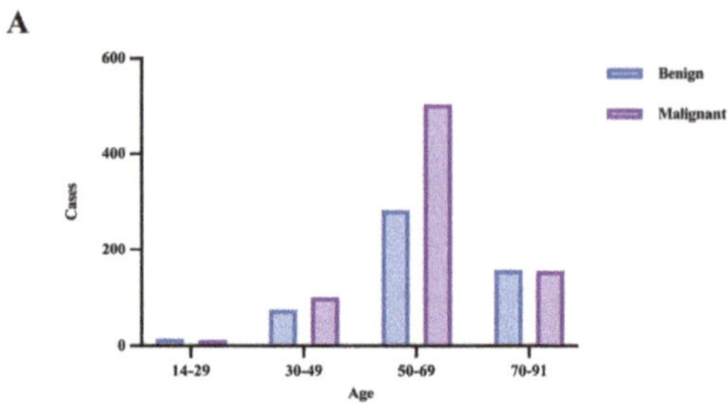
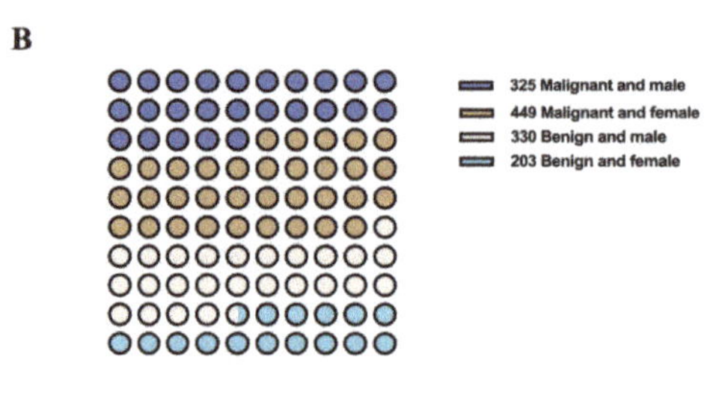

Figure 3. (**A**) The distribution of benign and malignant pleural effusions by age. (**B**) The distribution of benign and malignant pleural effusions by sex.

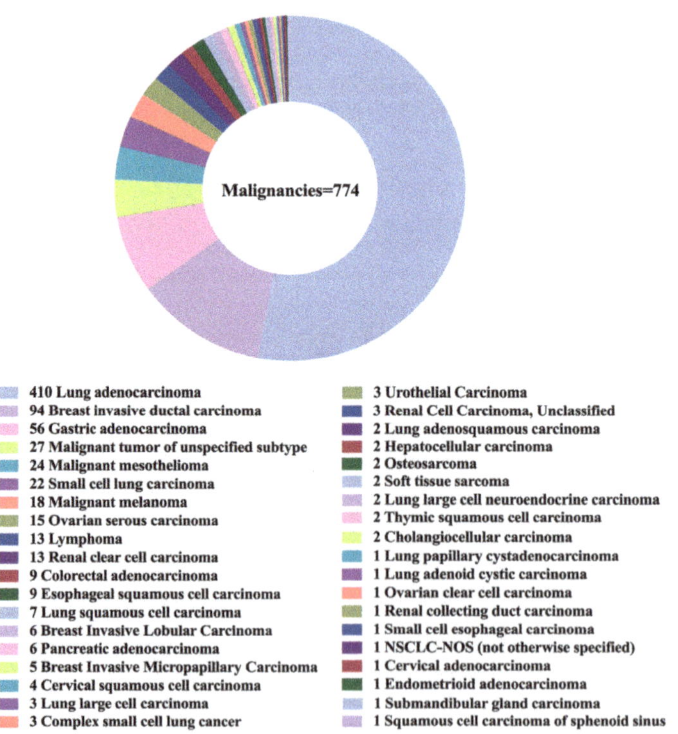

Figure 4. The specific pathological categories of 774 cases of malignant pleural effusion.

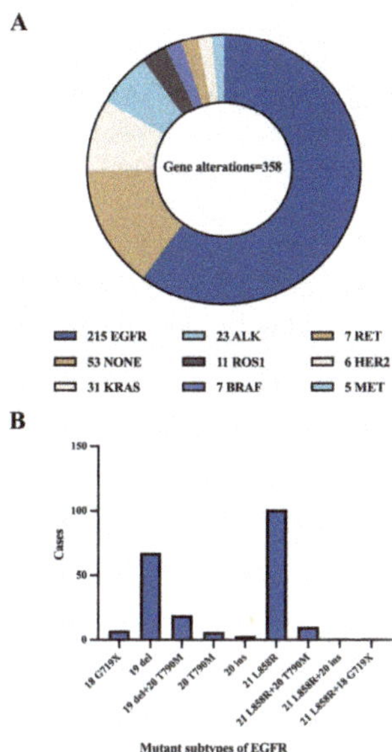

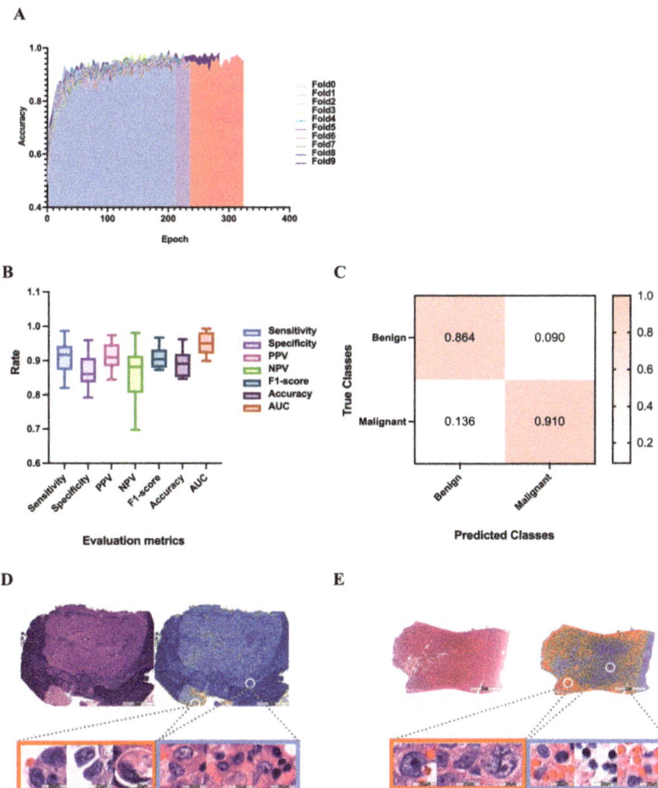

Figure 5. (**A**) The specific types of gene alterations. (**B**) Number of mutant subtypes of EGFR.

Figure 6. Results of the deep learning model for differentiating benign and malignant pleural effusion. (**A**) The learning curve of the deep learning model in distinguishing between benign and malignant

WSI images. (**B**) Evaluation metrics in test set of benign vs. malignant dataset. (**C**) Confusion matrix in test set of benign vs. malignant dataset. (**D**) A HE-stained image of breast invasive ductal carcinoma, corresponding heatmap and magnified pictures of different attention regions in the heatmap. The redder the color, the higher the confidence of the malignancy. (**E**) Another example of an HE-stained image of gastric adenocarcinoma, corresponding heatmap and magnified pictures of different attention regions in the heatmap.

3.3. Deep Learning Models in the Identification of the Primary Site of Metastatic Cancer

Next, we evaluated the performance of our method on the more difficult task of differentiating between pleural effusions of common metastatic carcinomas. The task was trained on 224 cases of breast carcinoma (augmented by 56 cases of breast carcinoma) 238 cases of gastric adenocarcinoma (augmented by 34 cases of gastric adenocarcinoma), and 246 cases of lung adenocarcinoma. It was validated on 19 cases of breast carcinoma 11 cases of gastric adenocarcinoma, and 82 cases of lung adenocarcinoma, and tested on 19 cases of breast carcinoma, 11 cases of gastric adenocarcinoma, and 82 cases of lung adenocarcinoma. Because the conventional MIL method, which was intended and widely implemented for weakly supervised positive/negative binary classification (for example cancer versus normal), was not suitable for this three-category task, we performed learning with the mMIL method [22], which shows a good classification effect in multiclassification tasks. In the test set, this process resulted in an average AUC of 0.910 (range: 0.879 to 0.960) and an average accuracy of 0.810 (range: 0.750 to 0.884), as shown in Figure 7A Among these, the average accuracy rates were 0.955 for gastric adenocarcinoma, 0.737 for breast invasive ductal carcinoma, and 0.807 for lung adenocarcinoma (Figure 7B). Figure 7C shown the confusion matrix of this task.

It is also important that in this classification task, the high attention regions of the deep learning model were consistent with the areas that cytopathologists focus on when making a diagnosis. For example, the trained model for gastric adenocarcinoma highlights predominantly scattered isolated malignant cells (Figure 7D) and uses them as strong evidence (high attention) for gastric adenocarcinoma, whereas the trained model for breast invasive ductal carcinoma emphasizes acini/glands or round cell groups (Figure 7E). For lung adenocarcinoma, the model emphasizes clusters with unregular borders and cellular pleomorphism (Figure 7F), which are consistent with human pathology expertise.

Misclassified gastric adenocarcinomas were all predicted to be lung adenocarcinomas For misclassified breast cancers, 78% of them were predicted to be lung adenocarcinomas and 22% were predicted to be gastric adenocarcinomas. Of the misclassified lung adenocarcinomas, 33.5% were predicted to be gastric adenocarcinomas and 66.5% were predicted to be breast cancers.

3.4. Predicting Gene Alteration Status from Whole-Slide Images Using Deep Learning Models

Next, we focused on lung adenocarcinoma WSIs and examined whether deep learning can be trained to predict gene alterations using only images as the input. To ensure that the training and test sets comprised sufficient images of gene alteration cases, we only selected common gene alterations as the target of this classification task. We investigated each of the common genes individually using binary classification. The prediction results for each common gene are shown in Figure 8A. The confusion matrices of common gene alteration are shown in Figure 8B–E. The heatmaps of common gene alteration are shown in Figure 8F–I.

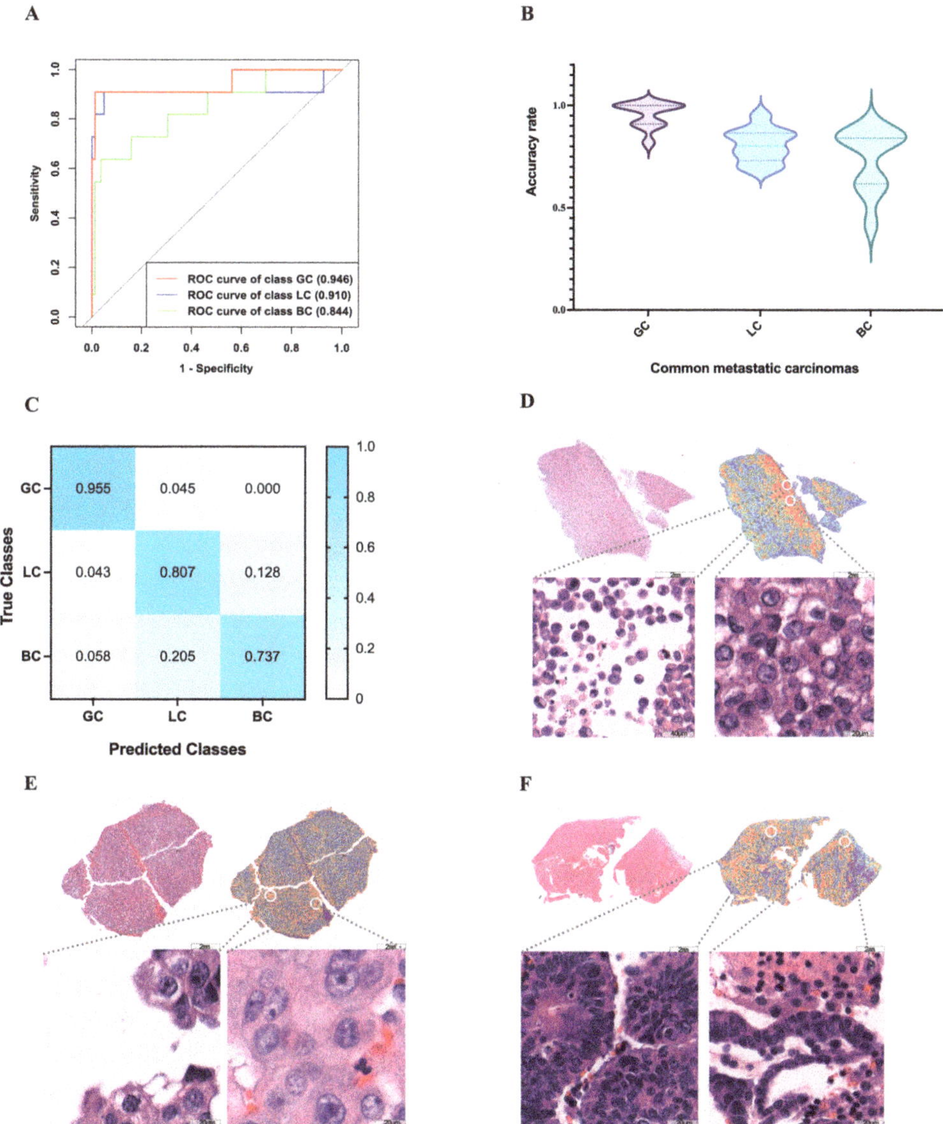

Figure 7. Results of the deep learning models in the identification of the primary site of metastatic carcinoma. (**A**) The ROC curve used for the common metastatic carcinomas on the test set. (**B**) The average accuracy rate of common metastatic carcinomas predicted by deep learning. (**C**) The confusion matrix used for the common metastatic carcinomas on the test set. (**D**–**F**) High-attention regions in the classification tasks. The gastric adenocarcinoma (**D**) highlights predominantly scattered isolated malignant cells, whereas the trained model for lung adenocarcinoma (**E**) emphasizes clusters with unregular borders and cellular pleomorphism. For breast invasive ductal carcinoma (**F**), the model emphasizes acini/glands or round cell groups. GC: gastric adenocarcinoma; LC: lung adenocarcinoma; BC: breast invasive ductal carcinoma.

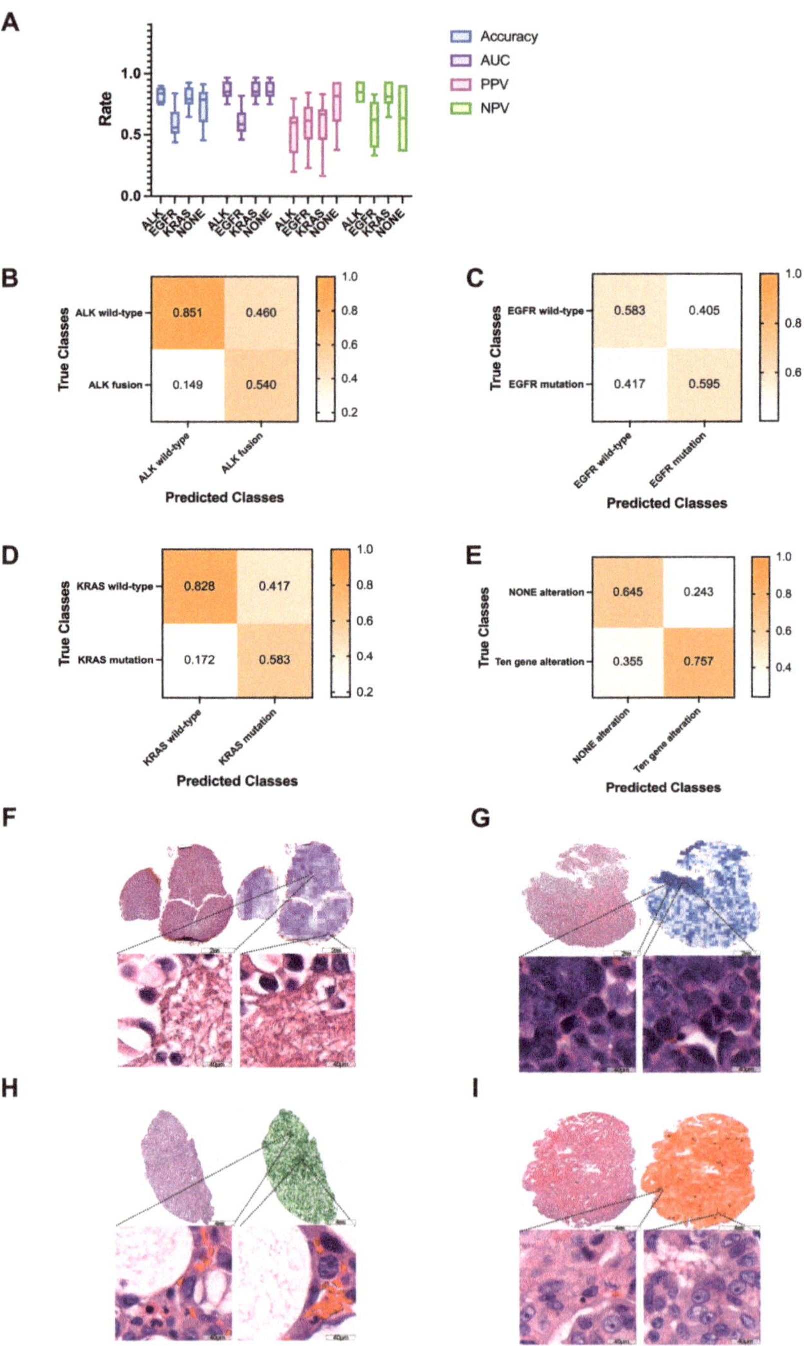

Figure 8. (**A**) The prediction results of common gene alterations in lung adenocarcinoma. (**B–E**) Th confusion matrices of common gene alteration. (**F–I**) Examples of the heatmaps of ALK fusion, EGFI mutation, KRAS mutation and NONE alteration, respectively. The darker the color, the higher th confidence of the corresponding gene alteration.

In the ALK dataset, the training set contained 345 ALK fusion (augmented by 23 ALK fusion) and 335 ALK wild-type WSIs, the validation set contained 13 ALK fusion and 201 ALK wild-type WSIs, and the test set contained 5 ALK fusion and 67 ALK wild-type WSIs. The results yielded an average AUC of 0.869 (range: 0.752 to 0.969), accuracy of 0.829 (0.750 to 0.903), PPV of 0.540 (0.200 to 0.800), and NPV of 0.850 (0.761 to 0.940) in the test set.

The KRAS dataset consisted of 31 KRAS mutations and 327 KRAS wild-type WSIs. After training on 190 KRAS mutations (augmented by 19 KRAS mutations) and 197 KRAS wild-type WSIs, the test set (6 KRAS mutation and 65 KRAS wild-type) yielded an average AUC of 0.804 (0.635 to 0.977), accuracy of 0.807 (0.648 to 0.930), PPV of 0.583 (0.166 to 0.833), and NPV of 0.828 (0.646 to 0.938).

For EGFR mutation or wild-type, the test set contained 43 EGFR mutations and 29 EGFR wild-type WSIs. The average AUC, accuracy, PPV, and NPV were 0.644 (0.468 to 0.821), 0.592 (0.480 to 0.840), 0.600 (0.231 to 0.846), and 0.583 (0.333 to 0.833), respectively. For ten gene alterations and NONE alterations, the test set contained 11 NONE-alteration and 61 gene-alteration WSIs. The results yielded average AUC, accuracy, PPV, and NPV values of 0.774 (0.615 to 0.879), 0.740 (0.458 to 0.917), 0.757 (0.377 to 0.934) and 0.645 (0.364 to 0.909), respectively.

4. Discussion

To the best of our knowledge, this is the largest study to date evaluating the application of deep learning to the cytological diagnosis of pleural effusion cell blocks. Deep learning focused in the direction of cytopathology is less available and has not been applied in more complex clinical scenarios. The majority of studies have only performed the differentiation of benign and malignant pleural fluid [23–25], and some research has studied only the most frequently mutated genes in lung adenocarcinoma [21], although these genes have no guiding meaning in target therapy. In addition, there are few publications on the use of deep learning to forecast the primary site of metastatic tumors. Identifying the primary site of metastatic cancer is a critical diagnostic task. Many patients present with pleural effusion as their first symptom, and different primary sites can lead to very different treatments [26].

In this study, we evaluated the use of deep learning to distinguish benign pleural effusion from malignant pleural effusion. Due to cytomorphologic overlap, proliferating mesothelial cells are frequently difficult to distinguish from cancer cells in routine cytopathological diagnostic procedures [27]. Our study demonstrates that a deep learning model can be used to aid in diagnostic work; it can classify normal and malignant effusions with an AUC of 0.932, an accuracy of 0.891, a sensitivity of 0.911, a specificity of 0.870, a PPV of 0.910, an NPV of 0.864, and an F1-score of 0.909. In institutions where there is a low level of diagnostic expertise, there is benefit to be gained from the use of deep learning systems. We further analyzed the cases in which the deep learning model misjudged, and found that the main reason for the false negative may be that the WSI contains very few tumor cells (accounting for 59.3%), and the deep learning model fails to make correct judgments based on these few tumor cells. The main causes of false positives are proliferating lymph node cells and hyperplastic mesothelial cells (48.6%), proliferating lymphocytes may not be well distinguished from lymphoma, small cell carcinoma, etc., and hyperplastic mesothelial cells may be indistinguishable from some malignant tumors with similar morphology, such as mesothelioma, gastric cancer, etc., resulting in incorrect prediction by deep learning systems.

Second, we examined the applicability of our model to the determination of the primary site of metastatic cancer in pleural fluid. Significant cytomorphologic overlap exists between carcinomas of different primary origins, and immunohistochemistry is frequently needed to determine the primary site. However, some pleural effusions of unknown primary may not be confirmed by limited immunohistochemical items due to the low cellularity in the cell block [28,29]. In our study, after training and testing on 560 WSIs (94 breast invasive ductal carcinomas, 56 gastric adenocarcinomas, and 410 lung adenocarcinomas), we discovered that our deep learning model can distinguish between the three types of metastatic cancer (0.910 AUC), which is of great benefit to patients with only pleural effusion specimens available when the cell block contains sparse tumor cells. According to the predictions of deep learning, the primary site is likely to be verified by only two or three immunohistochemistry markers; deep learning can considerably improve the confirmation rate of the primary site. It is worth mentioning that in clinical work, the prediction of rare tumors is more clinically significant, because the primary lesions of rare tumors often require a larger immunohistochemical panel to be clear, and the prediction of rare tumors by deep learning prediction systems can reduce the workload and patient costs. In our study, we did not make predictions for rare tumors due to the small number of rare tumors and the lack of public datasets available for pleural effusion cell blocks, but we shared our WSIs data in the hope that future studies can integrate our data to complete deep learning predictions for rare metastatic cancer in pleural fluid.

When analyzing the misclassified cases, gastric adenocarcinoma achieved a good accuracy, which may be caused by the relatively single morphology of metastatic gastric adenocarcinoma. In our study, metastatic gastric adenocarcinoma cells in pleural fluid were all scattered isolated malignant cells, and only 0.045% of gastric adenocarcinoma was misclassified as lung adenocarcinoma. The main cause of breast carcinoma misclassification (60%) is that some poorly differentiated breast invasive ductal carcinomas exhibiting as three-dimensional round cell groups were misclassified as solid lung adenocarcinoma. Lung adenocarcinoma has a variety of morphologies, including three-dimensional groups in papillary configurations, proliferation spheres or single-cell scattered forms, or a combination of multiple forms, and the varied morphology may cause its morphological overlap with breast and gastric adenocarcinoma, thereby reducing the accuracy of predictions.

We studied the feasibility of predicting targeted mutations using WSI images of pleural effusion cell blocks. Numerous clinical investigations have demonstrated that gene alteration status is a major predictor of the success of targeted therapy. The presence of ALK and ROS1 gene fusions correlates with the efficacy of ALK/MET inhibitor therapy. Patients with RET fusion could benefit from MET/RET/VEGFR inhibitors. BRAF-mutated patients benefit from BRAF inhibitor therapy, and KRAS/NRAS/HER2/PIK3CA mutation status is associated with the prognosis of some targeted drugs [30,31]. According to the National Comprehensive Cancer Network Guidelines for non-small cell lung cancer, gene mutation testing is essential prior to targeted therapy, and multitarget testing is strongly suggested for the most effective precision oncology treatment. Given the significance and impact of these genetic abnormalities, the ability to anticipate genetic alterations from pathology images rapidly and affordably may aid in the treatment of cancer patients. However, pleural effusion samples are not usually sufficient for a comprehensive analysis of targeted mutations [32]. Although some gene mutations have certain morphological characteristics [33], the positive rate of gene mutation prediction based on morphology is still low. To improve the success rate of gene detection, we used deep learning to predict the gene of most likely change.

In general, the results of our model in predicting targeted mutations are reasonable. Our model acquired good performance in predicting ALK fusion and KRAS mutation (0.869 AUC and 0.804 AUC), respectively, whereas the performance was relatively poor in predicting EGFR and NONE alterations. We suggest that the reason for the poor prediction accuracy of EGFR and NONE alterations is that they have more mutant subtypes. In our study, the mutant subtypes of EGFR are depicted in Figure 5B. Different mutant subtypes may result in different morphological alterations, thereby decreasing prediction accuracy. In addition, only ten genes were identified in our study, and cases with NONE alterations may contain altered genes other than the ten genes. The genetic complexity may further lead to the confounding of the corresponding morphology, consequently diminishing the accuracy of the prediction. In addition, the small amount of training data is a limiting factor for achieving higher accuracy. Therefore, we have shared the WSIs and corresponding genetic alteration information from this study so that future investigations based on using genetic subtypes and increasing the number of training instances may be able to further improve the prediction accuracy.

Moreover, we found that the attention heatmaps exhibited a high level of agreement with the cytologist's target region when tested on the benign vs. malignant dataset and primary site dataset, which gives us great hope and demonstrates the interpretability and dependability of our model. Although the accuracy of our model in predicting gene alterations is not very high, it is possible to detect morphological features associated with gene abnormalities if future research improves prediction accuracy.

Our research has the following limitations: (1) This was a single center, data-based study. There is no publicly available dataset for pleural effusion WSI images. Although we took clinically consecutive instances and did not normalize our WSI images to improve the robustness of the model, our model needs to be verified on a multicenter and larger dataset. In an effort to address the paucity of data on pleural effusion, we have made the data from this study available to researchers interested in additional investigation. (2) Due to the small number of cases with uncommon metastatic cancer and unusual mutations, our analysis was unable to forecast these conditions. More rare cases should be included in future studies to develop a more clinically applicable deep learning model.

5. Conclusions

Overall, this study implies that a deep learning model based on the pleural effusion cell blocks may be an effective diagnostic tool for cytopathologists. In addition to being able to differentiate between benign and malignant pleural effusion, the model can also identify the primary site of common metastatic malignancies, allowing for more precise medical treatment of patients. Precision medicine focuses increasingly on the genetic alterations of the disease, and different mutations result in different targeted therapies. Our model performs well in predicting KRAS mutations and ALK fusions, but further improvements are needed in predicting EGFR mutations and NONE mutations. Future research could analyze the subtypes of targeted mutations and collect more rare pleural effusion metastatic carcinomas and rare mutations to improve the model's accuracy. Although deep learning cannot be the gold standard for diagnosis, it can be a useful tool to aid in cytology diagnosis. Immunohistochemistry and genetic tests can achieve their goals more efficiently with the guidance of deep learning.

Author Contributions: W.R. and D.L. conceived the idea of this work. W.R. designed the methodology and the software. W.R. and Y.Z. carried out the validation of the methodology. W.R. performed the formal analysis. Q.W., H.J. and Y.G. performed the investigation. W.R., Y.Z. and Q.W. participated in annotation or curation of the dataset. W.R. and Y.Z. prepared and wrote the manuscript. D.L. reviewed and revised the manuscript. W.R. prepared the visualization of the manuscript. D.L. supervised this work. W.R. and D.L. administered this work. All authors have read and agreed to the published version of the manuscript.

Funding: This research received no external funding.

Institutional Review Board Statement: This study was approved by the institutional review boards of the Peking University Cancer Hospital (NO. 2018KT94). The study protocol was in accordance with the ethical guidelines established in the 1975 Declaration of Helsinki and complies with the procedures of the local ethical committee of the Humanitas Clinical and Research Center.

Informed Consent Statement: Informed consent was waived by the institutional review boards because the retrospective archival materials were anonymized and the results did not have any impact on patient management.

Data Availability Statement: The datasets generated and/or analyzed during the current study are available online at https://www.aliyundrive.com/s/cUnri7B82Qp (accessed on 1 January 2023).

Conflicts of Interest: The authors declare no conflict of interest.

References

1. Lepus, C.M.; Vivero, M. Updates in Effusion Cytology. *Surg. Pathol. Clin.* **2018**, *11*, 523–544. [CrossRef] [PubMed]
2. Pinto, D.; Chandra, A.; Crothers, B.A.; Kurtycz, D.F.I.; Schmitt, F. The international system for reporting serous fluid cytopathology: diagnostic categories and clinical management. *J. Am. Soc. Cytopathol.* **2020**, *9*, 469–477. [CrossRef] [PubMed]
3. Gayen, S. Malignant Pleural Effusion: Presentation, Diagnosis, and Management. *Am. J. Med.* **2022**, *135*, 1188–1192. [CrossRef] [PubMed]
4. Asciak, R.; Rahman, N.M. Malignant Pleural Effusion: From Diagnostics to Therapeutics. *Clin. Chest Med.* **2018**, *39*, 181–193. [CrossRef] [PubMed]
5. Ebata, T.; Okuma, Y.; Nakahara, Y.; Yomota, M.; Takagi, Y.; Hosomi, Y.; Asami, E.; Omuro, Y.; Hishima, T.; Okamura, T.; et al. Retrospective analysis of unknown primary cancers with malignant pleural effusion at initial diagnosis. *Thorac. Cancer* **2016**, *7*, 39–43. [CrossRef] [PubMed]
6. Addala, D.N.; Kanellakis, N.I.; Bedawi, E.O.; Dong, T.; Rahman, N.M. Malignant pleural effusion: Updates in diagnosis, management and current challenges. *Front. Oncol.* **2022**, *12*, 1053574. [CrossRef]
7. Arnold, D.T.; De Fonseka, D.; Perry, S.; Morley, A.; Harvey, J.E.; Medford, A.; Brett, M.; Maskell, N.A. Investigating unilateral pleural effusions: The role of cytology. *Eur. Respir. J.* **2018**, *52*, 1801254. [CrossRef]
8. Porcel, J.M. Diagnosis and characterization of malignant effusions through pleural fluid cytological examination. *Curr. Opin. Pulm. Med.* **2019**, *25*, 362–368. [CrossRef]
9. Roy-Chowdhuri, S.; Dacic, S.; Ghofrani, M.; Illei, P.B.; Layfield, L.J.; Lee, C.; Michael, C.W.; Miller, R.A.; Mitchell, J.W.; Nikolic, B.; et al. Collection and Handling of Thoracic Small Biopsy and Cytology Specimens for Ancillary Studies: Guideline From the College of American Pathologists in Collaboration With the American College of Chest Physicians, Association for Molecular Pathology, American Society of Cytopathology, American Thoracic Society, Pulmonary Pathology Society, Papanicolaou Society of Cytopathology, Society of Interventional Radiology, and Society of Thoracic Radiology. *Arch. Pathol. Lab. Med.* **2020**, *144*, 933–958. [CrossRef]
10. Niazi, M.K.K.; Parwani, A.V.; Gurcan, M.N. Digital pathology and artificial intelligence. *Lancet Oncol.* **2019**, *20*, e253–e261. [CrossRef]
11. Pallua, J.D.; Brunner, A.; Zelger, B.; Schirmer, M.; Haybaeck, J. The future of pathology is digital. *Pathol. Res. Pract.* **2020**, *216*, 153040. [CrossRef] [PubMed]
12. Bera, K.; Schalper, K.A.; Rimm, D.L.; Velcheti, V.; Madabhushi, A. Artificial intelligence in digital pathology—New tools for diagnosis and precision oncology. *Nat. Rev. Clin. Oncol.* **2019**, *16*, 703–715. [CrossRef]
13. Alam, M.R.; Abdul-Ghafar, J.; Yim, K.; Thakur, N.; Lee, S.H.; Jang, H.J.; Jung, C.K.; Chong, Y. Recent Applications of Artificial Intelligence from Histopathologic Image-Based Prediction of Microsatellite Instability in Solid Cancers: A Systematic Review. *Cancers* **2022**, *14*, 2590. [CrossRef]
14. Thakur, N.; Alam, M.R.; Abdul-Ghafar, J.; Chong, Y. Recent Application of Artificial Intelligence in Non-Gynecological Cancer Cytopathology: A Systematic Review. *Cancers* **2022**, *14*, 3529. [CrossRef] [PubMed]
15. Alrafiah, A.R. Application and performance of artificial intelligence technology in cytopathology. *Acta Histochem.* **2022**, *124*, 151890. [CrossRef]
16. Bankhead, P.; Loughrey, M.B.; Fernández, J.A.; Dombrowski, Y.; McArt, D.G.; Dunne, P.D.; McQuaid, S.; Gray, R.T.; Murray, L.J.; Coleman, H.G.; et al. QuPath: Open source software for digital pathology image analysis. *Sci. Rep.* **2017**, *7*, 16878. [CrossRef]
17. Szegedy, C.; Ioffe, S.; Vanhoucke, V.; Alemi, A.A. Inception-v4, inception-ResNet and the impact of residual connections on learning. In Proceedings of the AAAI-17: Thirty-First AAAI Conference on Artificial Intelligence, San Francisco, CA, USA, 4–9 February 2017.
18. Campanella, G.; Hanna, M.G.; Geneslaw, L.; Miraflor, A.; Werneck Krauss Silva, V.; Busam, K.J.; Brogi, E.; Reuter, V.E.; Klimstra, D.S.; Fuchs, T.J. Clinical-grade computational pathology using weakly supervised deep learning on whole slide images. *Nat. Med.* **2019**, *25*, 1301–1309. [CrossRef]
19. Lee, H.; Kim, S. Explaining Neural Networks Using Attentive Knowledge Distillation. *Sensors* **2021**, *21*, 1280. [CrossRef]
20. Shin, H.C.; Roth, H.R.; Gao, M.; Lu, L.; Xu, Z.; Nogues, I.; Yao, J.; Mollura, D.; Summers, R.M. Deep Convolutional Neural Networks for Computer-Aided Detection: CNN Architectures, Dataset Characteristics and Transfer Learning. *IEEE Trans. Med. Imaging* **2016**, *35*, 1285–1298. [CrossRef]
21. Coudray, N.; Ocampo, P.S.; Sakellaropoulos, T.; Narula, N.; Snuderl, M.; Fenyö, D.; Moreira, A.L.; Razavian, N.; Tsirigos, A. Classification and mutation prediction from non-small cell lung cancer histopathology images using deep learning. *Nat. Med.* **2018**, *24*, 1559–1567. [CrossRef]
22. Lu, M.Y.; Williamson, D.F.K.; Chen, T.Y.; Chen, R.J.; Barbieri, M.; Mahmood, F. Data-efficient and weakly supervised computational pathology on whole-slide images. *Nat. Biomed. Eng.* **2021**, *5*, 555–570. [CrossRef]
23. Xie, X.; Fu, C.C.; Lv, L.; Ye, Q.; Yu, Y.; Fang, Q.; Zhang, L.; Hou, L.; Wu, C. Deep convolutional neural network-based classification of cancer cells on cytological pleural effusion images. *Mod. Pathol.* **2022**, *35*, 609–614. [CrossRef] [PubMed]
24. Win, K.Y.; Choomchuay, S.; Hamamoto, K.; Raveesunthornkiat, M.; Rangsirattanakul, L.; Pongsawat, S. Computer Aided Diagnosis System for Detection of Cancer Cells on Cytological Pleural Effusion Images. *Biomed. Res. Int.* **2018**, *2018*, 6456724. [CrossRef]

25. Wang, X.; Chen, H.; Gan, C.; Lin, H.; Dou, Q.; Tsougenis, E.; Huang, Q.; Cai, M.; Heng, P.A. Weakly Supervised Deep Learning for Whole Slide Lung Cancer Image Analysis. *IEEE Trans Cybern.* **2020**, *50*, 3950–3962. [CrossRef] [PubMed]

26. Awadallah, S.F.; Bowling, M.R.; Sharma, N.; Mohan, A. Malignant pleural effusion and cancer of unknown primary site: A review of literature. *Ann. Transl. Med.* **2019**, *7*, 353. [CrossRef]

27. Shidham, V.B.; Layfield, L.J. Approach to Diagnostic Cytopathology of Serous Effusions. *Cytojournal* **2021**, *18*, 32. [CrossRef]

28. Harbhajanka, A.; Brickman, A.; Park, J.W.; Reddy, V.B.; Bitterman, P.; Gattuso, P. Cytomorphology, clinicopathologic, and cytogenetics correlation of myelomatous effusion of serous cavities: A retrospective review. *Diagn. Cytopathol.* **2016**, *44*, 742–747. [CrossRef]

29. Kuenen-Boumeester, V.; van Loenen, P.; de Bruijn, E.M.; Henzen-Logmans, S.C. Quality control of immunocytochemical staining of effusions using a standardized method of cell processing. *Acta Cytol.* **1996**, *40*, 475–479. [CrossRef]

30. Alexander, M.; Kim, S.Y.; Cheng, H. Update 2020: Management of Non-Small Cell Lung Cancer. *Lung* **2020**, *198*, 897–907. [CrossRef]

31. Chen, J.Y.; Cheng, Y.N.; Han, L.; Wei, F.; Yu, W.W.; Zhang, X.W.; Cao, S.; Yu, J.P. Predictive value of K-ras and PIK3CA in non-small cell lung cancer patients treated with EGFR-TKIs: A systemic review and meta-analysis. *Cancer Biol. Med.* **2015**, *12*, 126–139. [CrossRef]

32. DeMaio, A.; Clarke, J.M.; Dash, R.; Sebastian, S.; Wahidi, M.M.; Shofer, S.L.; Cheng, G.Z.; Li, X.; Wang, X.; Mahmood, K. Yield of Malignant Pleural Effusion for Detection of Oncogenic Driver Mutations in Lung Adenocarcinoma. *J. Bronchol. Interv. Pulmonol.* **2019**, *26*, 96–101. [CrossRef] [PubMed]

33. Li, Z.; Dacic, S.; Pantanowitz, L.; Khalbuss, W.E.; Nikiforova, M.N.; Monaco, S.E. Correlation of cytomorphology and molecular findings in EGFR+, KRAS+, and ALK+ lung carcinomas. *Am. J. Clin. Pathol.* **2014**, *141*, 420–428. [CrossRef] [PubMed]

Article

A Machine Learning-Based Online Prediction Tool for Predicting Short-Term Postoperative Outcomes Following Spinal Tumor Resections

Mert Karabacak and Konstantinos Margetis *

Department of Neurosurgery, Mount Sinai Health System, New York, NY 10029, USA
* Correspondence: konstantinos.margetis@mountsinai.org

Simple Summary: The overall incidence of spinal tumors in the United States was estimated to be 0.62 per 100,000 people. Surgical resection of spinal tumors intends to improve functional status, reduce pain, and, in some patients with isolated metastases or primary tumors, increase survival. Machine learning algorithms show great promise for predicting short-term postoperative outcomes in spinal tumor surgery. With this study, we aim to develop machine learning algorithms for predicting short-term postoperative outcomes and implement these models in an open-source web application.

Abstract: *Background:* Preoperative prediction of short-term postoperative outcomes in spinal tumor patients can lead to more precise patient care plans that reduce the likelihood of negative outcomes. With this study, we aimed to develop machine learning algorithms for predicting short-term postoperative outcomes and implement these models in an open-source web application. *Methods:* Patients who underwent surgical resection of spinal tumors were identified using the American College of Surgeons, National Surgical Quality Improvement Program. Three outcomes were predicted: prolonged length of stay (LOS), nonhome discharges, and major complications. Four machine learning algorithms were developed and integrated into an open access web application to predict these outcomes. *Results:* A total of 3073 patients that underwent spinal tumor resection were included in the analysis. The most accurately predicted outcomes in terms of the area under the receiver operating characteristic curve (AUROC) was the prolonged LOS with a mean AUROC of 0.745 The most accurately predicting algorithm in terms of AUROC was random forest, with a mean AUROC of 0.743. An open access web application was developed for getting predictions for individual patients based on their characteristics and this web application can be accessed here: huggingface.co/spaces/MSHS-Neurosurgery-Research/NSQIP-ST. *Conclusion:* Machine learning approaches carry significant potential for the purpose of predicting postoperative outcomes following spinal tumor resections. Development of predictive models as clinically useful decision-making tools may considerably enhance risk assessment and prognosis as the amount of data in spinal tumor surgery continues to rise.

Keywords: spine surgery; spinal tumors; artificial intelligence; machine learning; NSQIP; prediction; online prediction tool

Citation: Karabacak, M.; Margetis, K. A Machine Learning-Based Online Prediction Tool for Predicting Short-Term Postoperative Outcomes Following Spinal Tumor Resections. *Cancers* **2023**, *15*, 812. https://doi.org/10.3390/cancers15030812

Academic Editors: Hamid Khayyam, Ali Madani, Rahele Kafieh and Ali Hekmatnia

Received: 4 January 2023
Revised: 24 January 2023
Accepted: 26 January 2023
Published: 28 January 2023

1. Introduction

The overall incidence of spinal tumors in the US was estimated to be 0.62 per 100,000 people [1,2]. The majority of spinal tumors (up to 70%) are metastatic tumors. According to their location, spinal tumors are further divided into extradural (55%), intradural extramedullary (40%), and intramedullary (5%) [3,4]. Surgical resection of spinal tumors intends to improve functional status, reduce pain, and, in some patients with isolated metastases or primary tumors, increase survival [5–7]. Similar to other patients undergoing spine surgery, there is growing interest in finding the most effective ways to lower postoperative complications, length of hospital stays, and rate of nonhome discharges in the population of patients with spinal tumors [8,9]. Postoperative complications have a negative effect on a patient's short-term quality of life, can lengthen their hospital stay, and can raise the cost of their medical care [10,11]. Several preoperative risk factors, such as preoperative functional status, disseminated malignancy, and poor baseline health, have been shown to predict higher complications and length of stay (LOS) [12,13].

In order to track and determine risk-adjusted estimates for these outcomes, emphasis is being placed on registries and databases as part of growing efforts to bend the healthcare cost curve. As a result, clinicians nowadays must manage vast amounts of complex data, which necessitates the employment of strong analytical techniques [14]. Machine learning (ML) algorithms can utilize high-dimensional clinical data to create precise patient risk assessment models, contribute to the formation of smart guidelines, and influence healthcare decisions by tailoring care to patient needs. In comparison to traditional prognostic models, which usually incorporate logistic regression, ML provides significant advantages. First, ML hardly ever requires prior knowledge of key predictors [15]. Second, compared to logistic regression, ML often has fewer restrictions on the number of predictors used for a given dataset. In large datasets with a considerable number of predictors, ML is useful since associations between predictors and outcomes may not always be instantly evident. Third, complex, nonlinear correlations in datasets that are more challenging to express and interpret using logistic regression can be discovered through ML [16]. These benefits often lead to ML being more accurate and robust than logistic regression techniques on the same dataset [17,18].

Based on our literature search, no study has explored the ability of ML algorithms to predict prolonged LOS, nonhome discharges, and postoperative complications in a single study following surgery for spinal tumors, without dividing into subtypes. This study aimed to assess the efficacy of machine learning algorithms in predicting postoperative outcomes after spinal tumor resection and create a user-friendly and accessible predictive tool for this purpose.

2. Materials and Methods

2.1. Data Source

Data for this study is from the American College of Surgeons (ACS) National Surgical Quality Improvement Program (NSQIP) database, which was queried to identify spinal tumor patients who were surgically treated from 2016 to 2020. We chose the most recent five years of data to take into account the advances in medicine. The ACS-NSQIP database is a national surgical registry with over 700 participating medical centers across the US for adult patients who underwent major surgical procedures across all subspecialties, except for trauma and transplant [19,20]. The data for each case, including demographics, preoperative comorbidities, operative variables, and 30-day postoperative outcomes, are being gathered by trained, skilled clinical reviewers [21]. Regular database auditing guarantees high-quality data with a previously reported interobserver disagreement rate of less than 2% in 2020 [22]. Detailed information about the database and data collection methods have been provided elsewhere [23].

2.2. Guidelines

We followed Transparent Reporting of Multivariable Prediction Models for Individual Prognosis or Diagnosis (TRIPOD) [24] and Journal of Medical Internet Research (JMIR) Guidelines for Developing and Reporting Machine Learning Predictive Models in Biomedical Research [25]. This was a retrospective machine learning classification study (outcomes were binary categorical) for prognostication in spinal tumors.

2.3. Study Population

We queried the NSQIP database to identify patients in whom the following inclusion criteria were met: (1) elective surgery, (2) inpatient operation, (3) current procedural terminology (CPT) codes for surgical resection of spinal tumors, (4) operation under general anesthesia, (5) surgical subspecialty neurosurgery or orthopedics. CPT codes we used to define our cohort are provided in Table S1. We excluded patients with the following criteria: (1) emergency surgery, (2) patients with preoperative ventilator dependence, (3) patients with any unclean wounds (defined by wound classes 2 to 4), (4) patients with sepsis/shock/systemic inflammatory response syndrome 48 h before surgery, (5) patients with ASA physical status classification score of 4 and 5 or non-assigned.

2.4. Predictor Variables

Predictor variables included variables within the NSQIP database that were deemed to be known prior to the occurrence of the outcome of interest. These included (1) demographic information: age, sex, race/ethnicity, BMI (calculated from the height and weight), transfer status; (2) comorbidities and disease burden: diabetes mellitus, current smoke within one year, dyspnea, history of severe chronic obstructive pulmonary disease (COPD), ascites within 30 days prior to surgery, congestive heart failure within 30 days prior to surgery, hypertension requiring medication, acute renal failure, currently requiring or on dialysis, disseminated cancer, steroid or immunosuppressant for a chronic condition, >10% loss of body weight in last 6 months, bleeding disorders, preoperative transfusion of ≥ 1 unit of whole/packed RBCs within 72 h prior to surgery, ASA classification, functional status prior to surgery; (3) preoperative laboratory values: serum sodium, blood urea nitrogen (BUN), serum creatinine, serum albumin, total bilirubin, serum glutamic-oxaloacetic transaminase (SGOT), alkaline phosphatase, white blood cell (WBC) count, hematocrit, platelet count, partial thromboplastin time (PTT), International Normalized Ratio of pro thrombin time (PT) values, PT; (4) operative variables: surgical specialty, days from hospital admission to operation, CPT code for the procedure; (5) spinal tumor variables: tumor location (extradural, intradural). Definitions of these predictor variables are provided in the ACS-NSQIP PUF User Guides (https://www.facs.org/quality-programs/data-and-registries/acs-nsqip/participant-use-data-file/, accessed on 1 January 2023). For transfer status, the variable values other than 'Not transferred (admitted from home)' were grouped as 'Transferred'; for diabetes, the variable values' Non-Insulin' and 'Insulin' were grouped as 'Yes'; for dyspnea, the variable values 'Moderate Exertion' and 'At rest' were grouped as 'Yes'. Race and ethnicity variables were aggregated into one column, 'Race'. If the patients' Hispanic ethnicity values were 'Yes', their 'Race' values were assigned as 'Hispanic' regardless of their original values.

2.5. Outcome of Interest

The primary outcomes were prolonged length of stay, which we defined as total length of stay greater than 75% of the included patient population, nonhome discharge, and major complications. We defined nonhome discharge by dichotomizing the variable discharge destination. If patients required additional levels of care upon discharge, a nonhome discharge destination was identified and included 'Rehab', 'Skilled Care, Not Home', and 'Separate Acute Care'. Patients with unknown discharge destinations, hospice discharges, discharges to unskilled facilities, and patients who expired were not included. We defined major complications, based on the previous literature [26–28], as having one of these events post-operatively: deep incisional surgical site infection (SSI), organ/space SSI, wound disruption, unplanned reintubation, pulmonary embolism, being on a ventilator for more than 48 h, renal insufficiency, acute renal failure needing dialysis, cardiac arrest, myocardial infarction, bleeding requiring blood transfusions, deep vein thrombosis, sepsis, and septic shock. We did not include complications involving less serious events to major complications, such as superficial wound infection, pneumonia, and urinary tract infection.

2.6. Data Preprocessing

In order not to introduce bias with the exclusion of patients with missing values, we utilized imputation. Fifteen continuous variables contained at least one missing value.

After excluding variables with missing values for more than 25% of the patient population, missing values for continuous variables were imputed using the nearest neighbor (NN) imputation algorithm [29]. A value generated from cases in the entire dataset is used to replace each missing value for cases with missing values using NN imputation algorithms [30]. The only categorical variable that contained missing values was the variable race, and its missing values were imputed as 'Unknown'.

The robust scaler was utilized to scale continuous variables to account for outliers [31]. Additionally, normalization is essential for ensuring that all feature values are on the same scale and assigned the same weight. Each continuous variable (e.g., BMI, laboratory values) was put on the (0, 1) range using a min–max normalization [32]. Categorical nonbinary variables (e.g., race, CPT codes) were one-hot-encoded, and variables with ordinal characteristics (e.g., ASA classification, functional status) were coded with the ordinal encoder [33].

The adaptive synthetic sampling (ADASYN) approach for imbalanced learning was used to artificially generate cases of positive outcomes of interest (i.e., prolonged LOS, non-home discharges, major postoperative complications) based on the training and validation sets in order to overcome the class imbalance for a positive outcome of interest [34]. In order to enhance model learning and generalizability, ADASYN uses instances from the minority class that are difficult to learn and creates synthetic new cases based on these instances [35].

2.7. Training, Validation, and Test Sets

Data was split into training, validation, and test sets. The training set was used to develop the models, the validation set to adjust hyperparameters, and the test set to assess model performance. Data from 2015 to 2020 was split into training, validation, and test sets in a 60:20:20 ratio.

2.8. Modeling

Four supervised ML algorithms were utilized using the predictor variables to predict the outcomes: XGBoost, LightGBM, CatBoost, and random forest. We used the Optuna optimization library, where the optimized metric was the area under the receiver operating characteristic curve (AUROC). Optuna is a software framework for hyperparameter optimization that makes it simple to apply various state-of-the-art optimization techniques to carry out hyperparameter optimization quickly and effectively. To generate AUROC estimates that would serve as a guide for the optimization process, Tree-Structured Parzen Estimator Sampler (TPESampler) was employed as the Bayesian optimization algorithm. The final models for the outcomes were then built using the whole training set along with the optimized hyperparameters. ML analyses were performed in Python version 3.7.15.

2.9. Performance Evaluation

Models were evaluated graphically with receiver operating characteristic (ROC) curve, precision–recall curve (PRC), and calibration plots; and numerically with AUROC, area under PRC (AUPRC), accuracy, precision, recall, and Matthew's correlation coefficient (MCC).

The ability of a binary classifier system to discriminate between positive and negative cases is shown graphically in a ROC curve, and the AUROC summarizes the model's ability to do so. An AUROC of 1.0 indicates a perfect discriminator, whereas values of 0.90 to 0.99 are regarded as excellent, 0.80 to 0.89 as good, 0.70 to 0.79 as fair, and 0.51 to 0.69 as poor [36].

The model's ability to detect all positive cases without recognizing false positives is shown graphically in a PRC, which plots recall (sensitivity) against precision (positive predictive value) and is summarized by the AUPRC. AUPRC can be a more responsive metric when used with datasets where the positive class is relatively uncommon because PRCs assess the proportion of correct predictions among the positive predictions [37].

In addition to the performance plots and metrics, we also utilized Shapley additive explanations (SHAP) to investigate the relative importance of predictor variables. SHAP is a visualization method frequently used in ML to comprehend how models make predictions

2.10. Online Prediction Tool

We created a web application for getting predictions for individual patients based on their characteristics (Figure 1). This application is based on the models presented in this study with a few differences in implementation. The application and its source code are accessible on a platform that allows users to share ML models, Hugging Face (https://huggingface.co/spaces/MSHS-Neurosurgery-Research/NSQIP-ST, accessed on 1 January 2023).

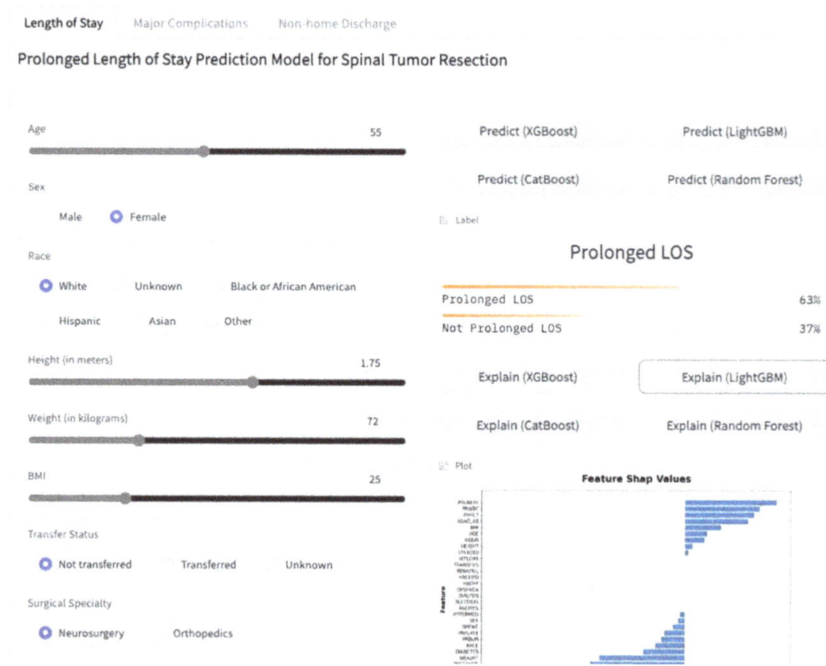

Figure 1. A screenshot of the online web application.

2.11. Statistical Analysis

The descriptive analyses were reported as means ($\pm$standard deviations) for normally distributed continuous variables, medians (interquartile ranges) for non-normally distributed continuous variables, and number of patients (% percentages). Group differences in outcomes were tested with the independent t-test for normally distributed continuous variables with equal variances, the Welch's t-test for normally distributed continuous variables with unequal variances, the Mann–Whitney U test for non-normally distributed continuous variables, and the Pearson's chi-squared test for categorical variables. Normality was evaluated with the Shapiro–Wilk test, and Levene's test was used to assess the equality of variances for a variable. The differences were considered to be statistically significant at $p < 0.05$. All statistical analyses were performed in Python version 3.7.15.

3. Results

Initially, a total of 6060 patients were identified via CPT codes. Inclusion and exclusion criteria were applied in a sequential manner. A total of 2449 were excluded due to non-elective surgeries, 145 due to outpatient surgeries, 17 due to anesthesia techniques other than general anesthesia, 42 due to surgical specialties other than neurosurgery or orthopedic surgery, 5 due to emergency surgeries, 8 due to preoperative ventilator dependency, 55 due to unclean wounds, 67 due to preoperative SIRS or sepsis, 165 due to ASA class 4, 5 or none assigned, 20 due to unknown LOS, 3 due to unknown major complication status and 11 due to discharge destination (Figure 2). After exclusion, 3073 patients were left in the analysis. There were 752 patients with prolonged LOS, 718 with nonhome discharges, and 379 with major complications. Characteristics of the patient population, both among the groups and in total, are presented in Tables S2–S4.

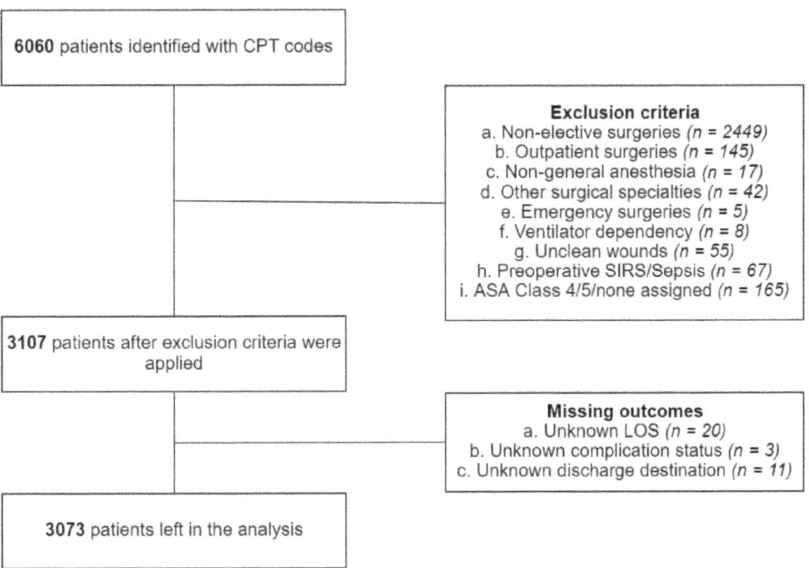

Figure 2. Patient selection process.

The most accurately predicted outcomes in terms of AUROC and accuracy were the prolonged LOS with a mean AUROC of 0.745 and accuracy of 0.804, and the major complications with a mean AUROC of 0.730 and accuracy of 0.856. The most accurately predicting algorithm in terms of AUROC was random forest, with a mean AUROC of 0.743, followed by LightGBM, with a mean AUROC of 0.729. The mean AUROCs for CatBoost and XGBoost were 0.726 and 0.704, respectively. Detailed metrics regarding the algorithms' performances are presented in Table 1. AUROC and AUPRC curves for the three outcomes are shown in Figures 3 and 4.

Table 1. Metrics regarding the algorithms' performances.

Outcome	Algorithm	P	R	F1	MCC	AUPRC	ACC	AUROC
	XGB	0.503	0.565	0.532	0.398	0.609	0.789	0.744
	LGB	0.449	0.641	0.528	0.423	0.621	0.808	0.748
LOS	CB	0.469	0.645	0.543	0.437	0.591	0.811	0.726
	RF	0.490	0.621	0.548	0.431	0.586	0.807	0.760
	Mean	**0.478**	**0.618**	**0.538**	**0.422**	**0.602**	**0.804**	**0.745**

Table 1. *Cont.*

Outcome	Algorithm	P	R	F1	MCC	AUPRC	ACC	AUROC
NHD	XGB	0.307	0.381	0.340	0.173	0.368	0.728	0.650
	LGB	0.343	0.475	0.398	0.262	0.410	0.764	0.712
	CB	0.436	0.477	0.455	0.304	0.454	0.763	0.725
	RF	0.414	0.436	0.425	0.261	0.402	0.745	0.719
	Mean	**0.375**	**0.442**	**0.405**	**0.250**	**0.408**	**0.750**	**0.701**
MC	XGB	0.192	0.405	0.261	0.212	0.293	0.862	0.718
	LGB	0.192	0.375	0.254	0.197	0.305	0.857	0.726
	CB	0.244	0.373	0.295	0.222	0.321	0.852	0.728
	RF	0.256	0.377	0.305	0.231	0.318	0.852	0.749
	Mean	**0.221**	**0.383**	**0.279**	**0.216**	**0.309**	**0.856**	**0.730**

P, precision; R, recall; MCC, Matthew's correlation coefficient; AUPRC, area under the precision recall curve; ACC, accuracy; AUROC, area under the receiver operating characteristic curve; LOS, length of stay; NHD, non-home discharge; MC major complications; XGB, XGBoost; LGB, LightGBM; CB, CatBoost; RF, Random Forest.

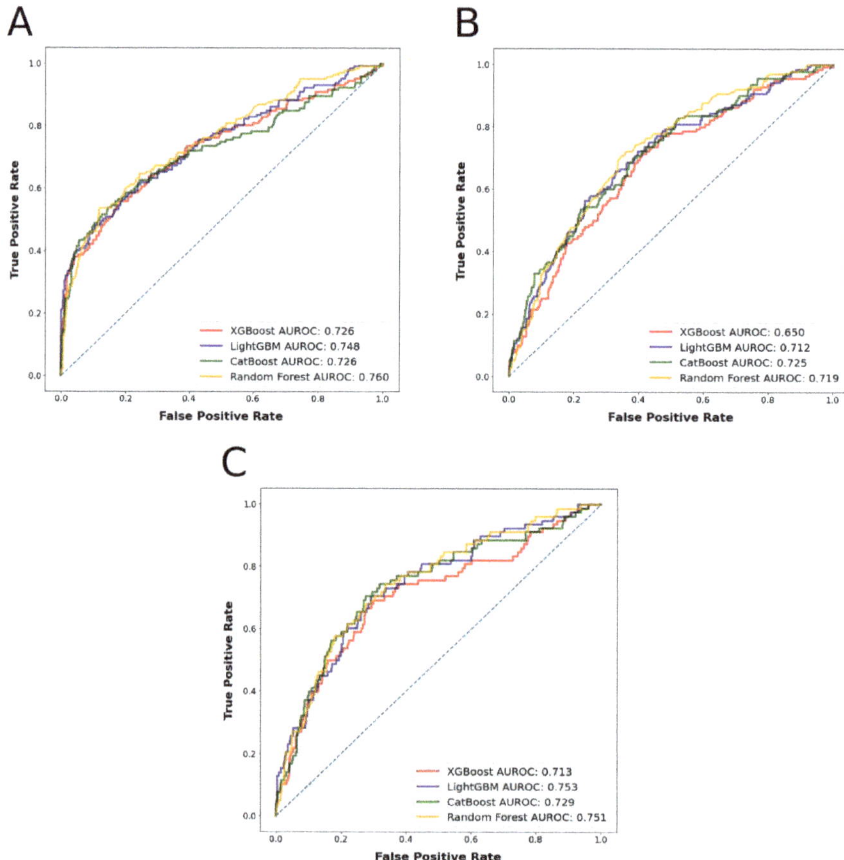

Figure 3. (**A**): Algorithms' receiver operator curves for the outcome prolonged length of sta (**B**): Algorithms' receiver operator curves for the outcome nonhome discharges. (**C**): Algorithm receiver operator curves for the outcome major complications.

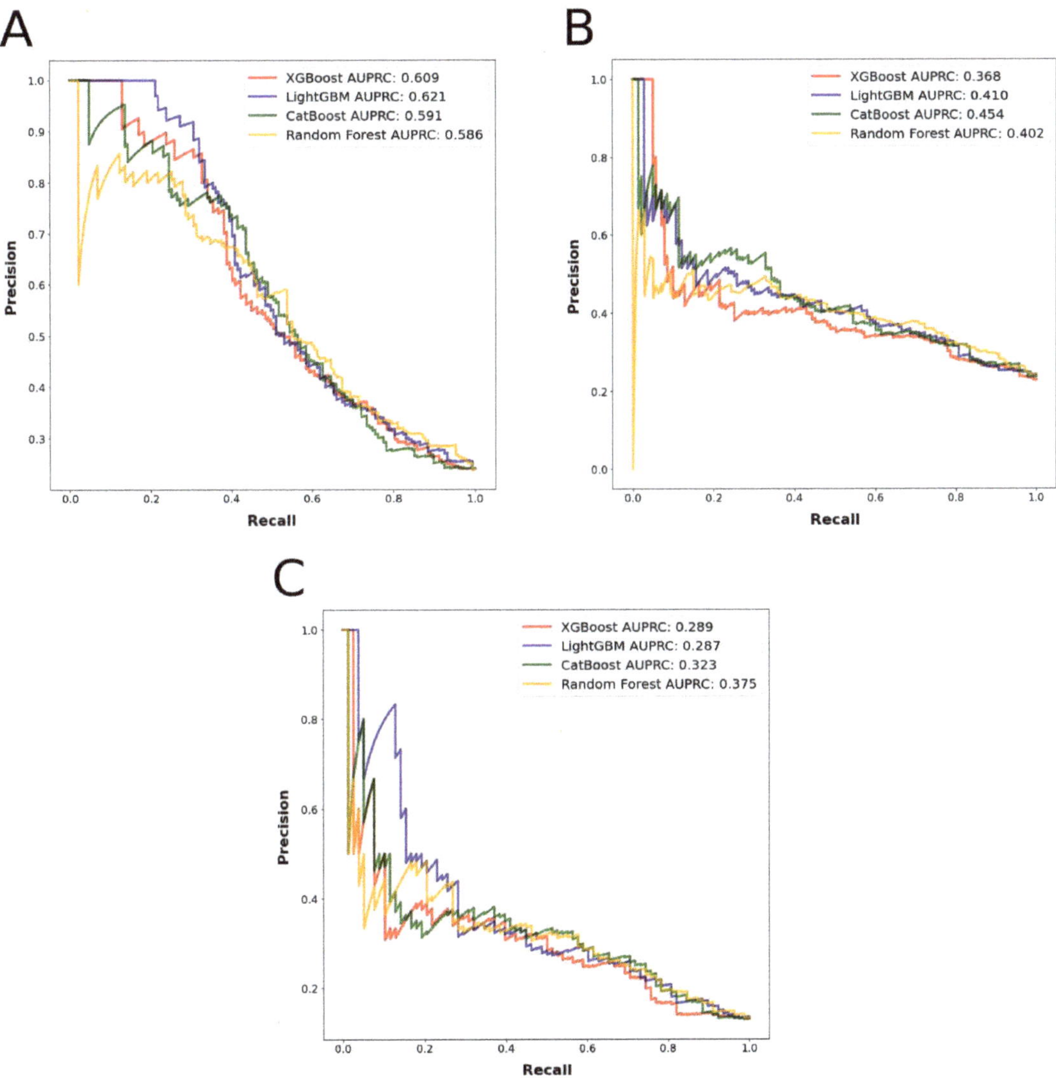

Figure 4. (**A**): Algorithms' precision–recall curves for the outcome prolonged length of stay. (**B**): Algorithms' precision–recall curves for the outcome nonhome discharges. (**C**): Algorithms' precision–recall curves for the outcome major complications.

SHAP plot of the XGBoost model for the outcome prolonged LOS, the CatBoost model for the outcome nonhome discharges, and the random forest model for the outcome major complications are presented in Figure 5. The other SHAP plots can be seen in Figure S1.

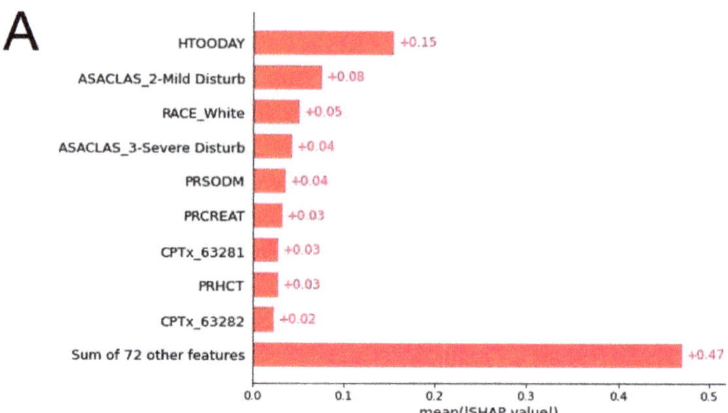

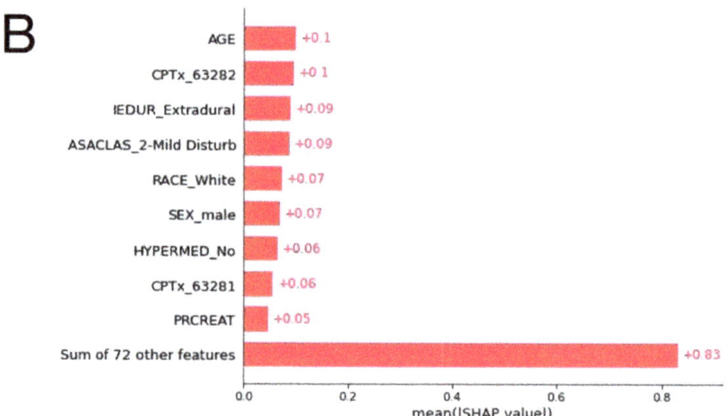

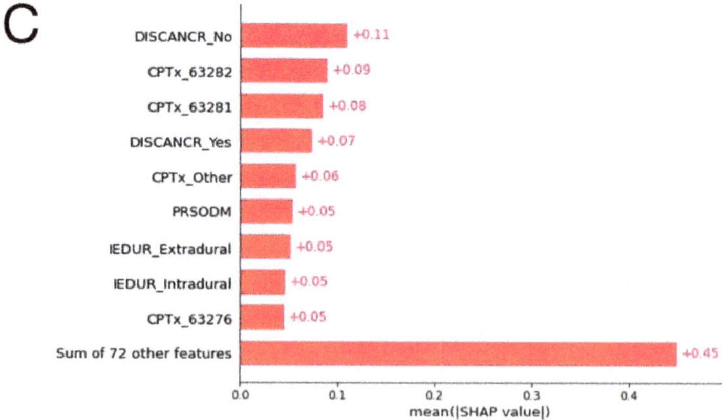

Figure 5. (**A**): The ten most important features and their mean SHAP values for the model predicting prolonged length of stay with the LightGBM algorithm. (**B**): The ten most important features and their mean SHAP values for the model predicting nonhome discharges with the CatBoost algorithm. (**C**): The ten most important features and their mean SHAP values for the model predicting major complications stay with the random forest algorithm.

4. Discussion

This study presents a set of ML algorithms that can preoperatively predict prolonged LOS, nonhome discharges, and major complications for patients undergoing spinal tumor resection. The results of the study here demonstrate significant potential for the prediction of surgical outcomes and may help in the risk stratification process for spinal tumor resections. Patients who are at risk of unfavorable outcomes after spinal tumor resection can be better informed about the risks of surgery, and providers can better customize patient care plans to reduce the risk of these unfavorable outcomes. This paper contributes to the literature by demonstrating the efficacy and significance of incorporating machine learning into clinical settings to predict postoperative spine surgery outcomes [38].

The ML algorithms were able to predict between 78.9% and 81.1% of the patients who had prolonged LOS accurately with AUROC values between 0.726 and 0.760; between 72.8% and 76.4% of the patients who had nonhome discharges accurately with AUROC values between 0.650 and 0.725, and between 85.2% and 86.2% of the patients who had major complications accurately with AUROC values between 0.718 and 0.749 in the test set. Based on prediction mean accuracies and AUROC values for the different outcomes, the random forest algorithm was found to have performed the best among all the algorithms tested. These results can be deemed as fair classification performance, as previously explained.

In addition to reporting our methods and results in this paper, we created a web application accessible to physicians worldwide. Our web application not only allows users to see predictions for the three investigated outcomes by the four different algorithms utilized in this study, but it also allows them to see visual explanations of the predictions by SHAP values and figures. This additional interpretability might be of benefit in clinical settings, where providers can address the individual risk factors to achieve the best possible postoperative outcomes for their patients. To the best of our knowledge, this is the first available ML-based web application that enables users to have predictions with explanations for postoperative outcomes after spinal tumor resections.

Yang et al. posted an online calculator based on their regression-based nomogram for spinal cord astrocytomas using patient data in the Surveillance, Epidemiology, and End Results (SEER) Program of the National Cancer Institute [39]. First of all, the patients included in this study were diagnosed between 1975 and 2016. It is a very broad timespan and it is not reported how the year of diagnosis has impact on the individual survival predictions because the online calculator does not have an input for the year of diagnosis. Despite achieving comparable results with our study in terms of classification performance, this online calculator does not incorporate advanced analytical techniques, such as the ML algorithms we utilized in our study. Our web application provides predictions by four ML algorithms, which allows users to have multiple insights for a single patient. Moreover, the input for the mentioned tool includes variables like 'histologic type', 'WHO grade' and 'postoperation radiotherapy' which would not be known prior to surgery. This approach would not make personalized treatment plans possible preoperatively. With this tool, users can have overall and cancer-specific survival predictions, while users can have predictions for 30-day postoperative outcomes with our web application. Previously, Karhade et al. incorporated only the best ML algorithm across the model performance metrics for predicting 30-day mortality after surgery for spinal metastases into an interactive interface web application [40]. This application only allows seven variables as input and predicts one outcome, thirty-day mortality. Although Karhade et al. used ML algorithms as classifiers, they did not mention the details of implementation in the paper. Preprocessing of the data was not mentioned except the imputation of missing data and, the source code for the classification algorithms and online application were not shared. The above factors limit the reproducibility of the results. Moreover, both of these tools lack elaboration of the predictions, unlike our application which provides that via SHAP values and figures.

The measures presented here for the machine learning algorithms are congruent with the current literature. The outcomes we picked for this investigation have not been studied in a single study using ML algorithms; nevertheless, a few publications have investigated the classification performance of machine learning algorithms in predicting postoperative outcomes in spinal tumor patients employing different data sources. Using institutional data, Masaad et al. compared the performance of the metastatic spinal tumor frailty index (MSTFI) with ML methods in identifying measures of frailty as predictors of outcomes [41]. The random forest algorithm performed best in the study and had an AUROC value of 0.62 for postoperative complications. Jin et al. queried IBM MarketScan Claims Database for adult patients receiving surgery for intradural tumors, with their primary outcome of interest being nonhome discharges and 90-day post-discharge admissions [42]. Their classification models were developed using a logistic regression approach regularized by the least absolute shrinkage and selection operator (LASSO) penalty, and they obtained AUROC values of 0.786 for nonhome discharges and 0.693 for 90-day readmissions.

Although a few studies using the NSQIP database that analyzed the accuracy of machine learning algorithms in predicting postoperative outcomes included, we did not include some of the available variables that would not be known prior to the surgery as input to our models, like total operative time [43,44]. The length of the procedure may be a consequence of unfavorable outcomes rather than its cause [45]. The study, in which Kalagara et al. analyzed the NSQIP database for readmissions following lumbar laminectomy and developed predictive models to identify readmitted patients, reported an overall accuracy of 95.9% and an AUROC value of 0.806 with a gradient boosting machine (GBM) model using all patient variables [46]. The most important variables that made this model achieve such good results included post-discharge complications and discharge destinations. A second GBM model to predict readmission utilizing only information known prior to readmission had an accuracy of 79.6% and an AUROC of 0.690. Still, this model included postoperative variables such as discharge destination and total hospital LOS, and those were among the most important features.

The SHAP analysis results are in line with the current literature on regression analysis for the relative importance of predictor variables. A study on predictors of discharge disposition following laminectomy for intradural extramedullary spinal tumors identified age over 65 years, ASA classification over three, and dependent functional status as predictors of nonhome discharge [13]. These variables were among the most important features of our machine learning models, as can be seen from the SHAP plots.

The study does have some potential limitations despite the strength of the methodology described. First, the sample of patients undergoing spinal tumor resection may not have accurately represented all the patients who undergo spinal tumor resection. The NSQIP dataset depends on reporting from participating hospitals. As a result, patients from hospitals with the infrastructure to uphold NSQIP reporting requirements will be overrepresented in the sample of spine tumor patients between 2016 and 2020. In addition, coding errors and other inaccuracies always affect studies using an extensive clinical database. Even though the NSQIP database is frequently used, there have been a few studies evaluating its actual accuracy. Neurosurgical procedure CPT codes contain numerous internal inconsistencies, according to Rolston et al. [47]. Second, NSQIP data do not include specific factors that might be associated with a patient's risk of unfavorable postoperative outcomes. For example, we could not assess the effect on outcomes of tumor-specific variables, such as histologic type or tumor size, because we lacked access to more granular data. While the current mean AUROCs between 0.703 and 0.734 are fair, adding these and other relevant variables may enhance the algorithm's performance.

5. Conclusions

Machine learning algorithms show great promise for predicting postoperative outcomes in spinal tumor surgery. These algorithms can be incorporated into clinically practical decision-making tools. The development of predictive models and the use of these models as accessible tools may significantly improve risk management and prognosis. Herein, we present and make publicly available a predictive algorithm for spinal tumor surgery aiming at the above goals.

Supplementary Materials: The following supporting information can be downloaded at: https://www.mdpi.com/article/10.3390/cancers15030812/s1. Table S1: CPT codes we used to define our cohort. Table S2: Characteristics of the patient population, both among the non-prolonged LOS and prolonged LOS groups and in total. Table S3: Characteristics of the patient population, both among the home discharge and nonhome discharge groups and in total. Table S4: Characteristics of the patient population, both among the no major complications and major complication groups and in total. Figure S1(a): The ten most important features and their mean SHAP values for the model predicting prolonged length of stay with the XGBoost algorithm. (b): The ten most important features and their mean SHAP values for the model predicting prolonged length of stay with the CatBoost algorithm. (c): The ten most important features and their mean SHAP values for the model predicting prolonged length of stay with the random forest algorithm. (d): The ten most important features and their mean SHAP values for the model predicting nonhome discharges with the XGBoost algorithm. (e): The ten most important features and their mean SHAP values for the model predicting nonhome discharges with the LightGBM algorithm. (f): The ten most important features and their mean SHAP values for the model predicting nonhome discharges with the Random Forest algorithm. (g): The ten most important features and their mean SHAP values for the model predicting major complications stay with the XGBoost algorithm. (h): The ten most important features and their mean SHAP values for the model predicting major complications stay with the LighthGBM algorithm. (i): The ten most important features and their mean SHAP values for the model predicting major complications stay with the random forest algorithm.

Author Contributions: Conceptualization, M.K. and K.M.; Methodology, M.K. and K.M.; Software, M.K.; Formal Analysis, M.K.; Data Curation, M.K.; Writing—Original Draft Preparation, M.K.; Writing—Review and Editing, K.M.; Visualization, M.K.; Supervision, K.M.; Project Administration, M.K. and K.M. All authors have read and agreed to the published version of the manuscript.

Funding: This research received no external funding.

Institutional Review Board Statement: This study was deemed exempt from approval by the Icahn School of Medicine at Mount Sinai institutional review board because it involved analysis of deidentified patient data (STUDY-22-01302, 29 December 2022).

Informed Consent Statement: Patient consent was waived due to analysis of deidentified patient data (STUDY-22-01302, 29 December 2022).

Data Availability Statement: Restrictions apply to the availability of these data. Data were obtained from American College of Surgeons National Surgical Quality Improvement Program and are available https://www.facs.org/quality-programs/data-and-registries/acs-nsqip/ (accessed on 1 January 2023) with the permission of American College of Surgeons.

Conflicts of Interest: The authors declare no conflict of interest.

Statement: The American College of Surgeons National Surgical Quality Improvement Program and the hospitals participating in the ACS NSQIP are the source of the data used herein; they have not verified and are not responsible for the statistical validity of the data analysis or the conclusions derived by the authors.

Source Code: The source code for preprocessing and analyzing the data is available on GitHub (https://github.com/mertkarabacak/NSQIP-ST, accessed on 1 January 2023).

References

1. Duong, L.M.; McCarthy, B.J.; McLendon, R.E.; Dolecek, T.A.; Kruchko, C.; Douglas, L.L.; Ajani, U.A. Descriptive Epidemiology of Malignant and Nonmalignant Primary Spinal Cord, Spinal Meninges, and Cauda Equina Tumors, United States, 2004–2007. *Cancer* **2012**, *118*, 4220–4227. [CrossRef] [PubMed]
2. Sharma, M.; Sonig, A.; Ambekar, S.; Nanda, A. Discharge Dispositions, Complications, and Costs of Hospitalization in Spinal Cord Tumor Surgery: Analysis of Data from the United States Nationwide Inpatient Sample, 2003–2010. *J. Neurosurg. Spine* **2014**, *20*, 125–141. [CrossRef] [PubMed]
3. Kaloostian, P.E.; Zadnik, P.L.; Etame, A.B.; Vrionis, F.D.; Gokaslan, Z.L.; Sciubba, D.M. Surgical Management of Primary and Metastatic Spinal Tumors. *Cancer Control* **2014**, *21*, 133–139. [CrossRef] [PubMed]

4. Schairer, W.W.; Carrer, A.; Sing, D.C.; Chou, D.; Mummaneni, P.V.; Hu, S.S.; Berven, S.H.; Burch, S.; Tay, B.; Deviren, V.; et al. Hospital Readmission Rates after Surgical Treatment of Primary and Metastatic Tumors of the Spine. *Spine* **2014**, *39*, 1801–1808. [CrossRef]

5. Galgano, M.; Fridley, J.; Oyelese, A.; Telfian, A.; Kosztowski, T.; Choi, D.; Gokaslan, Z.L. Surgical Management of Spinal Metastases. *Expert Rev. Anticancer. Ther.* **2018**, *18*, 463–472. [CrossRef] [PubMed]

6. Li, J.; Wei, W.; Xu, F.; Wang, Y.; Liu, Y.; Fu, C. Clinical Therapy of Metastatic Spinal Tumors. *Front. Surg.* **2021**, *8*, 626873. [CrossRef] [PubMed]

7. Barzilai, O.; Robin, A.M.; O'Toole, J.E.; Laufer, I. Minimally Invasive Surgery Strategies. *Neurosurg. Clin. North Am.* **2020**, *31*, 201–209. [CrossRef] [PubMed]

8. Camino Willhuber, G.; Elizondo, C.; Slullitel, P. Analysis of Postoperative Complications in Spinal Surgery, Hospital Length of Stay, and Unplanned Readmission: Application of Dindo-Clavien Classification to Spine Surgery. *Glob. Spine J.* **2019**, *9*, 279–286. [CrossRef]

9. Slattery, C.; Verma, K. Outcome Measures in Adult Spine Surgery: How Do We Identify the Outcomes of Greatest Utility for Research? *Clin. Spine Surg. A Spine Publ.* **2019**, *32*, 164–165. [CrossRef] [PubMed]

10. Kumar, N.; Patel, R.S.; Wang, S.S.Y.; Tan, J.Y.H.; Singla, A.; Chen, Z.; Ravikumar, N.; Tan, A.; Kumar, N.; Hey, D.H.W.; et al. Factors Influencing Extended Hospital Stay in Patients Undergoing Metastatic Spine Tumour Surgery and Its Impact on Survival. *J. Clin. Neurosci.* **2018**, *56*, 114–120. [CrossRef] [PubMed]

11. Pennington, Z.; Sundar, S.J.; Lubelski, D.; Alvin, M.D.; Benzel, E.C.; Mroz, T.E. Cost and Quality of Life Outcome Analysis of Postoperative Infections after Posterior Lumbar Decompression and Fusion. *J. Clin. Neurosci.* **2019**, *68*, 105–110. [CrossRef]

12. Zhou, R.P.; Mummaneni, P.V.; Chen, K.-Y.; Lau, D.; Cao, K.; Amara, D.; Zhang, C.; Dhall, S.; Chou, D. Outcomes of Posterior Thoracic Corpectomies for Metastatic Spine Tumors: An Analysis of 90 Patients. *World Neurosurg.* **2019**, *123*, e371–e378. [CrossRef] [PubMed]

13. Ahn, A.; Phan, K.; Cheung, Z.B.; White, S.J.W.; Kim, J.S.; Cho, S.K.-W. Predictors of Discharge Disposition Following Laminectomy for Intradural Extramedullary Spinal Tumors. *World Neurosurg.* **2019**, *123*, e427–e432. [CrossRef]

14. Kim, J.S.; Merrill, R.K.; Arvind, V.; Kaji, D.; Pasik, S.D.; Nwachukwu, C.C.; Vargas, L.; Osman, N.S.; Oermann, E.K.; Caridi, J.M.; et al. Examining the Ability of Artificial Neural Networks Machine Learning Models to Accurately Predict Complications Following Posterior Lumbar Spine Fusion. *Spine* **2018**, *43*, 853–860. [CrossRef]

15. Cruz, J.A.; Wishart, D.S. Applications of Machine Learning in Cancer Prediction and Prognosis. *Cancer Inf.* **2007**, *2*, 59–77. [CrossRef]

16. Kuhle, S.; Maguire, B.; Zhang, H.; Hamilton, D.; Allen, A.C.; Joseph, K.S.; Allen, V.M. Comparison of Logistic Regression with Machine Learning Methods for the Prediction of Fetal Growth Abnormalities: A Retrospective Cohort Study. *BMC Pregnancy Childbirth* **2018**, *18*, 333. [CrossRef] [PubMed]

17. Oermann, E.K.; Rubinsteyn, A.; Ding, D.; Mascitelli, J.; Starke, R.M.; Bederson, J.B.; Kano, H.; Lunsford, L.D.; Sheehan, J.P.; Hammerbacher, J.; et al. Using a Machine Learning Approach to Predict Outcomes after Radiosurgery for Cerebral Arteriovenous Malformations. *Sci. Rep.* **2016**, *6*, 21161. [CrossRef]

18. Lee, S.-I.; Celik, S.; Logsdon, B.A.; Lundberg, S.M.; Martins, T.J.; Oehler, V.G.; Estey, E.H.; Miller, C.P.; Chien, S.; Dai, J.; et al. A Machine Learning Approach to Integrate Big Data for Precision Medicine in Acute Myeloid Leukemia. *Nat. Commun.* **2018**, *9*, 42. [CrossRef] [PubMed]

19. Khuri, S.F.; Henderson, W.G.; Daley, J.; Jonasson, O.; Jones, S.R.; Campbell, D.A.J.; Fink, A.S.; Mentzer, R.M.J.; Steeger, J.E.; Study P.S.I. of the P.S. in S. The Patient Safety in Surgery Study: Background, Study Design, and Patient Populations. *J. Am. Coll. Surg.* **2007**, *204*, 1089–1102. [CrossRef]

20. Hall, B.L.; Hamilton, B.H.; Richards, K.; Bilimoria, K.Y.; Cohen, M.E.; Ko, C.Y. Does Surgical Quality Improve in the American College of Surgeons National Surgical Quality Improvement Program: An Evaluation of All Participating Hospitals. *Ann. Surg.* **2009**, *250*, 363–376. [CrossRef]

21. Ingraham, A.M.; Richards, K.E.; Hall, B.L.; Ko, C.Y. Quality Improvement in Surgery: The American College of Surgeons National Surgical Quality Improvement Program Approach. *Adv. Surg.* **2010**, *44*, 251–267. [CrossRef] [PubMed]

22. Ingraham, A.M.; Cohen, M.E.; Bilimoria, K.Y.; Dimick, J.B.; Richards, K.E.; Raval, M.V.; Fleisher, L.A.; Hall, B.L.; Ko, C.Y. Association of Surgical Care Improvement Project Infection-Related Process Measure Compliance with Risk-Adjusted Outcomes: Implications for Quality Measurement. *J. Am. Coll. Surg.* **2010**, *211*, 705–714. [CrossRef] [PubMed]

23. About ACS NSQIP. Available online: https://www.facs.org/quality-programs/data-and-registries/acs-nsqip/about-acs-nsqip/ (accessed on 29 September 2022).

24. Collins, G.S.; Reitsma, J.B.; Altman, D.G.; Moons, K.G.M. Transparent Reporting of a Multivariable Prediction Model for Individual Prognosis or Diagnosis (TRIPOD): The TRIPOD Statement. *BMC Med.* **2015**, *13*, 1. [CrossRef]

25. Luo, W.; Phung, D.; Tran, T.; Gupta, S.; Rana, S.; Karmakar, C.; Shilton, A.; Yearwood, J.; Dimitrova, N.; Ho, T.B.; et al. Guidelines for Developing and Reporting Machine Learning Predictive Models in Biomedical Research: A Multidisciplinary View. *J. Med. Internet Res.* **2016**, *18*, e323. [CrossRef]

26. Bovonratwet, P.; Ondeck, N.T.; Nelson, S.J.; Cui, J.J.; Webb, M.L.; Grauer, J.N. Comparison of Outpatient vs Inpatient Total Knee Arthroplasty: An ACS-NSQIP Analysis. *J. Arthroplast.* **2017**, *32*, 1773–1778. [CrossRef]

7. Basques, B.A.; Ibe, I.; Samuel, A.M.; Lukasiewicz, A.M.; Webb, M.L.; Bohl, D.D.; Grauer, J.N. Predicting Postoperative Morbidity and Readmission for Revision Posterior Lumbar Fusion. *Clin. Spine Surg.* **2017**, *30*, E770. [CrossRef]
8. Sood, A.; Abdollah, F.; Sammon, J.D.; Kapoor, V.; Rogers, C.G.; Jeong, W.; Klett, D.E.; Hanske, J.; Meyer, C.P.; Peabody, J.O.; et al. An Evaluation of the Timing of Surgical Complications Following Nephrectomy: Data from the American College of Surgeons National Surgical Quality Improvement Program (ACS-NSQIP). *World J. Urol.* **2015**, *33*, 2031–2038. [CrossRef] [PubMed]
9. Sklearn.Impute.KNNImputer. Available online: https://scikit-learn/stable/modules/generated/sklearn.impute.KNNImputer.html (accessed on 29 September 2022).
10. Beretta, L.; Santaniello, A. Nearest Neighbor Imputation Algorithms: A Critical Evaluation. *BMC Med. Inform. Decis. Mak.* **2016**, *16*, 74. [CrossRef]
11. Sklearn.Preprocessing.RobustScaler. Available online: https://scikit-learn/stable/modules/generated/sklearn.preprocessing.RobustScaler.html (accessed on 29 September 2022).
12. Sklearn.Preprocessing.MinMaxScaler. Available online: https://scikit-learn/stable/modules/generated/sklearn.preprocessing.MinMaxScaler.html (accessed on 29 September 2022).
13. Sklearn.Preprocessing.OrdinalEncoder. Available online: https://scikit-learn/stable/modules/generated/sklearn.preprocessing.OrdinalEncoder.html (accessed on 29 September 2022).
14. ADASYN—Version 0.9.1. Available online: https://imbalanced-learn.org/stable/references/generated/imblearn.over_sampling.ADASYN.html (accessed on 29 September 2022).
15. He, H.; Bai, Y.; Garcia, E.A.; Li, S. ADASYN: Adaptive Synthetic Sampling Approach for Imbalanced Learning. In Proceedings of the 2008 IEEE International Joint Conference on Neural Networks (IEEE World Congress on Computational Intelligence), Hong Kong, China, 1–6 June 2008; pp. 1322–1328.
16. Hanley, J.A.; McNeil, B.J. The Meaning and Use of the Area under a Receiver Operating Characteristic (ROC) Curve. *Radiology* **1982**, *143*, 29–36. [CrossRef]
17. Saito, T.; Rehmsmeier, M. The Precision-Recall Plot Is More Informative than the ROC Plot When Evaluating Binary Classifiers on Imbalanced Datasets. *PLoS ONE* **2015**, *10*, e0118432. [CrossRef]
18. Galbusera, F.; Casaroli, G.; Bassani, T. Artificial Intelligence and Machine Learning in Spine Research. *JOR Spine* **2019**, *2*, e1044. [CrossRef] [PubMed]
19. Yang, S.; Yang, X.; Wang, H.; Gu, Y.; Feng, J.; Qin, X.; Feng, C.; Li, Y.; Liu, L.; Fan, G.; et al. Development and Validation of a Personalized Prognostic Prediction Model for Patients with Spinal Cord Astrocytoma. *Front. Med.* **2022**, *8*, 802471. [CrossRef]
20. Karhade, A.V.; Thio, Q.C.B.S.; Ogink, P.T.; Shah, A.A.; Bono, C.M.; Oh, K.S.; Saylor, P.J.; Schoenfeld, A.J.; Shin, J.H.; Harris, M.B.; et al. Development of Machine Learning Algorithms for Prediction of 30-Day Mortality After Surgery for Spinal Metastasis. *Neurosurgery* **2019**, *85*, E83–E91. [CrossRef] [PubMed]
21. Massaad, E.; Williams, N.; Hadzipasic, M.; Patel, S.S.; Fourman, M.S.; Kiapour, A.; Schoenfeld, A.J.; Shankar, G.M.; Shin, J.H. Performance Assessment of the Metastatic Spinal Tumor Frailty Index Using Machine Learning Algorithms: Limitations and Future Directions. *Neurosurg. Focus* **2021**, *50*, E5. [CrossRef] [PubMed]
22. Jin, M.C.; Ho, A.L.; Feng, A.Y.; Medress, Z.A.; Pendharkar, A.V.; Rezaii, P.; Ratliff, J.K.; Desai, A.M. Prediction of Discharge Status and Readmissions after Resection of Intradural Spinal Tumors. *Neurospine* **2022**, *19*, 133–145. [CrossRef]
23. Zhong, H.; Poeran, J.; Gu, A.; Wilson, L.A.; Gonzalez Della Valle, A.; Memtsoudis, S.G.; Liu, J. Machine Learning Approaches in Predicting Ambulatory Same Day Discharge Patients after Total Hip Arthroplasty. *Reg. Anesth. Pain Med.* **2021**, *46*, 779–783. [CrossRef]
24. Lopez, C.D.; Gazgalis, A.; Peterson, J.R.; Confino, J.E.; Levine, W.N.; Popkin, C.A.; Lynch, T.S. Machine Learning Can Accurately Predict Overnight Stay, Readmission, and 30-Day Complications Following Anterior Cruciate Ligament Reconstruction. *Arthroscopy* **2022**, *in press*. [CrossRef]
25. Harris, A.H.S.; Trickey, A.W.; Eddington, H.S.; Seib, C.D.; Kamal, R.N.; Kuo, A.C.; Ding, Q.; Giori, N.J. A Tool to Estimate Risk of 30-Day Mortality and Complications After Hip Fracture Surgery: Accurate Enough for Some but Not All Purposes? A Study From the ACS-NSQIP Database. *Clin. Orthop. Relat. Res.* **2022**, *480*, 2335–2346. [CrossRef]
26. Kalagara, S.; Eltorai, A.E.M.; Durand, W.M.; DePasse, J.M.; Daniels, A.H. Machine Learning Modeling for Predicting Hospital Readmission Following Lumbar Laminectomy. *J. Neurosurg. Spine* **2019**, *30*, 344–352. [CrossRef]
27. Rolston, J.D.; Han, S.J.; Chang, E.F. Systemic Inaccuracies in the National Surgical Quality Improvement Program Database: Implications for Accuracy and Validity for Neurosurgery Outcomes Research. *J. Clin. Neurosci.* **2017**, *37*, 44–47. [CrossRef]

Current Oncology

Review

Artificial Intelligence for Cancer Detection—A Bibliometric Analysis and Avenues for Future Research

Erik Karger [1,*] and Marko Kureljusic [2]

[1] Information Systems and Strategic IT Management, University of Duisburg-Essen, 45141 Essen, Germany
[2] International Accounting, University of Duisburg-Essen, 45141 Essen, Germany
* Correspondence: erik.karger@uni-due.de

Abstract: After cardiovascular diseases, cancer is responsible for the most deaths worldwide. Detecting a cancer disease early improves the chances for healing significantly. One group of technologie that is increasingly applied for detecting cancer is artificial intelligence. Artificial intelligence ha great potential to support clinicians and medical practitioners as it allows for the early detectio of carcinomas. During recent years, research on artificial intelligence for cancer detection grew lot. Within this article, we conducted a bibliometric study of the existing research dealing with th application of artificial intelligence in cancer detection. We analyzed 6450 articles on that topic tha were published between 1986 and 2022. By doing so, we were able to give an overview of this researc field, including its key topics, relevant outlets, institutions, and articles. Based on our findings, w developed a future research agenda that can help to advance research on artificial intelligence fo cancer detection. In summary, our study is intended to serve as a platform and foundation fo researchers that are interested in the potential of artificial intelligence for detecting cancer.

Keywords: cancer detection; artificial intelligence; machine learning; deep learning; bibliome ric study

Citation: Karger, E.; Kureljusic, M. Artificial Intelligence for Cancer Detection—A Bibliometric Analysis and Avenues for Future Research. *Curr. Oncol.* **2023**, *30*, 1626–1647. https://doi.org/10.3390/ curroncol30020125

Received: 27 December 2022
Revised: 18 January 2023
Accepted: 27 January 2023
Published: 29 January 2023

1. Introduction

Living cells are the basic elements of all plants and animals. These cells constantl divide to replace destroyed cells or to enable the individual to grow. Although this i usually a balanced and controlled process, this genetic control can be damaged, possibl resulting in cancer [1]. Cancer is a disease that can affect most cell-based life. It befall mankind as long as it has existed and was already recognized and acknowledged by th ancient Egyptians [2]. After cardiovascular diseases, cancer is responsible for the mos deaths worldwide [3]. In 2018, there were more than 18 million new estimated cance cases and 9.6 million cancer deaths worldwide [4]. Given the threat that cancer constitutes researchers have already tried to understand for a long time how to cure this group o diseases in the best way.

Apart from treatments once cancer occurs, it is important to recognize the disease a soon as possible to increase the chances of recovery [5–8]. One reason why lung cance is the deadliest cancer type is that it is difficult to detect in early stages and hard t cure in an advanced stage [9,10]. Given the high benefits of detecting cancer in earl stages, new approaches are steadily being developed to support an early cancer diagnosi Mammography was introduced in 1960 [11] and is nowadays one of the most commo tools to detect breast cancer [12]. With digitalization and advances in computing powe computers have been increasingly used to support clinical practitioners with making medical diagnosis. Computer systems that help with the detection of cancer (compute aided detection, CAD) are an opportunity to support radiologists to achieve better detectio performance [13].

One technology that receives increasing attention in recent years is artificial intelligenc (AI). AI is a broad term that covers many different technologies and developments, suc as machine learning (ML) or deep learning (DL) [14]. In recent years, AI has been applied in medicine for several purposes, for example, to support medical practitioners with their decision-making [15]. In the context of oncology, AI is increasingly investigated and used for several different purposes [14]. One promising application is the detection and diagnosis of cancer. Due to its potential to effectively screen or diagnose cancer or polyps [14,16], AI might be a gamechanger in the early detection of cancer diseases and is the next step in the evolution of CAD.

Not only in clinical practice but also as a research field, AI for cancer detection and diagnosis grew rapidly over the past years. Since the 2010s, the annual research on AI-supported cancer diagnosis has been steadily increasing. It is nowadays a research field with contributions from different fields, such as medicine, computer science, mathematics, and engineering. Despite the fact that there are many reviews about AI on cancer [17–19], there is no comprehensive study that aims to give an overview of the research field of AI in cancer detection as a whole. This is surprising, since due to the wealth of research and publications, AI for cancer detection is nowadays a huge field that is hard to oversee. This makes it difficult for interested researchers and practitioners to obtain an impression of this field, its key publications, and the main topics addressed. Given that, we aim to close this research gap by giving an overview of the literature on AI for cancer detection. The first research question we aim to address is follows:

RQ1: What are the key topics of research on AI-supported cancer detection, and what are the most contributing research constituents and articles?

To answer our research question, we conduct a bibliometric study. A bibliometric study is a quantitative and statistical analysis of literature and allows for analyzing much larger bibliographic datasets than systematic literature reviews that follow a qualitative approach [20]. Due to their benefits, bibliometric studies have gained in popularity in recent years. Bibliometric approaches have been used in many different areas and disciplines, including pharmacy [21–23], oncology [24], or business and management [25]. By collecting and analyzing prior research, a bibliometric study can help to advance a field by systematically summarizing existing results. By doing so, reviews of the existing literature can also help to outline promising future research avenues and thus serve as a platform for interested scholars [26]. We follow this assumption and aim to derive future research avenues from our findings. Hence, our second research question is as follows:

RQ2: What are promising future research avenues that can help to advance the research on AI-based cancer detection?

The remainder of this article is structured as follows. In the next section, we will give an overview of AI and some foundational key terms. With that, we aim to equip readers that are not familiar with AI with basic knowledge and foundations about that technology. After that, we will explain our bibliometric approach in the third section. The bibliometric approach is divided into two phases, data collection and data analysis. Both phases are explained in more detail in two different subsections. In the fourth section, we will present the results of the bibliometric study. This is followed by a future research agenda in the fifth section. Finally, the sixth section consists of a discussion of this study's limitations and implications, while the seventh section contains concluding remarks.

2. Foundations of Artificial Intelligence

The beginning of AI can be dated to the year 1943 [27] when the first concept of an artificial neuron was proposed by [28]. Thirteen years later, at the Darthmouth Conference, the term artificial intelligence was used for the first time [29]. As such, AI is one of the newest fields that is investigated in science and engineering [29] and is nowadays a complex and thriving field with numerous research topics and many use-cases and applications for companies and in practice [30–32]. Especially in recent years, AI has experienced extensive growth and is viewed with interest from society and practice. The main reasons are advances in computing power and increasingly more data that are available to train AI systems [33]. It is important to note that AI is a multidisciplinary field, however, that is investigated in several research fields and disciplines, including neuroscience, psychology, computer science, and mathematics [34,35].

AI is an umbrella term that comprises a lot of different algorithms and technologies. One of the most frequently used AI technologies are artificial neural networks. If artificial neural networks are multilayered and consist of several hidden layers, they are also referred to as deep learning [30,31]. Artificial neural networks aim to simulate how humans and other biological organisms learn [36]. As such, artificial neural networks are inspired by the brains of living organisms and consist of processing units, called neurons, that are connected to each other [31]. These neurons receive inputs, which then are processed according to specific rules, resulting in an output of the neuron. Often, these neurons are arranged in different modules or layers. In this context, the term deep learning describes different types of complex neural networks that consist of a large number of neurons and layers. There are several other technologies that belong to AI, such as random forests [37,38] or support vector machines [39,40]. The explanation of these technologies, however, would go beyond the scope of this paper and is not necessary to understand the further results of this study.

Although modern AI systems have a lot of capabilities, they are not intelligent in the narrow sense. To describe the capabilities of AI, [41] was the first to differentiate between two forms of AI, namely strong and weak AI. Weak AI systems are only developed for single tasks and are not generally intelligent. Additionally, they lack other human characteristics like emotions, feelings, or a conscious mind [34,41]. Although weak AI systems often seem like they would be intelligent, they only behave like that [29,42]. In contrast, strong AI, also called artificial general intelligence (AGI), describes AI systems that have the intelligence or capabilities of humans [43,44]. This not only includes the intelligence but can also mean that these systems have emotions or feelings [34]. All of today's AI system belong to weak AI, while strong AI is not yet realized [45]. There are many assumptions about the time when a strong AI will be realized, with some researchers arguing that a strong AI might be never achieved [46].

3. Method

In this section, we explain our bibliometric approach. The conduction of a bibliometric study can be roughly divided into two steps. First, the data to be analyzed have to be collected. This step is described in the first subsection. The step of data collection is followed by the actual analysis of the data. This process is outlined in the second subsection.

3.1. Collection of Data

The first step was to collect the bibliometric data for our analysis. For the collection of bibliometric data, several databases exist, nowadays, with Scopus and Web of Science being among the most popular [47,48]. These databases differ in terms of their features and functionalities [49]. We decided to follow the recommendation of [20] to collect bibliometric data only from one database. We chose Scopus as the scientific database for our data collection. Scopus is a well-known database that has been used by several other bibliometric studies in the past [21,47,48,50–52]. Additionally, Scopus covers more journals than Web of Science and was therefore found to be suitable to identify as much research as possible [26]. Although there are other databases like Google Scholar and PubMed, we decided not to use these databases. First, Scopus has the option to develop a detailed search string and automatically download all bibliometric metadata, which is not possible with Google Scholar. Second, in comparison to PubMed, Scopus covers much more interdisciplinary research. As AI-based cancer detection is a multidisciplinary research topic, we found Scopus to be the most suitable database for conducting a bibliometric analysis.

For the creation of our search string, we oriented ourselves to other recent bibliometric studies that investigated AI within medicine [53] and pharmacy [21]. Our search string consists of two parts, one that covers the technical terms and another that consists of the application domain. The technical part consists of general technical terms like "artificial intelligence" or "machine learning". To search more broadly, we additionally searched for specific technologies, such as "artificial neural network", "deep learn*", fuzzy expert system", or "evolutionary computation". The applicational terms consisted of "cancer detect* and "cancer diagnos*". The use of * symbol is due to the syntax of Scopus and allows to search for all possible word endings of the search term. This led to the following search string that was applied:

((("artificial intelligence" OR "machine intelligence" OR "artificial neural network*" OR "machine learn*" OR "deep learn*" OR "thinking computer system" OR "fuzzy expert system*" OR "evolutionary computation" OR "hybrid intelligent system*") AND ("cancer detect*" OR "cancer diagnos*")).

The search was conducted in title, abstracts, and keywords on 23 September 2022. The initial results consisted of 7206 documents. We did several exclusion steps to refine the data collection and to come to our final sample. First, we limited our search to 2022 as the latest year of publication. This led to an elimination of eight articles. After that, we eliminated articles based on their document type. Herein, the only documents that remained were journal articles, conference papers, or reviews. This step led to the elimination of 604 publications, with 6594 articles remaining. After that, we excluded 137 non-English articles. As a last step, we eliminated seven articles with undefined authors. In summary, this led to an elimination of 756 publications, leaving a final sample of 6450 publications. Figure 1 shows an overview of the research process, the applied exclusion criteria, and the respective numbers of eliminated publications.

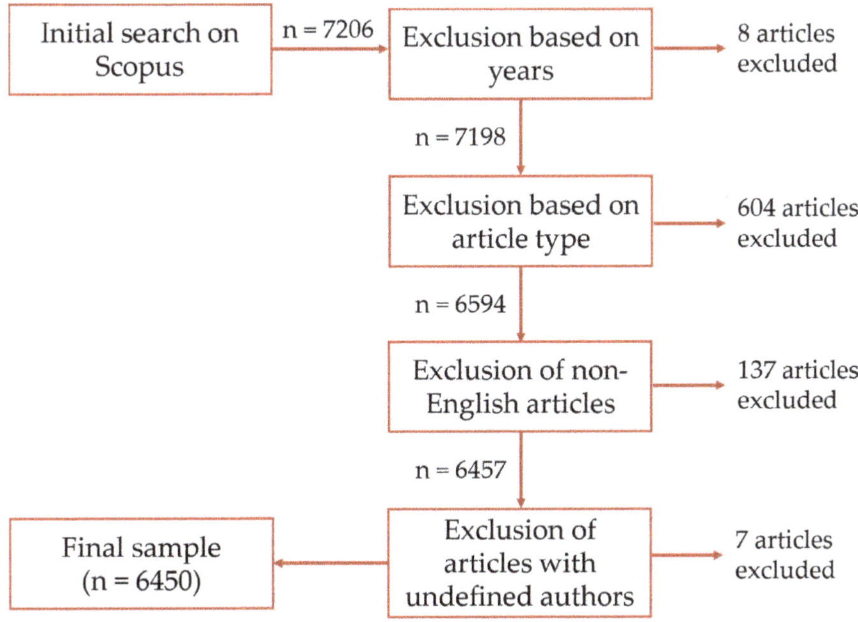

Figure 1. Overview of the literature collection and the exclusion criteria.

3.2. Data Analysis

In recent years, many tools that can help to analyze bibliometric data appeared [20]. In our study, we used two tools in combination, namely Bibliometrix/Bilioshiny and VOSviewer. First, Bibliometrix is an open source R package developed by [54]. It allows for a broad variety of different forms of analysis on bibliometric data [49]. We additionally complemented Bibliometrix with Biblioshiny. Biblioshiny enables the better creation of visualizations of bibliometric data [49]. We additionally complemented Biblioshiny and Bibliometrix with VOSviewer. VOSviewer is a tool for the visualization of bibliometric data. It was developed at Leiden University in the Netherlands by the Centre for Science and Technology Studies [49,55]. VOSviewer was applied in several bibliometric studies and enables the construction of bibliometric networks that show relationships between, among others, publications, outlets, keywords, or researchers. Additionally, VOSviewer supports the creation of co-citation, bibliographic coupling, and co-authorship analysis [49,55]. Although Biblioshiny stands out in terms of statistical functionalities, we found VOSviewer a suitable tool to visualize keyword co-occurrences.

4. Findings

The following three sections contain the results of our bibliometric analysis. First, we will give a general overview of the sample we collected and show of the fundamental key metrics. After that, we will show the results of our performance analysis. This first contains an overview of the sources with the most publications dealing with AI for cancer detection. Second, we present the most contributing countries, funding sponsors, and affiliations. After the performance analysis, we present a thematic analysis of the most relevant topics and key themes.

4.1. General Metrics and Overview

In this first subsection, we will present an overview of our sample and present some general metrics, such as annual production, document types, and information about the contributing authors. Table 1 shows an overview of the basic metrics of our final sample. In total, the sample consists of 6450 unique documents. These documents have been authored and co-authored by 23,854 different scholars, which is equal to 0.270 documents per author. In total, 247,762 references were cited and 9321 author's keywords appear. Additionally, 21,192 keywords plus were identified. The 6450 documents were published in 2018 different sources and received 19.87 citations on average. Of the 6232 multi-authored articles, around 25% were developed with an international team. The timeliness of this research topic is underpinned by the fact that the average document age is only 3.72 years old. This indicates that the majority of research has been published in the last 4 years.

Table 1. Main information and general metrics.

Metric	Value
Main information	
Timespan of publications	1986–2022
Sources (conferences and journals)	2018
Documents	6450
Average citations per document	19.87
Average document age	3.72
Total number of references	247,762
Number of author's keywords	9321
Number of keywords plus	21,192
Document types	
Journal article	4016
Conference article	1729
Review	708
Authors and collaboration	
Number of different AI-cancer authors	23,854
Documents per AI-cancer author	0.270
Single-authored documents	218
Multi-authored documents	6232
Authors of multi-authored documents	23,651
Co-authors per document	5.89
Collaboration index	3.8
International co-authorship	24.97%

We compared our bibliometric data with other bibliometric studies on different topics (for an overview, see Table 2). First, it is striking that a comparatively small number of publications on AI for cancer detection have been single-author documents. Only 218 of the 6450 documents were single-authored articles, which is equal to 3.38%. This might be an indicator of the very high complexity of this topic that makes it necessary to work together in large author teams. This assumption is further underpinned by the high collaboration index for our study. The collaboration index is often used to measure the cooperation between researchers and is calculated by dividing the total number of authors that contributed to multi-authored documents by the total number of multi-authored articles [56,57]. The number of documents per author is the lowest compared to the other bibliometric studies. This shows that a lot of different researchers contribute to the field of AI for cancer detection and that this field is not dominated by only a few researchers.

Table 2. Comparison of different bibliometric studies.

Study	[58]	[59]	[21]	[48]	This Study
Topic	Data quality	Blockchain in accounting	AI for drug discovery	Data governance	AI for cancer detection
Documents	159	93	3884	780	6450
Documents per author	0.305	0.443	0.322	0.367	0.27
Collaboration index	3.60	2.83	3.26	3.26	3.8
Single-authored documents	-	29%	6.7%	22.18%	3.4%

Figure 2 shows the annual production of research dealing with AI for cancer detection. The first research dealing with that topic was published in the 1980s. The first article can be dated to 1986. In this article, an expert system for the early detection of cervical cancer was proposed [60]. Until 1995, AI for cancer detection only experienced small growth in terms of annual production. In 1988 and 1990, no articles on this topic were published at all. In the following years, the number of publications only grew slowly. With 111 publications, the annual productions first topped the hundred mark in 2014. As the importance and potential of AI in general have increased, so has AI gained relevance in the field of cancer detection. As a result, most publications have been published in recent years (2019–2021). In 2022, 1213 publications had already been published before we collected the data for our study (23 September). Since it appears like statistically more publications are published in the last months of a year [21], we assume that the trend of increasing publications will be ongoing in 2022. Based on an extrapolation, we assume the total number of publications for 2022 will be 1872, with an estimate of 659 articles published after 23 September.

Figure 3 displays the distribution of disciplines among the publications. The data of Figure 3 were derived from Scopus wherein a publication is assigned to a discipline based on the outlet it was published in. However, some journals or conferences can belong to more than one discipline. Not surprisingly, we see that medicine and computer science outlets are the most popular ones within AI for cancer detection. A total of 23% and 21% of all articles have been published in outlets that belong to these disciplines. Medicine and computer science are followed by biochemistry, genetics, and molecular biology (11%); pharmacology, toxicology, and pharmaceutics (9%); and chemistry (8%). The dominance of medicine and computer science is not surprising, since oncology and the detection and treatment of cancer is one of the central disciplines in medicine, while AI is traditionally rooted within computer science.

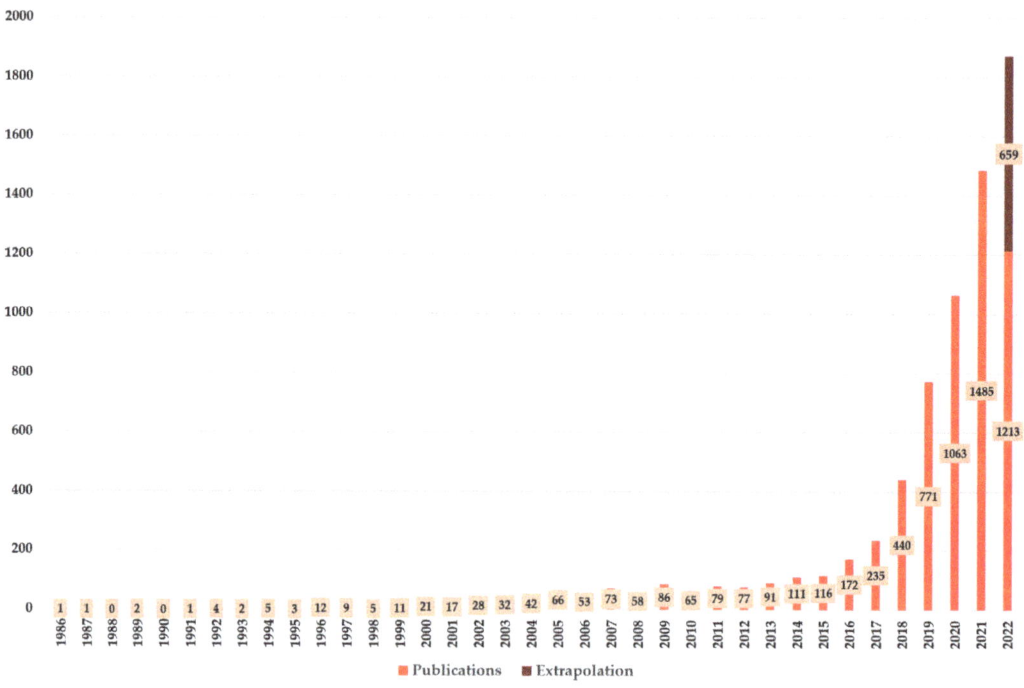

Figure 2. Overview of the annual production.

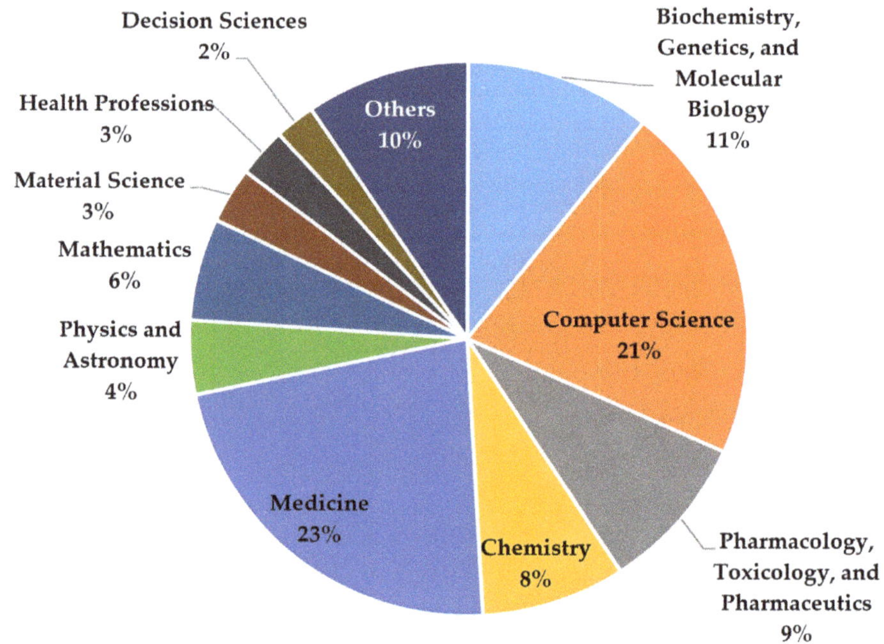

Figure 3. Overview of the most contributing disciplines.

4.2. Sources, Countries, and Affiliations

In this subsection, we show the results of our performance analysis, wherein we focus on the contributions of different research constituents. First, we take a look at the most relevant sources in terms of their absolute publication count within our sample. Table 3 shows the 20 sources with the most publications on AI for cancer detection. In total, the 6450 articles of our sample were published in 2018 different sources, which is equal to 3.2 publications per source. With 169 publications, Lecture Notes in Computer Science is the most important outlet in terms of absolute publication count. Lecture Notes in Computer Science is followed by Progress In Biomedical Optics And Imaging Proceedings Of SPIE (110 publications), Cancers (94 publications), and Computers In Biology And Medicine (88 publications).

Table 3. Overview of the sources with the most publications.

Rank	Source	Publications
01	Lecture Notes in Computer Science	169
02	Progress In Biomedical Optics and Imaging Proceedings Of SPIE	110
03	Cancers	94
04	Computers in Biology and Medicine	88
05	Computer Methods and Programs in Biomedicine	81
06	Scientific Reports	79
07	Plos One	71
08	European Radiology	69
09	Diagnostics	68
10	IEEE Access	64
11	Artificial Intelligence in Medicine	61
12	Medical Image Analysis	59
13	Proceedings Of SPIE The International Society for Optical Engineering	58
14	Frontiers in Oncology	55
15	Medical Physics	55
16	IEEE Transactions on Medical Imaging	53
17	Advances in Intelligent Systems and Computing	50
18	Computerized Medical Imaging and Graphics	47
19	Biomedical Signal Processing and Control	46
20	ACM International Conference Proceeding Series	45

In Table 4, we show the 20 most productive countries within AI for cancer detection in terms of absolute publication count. An article is assigned to a county when one of its authors is affiliated with one institution or company that is located within that country. Due to international collaboration, one article can therefore be assigned to more than one country. Hence, the total number of articles in Table 4 exceeds the total number of publications within our sample. Next to the total number of published articles, we also show the average age of the documents, as well as the average number of citations each document has received. Additionally, Table 4 shows the percentage of international co-authorship for every country. For example, an international co-authorship percentage of 50% would mean that 50% of the articles of one country have at least one author of another country.

In total, authors from 118 countries have contributed to research on AI for cancer detection. This very high number of contributing countries underlines the global importance of this topic. With 1627 articles, authors from the United States were the most productive ones. The United States are followed by China with 1202 contributions and India (1079 publications). The United Kingdom follows with a large gap (411 articles), Canada (264 publications) is in the fifth place. With 262 articles, the first European country to appear in the list is Germany in the sixth rank. Next to Germany, four other countries of the European Union are among the 20 most contributing nations, namely Italy, Spain, the Netherlands, and France.

Table 4. Overview of the countries with the most publications.

Rank	Country	Articles	Avg. Age (Years)	Avg. Cit.	Int. Co-Authorship
01	United States	1627	4.76	36.32	44.26%
02	China	1202	2.22	16.17	32.36%
03	India	1079	2.39	9.179	16.70%
04	United Kingdom	411	3.9	33.05	66.35%
05	Canada	264	3.52	33.94	55.88%
06	Germany	262	4.6	47.50	59.33%
07	Italy	248	3.73	28.39	55.21%
08	South Korea	221	2.53	23.63	40.95%
09	Japan	208	3.35	29.21	36.70%
10	Saudi Arabia	196	1.24	10.53	74.64%
11	Australia	190	3.36	32.81	70.47%
12	Spain	178	4.06	34.12	57.22%
13	Netherlands	177	2.8	41.17	64.90%
14	France	165	3.44	55.95	63.10%
15	Egypt	144	2.33	15.53	44.67%
16	Turkey	137	3.57	39.62	31.62%
17	Malaysia	134	3.36	17.31	49.50%
18	Iran	131	4.1	15.85	30.83%
19	Pakistan	123	1.69	13.78	67.42%
20	Taiwan	119	4.33	28.26	39.34%

When we look at the average age of the articles, it is striking that the United States not only has the most articles but also the oldest ones. In average, contributions from the United States have an age of 4.76 years. This is more than one year above the average age of the total sample. Among the top 20 countries, only Germany (4.6 years), Taiwan (4.33 years), Iran (4.1 years), and Spain (4.06 years) have an average article age of more than 4 years. This shows that these five countries are traditional contributors within the field of AI for cancer detection. It is striking that the contributions of China, the country with the second-most publications within our sample, are significantly younger. In average, Chinese contributions were 2.22 years old. This indicates that Chinese authors have contributed a lot, especially in the last few years. Only Saudi Arabia (1.24 years) and Pakistan (1.69 years) have younger articles on average.

Additionally, Table 4 shows the average citations the publications from a given country have received. The highest average citation numbers can be found for articles authored by authors from France (55.95 citations), Germany (47.50 citations), and the Netherlands (41.17 citations). The United States has received 36.32 citations on average, and Chinese publications have received 16.17. However, large parts of the different average citation counts can be explained with the average age of the articles. The average number of citations per document correlates with the average age of the articles, since recent articles have not had time to receive a high number of citations [47,61]. Additionally, it is interesting to observe that Indian articles received a much lower number of citations on average than Chinese ones (9.179 vs. 16.17), although the average age of the publications is relatively close to each other. However, it might be possible to explain this by the percentage of international co-authorship. While China has an international co-authorship ratio of 32.36%, this value is significantly lower for India (16.70%). Given that, it can be assumed that Indian research is much more isolated and probably not so much known in other countries, leading to a lower citation score.

The highest ratios of international co-authorship can be found for articles authored or co-authored by researchers from Saudi Arabia (74.64%), Australia (70.47%), Pakistan (67.42%), and the United Kingdom (66.35%). The lowest scores can be found for Indian (16.70%), Iranian (30.83%), Turkish (31.62%), and Chinese (32.36%) contributions. Furthermore, we can see that the average percentage of international co-authorship of the 20 most contributing countries is much higher than this value for the whole sample (24.97%

Likewise, this shows that many countries with only a few contributions tend to have a comparatively low amount of international co-authorship.

To further illustrate the international collaboration, Figure 4 shows an international collaboration map. Herein, collaborations between different countries are depicted with red lines. The thicker a red line between two countries is, the more collaboration took place among researchers of these two nations. To not overload it, only relationships with at least three contributions between two countries are depicted in Figure 4. Additionally, the countries' color represents their number of publications. The darker the blue is, the more publications have been contributed from researchers a specific country. Herein, we can see three large centers of collaboration, namely in the United States, China, and the European Union. These three areas have a lot of different collaborations with many different countries.

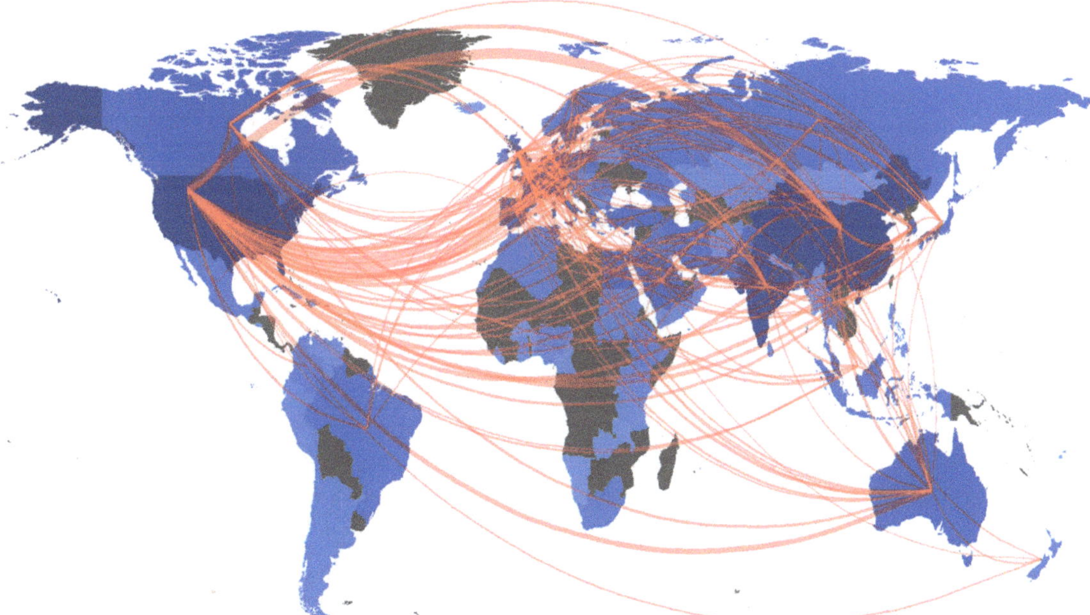

Figure 4. International co-collaboration map (generated with Biblioshiny).

Table 5 presents the 20 institutions and organizations that funded the most articles. With 539 publications, the National Natural Science Foundation of China has funded the most articles on AI for cancer detection. It is followed by the National Institutes of Health (408 publications), the National Cancer Institute (336 publications), and the National Science Foundation (113 publications). It is noteworthy that the top three funding sponsors together funded 1283 articles, which is almost equivalent to 20% of all publications dealing with AI for cancer detection. Both China and USA are most often represented, each with six funding sponsors among the top 20. They are followed by the European Union with three and Canada and UK with two funding sponsors.

Finally, Table 6 shows the 20 affiliations that authored the most publications within the field of AI for cancer detection. An article is assigned to one affiliation based on the contributing authors. Since an article can be authored or co-authored by researchers from different institutions, certain articles can be linked to more than one affiliation. With 219 articles, researchers from Sichuan University contributed to the most publications dealing with AI for cancer detection. The Sichuan University is followed by three affiliations located in the United States, namely the University of California (199 publications), the Memorial Sloan Kettering Cancer Center (195 publications), and the Stanford University

(170 publications). Of the 20 most contributing affiliations, eight are located in China and eight in the United States. Additionally, one affiliation is from the Netherlands (Radboud University Medical Center, 145 publications), Japan (The University Of Tokyo, 95 publications), the United Kingdom (University Of Cambridge, 90 publications), and Canada (University Of Toronto, 90 publications).

Table 5. Overview of the funding sponsors with the most funded publications.

Rank	Funding Sponsor	Country/Region	Quantity
01	National Natural Science Foundation of China	China	539
02	National Institutes of Health	USA	408
03	National Cancer Institute	USA	336
04	National Science Foundation	USA	113
05	National Key Research and Development Program of Chinas	China	106
06	U.S. Department of Health and Human Services	USA	89
07	Fundamental Research Funds for the Central Universities	China	79
08	National Research Foundation of Korea	South Korea	67
09	Natural Sciences and Engineering Research Council of Canada	Canada	60
10	European Regional Development Fund	EU	58
11	European Commission	EU	57
12	National Institute of Biomedical Imaging and Bioengineering	USA	50
13	Japan Society for the Promotion of Science	Japan	48
14	Ministry of Education of the People's Republic of China	China	40
15	Ministry of Science and Technology of the People's Republic of China	China	40
16	Canadian Institutes of Health Research	Canada	39
17	Cancer Research UK	UK	36
18	Science and Technology Commission of Shanghai Municipality	China	36
18	National Institute for Health Research	UK	34
19	Horizon 2020 Framework Programme	EU	33
20	Nvidia	USA	32

Table 6. Overview of the affiliations with the most publications.

Rank	Affiliation	Country/Region	Articles
01	Sichuan University	China	219
02	University of California	USA	199
03	Memorial Sloan Kettering Cancer Center	USA	195
04	Stanford University	USA	170
05	Fudan University	China	165
06	Shanghai Jiao Tong University	China	151
07	Harvard Medical School	USA	147
08	Huazhong University of Science and Technology	China	145
09	Radboud University Medical Center	Netherlands	145
10	University of Pennsylvania	USA	132
11	Southern Medical University	China	130
12	National Cancer Institute	USA	107
13	University of British Columbia	USA	104
14	Renmin Hospital of Wuhan University	China	101
15	Zhejiang University	China	101
16	The University of Tokyo	Japan	95
17	Emory University	USA	94
18	Sun Yat-Sen University Cancer Center	China	93
19	University of Cambridge	UK	90
20	University of Toronto	Canada	90

4.3. Content Analysis

In this section, we will thematically dive into the topics that are dealt with in AI for cancer detection research. First, Table 7 shows the 25 most frequently used keywords in our sample. This does not only include author keywords but also indexed keywords from Scopus. The keywords "human" and "humans" were most-often used, which indicates that most of the research belong to human medicine, specifically, cancer that affects humans. This is followed by "cancer diagnosis" and "diseases". The most frequently used technical keywords and terms in our sample were "deep learning" (2275 appearances), "machine learning" (2163 appearances), and "artificial intelligence", which appeared 1735 times. Other frequently used technologies according to the most often used keywords are "convolutional neural networks" (1021 appearances) and "artificial neural networks" (903 appearances).

Table 7. Overview of most frequently used keywords.

Rank	Keyword	Quantity
01	Human	3585
02	Humans	2685
03	Cancer Diagnosis	2621
04	Diseases	2521
05	Deep Learning	2275
06	Machine Learning	2163
07	Artificial Intelligence	1735
08	Female	1648
09	Breast Cancer	1498
10	Sensitivity and Specificity	1407
11	Controlled Study	1336
12	Diagnosis	1325
13	Diagnostic Accuracy	1273
14	Diagnostic Imaging	1245
15	Major Clinical Study	1199
16	Procedures	1123
17	Male	1092
18	Priority Journal	1088
19	Medical Imaging	1081
20	Adult	1061
21	Convolutional Neural Network	1021
22	Algorithm	1016
23	Computer Aided Diagnosis	909
24	Artificial Neural Network	903
25	Learning Systems	882

When we specifically focus on cancer types, breast cancer is most frequently addressed in the articles dealing with AI for cancer detection. With 1498 appearances, breast cancer is the ninth of the most often-used keywords. This is not surprising, since breast cancer is the most common carcinoma among women globally and comes with a low survival rate [62]. Breast cancer is followed by lung cancer (598 appearances), which causes the most cancer-related deaths worldwide [63]. Breast and lung cancer are followed by prostate cancer (425 appearances) and melanoma (skin cancer, 247 appearances).

To obtain a deeper understanding of the topics dealt with, Figure 5 shows a word cloud of the most frequently used keywords plus. Keywords plus are another way to analyze a document's content and are automatically generated out of words or phrases that are frequently used in the titles of an article's references [64,65]. In Figure 5, the size of words is determined based on their frequency in the keywords plus. Herein, many of the most frequently identified words are closely related to the cancer types that are most often addressed (e.g., "mammography", "lung cancer", "breast tumor", or "melanoma"), which is not a surprising result.

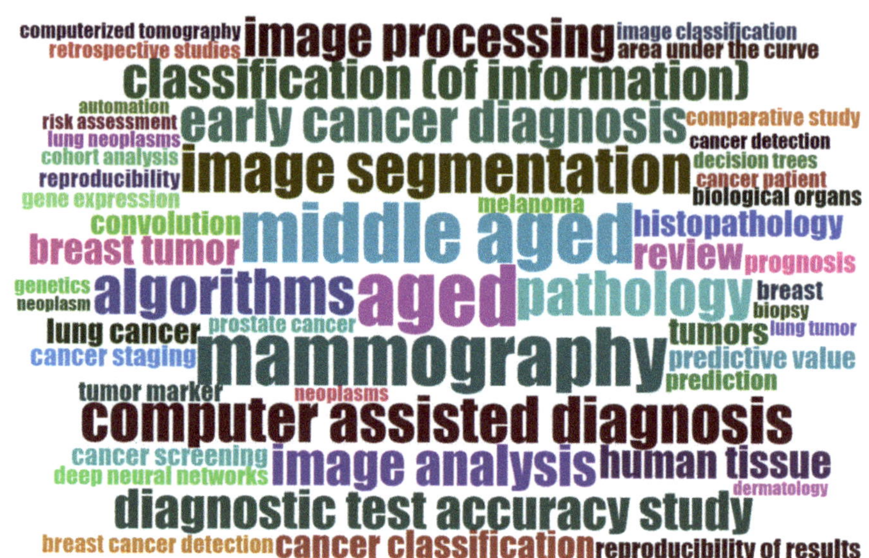

Figure 5. Word cloud of the most-frequently appearing keywords plus (generated with Biblioshiny).

Additionally, Figure 6 shows a keyword co-occurrence network of author keyword and indexed keywords of our sample. Like in the word cloud in Figure 5, the font size depends on the frequency a term is used. Terms that frequently appear together are linked with lines and are arranged in clusters of the same color. Terms that appear in the center of the network, such as "deep learning"; "machine learning", "artificial intelligence", or "machine learning", are connected with many other words in the network. It is noteworthy that it is hard to distinguish clear thematical clusters based on the color in Figure 6. Although a red and a green cluster are visible, the keywords that belong to these clusters have many relations to terms that do not belong to these clusters. Keywords in yellow, blue, or purple for example, are spread in the whole network and to not represent clearly distinguishable thematic fields. Despite the fact that AI for cancer detection is a multidisciplinary field, we can conclude from Figure 6 that knowledge and research on that topic is not fragmented. Although different clusters can be identified, these are not isolated from other research streams, which shows the overall coherence within that research field.

Finally, Table 8 shows the 30 most-cited articles in our sample. A total of 13 of the 30 articles do have a general focus on AI's potential for drug discovery and do not focus on a single cancer type. Among the other articles, breast cancer (10 publications) is the cancer type that is most often addressed, followed by brain tumors (3 publications). With 2136 citations at the point of time our data were collected, the article "Classification and diagnostic prediction of cancers using gene expression profiling and artificial neural networks" is the most often cited publications in our sample. In their article, the authors show the potential and applications of artificial neural networks for diagnosing cancer and the identification of candidate targets for therapy. Although this article is comparatively old and has been published in 2001, the results were already promising and showed the great potential of artificial neural networks. In rank two, the article "The evaluation of tumor infiltrating lymphocytes (TILs) in breast cancer: recommendations by an International TILs Working Group 2014" follows with 1533 citations. Although AI and ML is only partly covered in this article, the authors mention ML to be a promising tool for the future assessment of TILs [66] (p. 269).

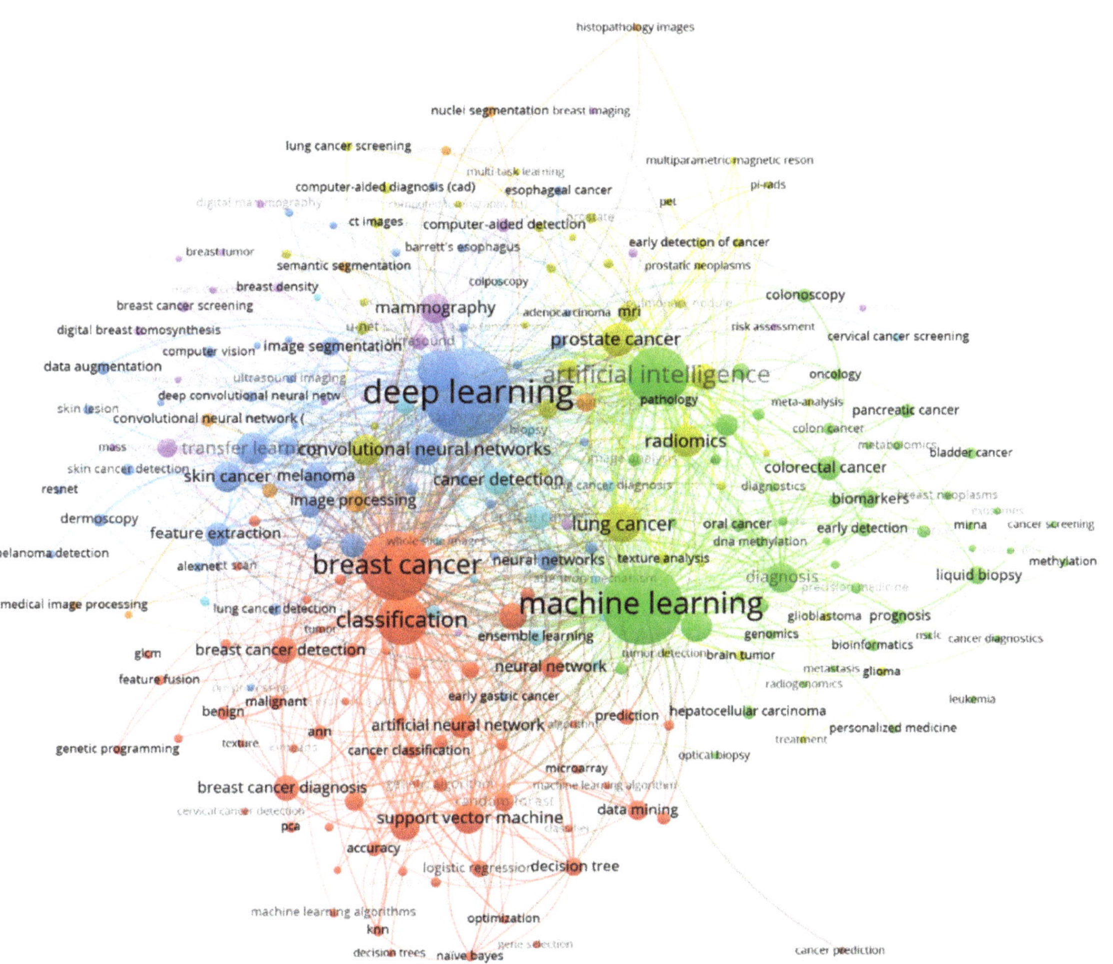

Figure 6. Network of keyword co-occurences.

Table 8. Overview of the 30 most-often cited articles.

Rank	Authors	Year	Focus	Citations	Reference
01	Khan et al.	2001	General investigation	2136	[67]
02	Salgado et al.	2015	Breast cancer	1533	[66]
03	Kourou et al.	2015	General investigation	1426	[68]
04	Bejnordi et al.	2017	Breast cancer/lymph node metastases	1305	[69]
05	Lu and Fei	2014	General investigation	1252	[70]
06	Coudray et al.	2018	Lung cancer	1018	[71]
07	McKinney et al.	2020	Breast cancer	774	[72]
08	Johnson et al.	2019	General investigation	720	[73]
09	Cruz and Wishart	2006	General investigation	693	[74]
10	Statnikov et al.	2005	General investigation	644	[75]
11	Spanhol et al.	2016	Breast cancer	626	[76]
12	Haenssle et al.	2018	Skin cancer	588	[77]
13	Litjens et al.	2016	Breast cancer/prostate cancer	581	[78]
14	Mazurowski et al.	2008	Breast cancer	557	[79]

Table 8. *Cont.*

Rank	Authors	Year	Focus	Citations	Reference
15	Akay	2009	Breast cancer	554	[80]
16	Bi et al.	2019	General investigation	553	[81]
17	Zacharaki et al.	2009	Brain tumors	542	[82]
18	Shrestha and Mahmood	2019	General investigation	525	[83]
19	Tang et al.	2009	Breast cancer	488	[84]
20	Statnikov et al.	2008	General investigation	467	[85]
21	Irshad et al.	2014	General investigation	450	[86]
22	Zhao et al.	2018	Brain tumors	428	[87]
23	Dou et al.	2017	General investigation	421	[88]
24	Zheng et al.	2014	Breast cancer	374	[89]
25	Lee et al.	2008	General investigation	372	[90]
26	Limkin et al.	2017	General investigation	364	[91]
27	Albarqouni et al.	2016	Breast cancer	360	[92]
28	Urban et al.	2018	Polyps/Colorectal cancer	347	[16]
29	Ribli et al.	2018	Breast cancer	346	[93]
30	Işın et al.	2016	Brain tumor	345	[94]

5. Future Research Agenda

In the prior sections, we presented the results of our bibliometric study. Based on our findings, we will present promising avenues for future research in this section. These have the purpose to serve as an orientation for interested scholars.

First, considering the word cloud in Figure 5 and the focus of the most-cited studies, it becomes evident that the current state of research mainly focuses on the predictive performance of a limited number of applied AI algorithms. The interaction between the computer system and the humans involved, also referred to as human–computer interaction, is a topic addressed much more rarely. It is important to investigate how the interaction between AI and the humans may or should look in the context of cancer diagnosis. In general, there are different conceivable scenarios, namely substitution, augmentation, and assemblage [95,96]. Augmentation refers to the scenario that AI and humans augment each other, while assemblage means that the AI and humans are brought together dynamically to function as a unit. Finally, substitution means that the human is completely replaced by the AI system [96]. Future research needs to investigate which form of cooperation between AI and humans is most suitable in the context of cancer diagnosis. This involves the question of whether a substitution is possible and, especially, if it is desirable, at all. There are already a few promising studies available that investigate human–computer interaction in the health industry [97,98]. Therefore, these studies can be used as a foundation for future studies that address the relationship between AI and humans. Additionally, trust between the AI cancer detection model and humans involved is an important factor. Although AI systems often have accuracy that surpasses that of human experts, there is a lack of trust in the predictions generated by AI systems [99]. It should be therefore investigated what reasons exist for a lack of trust and how trust in the AI system can be improved. This also holds true for patients who might be subject to treatments that are mainly based on the results of an AI system. Explainable artificial intelligence (XAI, see below) might be one way to increase the trust in an AI system.

One important aspect is also the security and robustness of the AI models. Many AI models that are described in the literature were evaluated only on one dataset. Therefore, it might remain unclear if the AI model can be transferred to input data that stems from different scanning machines. Therefore, it would be worthwhile to investigate how AI models must be designed to ensure their transferability [100–102]. In this context, it also might make sense to evaluate AI models using several datasets generated by different sensors or different manufacturers. As outlined above, AI systems require a large amount of data to learn and to develop robust models. When it comes to data, it is additionally important to ensure the trustworthiness, reliability, and security of the sources or platforms the data stem from [103,104]. If malicious actors succeed in manipulating or changing the data that are used as an input for the AI system, this might affect the AI system's result. Therefore, these results are not reliable anymore and might endanger the patient's health due to the risk of wrong results. Data storage is an especially important aspect, as medical data is subject to special data-protection regulations. Therefore, it should be examined what storage solutions are compliant with regulations, such as the GDPR or HIPAA, and how to ensure that the data is not traceable. In this context, future research should also verify whether the pseudonymization of the data is sufficient or whether complete anonymization is required. Different researchers also examine whether new technologies for the distributed storage and management of data, such as the blockchain, might be suitable for medical data [105–107]. Future research could therefore take a critical look if a blockchain would make sense for the purpose of managing and storing medical data or if other technologies and databases are more suitable. Moreover, it is noteworthy that there are already a few studies available that investigate security and robustness aspects of AI models for cancer detection. Approaches such as the external validation of AI algorithms [108] and robustness tests against adversarial images [109], as well as comprehensive data preprocessing [110,111], are promising to achieve robustness and security goals and should therefore be investigated in more detail. In this context, the application of design science research could also be a way to iteratively address specific security problems in order to find an efficient solution. Examples of design science research can be found in business administration [31,112] and information systems [113,114].

As mentioned above, the explainability of an AI model is an important factor to ensure the acceptance for and trust in an AI model. With an AI system's advancing complexity, it is increasingly difficult to understand how it comes up with its results and predictions. This holds true for the most of today's AI and machine learning algorithms, which are very complex and [21,115] considered black boxes. Although XAI is hard to achieve, it is necessary for certain use-cases in critical areas like law or medicine [116–118]. For cancer detection, XAI can be considered very relevant. This might not be the case as long as the AI system's results are doublechecked by doctors or oncologist. However, before AI can be used independently, explainability is an important challenge that needs to be addressed [119]. Recent reviews and surveys demonstrate that XAI in medicine is still one of the most signifcant research gaps and remains largely unanswered [102,120]. Future research should therefore investigate how AI systems for cancer detection can be made transparent enough that their results are understandable. It might make sense to collaborate with AI researchers or scholars from other disciplines since some promising XAI applications and developments might not yet be applied in the context of cancer detection.

Table 9 below provides an overview of our future research agenda and presents possible future research questions that might help advance the field of AI for cancer detection.

Table 9. Future research agenda.

Focus	Possible Research Questions
Human Computer Interaction	How can the interaction between doctors and AI models be designed efficiently? What is the current state of trust towards AI based models in medicine? How can trust in AI be built for doctors and patients? How can AI experts and clinical practitioners cooperate and work together in the best way? What is the role of explainable AI for building trust?
Robustness and security	How reliable are trained AI models on other cancer datasets (e.g., generated by other sensors)? Can adversarial attacks outsmart AI models in medicine? How should an AI system for cancer detection be designed to make it robust and secure against adversarial attacks and actors? Could a cancer detection algorithm be applied to other types of cancer?

Table 9. *Cont.*

Focus	Possible Research Questions
Explainable AI	Should explainable AI models be preferred instead of the most accurate one? How should explainable AI be designed to increase the trust in the AI system and its decisions? What are the promising approaches in XAI that have not yet been applied in the medical field?
Data Storage	Where should the data of the scans be stored to ensure data privacy rights? Do new technologies like blockchain have a potential for the storage and management of medical data? Should patient data be irreversibly anonymized or only pseudonymized?

6. Discussion

In this study, we conducted a bibliometric analysis of 6450 articles dealing with the potential and application of AI for cancer detection and diagnosis. To the best of our knowledge, this is the first study that uses a bibliometric approach to analyze research on AI for cancer detection. This study has several implications and benefits for both researchers and practitioners. First, interested researchers can use the study at hand to obtain an initial overview of research on AI for cancer diagnosis. This involves information about the scientific landscape and the most influential articles, as well as core topics and key themes investigated. As such, this study can help to equip interested scholars with an initial understanding of the research field dealing with AI-based cancer detection. Our research agenda furthermore can serve as a foundation for future research to build on to further develop this exciting field. Additionally, both clinical as well as commercial practitioners can use our study to obtain an initial insight about the potential of AI for supporting the diagnosis of cancer.

Our study is subject to certain limitations. First, we used Scopus as the only scientific database for the collection of our bibliometric data. Although, as outlined above, Scopus covers a huge number of different conferences and journals, it is likely that different publications were not covered by our research. Bibliometric studies on AI for cancer detection that use other databases for data collection might therefore lead to slightly different results. However, we believe that the most of our key results, especially the most important topics and key themes, are likely to maintain constant even if other databases would be used. Furthermore, the application of other bibliometric tools and analysis methods like citation analysis [121] or bibliographic coupling [122,123] might lead to additional results that were not part of this study. Additionally, AI is a fast-evolving field. New research on AI for the purpose of cancer detection is published every month. This study's results are therefore only able to show the current state of the art.

7. Conclusions

AI is a promising technology that is increasingly applied to detect or diagnose cancer. In recent years, research on AI for cancer detection grew rapidly, resulting in a high number of research articles on that topic. Due to the large amount of research that is available, it is hard for interested scholars or clinical practitioners to obtain an initial understanding of this field. Against this backdrop, we aimed to provide researchers with an overview and analysis of the research field of AI for cancer detection. For this purpose, we conducted a bibliometric study of the existing research on that topic. In total, we identified and analyzed 6450 articles published between 1986 and 2022.

Our analysis consisted of different parts. First, we gave a general overview of our sample and presented the development of scientific production over the year and which disciplines contributed to it. After that, we conducted a performance analysis. Herein, we identified the most productive institutions and countries. Additionally, we gave an overview of the most relevant outlets and the international collaboration. Finally, we thematically analyzed the sample and identified key topics and the most-cited publications. We found that breast and lung cancer are cancer types most often addressed by recent research. Based on these findings, we developed a future research agenda that is supposed to guide researchers to further advance the field of AI-based cancer diagnosis. We believe that we provide a systematic and holistic overview of this exciting field of research and hope that our study will serve interested scholars and practitioners as a valuable overview.

Author Contributions: Conceptualization, E.K. and M.K.; methodology, E.K. and M.K.; software, E.K. and M.K.; validation, E.K. and M.K.; formal analysis, E.K. and M.K.; investigation, E.K. and M.K.; resources, E.K. and M.K.; data curation, E.K. and M.K.; writing—original draft preparation, E.K. and M.K.; writing—review and editing, E.K. and M.K.; visualization, E.K. and M.K.; supervision, E.K.; project administration, M.K.; funding acquisition, not applicable. All authors have read and agreed to the published version of the manuscript.

Funding: This research received no external funding.

Institutional Review Board Statement: Not applicable.

Informed Consent Statement: Not applicable.

Data Availability Statement: Not applicable.

Acknowledgments: We acknowledge support by the Open Access Publication Fund of the University of Duisburg-Essen.

Conflicts of Interest: The authors declare no conflict of interest.

References

1. Stephens, F.O.; Aigner, K.R. *Basics of Oncology*, 2nd ed.; Springer International Publishing: Cham, Switzerland, 2015; ISBN 978-3-319-23368-0.
2. Shaikh, K.; Krishnan, S.; Thanki, R. *Artificial Intelligence in Breast Cancer Early Detection and Diagnosis*; Springer International Publishing AG: Cham, Switzerland, 2021; ISBN 978-3-030-59207-3.
3. Sudhakar, A. History of Cancer, Ancient and Modern Treatment Methods. *J. Cancer Sci. Ther.* **2009**, *1*, 1. [CrossRef] [PubMed]
4. Bray, F.; Ferlay, J.; Soerjomataram, I.; Siegel, R.L.; Torre, L.A.; Jemal, A. Global cancer statistics 2018: GLOBOCAN estimates of incidence and mortality worldwide for 36 cancers in 185 countries. *CA Cancer J. Clin.* **2018**, *68*, 394–424. [CrossRef] [PubMed]
5. Leddin, D.J.; Enns, R.; Hilsden, R.; Plourde, V.; Rabeneck, L.; Sadowski, D.C.; Singh, H. Canadian Association of Gastroenterology Position Statement on Screening Individuals at Average Risk for Developing Colorectal Cancer: 2010. *Can. J. Gastroenterol.* **2010**, *24*, 705–714. [CrossRef] [PubMed]
6. Ghebrial, M.; Aktary, M.L.; Wang, Q.; Spinelli, J.J.; Shack, L.; Robson, P.J.; Kopciuk, K.A. Predictors of CRC Stage at Diagnosis among Male and Female Adults Participating in a Prospective Cohort Study: Findings from Alberta's Tomorrow Project. *Curr. Oncol.* **2021**, *28*, 4938–4952. [CrossRef] [PubMed]
7. Schoen, R.E.; Pinsky, P.F.; Weissfeld, J.L.; Yokochi, L.A.; Church, T.R.; Laiyemo, A.O.; Bresalier, R.; Andriole, G.L.; Buys, S.S.; Crawford, E.D.; et al. Colorectal-Cancer Incidence and Mortality with Screening Flexible Sigmoidoscopy. *N. Engl. J. Med.* **2012**, *366*, 2345–2357. [CrossRef]
8. Fahim, C.; Huyer, L.D.; Lee, T.T.; Prashad, A.; Leonard, R.; Khare, S.R.; Stiff, J.; Chadder, J.; Straus, S.E. Implementing and Sustaining Early Cancer Diagnosis Initiatives in Canada: An Exploratory Qualitative Study. *Curr. Oncol.* **2021**, *28*, 4341–4356. [CrossRef]
9. Parveen, S.S.; Kavitha, C. Detection of lung cancer nodules using automatic region growing method. In Proceedings of the 2013 Fourth International Conference on Computing, Communications and Networking Technologies (ICCCNT), Tiruchengode, India, 4–6 July 2013; pp. 1–6, ISBN 978-1-4799-3926-8.
10. Patel, D.; Shah, Y.; Thakkar, N.; Shah, K.; Shah, M. Implementation of Artificial Intelligence Techniques for Cancer Detection. *Augment. Hum. Res.* **2019**, *5*, 6. [CrossRef]
11. Kalaf, J.M. Mammography: A history of success and scientific enthusiasm. *Radiol. Bras.* **2014**, *47*, VII–VIII. [CrossRef]
12. Mahoro, E.; Akhloufi, M.A. Applying Deep Learning for Breast Cancer Detection in Radiology. *Curr. Oncol.* **2022**, *29*, 8767–8793. [CrossRef]
13. Rodríguez-Ruiz, A.; Krupinski, E.; Mordang, J.-J.; Schilling, K.; Heywang-Köbrunner, S.H.; Sechopoulos, I.; Mann, R.M. Detection of Breast Cancer with Mammography: Effect of an Artificial Intelligence Support System. *Radiology* **2019**, *290*, 305–314. [CrossRef]
14. Qiu, H.; Ding, S.; Liu, J.; Wang, L.; Wang, X. Applications of Artificial Intelligence in Screening, Diagnosis, Treatment, and Prognosis of Colorectal Cancer. *Curr. Oncol.* **2022**, *29*, 1773–1795. [CrossRef] [PubMed]
15. Secasan, C.C.; Onchis, D.; Bardan, R.; Cumpanas, A.; Novacescu, D.; Botoca, C.; Dema, A.; Sporea, I. Artificial Intelligence System for Predicting Prostate Cancer Lesions from Shear Wave Elastography Measurements. *Curr. Oncol.* **2022**, *29*, 4212–4223. [CrossRef] [PubMed]

16. Urban, G.; Tripathi, P.; Alkayali, T.; Mittal, M.; Jalali, F.; Karnes, W.; Baldi, P. Deep Learning Localizes and Identifies Polyps in Real Time With 96% Accuracy in Screening Colonoscopy. *Gastroenterology* **2018**, *155*, 1069–1078.e8. [CrossRef] [PubMed]

17. Goldenberg, S.L.; Nir, G.; Salcudean, S.E. A new era: Artificial intelligence and machine learning in prostate cancer. *Nat. Rev. Urol.* **2019**, *16*, 391–403. [CrossRef]

18. Dlamini, Z.; Francies, F.Z.; Hull, R.; Marima, R. Artificial intelligence (AI) and big data in cancer and precision oncology. *Comput. Struct. Biotechnol. J.* **2020**, *18*, 2300–2311. [CrossRef]

19. Huang, S.; Yang, J.; Fong, S.; Zhao, Q. Artificial intelligence in cancer diagnosis and prognosis: Opportunities and challenges. *Cancer Lett.* **2019**, *471*, 61–71. [CrossRef]

20. Donthu, N.; Kumar, S.; Mukherjee, D.; Pandey, N.; Lim, W.M. How to conduct a bibliometric analysis: An overview and guidelines. *J. Bus. Res.* **2021**, *133*, 285–296. [CrossRef]

21. Karger, E.; Kureljusic, M. Using Artificial Intelligence for Drug Discovery: A Bibliometric Study and Future Research Agenda. *Pharmaceuticals* **2022**, *15*, 1492. [CrossRef]

22. Sampietro, A.; Pérez-Areales, F.J.; Martínez, P.; Arce, E.M.; Galdeano, C.; Muñoz-Torrero, D. Unveiling the Multitarget Anti-Alzheimer Drug Discovery Landscape: A Bibliometric Analysis. *Pharmaceuticals* **2022**, *15*, 545. [CrossRef]

23. Chiari, W.; Damayanti, R.; Harapan, H.; Puspita, K.; Saiful, S.; Rahmi, R.; Rizki, D.R.; Iqhrammullah, M. Trend of Polymer Research Related to COVID-19 Pandemic: Bibliometric Analysis. *Polymers* **2022**, *14*, 3297. [CrossRef]

24. Franco, P.; Segelov, E.; Johnsson, A.; Riechelmann, R.; Guren, M.G.; Das, P.; Rao, S.; Arnold, D.; Spindler, K.-L.G.; Deutsch, E.; et al. A Machine-Learning-Based Bibliometric Analysis of the Scientific Literature on Anal Cancer. *Cancers* **2022**, *14*, 1697. [CrossRef] [PubMed]

25. Danvila-Del-Valle, I.; Estévez-Mendoza, C.; Lara, F.J. Human resources training: A bibliometric analysis. *J. Bus. Res.* **2019**, *101*, 627–636. [CrossRef]

26. Paul, J.; Criado, A.R. The art of writing literature review: What do we know and what do we need to know? *Int. Bus. Rev.* **2020**, *29*, 101717. [CrossRef]

27. Russell, S.J.; Norvig, P. *Artificial Intelligence: A Modern Approach*, 4th ed.; Pearson: London, UK, 2020.

28. Mcculloch, W.S.; Pitts, W.H. A logical calculus of the ideas immanent in nervous activity. *Bull. Math. Biophys.* **1943**, *5*, 115–133. [CrossRef]

29. Russell, S.J.; Norvig, P.; Davis, E.; Edwards, D. *Artificial Intelligence: A Modern Approach*, 3rd ed.; Pearson: London, UK, 2016; ISBN 9781292153964.

30. Goodfellow, I.; Bengio, Y.; Courville, A. *Deep Learning*; MIT Press: Cambridge, MA, USA, 2016.

31. Kureljusic, M.; Reisch, L. Revenue forecasting for European capital market-oriented firms: A comparative prediction study between financial analysts and machine learning models. *Corp. Ownersh. Control* **2022**, *19*, 159–178. [CrossRef]

32. Leitner-Hanetseder, S.; Lehner, O.M.; Eisl, C.; Forstenlechner, C. A profession in transition: Actors, tasks and roles in AI-based accounting. *J. Appl. Account. Res.* **2021**, *22*, 539–556. [CrossRef]

33. Haenlein, M.; Kaplan, A. A Brief History of Artificial Intelligence: On the Past, Present, and Future of Artificial Intelligence. *Calif. Manag. Rev.* **2019**, *61*, 5–14. [CrossRef]

34. Taulli, T. *Artificial Intelligence Basics: A Non-Technical Introduction, 2019*, 1st ed.; Apress: Berkeley, CA, USA, 2019; ISBN 978-1-4842-5028-0.

35. Shalev-Shwartz, S.; Ben-David, S. *Understanding Machine Learning: From Theory to Algorithms*; Cambridge University Press: Cambridge, UK, 2014; ISBN 9781107057135.

36. Aggarwal, C.C. *Neural Networks and Deep Learning*; Springer International Publishing: Cham, Switzerland, 2018; ISBN 978-3-319-94462-3.

37. Murugan, A.; Nair, S.H.; Kumar, K.P.S. Detection of Skin Cancer Using SVM, Random Forest and kNN Classifiers. *J. Med. Syst.* **2019**, *43*, 269. [CrossRef]

38. Nuklianggraita, T.N.; Adiwijaya, A.; Aditsania, A. On the Feature Selection of Microarray Data for Cancer Detection based on Random Forest Classifier. *J. INFOTEL* **2020**, *12*, 89–96. [CrossRef]

39. Ragab, D.A.; Sharkas, M.; Marshall, S.; Ren, J. Breast cancer detection using deep convolutional neural networks and support vector machines. *PeerJ* **2019**, *7*, e6201. [CrossRef]

40. Sweilam, N.; Tharwat, A.; Moniem, N.A. Support vector machine for diagnosis cancer disease: A comparative study. *Egypt. Inform. J.* **2010**, *11*, 81–92. [CrossRef]

41. Searle, J.R. Minds, brains, and programs. *Behav. Brain Sci.* **1980**, *3*, 417–424. [CrossRef]

42. Franklin, S. History, motivations, and core themes. In *The Cambridge Handbook of Artificial Intelligence*; Frankish, K., Ramsey, W.M. Eds.; Cambridge University Press: Cambridge, UK, 2014; pp. 15–33. ISBN 9781139046855.

43. Adams, S.; Arel, I.; Bach, J.; Coop, R.; Furlan, R.; Goertzel, B.; Hall, J.S.; Samsonovich, A.; Scheutz, M.; Schlesinger, M.; et al. Mapping the Landscape of Human-Level Artificial General Intelligence. *AI Mag.* **2012**, *33*, 25–42. [CrossRef]

44. Van Gerven, M. Computational Foundations of Natural Intelligence. *Front. Comput. Neurosci.* **2017**, *11*, 112. [CrossRef] [PubMed]

45. Dingli, A.; Haddod, F.; Klüver, C. *Artificial Intelligence in Industry 4.0*; Springer International Publishing: Cham, Switzerland, 2021; ISBN 978-3-030-61044-9.

46. Braga, A.; Logan, R.K. The Emperor of Strong AI Has No Clothes: Limits to Artificial Intelligence. *Information* **2017**, *8*, 156. [CrossRef]

47. Forliano, C.; De Bernardi, P.; Yahiaoui, D. Entrepreneurial universities: A bibliometric analysis within the business and management domains. *Technol. Forecast. Soc. Chang.* **2020**, *165*, 120522. [CrossRef]

48. Jagals, M.; Karger, E.; Ahlemann, F. Already grown-up or still in puberty? A bibliometric review of 16 years of data governance research. *Corp. Ownersh. Control* **2019**, *19*, 105–120. [CrossRef]

9.   Moral-Muñoz, J.A.; Herrera-Viedma, E.; Santisteban-Espejo, A.; Cobo, M.J. Software tools for conducting bibliometric analysis in science: An up-to-date review. *Prof. Inf.* **2020**, *29*, e290103. [CrossRef]

0.   Tandon, A.; Kaur, P.; Mäntymäki, M.; Dhir, A. Blockchain applications in management: A bibliometric analysis and literature review. *Technol. Forecast. Soc. Chang.* **2021**, *166*, 120649. [CrossRef]

1.   Donthu, N.; Kumar, S.; Pattnaik, D. Forty-five years of Journal of Business Research: A bibliometric analysis. *J. Bus. Res.* **2019**, *109*, 1–14. [CrossRef]

2.   Wulfert, T.; Karger, E. A bibliometric analysis of platform research in e-commerce: Past, present, and future research agenda. *Corp. Ownersh. Control* **2022**, *20*, 185–200. [CrossRef]

3.   Tran, B.X.; Vu, G.T.; Ha, G.H.; Vuong, Q.-H.; Ho, M.-T.; Vuong, T.-T.; La, V.-P.; Nghiem, K.-C.P.; Nguyen, H.L.T.; Latkin, C.A.; et al. Global Evolution of Research in Artificial Intelligence in Health and Medicine: A Bibliometric Study. *J. Clin. Med.* **2019**, *8*, 360. [CrossRef] [PubMed]

4.   Aria, M.; Cuccurullo, C. bibliometrix: An R-tool for comprehensive science mapping analysis. *J. Informetr.* **2017**, *11*, 959–975. [CrossRef]

5.   Van Eck, N.J.; Waltman, L. Software survey: VOSviewer, a computer program for bibliometric mapping. *Scientometrics* **2010**, *84*, 523–538. [CrossRef]

6.   Elango, B.; Rajendran, P. Authorship trends and collaboration pattern in the marine sciences literature: A scientometric study. *Int. J. Inf. Dissem. Technol.* **2012**, *2*, 166–169.

7.   Koseoglu, M.A. Mapping the institutional collaboration network of strategic management research: 1980–2014. *Scientometrics* **2016**, *109*, 203–226. [CrossRef]

8.   Secinaro, S.; Brescia, V.; Calandra, D.; Biancone, P. Data quality for health sector innovation and accounting man-agement: A twenty-year bibliometric analysis. *Econ. Aziend. Online* **2021**, *12*, 407–431. [CrossRef]

9.   Secinaro, S.; Mas, F.D.; Brescia, V.; Calandra, D. Blockchain in the accounting, auditing and accountability fields: A bibliometric and coding analysis. *Account. Audit. Account. J.* **2021**, *35*, 168–203. [CrossRef]

0.   Wied, G.L.; Bartels, P.H.; Bibbo, M.; Dytch, H.; Weber, J.E. Expert System Design under Uncertainty of Human Diagnosticians. In Proceedings of the Eighth Annual Conference of the IEEE/Engineering in Medicine and Biology Society, Fort Worth, TX, USA, 7–10 November 1986; pp. 757–760.

1.   Massaro, M.; Dumay, J.; Guthrie, J. On the shoulders of giants: Undertaking a structured literature review in accounting. *Account. Audit. Account. J.* **2016**, *29*, 767–801. [CrossRef]

2.   Alanazi, J.; Unnisa, A.; Alanazi, M.; Alharby, T.N.; Moin, A.; Rizvi, S.M.D.; Hussain, T.; Awadelkareem, A.M.; Elkhalifa, A.O.; Faiyaz, S.S.M.; et al. 3-Methoxy Carbazole Impedes the Growth of Human Breast Cancer Cells by Suppressing NF-κB Signaling Pathway. *Pharmaceuticals* **2022**, *15*, 1410. [CrossRef]

3.   Wu, G.X.; Raz, D.J. Lung Cancer Screening. In *Lung Cancer: Treatment and Research*; Reckamp, K.L., Ed.; Springer International Publishing: Cham, Switzerland, 2016; pp. 1–23. ISBN 978-3-319-40389-2.

4.   Garfield, E.; Sher, I.H. KeyWords Plus™—Algorithmic derivative indexing. *J. Am. Soc. Inf. Sci.* **1993**, *44*, 298–299. [CrossRef]

5.   Zhang, J.; Yu, Q.; Zheng, F.; Long, C.; Lu, Z.; Duan, Z. Comparing keywords plus of WOS and author keywords: A case study of patient adherence research. *J. Assoc. Inf. Sci. Technol.* **2015**, *67*, 967–972. [CrossRef]

6.   Salgado, R.; Denkert, C.; Demaria, S.; Sirtaine, N.; Klauschen, F.; Pruneri, G.; Wienert, S.; Van den Eynden, G.; Baehner, F.L.; Penault-Llorca, F.; et al. The evaluation of tumor-infiltrating lymphocytes (TILs) in breast cancer: Recommendations by an International TILs Working Group 2014. *Ann. Oncol.* **2015**, *26*, 259–271. [CrossRef] [PubMed]

7.   Khan, J.; Wei, J.S.; Ringnér, M.; Saal, L.H.; Ladanyi, M.; Westermann, F.; Berthold, F.; Schwab, M.; Antonescu, C.R.; Peterson, C.; et al. Classification and diagnostic prediction of cancers using gene expression profiling and artificial neural networks. *Nat. Med.* **2001**, *7*, 673–679. [CrossRef] [PubMed]

8.   Kourou, K.; Exarchos, T.P.; Exarchos, K.P.; Karamouzis, M.V.; Fotiadis, D.I. Machine learning applications in cancer prognosis and prediction. *Comput. Struct. Biotechnol. J.* **2015**, *13*, 8–17. [CrossRef] [PubMed]

9.   Bejnordi, B.E.; Veta, M.; Van Diest, P.J.; Van Ginneken, B.; Karssemeijer, N.; Litjens, G.; Van Der Laak, J.A.W.M.; Hermsen, M.; Manson, Q.F.; Balkenhol, M.; et al. Diagnostic Assessment of Deep Learning Algorithms for Detection of Lymph Node Metastases in Women with Breast Cancer. *JAMA* **2017**, *318*, 2199–2210. [CrossRef]

0.   Lu, G.; Fei, B. Medical hyperspectral imaging: A review. *J. Biomed. Opt.* **2014**, *19*, 010901. [CrossRef]

1.   Coudray, N.; Ocampo, P.S.; Sakellaropoulos, T.; Narula, N.; Snuderl, M.; Fenyö, D.; Moreira, A.L.; Razavian, N.; Tsirigos, A. Classification and mutation prediction from non–small cell lung cancer histopathology images using deep learning. *Nat. Med.* **2018**, *24*, 1559–1567. [CrossRef]

2.   McKinney, S.M.; Sieniek, M.; Godbole, V.; Godwin, J.; Antropova, N.; Ashrafian, H.; Back, T.; Chesus, M.; Corrado, G.S.; Darzi, A.; et al. International evaluation of an AI system for breast cancer screening. *Nature* **2020**, *577*, 89–94. [CrossRef]

3.   Johnson, J.M.; Khoshgoftaar, T.M. Survey on deep learning with class imbalance. *J. Big Data* **2019**, *6*, 27. [CrossRef]

4.   Cruz, J.A.; Wishart, D.S. Applications of Machine Learning in Cancer Prediction and Prognosis. *Cancer Inform.* **2006**, *2*, 59–77. [CrossRef]

5.   Statnikov, A.; Aliferis, C.F.; Tsamardinos, I.; Hardin, D.; Levy, S. A comprehensive evaluation of multicategory classification methods for microarray gene expression cancer diagnosis. *Bioinformatics* **2004**, *21*, 631–643. [CrossRef] [PubMed]

6.   Spanhol, F.A.; Oliveira, L.S.; Petitjean, C.; Heutte, L. A Dataset for Breast Cancer Histopathological Image Classification. *IEEE Trans. Biomed. Eng.* **2015**, *63*, 1455–1462. [CrossRef] [PubMed]

77. Haenssle, H.A.; Fink, C.; Schneiderbauer, R.; Toberer, F.; Buhl, T.; Blum, A.; Kalloo, A.; Hassen, A.B.H.; Thomas, L.; Enk, A.; et al. Man against machine: Diagnostic performance of a deep learning convolutional neural network for dermoscopic melanoma recognition in comparison to 58 dermatologists. *Ann. Oncol. Off. J. Eur. Soc. Med Oncol.* **2018**, *29*, 1836–1842. [CrossRef]
78. Litjens, G.; Sánchez, C.I.; Timofeeva, N.; Hermsen, M.; Nagtegaal, I.; Kovacs, I.; Hulsbergen-van de Kaa, C.; Bult, P.; Van Ginneken, B.; Van Der Laak, J. Deep learning as a tool for increased accuracy and efficiency of histopathological diagnosis. *Sci. Rep.* **2016**, *6*, 26286. [CrossRef] [PubMed]
79. Mazurowski, M.A.; Habas, P.A.; Zurada, J.M.; Lo, J.Y.; Baker, J.A.; Tourassi, G.D. Training neural network classifiers for medical decision making: The effects of imbalanced datasets on classification performance. *Neural Netw.* **2008**, *21*, 427–436. [CrossRef]
80. Akay, M.F. Support vector machines combined with feature selection for breast cancer diagnosis. *Expert Syst. Appl.* **2009**, *36*, 3240–3247. [CrossRef]
81. Bi, W.L.; Hosny, A.; Schabath, M.B.; Giger, M.L.; Birkbak, N.J.; Mehrtash, A.; Allison, T.; Arnaout, O.; Abbosh, C.; Dunn, I.F.; et al. Artificial intelligence in cancer imaging: Clinical challenges and applications. *CA A Cancer J. Clin.* **2019**, *69*, 127–157. [CrossRef] [PubMed]
82. Zacharaki, E.I.; Wang, S.; Chawla, S.; Yoo, D.S.; Wolf, R.; Melhem, E.R.; Davatzikos, C. Classification of brain tumor type and grade using MRI texture and shape in a machine learning scheme. *Magn. Reson. Med.* **2009**, *62*, 1609–1618. [CrossRef]
83. Shrestha, A.; Mahmood, A. Review of Deep Learning Algorithms and Architectures. *IEEE Access* **2019**, *7*, 53040–53065. [CrossRef]
84. Tang, J.; Rangayyan, R.M.; Xu, J.; El Naqa, I.; Yang, Y. Computer-Aided Detection and Diagnosis of Breast Cancer with Mammography: Recent Advances. *IEEE Trans. Inf. Technol. Biomed.* **2009**, *13*, 236–251. [CrossRef] [PubMed]
85. Statnikov, A.; Wang, L.; Aliferis, C.F. A comprehensive comparison of random forests and support vector machines for microarray-based cancer classification. *BMC Bioinform.* **2008**, *9*, 319. [CrossRef] [PubMed]
86. Irshad, H.; Veillard, A.; Roux, L.; Racoceanu, D. Methods for Nuclei Detection, Segmentation, and Classification in Digital Histopathology: A Review—Current Status and Future Potential. *IEEE Rev. Biomed. Eng.* **2013**, *7*, 97–114. [CrossRef]
87. Zhao, X.; Wu, Y.; Song, G.; Li, Z.; Zhang, Y.; Fan, Y. A deep learning model integrating FCNNs and CRFs for brain tumor segmentation. *Med. Image Anal.* **2017**, *43*, 98–111. [CrossRef] [PubMed]
88. Dou, Q.; Chen, H.; Yu, L.; Qin, J.; Heng, P.-A. Multilevel Contextual 3-D CNNs for False Positive Reduction in Pulmonary Nodule Detection. *IEEE Trans. Biomed. Eng.* **2016**, *64*, 1558–1567. [CrossRef]
89. Zheng, B.; Yoon, S.W.; Lam, S.S. Breast cancer diagnosis based on feature extraction using a hybrid of K-means and support vector machine algorithms. *Expert Syst. Appl.* **2014**, *41*, 1476–1482. [CrossRef]
90. Lee, E.; Chuang, H.-Y.; Kim, J.-W.; Ideker, T.; Lee, D. Inferring Pathway Activity toward Precise Disease Classification. *PLoS Comput. Biol.* **2008**, *4*, e1000217. [CrossRef]
91. Limkin, E.J.; Sun, R.; Dercle, L.; Zacharaki, E.I.; Robert, C.; Reuzé, S.; Schernberg, A.; Paragios, N.; Deutsch, E.; Ferté, C. Promises and challenges for the implementation of computational medical imaging (radiomics) in oncology. *Ann. Oncol.* **2017**, *28*, 1191–1206. [CrossRef] [PubMed]
92. Albarqouni, S.; Baur, C.; Achilles, F.; Belagiannis, V.; Demirci, S.; Navab, N. AggNet: Deep Learning from Crowds for Mitosis Detection in Breast Cancer Histology Images. *IEEE Trans. Med Imaging* **2016**, *35*, 1313–1321. [CrossRef]
93. Ribli, D.; Horváth, A.; Unger, Z.; Pollner, P.; Csabai, I. Detecting and classifying lesions in mammograms with Deep Learning. *Sci. Rep.* **2018**, *8*, 4165. [CrossRef]
94. Işın, A.; Direkoğlu, C.; Şah, M. Review of MRI-based Brain Tumor Image Segmentation Using Deep Learning Methods. *Procedia Comput. Sci.* **2016**, *102*, 317–324. [CrossRef]
95. Dellermann, D.; Ebel, P.; Söllner, M.; Leimeister, J.M. Hybrid Intelligence. *Bus. Inf. Syst. Eng.* **2019**, *61*, 637–643. [CrossRef]
96. Maedche, A.; Legner, C.; Benlian, A.; Berger, B.; Gimpel, H.; Hess, T.; Hinz, O.; Morana, S.; Söllner, M. AI-Based Digital Assistants. *Bus. Inf. Syst. Eng.* **2019**, *61*, 535–544. [CrossRef]
97. Tschandl, P.; Rinner, C.; Apalla, Z.; Argenziano, G.; Codella, N.; Halpern, A.; Janda, M.; Lallas, A.; Longo, C.; Malvehy, J.; et al. Human–computer collaboration for skin cancer recognition. *Nat. Med.* **2020**, *26*, 1229–1234. [CrossRef] [PubMed]
98. Cai, C.J.; Winter, S.; Steiner, D.; Wilcox, L.; Terry, M. "Hello AI": Uncovering the Onboarding Needs of Medical Practitioners for Human-AI Collaborative Decision-Making. *Proc. ACM Human-Computer Interact.* **2019**, *3*, 1–24. [CrossRef]
99. Schmidt, P.; Biessmann, F.; Teubner, T. Transparency and trust in artificial intelligence systems. *J. Decis. Syst.* **2020**, *29*, 260–278. [CrossRef]
100. Carter, S.M.; Rogers, W.; Win, K.T.; Frazer, H.; Richards, B.; Houssami, N. The ethical, legal and social implications of using artificial intelligence systems in breast cancer care. *Breast* **2019**, *49*, 25–32. [CrossRef] [PubMed]
101. Houssami, N.; Kirkpatrick-Jones, G.; Noguchi, N.; Lee, C.I. Artificial Intelligence (AI) for the early detection of breast cancer: a scoping review to assess AI's potential in breast screening practice. *Expert Rev. Med. Devices* **2019**, *16*, 351–362. [CrossRef]
102. Tjoa, E.; Guan, C. A Survey on Explainable Artificial Intelligence (XAI): Toward Medical XAI. *IEEE Trans. Neural Netw. Learn. Syst.* **2020**, *32*, 4793–4813. [CrossRef]
103. Karger, E.; Jagals, M.; Ahlemann, F. Blockchain for AI Data—State of the Art and Open Research. In Proceedings of the 42nd International Conference on Information Systems (ICIS), Austin, TX, USA, 12–15 December 2021.
104. Salah, K.; Rehman, M.H.U.; Nizamuddin, N.; Al-Fuqaha, A. Blockchain for AI: Review and Open Research Challenges. *IEEE Access* **2019**, *7*, 10127–10149. [CrossRef]
105. Xia, Q.; Sifah, E.B.; Smahi, A.; Amofa, S.; Zhang, X. BBDS: Blockchain-Based Data Sharing for Electronic Medical Records in Cloud Environments. *Information* **2017**, *8*, 44. [CrossRef]

06. Yue, X.; Wang, H.; Jin, D.; Li, M.; Jiang, W. Healthcare Data Gateways: Found Healthcare Intelligence on Blockchain with Novel Privacy Risk Control. *J. Med. Syst.* **2016**, *40*, 218. [CrossRef] [PubMed]
07. Mamoshina, P.; Ojomoko, L.; Yanovich, Y.; Ostrovski, A.; Botezatu, A.; Prikhodko, P.; Izumchenko, E.; Aliper, A.; Romantsov, K.; Zhebrak, A.; et al. Converging blockchain and next-generation artificial intelligence technologies to decentralize and accelerate biomedical research and healthcare. *Oncotarget* **2017**, *9*, 5665–5690. [CrossRef] [PubMed]
08. Martin-Noguerol, T.; Luna, A. External validation of AI algorithms in breast radiology: The last healthcare security checkpoint? *Quant. Imaging Med. Surg.* **2021**, *11*, 2888–2892. [CrossRef] [PubMed]
09. Zhou, Q.; Zuley, M.; Guo, Y.; Yang, L.; Nair, B.; Vargo, A.; Ghannam, S.; Arefan, D.; Wu, S. A machine and human reader study on AI diagnosis model safety under attacks of adversarial images. *Nat. Commun.* **2021**, *12*, 7281. [CrossRef]
10. Sajjadnia, Z.; Khayami, R.; Moosavi, M.R. Preprocessing Breast Cancer Data to Improve the Data Quality, Diagnosis Procedure, and Medical Care Services. *Cancer Inform.* **2020**, *19*, 1176935120917955. [CrossRef]
11. Kureljusic, M.; Karger, E. Data Preprocessing as a Service—Outsourcing der Datenvorverarbeitung für KI-Modelle mithilfe einer digitalen Plattform. *Inform. Spektrum* **2021**, *45*, 13–19. [CrossRef]
12. Muniz, E.C.L.; Dandolini, G.A.; Biz, A.A.; Ribeiro, A.C. Customer knowledge management and smart tourism destinations: A framework for the smart management of the tourist experience—SMARTUR. *J. Knowl. Manag.* **2020**, *25*, 1336–1361. [CrossRef]
13. vom Brocke, J.; Winter, R.; Hevner, A.; Maedche, A. Special Issue Editorial –Accumulation and Evolution of Design Knowledge in Design Science Research: A Journey Through Time and Space. *J. Assoc. Inf. Syst.* **2020**, *21*, 520–544. [CrossRef]
14. Kohli, R.; Melville, N.P. Digital innovation:Areview and synthesis. *Inf. Syst. J.* **2018**, *29*, 200–223. [CrossRef]
15. Bauer, K.; Hinz, O.; van der Aalst, W.; Weinhardt, C. Expl(AI)n It to Me—Explainable AI and Information Systems Research. *Bus. Inf. Syst. Eng.* **2021**, *63*, 79–82. [CrossRef]
16. Gunning, D.; Stefik, M.; Choi, J.; Miller, T.; Stumpf, S.; Yang, G.-Z. XAI—Explainable artificial intelligence. *Sci. Robot.* **2019**, *4*, eaay7120. [CrossRef] [PubMed]
17. Samek, W.; Montavon, G.; Vedaldi, A.; Hansen, L.K.; Müller, K.-R. *Explainable AI: Interpreting, Explaining and Visualizing Deep Learning*; Springer International Publishing: Cham, Switzerland, 2019; ISBN 978-3-030-28953-9.
18. Escalante, H.J.; Escalera, S.; Guyon, I.; Baró, X.; Güçlütürk, Y.; Güçlü, U.; van Gerven, M. *Explainable and Interpretable Models in Computer Vision and Machine Learning*; Springer International Publishing: Cham, Switzerland, 2018; ISBN 978-3-319-98130-7.
19. Baughan, N.; Douglas, L.; Giger, M.L. Past, Present, and Future of Machine Learning and Artificial Intelligence for Breast Cancer Screening. *J. Breast Imaging* **2022**, *4*, 451–459. [CrossRef]
20. Confalonieri, R.; Coba, L.; Wagner, B.; Besold, T.R. A historical perspective of explainable Artificial Intelligence. *WIREs Data Min. Knowl. Discov.* **2020**, *11*, e1391. [CrossRef]
21. Stremersch, S.; Verniers, I.; Verhoef, P.C.; Chan, T.H.; Tse, C.H.; Uddin, S.; Khan, A.; Fox, C.W.; Paine, C.E.; Sauterey, B.; et al. The Quest for Citations: Drivers of Article Impact. *J. Mark.* **2007**, *71*, 171–193. [CrossRef]
22. Kessler, M.M. Bibliographic coupling between scientific papers. *Am. Doc.* **1963**, *14*, 10–25. [CrossRef]
23. Weinberg, B.H. Bibliographic coupling: A review. *Inf. Storage Retr.* **1974**, *10*, 189–196. [CrossRef]

Article

MRI-Based Radiomics Combined with Deep Learning for Distinguishing IDH-Mutant WHO Grade 4 Astrocytomas from IDH-Wild-Type Glioblastomas

Seyyed Ali Hosseini [1,2,*], Elahe Hosseini [3], Ghasem Hajianfar [4], Isaac Shiri [5], Stijn Servaes [1,2], Pedro Rosa-Neto [1,2], Laiz Godoy [6], MacLean P. Nasrallah [7], Donald M. O'Rourke [8], Suyash Mohan [6] and Sanjeev Chawla [6,*]

[1] Translational Neuroimaging Laboratory, The McGill University Research Centre for Studies in Aging, Douglas Hospital, McGill University, Montréal, QC H4H 1R3, Canada

[2] Department of Neurology and Neurosurgery, Faculty of Medicine, McGill University, Montréal, QC H3A 2B4, Canada

[3] Department of Electrical and Computer Engineering, Kharazmi University, Tehran 15719-14911, Iran

[4] Rajaie Cardiovascular Medical and Research Center, Iran University of Medical Science, Tehran 19956-14331, Iran

[5] Division of Nuclear Medicine and Molecular Imaging, Geneva University Hospital, CH-1211 Geneva, Switzerland

[6] Department of Radiology, Perelman School of Medicine at the University of Pennsylvania, Philadelphia, PA 19104, USA

[7] Department of Pathology and Laboratory Medicine, Perelman School of Medicine, University of Pennsylvania, Philadelphia, PA 19104, USA

[8] Department of Neurosurgery, Perelman School of Medicine, University of Pennsylvania, Philadelphia, PA 19104, USA

* Correspondence: seyyed.ali.hosseini@mail.mcgill.ca (S.A.H.); sanjeev.chawla@pennmedicine.upenn.edu (S.C.); Tel.: +1-438-929-6575 (S.A.H.); +1-215-615-1662 (S.C.)

Citation: Hosseini, S.A.; Hosseini, E.; Hajianfar, G.; Shiri, I.; Servaes, S.; Rosa-Neto, P.; Godoy, L.; Nasrallah, M.P.; O'Rourke, D.M.; Mohan, S.; et al. MRI-Based Radiomics Combined with Deep Learning for Distinguishing IDH-Mutant WHO Grade 4 Astrocytomas from IDH-Wild-Type Glioblastomas. *Cancers* 2023, 15, 951. https://doi.org/10.3390/cancers15030951

Academic Editors: Ali Madani, Hamid Khayyam, Rahele Kafieh and Ali Hekmatnia

Received: 1 January 2023
Revised: 30 January 2023
Accepted: 30 January 2023
Published: 2 February 2023

Simple Summary: To differentiate IDH-mutant grade 4 astrocytomas from IDH-wild-type glioblastomas, two MRI sequences (post-contrast T1 and T2-FLAIR) were acquired from 57 patients. The images were resliced, resampled, and realigned. In the next step, tumors were segmented semi-automatically into subregions including whole tumor, edema region, core tumor, enhancing region, and necrotic region. A total of 105 radiomic features were extracted from each subregion. The data were divided randomly into training and testing sets. A deep learning-based data augmentation method (CTGAN) was implemented to synthesize 200 datasets. A total of 18 classifiers were used to distinguish two genotypes of grade 4 astrocytomas. The best discriminatory power was obtained from core tumor regions overlaid on post-contrast T1 using the K-best feature selection algorithm and a Gaussian naïve Bayes classifier.

Abstract: This study aimed to investigate the potential of quantitative radiomic data extracted from conventional MR images in discriminating IDH-mutant grade 4 astrocytomas from IDH-wild-type glioblastomas (GBMs). A cohort of 57 treatment-naïve patients with IDH-mutant grade 4 astrocytomas ($n = 23$) and IDH-wild-type GBMs ($n = 34$) underwent anatomical imaging on a 3T MR system with standard parameters. Post-contrast T1-weighted and T2-FLAIR images were co-registered. A semi-automatic segmentation approach was used to generate regions of interest (ROIs) from different tissue components of neoplasms. A total of 1050 radiomic features were extracted from each image. The data were split randomly into training and testing sets. A deep learning-based data augmentation method (CTGAN) was implemented to synthesize 200 datasets from the training sets. A total of 18 classifiers were used to distinguish two genotypes of grade 4 astrocytomas. From generated data using 80% training set, the best discriminatory power was obtained from core tumor regions overlaid on post-contrast T1 using the K-best feature selection algorithm and a Gaussian naïve Bayes classifier (AUC = 0.93, accuracy = 0.92, sensitivity = 1, specificity = 0.86, PR_AUC = 0.92). Similarly, high diagnostic performances were obtained from original and generated data using 50% and 30% training sets. Our findings suggest that conventional MR imaging-based radiomic features combined with machine/deep learning methods may be valuable in discriminating IDH-mutant grade 4 astrocytomas from IDH-wild-type GBMs.

Keywords: grade 4 astrocytoma; glioblastoma; isocitrate dehydrogenase mutation; conventional magnetic resonance imaging; radiomics; machine learning; deep learning

1. Introduction

Glioblastomas (GBMs) are devastating and universally fatal brain cancers in adults despite advancements in diagnostic and therapeutic strategies [1]. Approximately 14,000 new cases of GBM are diagnosed in the US each year, with an estimated incidence of 3.19 per 100,000 people [2]. In recent years, the emergence of molecular profiling in neuro-oncology has had a considerable bearing on the classification, diagnosis, prognosis, and clinical management of GBM patients [3]. The 2016 WHO classification system recognized the somatic mutation of the isocitrate dehydrogenase (IDH) gene in gliomas as a distinct entity regardless of histopathological features [4]. IDH mutation occurs in 50–70% of WHO grade 2/3 gliomas and 10% of GBMs [5], which has been considered as a new paradigm in determining the prognosis of these patients. The new 2021 WHO system has reclassified GBMs as IDH-mutant grade 4 astrocytomas or IDH-wild-type GBMs based on gene expression profiles [6]. It has been widely reported that glioma patients harboring IDH mutations demonstrate a better response to chemoradiation therapy and live longer than those with IDH-wild-type alleles [7,8]. Immunohistochemical analyses and exomic sequencing are considered the gold standard for determining IDH mutation status in gliomas [9,10]; however, several factors, such as tissue heterogeneity, partial sampling of tissue specimens, and presence of variable amounts of antigens constrain the utility of these methods in reliable detection of IDH mutation status [11]. Moreover, it is not always possible to perform neurosurgical interventions because of the eloquent locations of these neoplasms.

Therefore, non-invasive identification of IDH-mutant gliomas is vital for making informed decisions on therapeutic intervention and prognosticating these patients. IDH mutations confer the neomorphic activity of an enzyme leading to the conversion of alpha-ketoglutarate (α-KG) to 2-hydroxyglutarate (2HG) [12]. Prior studies [13–15] have reported the clinical utility of modified MR spectroscopy sequences in identifying IDH-mutant gliomas by detecting characteristic resonances of 2HG. However, not all IDH-mutant gliomas show the neomorphic activity of the 2-HG production [16]. Moreover, these sophisticated spectroscopic sequences are not readily available in routine clinical settings.

Conventional magnetic resonance imaging (MRI) remains the mainstay for determining tumor location, size, and structural features in neurooncology [17]. Radiomics is a rapidly evolving translational field that automatically produces mineable high-dimensionality data from positron emission tomography (PET) [18,19], computed tomography (CT), and MRI images with high precision [20–22]. Several previous studies have documented the clinical potential of quantitative radiomic features extracted from conventional MRI data in diagnosis, determining molecular signatures, assessing treatment response, and predicting survival outcomes in GBM patients [23–28]. Some other studies have also reported promising findings in identifying IDH-mutant grade 4 astrocytomas using conventional neuroimaging-based radiomic classification models with variable accuracies [29,30]. However, these studies were limited by the extraction of a sparse number of radiomic features (n = 31) [29] or by the inclusion of a small sample size of IDH-mutant grade 4 astrocytomas (n = 7) [30].

With these inadequacies in mind, the current study was designed to investigate the potential utility of radiomic features extracted from different tumor habitats as visible on widely available and universally acquired preoperative post-contrast T1 weighted and

T2-FLAIR images in differentiating IDH-mutant grade 4 astrocytomas from IDH-wild type GBMs.

2. Materials and Methods

2.1. Patient Population

This retrospective study was approved by the institutional review board and wa compliant with the Health Insurance Portability and Accountability Act. The inclusion criteria for enrollment in the present study were that all patients had (a) histopathologicall confirmed grade 4 astrocytoma according to the WHO classification system, (b) a known IDH mutation genotype using immunohistochemistry and/or gene sequencing, and (c available preoperative anatomical MR images acquired using identical data acquisition protocol. Based upon the inclusion criteria, a cohort of 57 patients (mean age = 57.7 $\pm$ 6. years, 39 males and 18 females) with newly diagnosed grade 4 astrocytoma and GBM wer recruited in this study. Of these 57 patients, 23 had the IDH-mutant genotype, and 34 hac the IDH-wild-type genotype.

2.2. Determination of IDH Mutational Status by Immunohistochemistry and Sequencing

Hematoxylin, eosin staining, and immunohistochemistry were conducted on 5-micror thick, formalin-fixed (10%), paraffin-embedded tissue sections mounted on Leica Surgipatl slides followed by drying for 60 min at 70 °C. In addition, immunohistochemistry to detec the IDH1 p.R132H variant was performed by using an anti-IDH1-R132H antibody (monc clonal mouse anti-human IDH1 (R132H), Dianova, DIA Clone H09) and DAB chromoger was performed on a Leica Bond III instrument using a bond polymer refine detection system (Leica Microsystems AR9800) following a 20-min heat-induced epitope retrieva with Epitope Retrieval 2, EDTA, pH 9.0. Appropriate positive and negative controls wer included.

In addition, massively parallel sequencing or RealTime polymerase chain reaction (PCR) was performed to confirm the immunohistochemical results and to interrogate othe IDH variants. For RealTime PCR, formalin-fixed, paraffin-embedded (FFPE) specimen with >20% tumor content were analyzed for IDH1 and IDH2 variants using Abbott Re alTime Assays (Abbott Molecular, Inc., Abbott Park, IL, USA) after extraction using the QIAamp DSP DNA FFPE Tissue Kit (Qiagen, Hilden, Germany). The Abbott RealTime IDH assay detects 5 single nucleotide variants (SNVs) in IDH1 (p.R132C, p.R132H, p.R132C p.R132S, and p.R132L). The Abbott RealTime IDH2 assay detects 9 SNVs in IDH2 (p.R140C p.R140L, p.R140G, p.R140W, p.R172K, p.R172M, p.R172G, p.R172S, and p.R172W). The Abbott m2000rt software performs variant calling, and results are qualitatively reported a positive or not detected. Tests were performed according to the manufacturer's instruction by adding a dilution step to the IDH2 assay. For massively parallel sequencing, the pane gives full gene coverage of 152 genes, using the Agilent Haploplex design with uniqu molecular identifiers as described previously [31]. Briefly, DNA was extracted from FFPI or specimens preserved in PreservCyt. Samples were multiplexed and sequenced on a HiSeq with total deduplicated reads of 6.5 million/sample; duplicate reads were removec based on incorporating unique molecular identifiers. All variants were identified usin an in-house data processing bioinformatics pipeline capable of detecting SNVs, insertion and/or deletions (indels), and copy number gains for a subset of genes based on increasec read depth. An experienced neuropathologist (MPN) reviewed cases from all patients t confirm the IDH status.

2.3. MRI Data Acquisition

All patients underwent an MRI on a 3T Tim Trio whole-body MR scanner (Siemen: Erlangen, Germany) equipped with a 12-channel phased array head coil. The anatomica imaging protocol included an axial 3D-T1-weighted magnetization-prepared rapid acquis tion of gradient echo (MPRAGE) imaging [repetition time (TR)/echo time (TE)/inversion time (TI) = 1760/3.1/950 ms]; in-plane resolution = 1×1 mm^2; slice thickness = 1 mm; the number of slices = 192; and axial T2-FLAIR imaging (TR/TE/TI = 9420/141/2500 ms, slice thickness = 3 mm; the number of slices = 60). The post-contrast T1-weighted images were acquired with the same parameters as the pre-contrast acquisition after administration of the standard dose of gadobenate dimeglumine (MultiHance, Bracco Imaging, Milano, Italy) intravenous contrast agent using a power injector (Medrad, Idianola, PA, USA).

2.4. Image Processing

The overview of the image processing and radiomics pipeline, which includes image registration, tissue segmentation, feature extraction, feature selection, and radiomics model building, is shown in Figure 1. An investigator (SAH) blinded to the IDH mutational status performed all image processing steps. Post-contrast T1-weighted images were resliced, resampled, and co-registered with T2-FLAIR images using a linear affine transformation. A semi-automatic segmentation approach was used to generate regions of interest (ROIs) on the anatomical images. Care was taken to exclude surrounding normal brain vessels. Manual inspections were performed by an experienced neuroradiologist to correct for any pixel anomalies present within the ROIs. Accordingly, these ROIs were modified manually by adding pixels for tumor regions not included in the initial ROIs or by removing pixels for non-tumor regions included in the initial ROIs. Post-contrast T1 weighted images were used to segment solid/contrast-enhancing regions, necrotic regions, and core tumors (solid + necrotic region). T2-FLAIR images were used to segment peri-tumoral edematous regions and whole tumor volumes. All tissue segmentations were performed using 3D slicer software. To maximize the characterization of tumors, these 5 segmented ROIs were overlaid on the source post-contrast T1-weighted images and T2-FLAIR images for the data analysis (Figures 2 and 3). A bias field correction using N4 and an image normalization using histogram matching were performed using the 3D slicer software on the MRI images before feature extraction to avoid any potential bias field distortions and data heterogeneity bias.

2.5. Radiomic Feature Extraction

From each segmented ROI, 105 original radiomic features from categories (shape, first-order statistical, second-order texture, and higher-order statistic) were extracted using the PyRadiomics package in python [32]. These original features can be sub-divided into 7 classes, including 13 shape features, 18 first-order statistical features, 23 gray level co-occurrence matrix (GLCM) features, 14 gray level dependence matrix (GLDM) features, 16 gray level size zone matrix (GLSZM) features, 16 gray level run length matrix (GLRLM) features, and 5 neighboring gray-tone difference matrix (NGTDM) features. Altogether, 525 radiomic features were extracted from 5 ROIs of each image for a total of 1050 features from post-contrast T1- and T2-FLAIR images. The radiomics features used comply with the standard described by the Imaging Biomarker Standardization Initiative (IBSI) [33]. A high-performance computer system with 16GB RAM and an Intel Core i7-7700 CPU processor @3.60 GHz was used for data processing. The feature extraction took an average of 2–3 min per patient image set. A list of all features is summarized in Supplementary Table S1.

2.6. Radiomics Feature Selection/Dimension Reduction

It is important to eliminate irrelevant or redundant variables that may cause data overfitting and may bias the performance of the prediction model. Multiple feature selection algorithms, including recursive feature elimination (RFE), minimum redundancy, maximum relevance (mRmR), and K-best were employed to select image features. Patients were divided into 2 mutually exclusive training (80%, 50%, and 30%) and testing (20%, 50%, and 70%) sets using the random shuffling method. Ten percent of the training set was split off to serve as the validation set. All data were normalized by MinMax normalization. The mRmR feature selection technique was used to select 15 features.

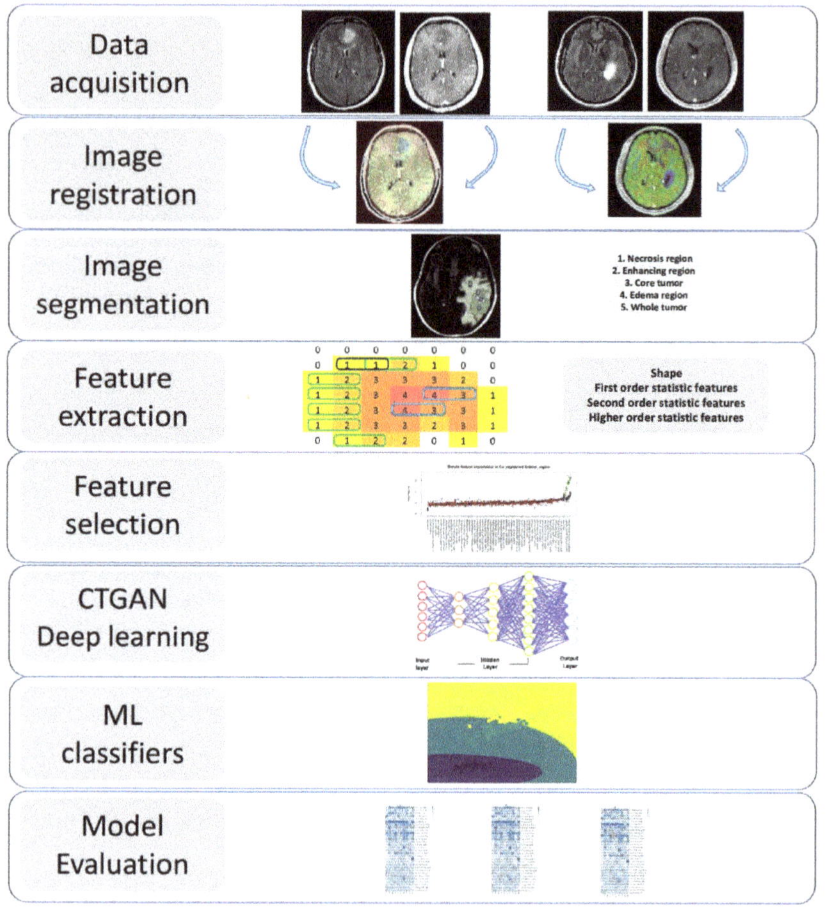

Figure 1. The overview of the image processing and radiomics pipeline.

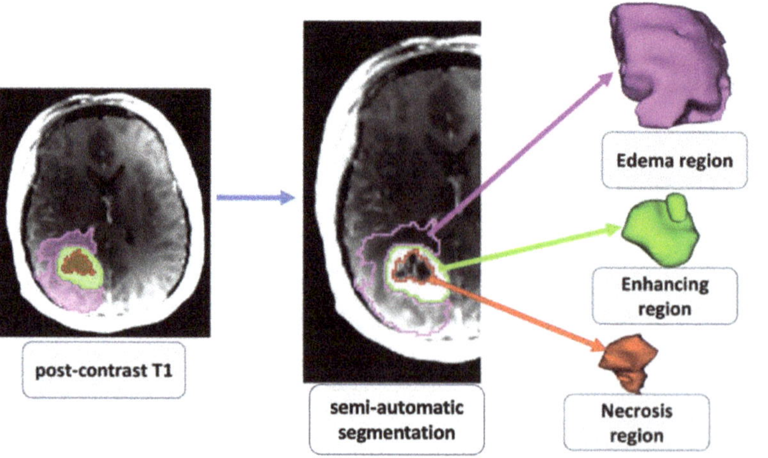

Figure 2. 2D and 3D visualization of various subregions of a grade 4 astrocytoma as visible on post-contrast T1-weighted image.

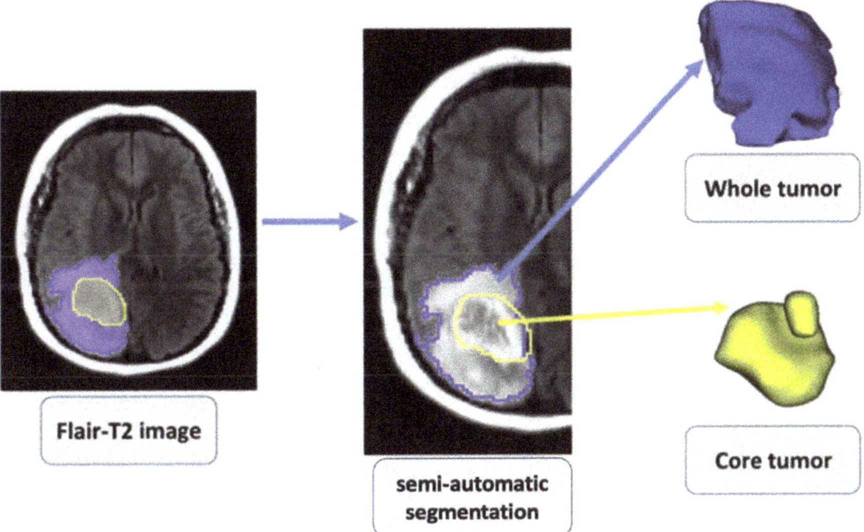

Figure 3. 2D and 3D visualization of various subregions of a grade 4 astrocytoma as visible on T2-FLAIR image.

2.7. Deep Learning Approach for Data Augmentation

The current study implemented a deep learning method based on generative adversarial networks (GAN) for data augmentation [34]. CTGAN is a GAN-based deep learning data synthesizer to increase the number of our datasets that can improve the reproducibility and discriminatory power of radiomics features [35–37]. After splitting the data set and selecting bold features using various feature selection algorithms, the selected radiomic features from each model with the highest number were used as the input value for CTGAN to synthesize 200 radiomic features. As a result, after splitting 80%, 50%, and 30% of 57 original data for the training sets, 245, 228, and 217 datasets (80%, 50%, and 30% of 57 + 200 = 245, 228, and 217), including original and generated data, were synthesized, respectively. Different splitting percentages were used to confirm our findings [38] and to prevent the impact of data leakage in our results [39]. Furthermore, a random noise (normal distribution, mean = 0.0, standard deviation = 0.05) [40] was added to the training set. The test sets were not generated, and the original datasets were used for the testing sets.

2.8. Machine Learning Classifiers for Prediction Model Building

To develop a prediction model for distinguishing IDH-mutant grade 4 astrocytomas from IDH-wild-type GBMs, a total of 18 single and ensembled machine learning classifiers [Bernoulli naïve Bayes (BNB), multilayer perceptron (MLP), support vector classifier (SVC), Gaussian naïve Bayes (GNB), quadratic discriminant analysis (QDA), bagging classifier, linear discriminant analysis (LDA), logistic regression (RG), ridge, ada boost (AD), hist gradient boosting (HGB), K-neighbors (KN) (K = 5), random forest (RF), gradient boosting (GB), extra trees (ET), decision tree (DT), nearest centroid (NC), and passive aggressive (PA] were employed using an in-house-developed python package. All cases in the training cohort (80%, 50%, and 30%) were used to train the classifiers, and an internal validation (cross-validation) was performed from the testing cohort (20%, 50%, and 70%). Receiver operative characteristic (ROC) curve analyses were performed to evaluate the diagnostic potentials of prediction models in distinguishing 2 groups (IDH-mutant grade 4 astrocytomas and IDH-wild-type GBMs). Area under the ROC curve (AUC), area under the precision-recall curve (PR_AUC), accuracy (ACC), sensitivity, specificity, and negative and positive predictive values (NPV and PPV, respectively) were determined for each prediction model as performance metrics.

3. Results

When original MRI data (*n* = 57) were used in discriminating IDH-mutant grade 4 astrocytomas from IDH-wild-type GBMs, the best discriminatory performance (AUC = 0.93, ACC = 0.92, sensitivity = 1, specificity = 0.86, PR_AUC = 0.92) was obtained from solid contrast enhancing, and core tumor (solid + necrotic region) overlaid on post-contrast T1-weighted images using various combinations of feature selection algorithms and machine learning classifiers. The predictive power, accuracy, sensitivity, specificity, and PR_AUC of the best 10 methods in distinguishing two genotypes of grade 4 astrocytomas are summarized in Table 1.

Table 1. Best 10 performances of multi-segmentation approaches, multi-machine learning classifiers and multi-feature selection algorithms in discriminating IDH-mutant grade 4 astrocytomas from IDH-wild-type GBMs using original (Or) data set.

Radiomic Feature Combination	AUC	Accuracy	Sensitivity	Specificity	PR_AUC
Or_PC_T1_Core_AB_Kbest	0.93	0.92	1	0.86	0.92
Or_PC_T1_Core_KN_Kbest	0.93	0.92	1	0.86	0.92
Or_PC_T1_Core_LR_Kbest	0.93	0.92	1	0.86	0.92
Or_PC_T1_Core_MLP_Kbest	0.93	0.92	1	0.86	0.92
Or_T2-FLAIR_Enhancing_DT_Kbest	0.93	0.92	1	0.86	0.92
Or_T2-FLAIR_Enhancing_DT_mRmR	0.93	0.92	1	0.86	0.92
Or_T2-FLAIR_Enhancing_GB_mRmR	0.93	0.92	1	0.86	0.92
Or_T2-FLAIR_Enhancing_RF_mRmR	0.93	0.92	1	0.86	0.92
Or_PC_T1_Enhancing_HGB_RFE	0.93	0.92	1	0.86	0.92
Or_PC_T1_Enhancing_HGB_mRmR	0.93	0.92	1	0.86	0.92

The relative importance of the best 10 methods in terms of predictive power, accuracy, sensitivity, specificity, and PR_AUC in discriminating two genotypes of grade 4 astrocytomas by using various combinations of feature selection algorithms, machine learning classifiers, and segmented tumor regions when 80%, 50%, and 30% of the generated data were used as training sets are summarized in Table 2, Table 3, and Table 4, respectively. From generated data using 80% as the training set (Table 2), the best discriminatory power (AUC = 0.93, accuracy = 0.92, sensitivity = 1, specificity = 0.86, and PR_AUC = 0.92) in distinguishing two genotypes of grade 4 astrocytomas was obtained from core regions overlaid on post-contrast T1 images when K-best and RFE feature selection algorthims and GNB and PA classifiers were applied. A similar high-diagnostic performance was obtained from enhancing regions overlaid on T2-FLAIR images when the K-best feature selection algorithm and DT and bagging classifiers were applied. From generated data using 50% as the training set (Table 3), necrotic regions of co-registered, post-contrast T1 images with mRmR feature selection and bagging and RF classifiers and the edematous region of the co-registered, post-contrast T1 image with mRmR feature selection and KN classifier provided the highest predictive power (AUC = 0.92, accuracy = 0.92, sensitivity = 0.91, specificity = 0.94, and PR_AUC = 0.93). From generated data using 30% as the training set (Table 4), the core regions of co-registered, post-contrast T1 images with K-best feature selection and LR classifier provided the highest predictive power (AUC = 0.91, accuracy = 0.92, sensitivity = 0.86, specificity = 0.96, and PR_AUC = 0.92).

Heatmaps of predictive power (AUC), predictive accuracy (ACC), sensitivity (SEN) and specificity (SPE) for discriminating IDH-mutant grade 4 astrocytomas from IDH-wild-type GBMs utilizing a variety of feature selections (training set equal to 80%), and machine learning algorithms applied to distinct subregions of neoplasms, are shown in Supplementary Figures S1–S4, respectively. In addition, the comprehensive findings from using a multi-segmentation approach, feature selection algorithms, and multi-machine learning classifiers in discriminating IDH-mutant grade 4 astrocytomas from IDH-wild-type GBMs in original and generated data with different training and testing sets are provided in the Supplementary File.

Table 2. Best 10 performances of multi-segmentation approaches, multi-machine learning classifiers, and multi-feature selection algorithms in discriminating IDH-mutant grade 4 astrocytomas from IDH-wild-type GBMs using generated (Ge) data with 80% training set.

Radiomic Feature Combination	AUC	Accuracy	Sensitivity	Specificity	PR_AUC
Ge_PC_T1_Core_GNB_Kbest	0.93	0.92	1	0.86	0.92
Ge_PC_T1_Core_PA_RFE	0.93	0.92	1	0.86	0.92
Ge_T2_FLAIR_Enhancing_Bagging_Kbest	0.93	0.92	1	0.86	0.92
Ge_T2_FLAIR_Enhancing_DT_Kbest	0.93	0.92	1	0.86	0.92
Ge_T2_FLAIR_Whole_AB_Kbest	0.90	0.92	0.80	1	0.94
Ge_PC_T1_Core_RF_Kbest	0.90	0.92	0.80	1	0.94
Ge_PC_T1_Core_RF_RFE	0.90	0.92	0.80	1	0.94
Ge_PC_T1_Core_HGB_Kbest	0.90	0.92	0.80	1	0.94
Ge_PC_T1_Edema_AB_Kbest	0.90	0.92	0.80	1	0.94
Ge_PC_T1_Edema_Bagging_Kbest	0.90	0.92	0.80	1	0.94

Table 3. Best 10 performances of multi-segmentation approaches, multi-machine learning classifiers, and multi-feature selection algorithms in discriminating IDH-mutant grade 4 astrocytomas from IDH-wild-type GBMs using generated (Ge) data with 50% training set.

Radiomic Feature Combination	AUC	Accuracy	Sensitivity	Specificity	PR_AUC
Ge_PC_T1_Necrosis_Bagging_mRmR	0.92	0.92	0.91	0.94	0.93
Ge_PC_T1_Necrosis_RF_mRmR	0.92	0.92	0.91	0.94	0.93
Ge_PC_T1_Edema_KN_mRmR	0.92	0.92	0.91	0.94	0.93
Ge_PC_T1_Necrosis_KN_RFE	0.89	0.89	0.91	0.87	0.89
Ge_PC_T1_Edema_HGB_RFE	0.89	0.89	0.91	0.87	0.89
Ge_PC_T1_Necrosis_KN_RFE	0.88	0.89	0.82	0.94	0.90
Ge_PC_T1_Necrosis_KN_RFE	0.88	0.89	0.82	0.94	0.90
Ge_PC_T1_Edema_HGB_RFE	0.88	0.89	0.82	0.94	0.90
Ge_PC_T1_Edema_HGB_RFE	0.88	0.89	0.82	0.94	0.90
Ge_PC_T1_Core_KN_RFE	0.88	0.89	0.82	0.94	0.90

Table 4. Best 10 performances of multi-segmentation approaches, multi-machine learning classifiers, and multi-feature selection algorithms in discriminating IDH-mutant grade 4 astrocytomas from IDH-wild-type GBMs using generated (Ge) data with 30% training set.

Radiomic Feature Combination	AUC	Accuracy	Sensitivity	Specificity	PR_AUC
Ge_PC_T1_Core_LR_Kbest	0.91	0.92	0.86	0.96	0.92
Ge_PC_T1_Core_Ridge_Kbest	0.89	0.89	0.86	0.92	0.88
Ge_PC_T1_Core_SVC_mRmR	0.86	0.89	0.71	1	0.91
Ge_PC_T1_Core_LDA_Kbest	0.84	0.84	0.86	0.83	0.83
Ge_T2_FLAIR_Core_HGB_Kbest	0.82	0.79	0.93	0.71	0.80
Ge_T2_FLAIR_Core_LR_Kbest	0.81	0.84	0.71	0.92	0.83
Ge_PC_T1_Edema_GB_Kbest	0.81	0.84	0.71	0.92	0.83
Ge_T2_FLAIR_Core_LDA_Kbest	0.81	0.81	0.78	0.83	0.80
Ge_T2_FLAIR_Core_Ridge_Kbest	0.81	0.81	0.78	0.83	0.80
Ge_PC_T1_Enhancing_QDA_Kbest	0.80	0.79	0.86	0.75	0.79

4. Discussion

In this study, we investigated the clinical utility of a conventional neuroimaging-based radiomics approach with deep learning in determining the IDH status of grade 4 astrocytomas. A total of 1050 radiomic features were extracted from different tumor habitats (solid/contrast enhancing, central necrotic, peritumoral edematous, core tumor and whole tumor regions), encompassing post-contrast T1-weighted and T2-FLAIR images. Our work is an extension of previous studies as we used a GAN-based algorithm to increase our sample size and used a large number of machine learning classifiers (n = 18) to build a reliable prediction model in distinguishing IDH-mutant grade 4 astrocytomas and IDH wild-type GBMs. In the testing cohort, our best prediction model consisted of a central necrotic region from post-contrast, T1-weighted images when a combination of the K-best feature selection algorithm and Gaussian naïve Bayes classifier were used together. This prediction model achieved a high diagnostic performance (AUC = 0.93, accuracy = 0.92, sensitivity = 1, specificity = 0.86, PR_AUC = 0.92) in discriminating two genotypes of grade 4 astrocytomas.

IDH mutation has been recognized as one of the most important molecular markers for diagnosis of gliomas and GBMs based on the 2016 WHO classification system. In addition, according to the recent 2021 WHO classification of tumors of the central nervous system (CNS) [6], previously called IDH-mutant GBM, is now designated as IDH-mutant grade 4 astrocytoma, and GBM is diagnosed in the setting of IDH-wild-type status. It has been reported that IDH mutational status is an independent favorable prognostic factor for conferring longer progression-free and overall survival in GBM patients [7,8]. Moreover, patients with IDH-mutant grade 4 gliomas have been shown to exhibit a better prognosis than those with IDH-wild-type grade 3 gliomas. Collectively, these clinical findings emphasize the importance of determining IDH-mutant status in grade 4 astrocytomas [41]. The immuno histochemical assay is the most commonly used method for assessing IDH mutational status following invasive surgical interventions, which are associated with operative risks [42,43]. Moreover, the possibility of sampling error is highly relevant to determining histological grade and molecular profiling [11,44]. For example, IDH sequencing may be falsely negative if there are few glioma cells present within a tumor specimen [44] or substantial genetic heterogeneity occurs within the tumor specimen [11]. In addition, some exome sequencing studies have reported that traditional immunohistochemical assays do not detect IDH mutant status in ~15% of gliomas [45]. Therefore, it is essential to develop non-invasive and objective imaging biomarkers for determining IDH mutational status in gliomas.

Mechanistically, wild-type IDH normally catalyzes the reversible, NADP+-dependent oxidative decarboxylation of isocitrate to alpha-ketoglutarate (α-KG) in the TCA cycle. However, IDH mutations confer a neomorphic enzyme activity converting α-KG to 2HG. Therefore, the oncometabolite 2HG has been proposed as a putative biomarker for IDH specific genetic profiles for gliomas. A few studies have employed modified spectroscopic sequences and post-processing tools for detecting spectral resonances of 2HG from IDH mutant gliomas [15,46–48]. However, the non-availability of these sequences and tools in the routine clinical setting renders these techniques less attractive. Moreover, diagnostic challenges may also arise due to the presence of a high degree of genetic heterogeneity within GBMs and partial sampling of these lesions, especially when single voxel spec troscopic methods are employed. In contrast, conventional MRI is a widely available, fast, easy-to-use, and economically affordable imaging modality that provides valuable information about brain tumor structural and morphological characteristics. Qualitative imaging features, such as frontal lobe tumor location, homogeneous signal intensity, sparse contrast enhancement within the tumor beds, and less intensive tumor infiltration are some of the imaging signatures that have been used to identify IDH-mutant gliomas with variable success [49–51]. However, all these qualitative associations were largely based on univariate analyses and hence, were prone to inter-observer variably. Therefore, a comprehensive analysis of imaging features is warranted for reliable prediction of IDH mutational status in spatially and temporally heterogeneous GBMs.

Radiomics is a quantitative analytical method of medical images that provides in formation that is generally difficult to perceive by visual inspection. Compared with conventional analytical approaches, radiomic analysis can provide a more efficient and unbiased quantification of imaging information. Readily interpretable and quantitative features, such as intensity distributions, spatial relationships, textural heterogeneity, and shape descriptors are extracted from a pre-defined ROI encompassing both solid and peritumoral regions of neoplasms in a typical fashion [52]. The training cohort is used to instruct the computer algorithm to detect patterns of features that are subsequently examined in a validation cohort to evaluate the algorithm's performance in correctly predicting the presence or absence of a feature and its association with an outcome. In the recent past, the field of radiogenomics has been established to study the relationship between imaging features and underlying molecular processes and characteristics. Recently, it has been widely reported that radiomics/radiogenomics aids in guiding clinical decision making in neuro-oncology, particularly for making an accurate diagnosis, prognosis, and response assessment [23–27,53].

IDH mutation occurs only in 10% of grade 4 astrocytomas, so we could only include data from 23 IDH-mutant cases in the present study. Due to this small sample size and imbalance in data distribution, our data was prone to overfitting. Furthermore, in situations with an insufficient number of training datasets, the model is often overtrained. Consequently, the model performs well during the training stage but comparatively poorly during the subsequent testing stage. To address this challenge of small sample size, we leveraged the use of a well-established GAN method for synthesizing high-quality images and, in turn, raised the total sample size from 57 to 200. GAN is a deep learning architecture in which two neural networks compete against each other in a zero-sum game framework [54]. A GAN model consists of two components: a generator and a discriminator. In the training stage, the datasets produced by the generator, along with real images, serve as inputs to the discriminator. This can be considered comparable to enlarging the training datasets for the discriminator, whose purpose is to differentiate the real from the generated images [55]. Consequently, the discriminator will not immediately succumb to overfitting through the competitive relationship between these two networks, even when a limited number of training samples are used.

In a previous study [56], IDH mutational status was determined from a mixed population of grade III and grade IV gliomas. In the present study, only a histologically homogenous population of gliomas (grade IV astrocytomas) was included. Moreover, numerous radiomics features and machine learning classifiers were applied to predict IDH mutational status. Tumor necrosis was recognized as an important imaging feature and contributed most to the prediction model for distinguishing IDH-mutant grade 4 astrocytomas from IDH-wild-type GBMs when the K-best radiomics feature algorithm and decision tree (DT) classifier were used together. This finding is in agreement with an earlier study [56] in which IDH mutation was associated with a smaller enhancing volume and a larger necrotic volume when multiparametric radiomic profiles were analyzed. Additionally, imaging features from whole tumor volumes were found to be associated with IDH mutation status when the K-best radiomics feature selection algorithm and AB classifier were used together (AUC = 0.93). This finding may be explained by the fact that IDH-mutant gliomas have a more heterogeneous imaging microenvironment because of their stepwise gliomagenesis [57]. Our findings are also consistent with previous studies that have reported a larger tumor volume [58] and a lower degree of cellularity [59] in IDH-mutant compared to those in IDH-wild-type gliomas. Taken together, our results and published findings indicate that quantitative radiomic features can predict the IDH mutation status of grade 4 astrocytomas with high diagnostic power. However, these findings warrant further validation in multicentric, prospective studies with larger patient populations.

5. Conclusions

In conclusion, a prediction model based on conventional MRI-extracted radiomic features achieved promising diagnostic power in distinguishing IDH-mutant grade 4 astrocytomas from IDH-wild-type GBMs.

Supplementary Materials: The following supporting information can be downloaded at: https://www.mdpi.com/article/10.3390/cancers15030951/s1, Figure S1: The area under the curve of AUC (predictive power) heatmap for differentiating IDH-mutant from IDH wild-type grade-4 astrocytoma employing a variety of feature selections and machine learning classifiers applied to distinct neoplasm subregions; Figure S2: Accuracy (ACC) heatmap for differentiating IDH-mutant from IDH wild-type grade-4 astrocytomas employing a variety of feature selections and machine learning classifier applied to distinct neoplasm subregions; Figure S3: Sensitivity (SEN) heatmap for differentiating IDH-mutant from IDH wild-type grade-4 astrocytomas employing a variety of feature selections and machine learning classifiers applied to distinct neoplasm; Figure S4: Specificity (SPE) heatmap for differentiating IDH-mutant from IDH wild-type grade-4 astrocytomas employing a variety of feature selections and machine learning classifiers applied to distinct neoplasm; Table S1. Radiomics feature name, set, and family extracted in this study; Table S2: The complete result of multi-machine learning algorithms, feature selection, and multi-segmentation approaches in discriminating IDH-mutant grade-4 astrocytomas from IDH wild-type GBMs of generated data with 80:20 training:testing set; Table S3. The complete result of multi-machine learning algorithms, feature selection, and multi-segmentation approaches in discriminating IDH-mutant grade-4 astrocytomas from IDH wild-type GBMs of original data.

Author Contributions: Conceptualization, S.C. and S.A.H.; methodology, S.C., S.A.H., L.G., M.P.N., D.M.O. and S.M.; software, S.A.H., E.H., G.H. and I.S.; validation, S.C., S.A.H., E.H., G.H., I.S., S.S., P.R.-N. and S.A.H.; investigation, S.A.H. and L.G.; resources, S.C.; data curation, S.A.H. and L.G.; writing—original draft preparation, S.C. and S.A.H.; writing—review and editing, S.C., S.A.H., L.G., M.P.N., D.M.O., S.M., E.H., G.H., I.S., S.S. and P.R.-N.; visualization, S.A.H.; supervision, S.C.; project administration, S.C. and S.A.H.; funding acquisition, S.C. All authors have read and agreed to the published version of the manuscript.

Funding: This work was partially supported by a grant obtained from the Research Foundation of the University of Pennsylvania, Philadelphia, PA, USA (PI = SC).

Institutional Review Board Statement: This study was approved by the institutional review board (protocol # 829645) and was compliant with the health insurance portability and accountability act (HIPPA).

Informed Consent Statement: Informed consent was obtained from all subjects involved in the study.

Data Availability Statement: The data supporting this study's findings and data processing algorithms will be available from the investigative team upon reasonable request.

Acknowledgments: We thank Lisa Desiderio, University of Pennsylvania Neuroradiology Clinical Research division, and the MRI technicians for their valuable contributions to this project.

Conflicts of Interest: The authors declare no conflict of interest.

References

1. Wen, P.Y.; Weller, M.; Lee, E.Q.; Alexander, B.M.; Barnholtz-Sloan, J.S.; Barthel, F.P.; Batchelor, T.T.; Bindra, R.S.; Chang, S.M.; Chiocca, E.A. Glioblastoma in adults: A Society for Neuro-Oncology (SNO) and European Society of Neuro-Oncology (EANO) consensus review on current management and future directions. *Neuro Oncol.* **2020**, *22*, 1073–1113. [CrossRef] [PubMed]
2. Alexander, B.M.; Cloughesy, T.F. *Platform Trials Arrive on Time for Glioblastoma*; Oxford University Press: New York, NY, USA, 2018; pp. 723–725.
3. Thakkar, J.P.; Dolecek, T.A.; Horbinski, C.; Ostrom, Q.T.; Lightner, D.D.; Barnholtz-Sloan, J.S.; Villano, J.L. Epidemiologic and Molecular Prognostic Review of GlioblastomaGBM Epidemiology and Biomarkers. *Cancer Epidemiol. Biomark. Prev.* **2014**, *23*, 1985–1996. [CrossRef] [PubMed]
4. Louis, D.N.; Perry, A.; Reifenberger, G.; Von Deimling, A.; Figarella-Branger, D.; Cavenee, W.K.; Ohgaki, H.; Wiestler, O.D.; Kleihues, P.; Ellison, D.W. The 2016 World Health Organization classification of tumors of the central nervous system: A summary. *Acta Neuropathol.* **2016**, *131*, 803–820. [CrossRef] [PubMed]
5. Vigneswaran, K.; Neill, S.; Hadjipanayis, C.G. Beyond the World Health Organization grading of infiltrating gliomas: Advances in the molecular genetics of glioma classification. *Ann. Transl. Med.* **2015**, *3*, 95.
6. Louis, D.N.; Perry, A.; Wesseling, P.; Brat, D.J.; Cree, I.A.; Figarella-Branger, D.; Hawkins, C.; Ng, H.; Pfister, S.M.; Reifenberger, G. The 2021 WHO classification of tumors of the central nervous system: A summary. *Neuro Oncol.* **2021**, *23*, 1231–1251. [CrossRef]

7.   Yan, W.; Zhang, W.; You, G.; Bao, Z.; Wang, Y.; Liu, Y.; Kang, C.; You, Y.; Wang, L.; Jiang, T. Correlation of IDH1 mutation with clinicopathologic factors and prognosis in primary glioblastoma: A report of 118 patients from China. *PLoS ONE* **2012**, *7*, e30339. [CrossRef]

8.   Zhang, C.-B.; Bao, Z.-S.; Wang, H.-J.; Yan, W.; Liu, Y.-W.; Li, M.-Y.; Zhang, W.; Chen, L.; Jiang, T. Correlation of IDH1/2 mutation with clinicopathologic factors and prognosis in anaplastic gliomas: A report of 203 patients from China. *J. Cancer Res. Clin. Oncol.* **2014**, *140*, 45–51. [CrossRef]

9.   Parsons, D.W.; Jones, S.; Zhang, X.; Lin, J.C.-H.; Leary, R.J.; Angenendt, P.; Mankoo, P.; Carter, H.; Siu, I.-M.; Gallia, G.L. An integrated genomic analysis of human glioblastoma multiforme. *Science* **2008**, *321*, 1807–1812. [CrossRef]

10.   Yan, H.; Parsons, D.W.; Jin, G.; McLendon, R.; Rasheed, B.A.; Yuan, W.; Kos, I.; Batinic-Haberle, I.; Jones, S.; Riggins, G.J. IDH1 and IDH2 mutations in gliomas. *N. Engl. J. Med.* **2009**, *360*, 765–773. [CrossRef]

11.   Preusser, M.; Wöhrer, A.; Stary, S.; Höftberger, R.; Streubel, B.; Hainfellner, J.A. Value and limitations of immunohistochemistry and gene sequencing for detection of the IDH1-R132H mutation in diffuse glioma biopsy specimens. *J. Neuropathol. Exp. Neurol.* **2011**, *70*, 715–723. [CrossRef]

12.   Dang, L.; White, D.W.; Gross, S.; Bennett, B.D.; Bittinger, M.A.; Driggers, E.M.; Fantin, V.R.; Jang, H.G.; Jin, S.; Keenan, M.C. Cancer-associated IDH1 mutations produce 2-hydroxyglutarate. *Nature* **2009**, *462*, 739–744. [CrossRef]

13.   Pope, W.B.; Prins, R.M.; Albert Thomas, M.; Nagarajan, R.; Yen, K.E.; Bittinger, M.A.; Salamon, N.; Chou, A.P.; Yong, W.H.; Soto, H. Non-invasive detection of 2-hydroxyglutarate and other metabolites in IDH1 mutant glioma patients using magnetic resonance spectroscopy. *J. Neuro-Oncol.* **2012**, *107*, 197–205. [CrossRef]

14.   Choi, C.; Ganji, S.K.; DeBerardinis, R.J.; Hatanpaa, K.J.; Rakheja, D.; Kovacs, Z.; Yang, X.-L.; Mashimo, T.; Raisanen, J.M.; Marin-Valencia, I. 2-hydroxyglutarate detection by magnetic resonance spectroscopy in IDH-mutated patients with gliomas. *Nat. Med.* **2012**, *18*, 624–629. [CrossRef]

15.   Verma, G.; Mohan, S.; Nasrallah, M.P.; Brem, S.; Lee, J.Y.; Chawla, S.; Wang, S.; Nagarajan, R.; Thomas, M.A.; Poptani, H. Non-invasive detection of 2-hydroxyglutarate in IDH-mutated gliomas using two-dimensional localized correlation spectroscopy (2D L-COSY) at 7 Tesla. *J. Transl. Med.* **2016**, *14*, 274. [CrossRef]

16.   Ichimura, K.; Pearson, D.M.; Kocialkowski, S.; Bäcklund, L.M.; Chan, R.; Jones, D.T.; Collins, V.P. IDH1 mutations are present in the majority of common adult gliomas but rare in primary glioblastomas. *Neuro Oncol.* **2009**, *11*, 341–347. [CrossRef]

17.   Villanueva-Meyer, J.E.; Mabray, M.C.; Cha, S. Current clinical brain tumor imaging. *Neurosurgery* **2017**, *81*, 397–415. [CrossRef]

18.   Hosseini, S.A.; Hajianfar, G.; Shiri, I.; Zaidi, H. Lymphovascular Invasion Prediction in Lung Cancer Using Multi-Segmentation PET Radiomics and Multi-Machine Learning Algorithms. In Proceedings of the 2021 IEEE Nuclear Science Symposium and Medical Imaging Conference (NSS/MIC), Piscataway, NJ, USA, 16–23 October 2021.

19.   Hosseini, S.A.; Hajianfar, G.; Shiri, I.; Zaidi, H. Lung Cancer Recurrence Prediction Using Radiomics Features of PET Tumor Sub-Volumes and Multi-Machine Learning Algorithms. In Proceedings of the 2021 IEEE Nuclear Science Symposium and Medical Imaging Conference (NSS/MIC), Piscataway, NJ, USA, 16–23 October 2021.

20.   Hosseini, S.A.; Shiri, I.; Hajianfar, G.; Bahadorzadeh, B.; Ghafarian, P.; Zaidi, H.; Ay, M.R. Synergistic impact of motion and acquisition/reconstruction parameters on 18F-FDG PET radiomic features in non-small cell lung cancer: Phantom and clinical studies. *Med. Phys.* **2022**, *49*, 3783–3796. [CrossRef]

21.   Hosseini, S.A.; Hajianfar, G.; Shiri, I.; Zaidi, H. PET Image Radiomics Feature Variability in Lung Cancer: Impact of Image Segmentation. In Proceedings of the 2021 IEEE Nuclear Science Symposium and Medical Imaging Conference (NSS/MIC), Piscataway, NJ, USA, 16–23 October 2021.

22.   Hosseini, S.A.; Shiri, I.; Hajianfar, G.; Ghafarian, P.; Karam, M.B.; Ay, M.R. The impact of preprocessing on the PET-CT radiomics features in non-small cell lung cancer. *Front. Biomed. Technol.* **2021**, *8*, 261–272. [CrossRef]

23.   Ruan, Z.; Mei, N.; Lu, Y.; Xiong, J.; Li, X.; Zheng, W.; Liu, L.; Yin, B. A Comparative and Summative Study of Radiomics-based Overall Survival Prediction in Glioblastoma Patients. *J. Comput. Assist. Tomogr.* **2022**, *46*, 470–479. [CrossRef]

24.   Aftab, K.; Aamir, F.B.; Mallick, S.; Mubarak, F.; Pope, W.B.; Mikkelsen, T.; Rock, J.P.; Enam, S.A. Radiomics for precision medicine in glioblastoma. *J. Neuro-Oncol.* **2022**, *156*, 217–231. [CrossRef]

25.   Baine, M.; Burr, J.; Du, Q.; Zhang, C.; Liang, X.; Krajewski, L.; Zima, L.; Rux, G.; Zhang, C.; Zheng, D. The potential use of radiomics with pre-radiation therapy MR imaging in predicting risk of pseudoprogression in glioblastoma patients. *J. Imaging* **2021**, *7*, 17. [CrossRef] [PubMed]

26.   Lu, Y.; Patel, M.; Natarajan, K.; Ughratdar, I.; Sanghera, P.; Jena, R.; Watts, C.; Sawlani, V. Machine learning-based radiomic, clinical and semantic feature analysis for predicting overall survival and MGMT promoter methylation status in patients with glioblastoma. *Magn. Reson. Imaging* **2020**, *74*, 161–170. [CrossRef] [PubMed]

27.   Sasaki, T.; Kinoshita, M.; Fujita, K.; Fukai, J.; Hayashi, N.; Uematsu, Y.; Okita, Y.; Nonaka, M.; Moriuchi, S.; Uda, T. Radiomics and MGMT promoter methylation for prognostication of newly diagnosed glioblastoma. *Sci. Rep.* **2019**, *9*, 14435. [CrossRef] [PubMed]

28.   Hosseini, S.A.; Shiri, I.; Hajianfar, G.; Bagley, S.; Nasrallah, M.; O'Rourke, D.M.; Mohan, S.; Chawla, S. MRI based Radiomics for Distinguishing IDH-mutant from IDH wild-type Grade-4 Astrocytomas. In Proceedings of the 31st Annual Meeting of ISMRM, London, UK, 7–12 May 2022.

29. Lee, M.H.; Kim, J.; Kim, S.-T.; Shin, H.-M.; You, H.-J.; Choi, J.W.; Seol, H.J.; Nam, D.-H.; Lee, J.-I.; Kong, D.-S. Prediction of IDH1 mutation status in glioblastoma using machine learning technique based on quantitative radiomic data. *World Neurosurg.* **2019**, *125*, e688–e696. [CrossRef]

30. Hsieh, K.L.-C.; Chen, C.-Y.; Lo, C.-M. Radiomic model for predicting mutations in the isocitrate dehydrogenase gene in glioblastomas. *Oncotarget* **2017**, *8*, 45888. [CrossRef]

31. Nasrallah, M.P.; Binder, Z.A.; Oldridge, D.A.; Zhao, J.; Lieberman, D.B.; Roth, J.J.; Watt, C.D.; Sukhadia, S.; Klinman, E.; Daber, R.D. Molecular neuropathology in practice: Clinical profiling and integrative analysis of molecular alterations in glioblastoma. *Acad. Pathol.* **2019**, *6*, 2374289519848353. [CrossRef]

32. Van Griethuysen, J.J.; Fedorov, A.; Parmar, C.; Hosny, A.; Aucoin, N.; Narayan, V.; Beets-Tan, R.G.; Fillion-Robin, J.-C.; Pieper, S.; Aerts, H.J. Computational radiomics system to decode the radiographic phenotype. *Cancer Res.* **2017**, *77*, e104–e107. [CrossRef]

33. Zwanenburg, A.; Leger, S.; Vallières, M.; Löck, S. Image biomarker standardisation initiative. *arXiv* **2016**, arXiv:1612.07003.

34. Gupta, A.; Bhatt, D.; Pandey, A. Transitioning from Real to Synthetic data: Quantifying the bias in model. *arXiv* **2021**, arXiv:2105.04144.

35. Dina, A.S.; Siddique, A.; Manivannan, D. Effect of Balancing Data Using Synthetic Data on the Performance of Machine Learning Classifiers for Intrusion Detection in Computer Networks. *arXiv* **2022**, arXiv:2204.00144. [CrossRef]

36. Pereira, M.; Kshirsagar, M.; Mukherjee, S.; Dodhia, R.; Ferres, J.L. An Analysis of the Deployment of Models Trained on Private Tabular Synthetic Data: Unexpected Surprises. *arXiv* **2021**, arXiv:2106.10241.

37. Stadler, T.; Oprisanu, B.; Troncoso, C. Synthetic data–anonymisation groundhog day. In Proceedings of the 31st USENIX Security Symposium (USENIX Security 22), Boston, MA, USA, 10–12 August 2022.

38. Sepehri, S.; Tankyevych, O.; Upadhaya, T.; Visvikis, D.; Hatt, M.; Cheze Le Rest, C. Comparison and Fusion of Machine Learning Algorithms for Prospective Validation of PET/CT Radiomic Features Prognostic Value in Stage II-III Non-Small Cell Lung Cancer. *Diagnostics* **2021**, *11*, 675. [CrossRef]

39. Hannun, A.; Guo, C.; van der Maaten, L. Measuring data leakage in machine-learning models with Fisher information. In Proceedings of the Thirty-First International Joint Conference on Artificial Intelligence (IJCAI-22), Vienna, Austria, 23–29 July 2022.

40. Dietterich, T.G. Ensemble methods in machine learning. In Proceedings of the International Workshop on Multiple Classifier Systems, Cagliari, Italy, 21–23 June 2000; Springer: Berlin/Heidelberg, Germany, 2000.

41. Hartmann, C.; Hentschel, B.; Wick, W.; Capper, D.; Felsberg, J.; Simon, M.; Westphal, M.; Schackert, G.; Meyermann, R.; Pietsch, T. Patients with IDH1 wild type anaplastic astrocytomas exhibit worse prognosis than IDH1-mutated glioblastomas, and IDH1 mutation status accounts for the unfavorable prognostic effect of higher age: Implications for classification of gliomas. *Acta Neuropathol.* **2010**, *120*, 707–718. [CrossRef]

42. Bhandari, A.P.; Liong, R.; Koppen, J.; Murthy, S.; Lasocki, A. Noninvasive determination of IDH and 1p19q status of lower-grade gliomas using MRI radiomics: A systematic review. *Am. J. Neuroradiol.* **2021**, *42*, 94–101. [CrossRef]

43. Akay, A.; Rüksen, M.; Islekel, S. Magnetic resonance imaging-guided stereotactic biopsy: A review of 83 cases with outcomes. *Asian J. Neurosurg.* **2019**, *14*, 90. [CrossRef]

44. Horbinski, C. What do we know about IDH1/2 mutations so far, and how do we use it? *Acta Neuropathol.* **2013**, *125*, 621–636. [CrossRef]

45. Gutman, D.A.; Dunn, W.D.; Grossmann, P.; Cooper, L.A.; Holder, C.A.; Ligon, K.L.; Alexander, B.M.; Aerts, H.J. Somatic mutations associated with MRI-derived volumetric features in glioblastoma. *Neuroradiology* **2015**, *57*, 1227–1237. [CrossRef]

46. Askari, P.; Dimitrov, I.E.; Ganji, S.K.; Tiwari, V.; Levy, M.; Patel, T.R.; Pan, E.; Mickey, B.E.; Malloy, C.R.; Maher, E.A. Spectral fitting strategy to overcome the overlap between 2-hydroxyglutarate and lipid resonances at 2.25 ppm. *Magn. Reson. Med.* **2021**, *86*, 1818–1828. [CrossRef]

47. Choi, C.; Raisanen, J.M.; Ganji, S.K.; Zhang, S.; McNeil, S.S.; An, Z.; Madan, A.; Hatanpaa, K.J.; Vemireddy, V.; Sheppard, C.A. Prospective longitudinal analysis of 2-hydroxyglutarate magnetic resonance spectroscopy identifies broad clinical utility for the management of patients with IDH-mutant glioma. *J. Clin. Oncol.* **2016**, *34*, 4030. [CrossRef]

48. An, Z.; Ganji, S.K.; Tiwari, V.; Pinho, M.C.; Patel, T.; Barnett, S.; Pan, E.; Mickey, B.E.; Maher, E.A.; Choi, C. Detection of 2-hydroxyglutarate in brain tumors by triple-refocusing MR spectroscopy at 3T in vivo. *Magn. Reson. Med.* **2017**, *78*, 40–48. [CrossRef]

49. Sonoda, Y.; Shibahara, I.; Kawaguchi, T.; Saito, R.; Kanamori, M.; Watanabe, M.; Suzuki, H.; Kumabe, T.; Tominaga, T. Association between molecular alterations and tumor location and MRI characteristics in anaplastic gliomas. *Brain Tumor Pathol.* **2015**, *32*, 99–104. [CrossRef] [PubMed]

50. Baldock, A.L.; Yagle, K.; Born, D.E.; Ahn, S.; Trister, A.D.; Neal, M.; Johnston, S.K.; Bridge, C.A.; Basanta, D.; Scott, J. Invasion and proliferation kinetics in enhancing gliomas predict IDH1 mutation status. *Neuro Oncol.* **2014**, *16*, 779–786. [CrossRef] [PubMed]

51. Qi, S.; Yu, L.; Li, H.; Ou, Y.; Qiu, X.; Ding, Y.; Han, H.; Zhang, X. Isocitrate dehydrogenase mutation is associated with tumor location and magnetic resonance imaging characteristics in astrocytic neoplasms. *Oncol. Lett.* **2014**, *7*, 1895–1902. [CrossRef] [PubMed]

52. Aerts, H.J.; Velazquez, E.R.; Leijenaar, R.T.; Parmar, C.; Grossmann, P.; Carvalho, S.; Bussink, J.; Monshouwer, R.; Haibe-Kains, B.; Rietveld, D. Decoding tumour phenotype by noninvasive imaging using a quantitative radiomics approach. *Nat. Commun.* **2014**, *5*, 4006. [CrossRef] [PubMed]

3. Hu, S.; Luo, M.; Li, Y. Machine Learning for the Prediction of Lymph Nodes Micrometastasis in Patients with Non-Small Cell Lung Cancer: A Comparative Analysis of Two Practical Prediction Models for Gross Target Volume Delineation. *Cancer Manag. Res.* **2021**, *13*, 4811. [CrossRef]
4. Zhu, L.; Chen, Y.; Ghamisi, P.; Benediktsson, J.A. Generative adversarial networks for hyperspectral image classification. *IEEE Trans. Geosci. Remote Sens.* **2018**, *56*, 5046–5063. [CrossRef]
5. Creswell, A.; White, T.; Dumoulin, V.; Arulkumaran, K.; Sengupta, B.; Bharath, A.A. Generative adversarial networks: An overview. *IEEE Signal Process. Mag.* **2018**, *35*, 53–65. [CrossRef]
6. Zhang, B.; Chang, K.; Ramkissoon, S.; Tanguturi, S.; Bi, W.L.; Reardon, D.A.; Ligon, K.L.; Alexander, B.M.; Wen, P.Y.; Huang, R.Y. Multimodal MRI features predict isocitrate dehydrogenase genotype in high-grade gliomas. *Neuro Oncol.* **2017**, *19*, 109–117. [CrossRef]
7. Lee, S.; Choi, S.H.; Ryoo, I.; Yoon, T.J.; Kim, T.M.; Lee, S.-H.; Park, C.-K.; Kim, J.-H.; Sohn, C.-H.; Park, S.-H. Evaluation of the microenvironmental heterogeneity in high-grade gliomas with IDH1/2 gene mutation using histogram analysis of diffusion-weighted imaging and dynamic-susceptibility contrast perfusion imaging. *J. Neuro-Oncol.* **2015**, *121*, 141–150. [CrossRef]
8. Metellus, P.; Coulibaly, B.; Colin, C.; de Paula, A.M.; Vasiljevic, A.; Taieb, D.; Barlier, A.; Boisselier, B.; Mokhtari, K.; Wang, X.W. Absence of IDH mutation identifies a novel radiologic and molecular subtype of WHO grade II gliomas with dismal prognosis. *Acta Neuropathol.* **2010**, *120*, 719–729. [CrossRef]
9. Xing, Z.; Yang, X.; She, D.; Lin, Y.; Zhang, Y.; Cao, D. Noninvasive assessment of IDH mutational status in World Health Organization grade II and III astrocytomas using DWI and DSC-PWI combined with conventional MR imaging. *Am. J. Neuroradiol.* **2017**, *38*, 1138–1144. [CrossRef] [PubMed]

Article

FabNet: A Features Agglomeration-Based Convolutional Neural Network for Multiscale Breast Cancer Histopathology Images Classification

Muhammad Sadiq Amin and Hyunsik Ahn *

Department of Robot System Engineering, Tongmyong University, Busan 48520, Republic of Korea
* Correspondence: hsahn@tu.ac.kr

Simple Summary: Histology sample images are usually diagnosed definitively based on the radiologist's extensive knowledge, yet, owing to the highly gritty visual appearance of such images, specialists sometimes differ on their evaluations. Automating the image diagnostic process and decreasing the analysis time may be achieved via the use of advanced deep learning algorithms. Diagnostic objectivity may be improved with the use of more effective and accurate automated technologies by lessening the differences between the humans. In this research, we propose a CNN model architecture for cancer image classification by accumulating layers closer together to further merge the semantic and spatial features. Regarding precision, our suggested cutting-edge model improves upon the current state-of-the-art approaches.

Abstract: The definitive diagnosis of histology specimen images is largely based on the radiologist's comprehensive experience; however, due to the fine to the coarse visual appearance of such images, experts often disagree with their assessments. Sophisticated deep learning approaches can help to automate the diagnosis process of the images and reduce the analysis duration. More efficient and accurate automated systems can also increase the diagnostic impartiality by reducing the difference between the operators. We propose a FabNet model that can learn the fine-to-coarse structural and textural features of multi-scale histopathological images by using accretive network architecture that agglomerate hierarchical feature maps to acquire significant classification accuracy. We expand on a contemporary design by incorporating deep and close integration to finely combine features across layers. Our deep layer accretive model structure combines the feature hierarchy in an iterative and hierarchically manner that infers higher accuracy and fewer parameters. The FabNet can identify malignant tumors from images and patches from histopathology images. We assessed the efficiency of our suggested model standard cancer datasets, which included breast cancer as well as colon cancer histopathology images. Our proposed avant garde model significantly outperform existing state-of-the-art models in respect of the accuracy, F1 score, precision, and sensitivity, with fewer parameters.

Keywords: artificial intelligence; deep learning; pattern recognition; computer-assisted diagnosis; convolutional neural networks; breast cancer; colon cancer; histopathological images

Citation: Amin, M.S.; Ahn, H. FabNet: A Features Agglomeration-Based Convolutional Neural Network for Multiscale Breast Cancer Histopathology Images Classification. *Cancers* **2023**, *15*, 1013. https://doi.org/10.3390/cancers15041013

Academic Editors: Hamid Khayyam, Ali Madani, Rahele Kafieh and Ali Hekmatnia

Received: 12 January 2023
Revised: 31 January 2023
Accepted: 31 January 2023
Published: 5 February 2023

1. Introduction

Breast cancer is the most prevalent types of cancer in women, affecting 2.1 million women annually, and it is responsible for the bulk of cancer-related deaths globally [1]. It has been estimated that the prevalence rates of breast cancer range from 19.3 per 100,000 African women to 89.7 per 100,000 European women [2]. Breast cancer is a fatal condition that can occur in nearly any bodily region or tissue when irregular cells abnormally spread, infiltrate, or move into adjacent tissues. The number of reported cases has increased in recent years, and it is projected to reach 27 million by 2030 [3–7]. Considering the high cancer mortality rate, colonoscopy and computer tomography are recommended for regular tests [8]. A biopsy examination is used to diagnose abnormalities in the breast and colon if suspicious cells are found. Hematoxylin and eosin (H&E) are often used to stain the isolated sample. When Hematoxylin interacts with Deoxyribonucleic Acid (DNA), it dyes the nuclei purple or blue, while Eosin stains other structures pink when it reacts with proteins [9].

The diagnosis of all cancer types, including breast and colon cancers, is based on histopathological images, which are considered to be essential. Histopathological examination, contrastingly, is a long-winded clinical practice, with the key impediment to successful image processing being a difference in the visibility in th H&E-colored regions. Various considerations, such as laboratory technique anomalies, discrepancies in sample positioning, operator-related heterogeneity, device diversity, and the usage of different fluorophores for staining, may all influence the diagnosis [10]. For even seasoned oncologists, recognizing and evaluating these discrepancies during a diagnosis could be challenging. As a result, there is a significant necessity for intelligent automated diagnostic systems to provide oncologists with reliable evaluations and improve the diagnostic performance.

Deep-learning-based approaches are currently the course of the research, and they have a profound impact on clinical trials and even the evolution and progress of targeted treatment methods. With the advancement in digital imaging technology, the automated diagnosis and detection of cancer types in whole slides images have received a great deal of interest. Several methods for analyzing histological images have been adopted, ranging from conventional to machine-learning-based ones [11]. Deep learning (DL) approaches have increasingly outperformed traditional machine learning (ML) algorithms in terms of end-to-end processing automation [12,13]. Deep learning-based techniques such as convolutional neural networks (CNN) have been successfully used in medical imaging to detect diabetic retinopathy [14], diagnose bone osteoarthritis [15], and for other purposes [16]. CNN-based histological image analysis methods have previously been shown to be effective for breast cancer diagnosis [17] and micro-level pathological image analysis [18,19].

The advent of the use convolutional neural networks as the basis of several visual tasks for different applications has made architecture searching a key driver in sustaining advancement with the right task extensions and data [20–22]. Because of the growing size and sophistication of networks, more effort is being put into developing the architecture motifs of nodes and nodes connectivity strategies that can be integrated systematically. This has resulted in wider and deeper networks; however, there is a need for more closely linked networks. To overcome these obstacles, various blocks or units have been integrated to match and change the network sizes, such as bottlenecks for reducing the dimensions [23,24] or residual, concatenated connections for features propagation [25,26].

In this paper, we suggest a CNN model design by accumulating layers that are even more close together to further fuse the semantic and spatial details for cancer image classification. Our accretive architecture incorporates more depth and sharing by expanding the existing approaches' "shallow" skip connections [27] and focuses on merging the features from all of the layers and channels. Our contributions to this research are as follows:

1. We proposed a FabNet model that can learn the fine-to-coarse structural and textural features of multi-scale histopathological images by accretive network architecture that agglomerate hierarchical feature maps to acquire significant classification accuracy.
2. To preserve and integrate the features, our model links convolutional blocks in a closely coupled tree-based architecture. This method employs every layer of the network from the shallowest to the deepest layers to learn about the rich patterns that occupy a large portion of the feature pile.
3. We assessed the FabNet model using two publicly available standard datasets that are related to breast cancer and colorectal cancer and noticed that it outperforms the current state-of-the-art models in terms of accuracy, F1 score, sensitivity, and precision

when we evaluated our model at different magnification scales of both binary and multi classification.

The rest of this article is structured as follows: Section 2 addresses the related work. Section 3 defines the design of the proposed FabNet model. We define the experimental setup, datasets, training, and implementation descriptions, and provide a detailed analysis of the performance in Section 4. The discussion, conclusions, and possible future research directions are all contained in Section 5.

2. Related Works

There has been extensive work that has been conducted in the literature to establish strategies for classifying and recognizing breast and colon cancers from histopathology images. The majority of the current approaches utilize computer-aided diagnosis (CAD) techniques to identify breast-cancer-related tumors that include benign and malignant ones. Before the deep learning breakthrough, the data were examined using conventional machine learning techniques based on supervised learning methods [28] to obtain the data features.

2.1. Conventional Learning Methods

The bulk of the research in this area has concentrated on a small data sample taken mostly from proprietary datasets. In 2013, several algorithms were used to classify the nuclei from a dataset containing five hundred images from fifty patients, including Gaussian mixture models and fuzzy C-means clustering techniques. This study reported 96% accuracy for two category classifications [29], suggesting that such machine learning based approaches allowed adequately comprehensive and precise research and were considered to be useful for supporting breast cancer diagnostics. Spanhol et al. [30] published yet another study in which they achieved 85.1 % accuracy on a breast cancer dataset. They applied support vector machines for a patient-level analysis. Using a database of ninety-two specimens, George et al. [31] proposed a breast cancer classification method by applying neural nets with a support vector machine, which achieved 94 percent accuracy. Zhang et al. [32] suggested a cascading approach with a refusal alternative. This procedure was evaluated on a dataset with 361 specimens [33]. This study [34] suggested the application of different classifiers such as support vector machines and the k-nearest neighbor for breast cancer histology image classification. They achieved 87 % accuracy by utilizing assembling voting using the mentioned techniques. In this study [35], adaptive sparse support vector machine-based techniques were applied on a dataset at a 40× magnification level. They reported 94.97% accuracy. There have been a couple of other studies on histopathological representations for carcinoma classification; these studies specifically explain the dichotomies and shortcomings of various publicly accessible benchmark data [36,37].

2.2. Deep Learning Approaches

Deep learning has ushered in a new era in the domain of general object classification and detection. The classification of cancer histopathological images (i.e., breast and colon) has been a significant field of study due to advances in medical computer vision and deep learning. Because of the elevated histopathological image resolutions, the conventional machine learning algorithms and deep neural network models used to explicitly view the WSI have resulted in very complex network designs that are a challenge to training [38]. The number of samples used in the classification cancer histopathology images is limited and the image size is large, making the training of CNN models challenging. Furthermore, image compression of the entire oncology image array to the CNN's input size would result in a loss of the richness of the detailed feature data. As a result, some researchers suggested the classification of images based on patches to alleviate the challenge. In this study [39], the author used a technique to achieve the arbitrary extraction of patches based on a window slithering approach to extract image patches from the BreakHis dataset. AlexNet [40] was trained on the extracted patches, and then, integrated the outcomes to classify into relevant categories. Another study by Arajo et al. [3] suggested a convolutional neural network for automatic feature extraction from a dataset that contained 512×512 size patches. The images were grouped into four classes during training, which were used for multi-classification, as well as two classes, which were used for binary classification.

Because of the image patches extraction process, CNN became capable of training whole slide images with reasonable details. This study [41] suggested a convolutional neural network with a two-level model for high-resolution WSIs classification. The first model is based on a minimal anomaly model that can distinguish between patterns automatically during training on image patches, and a second model that classifies the results by an SVM classifier. In another study, Alom et al. [2] suggested the merging of three models to classify breast cancer histology images. A CNN-based methodology achieved 77.8 percent accuracy for multi-classification, while it found an 83.3 percent accuracy for binary classification on the breast histology 2015 dataset [3].

Han et al. [42] recently suggested a class structure-based deep convolutional neural network that achieved 93.2 percent accuracy on the BreakHis dataset. Table 1 elaborates on the details of recent advancements in the cancer research domain.

Table 1. A review of supervised learning models. The staining abbreviations stand for H&E (hematoxylin and eosin); PHH3 (Phosphohistone-H3).

Reference	Local/Global	Cancer Type	Staining	Method	Dataset
Ceresin et al. (2013) [43]	Local-level	Breast	Hematoxylin and eosin	CNN	ICPR2012 (50 images)
Wang et al. (2014) [44]	Local-level	Breast	Hematoxylin and eosin	Rippled integration of CNN	ICPR2012 (50 images)
Raza et al. (2016) [45]	Local-level	Colorectal	Hematoxylin and eosin	Cell detection Spatially constrained CNN + handcrafted features	Private CRC dataset (15 images)
Tellez et al. (2019) [46]	Local-level	Breast	Hematoxylin and eosin; PHH3	CNN	TNBC (36 images); TUPAC (814 images)
Ehteshami et al. (2017) [47]	Global-level	Breast	Hematoxylin and eosin	Stacked CNN incorporating contextual information	Private set (221 images)
Ehteshami et al. (2018) [48]	Global-level	Breast	Hematoxylin and eosin	Integration of DHACNN & LSTM	BreakHis (7909 images)

Even though the preceding studies demonstrate that patch-based image classification approaches are commonly used in different breast cancer histopathology datasets, histopathology images contain a large number of fine details that need to be extracted with utmost accuracy and precision. We present FabNet, a CNN model that ensembles every fine-to-coarse detail for more accurate learning. This method employs every layer of network from the shallowest to the deepest layers to learn about the rich patterns that occupy a large portion of the feature pile.

3. FabNet: Features Agglomeration Approach

We define agglomeration as the combination or merging of network layers in a closely coupled manner. In the proposed model FabNet, as shown in Figure 1, we are particularly focused on the productive accumulation of depth, dimensions, and resolutions. We define an agglomeration sequence as deep if it is holistic, discrete, and the initial agglomerated layer moves features through several agglomerations. Since our network has multiple layers and connections, we designed modular architecture that tends to reduce the complexity by grouping and replication. The proposed network layers are subdivided into blocks, for example, B1, which are further subdivided into stages based on the feature resolution. This design is focused on agglomerating the blocks to preserve and combine the feature channels. In Figure 2, a conv block (i.e., B1) is shown, which comprises two convolutional layers with 5×5 and 3×3 filter window sizes. Both of the convolutional layer activation maps are concatenated, and then transferred to another convolutional layer with 1×1 filter size window to reduce the optimal channels. Agglomeration starts on the smallest, shallowest scale and gradually merges on the deeper, wider scales in a repetitive manner. In this manner, the shallow features are redefined as they progress over to deeper blocks of layers.

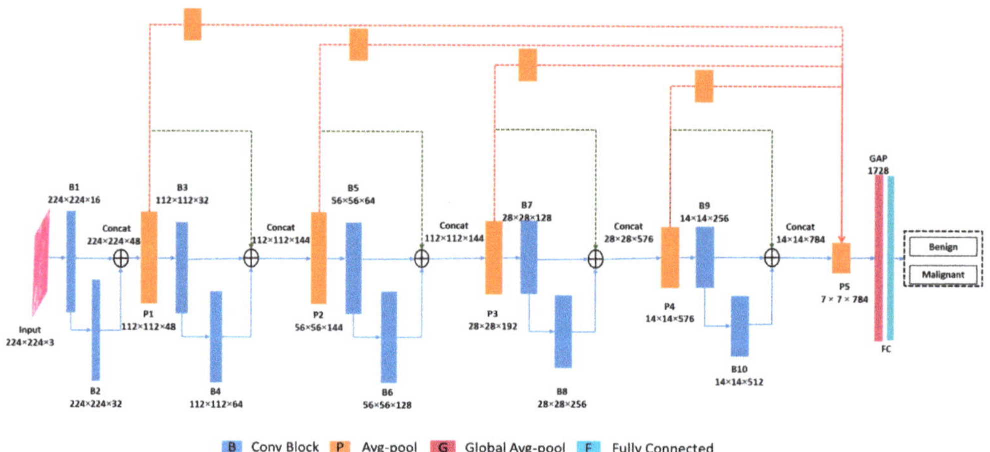

Figure 1. FabNet model: a detailed architectural overview.

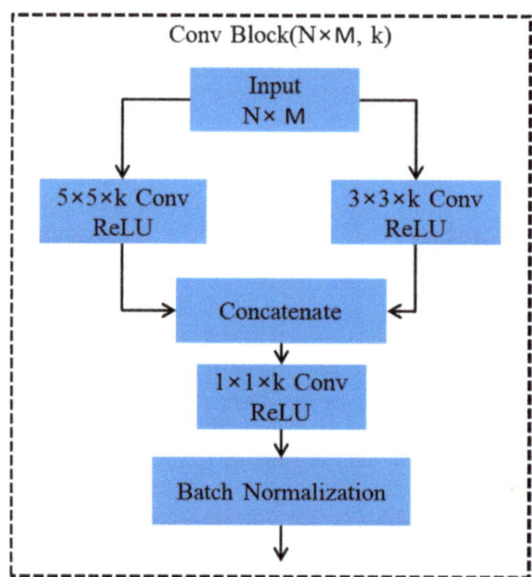

Figure 2. Internal architecture of conv blocks (i.e., B1). The input passes through two 5×5 and 3×3 convolutional layers; the output is concatenated in the proceeding step.

For a sequence of blocks $\{B1, B2, B3 \ldots .. Bn\}$, we formulated the function $\Re$ for such a repetition below.

$$\Re(B1, B2, B3 \ldots .. Bn) = \Re(\Sigma B1, B2, B3 \ldots .. Bn) \tag{1}$$

In Equation (1), n is the number of blocks. To increase the depth of the network and the performance, we merge or fuse blocks in a tree-like closely coupled structure. We pass an agglomerated node's feature map back to the baseline as the input feature map to the next sub-module, instead of forwarding intermediate agglomerations further up the tree. This spreads the agglomerations of all of the previous modules, rather than the preceding module only, to help the best preservation of features. We combine the parent and left child nodes of the same depth in the performance.

Our model consists of conv blocks, which are the basic building block of each node. The input of a conv block in the case of B1 accepts an input of $224 \times 224 \times 3$. This input is passed to two different convolutional layers simultaneously for convolutional operations to be performed. Both of the convolutional layers apply 16 kernels with filter window sizes of 3×3 and 5×5 each with nonlinearity (ReLU), which aims to alleviate the issue of vanishing gradients, as well as improve the network's training speed. To generate an optimal feature map, the feature maps of these two convolutional layers are combined, and thereafter, transferred to a 1×1 convolutional layer. In each convolutional layer that is discussed above, we use zero padding, which preserves the original image size, while also providing valuable knowledge about feature learning, which aids in the extraction of low-level features for the subsequent layers. Following that, we apply batch normalization, which balances the inferences of the preceding activation layer by subtracting the batch mean and dividing the batch division, thereby increasing the network stability.

The output of conv block B1 is fed into B2, which has a similar internal architecture to that of B1, as depicted in Figure 2, except for the number of kernels. Conv B2 contains 32 convolution filters. The feature maps of both of the conv blocks are then concatenated, which results in an enhanced collective feature map. We apply an average pooling operation with an average pooling layer with 2×2 patches of the feature map with a stride of two. This layer down-samples the estimation complexities and parameters from the evaluated image by dividing it into rectangular pooling window areas, which is proceeded by a mean value estimation for every region. The inference of the average pooled image propagates to the next block as an input to conv block B3, which is fed into the final stage C5. As it was mentioned earlier, B3 contains the same internal architecture as those of conv blocks B1 and B2, but the number of convolution filters is 32. The output feature map of B3 is fed into conv block B4 as an input. The internal convolutional layers of conv block B4 apply 64 convolution filters to learn the features. The feature maps of B3 and B4 are fused to generate an extended feature map, which is proceeding by average pooling for down-sampling. The average pooled value feeds into the next conv block B5. Conv block B6 is fed to B5 as an input. B6 utilizes 128 convolution filters.

The feature maps of conv block B5 are conv block B6 which is concatenated to fuse the feature, which results in an enhanced feature map with detailed data information. This step is preceded by an average pooling operation to obtain half of the image size. The result of the pooled value is fed into conv block B5. The network repeats the same operation until it reaches conv block B10. The only difference is between the blocks is the number of convolution filters, which is 256 for B8 and 512 for B10. Until it reaches B10, the feature maps of the entire network resulted in optimized propagation from the shallower to the deeper layers and blocks, which makes the proposed network compact and closely bind the entire network. The best features of every block and stage are collected and fused at stage C5 by the extensions from C1 to C5 and by bridging the adjacent blocks The C5 is subjected to the global average pooling function, which significantly reduces the number of data, and thus, the classification layers by measuring the average results of every feature map in the preceding layer. The output layer, which is the last dense layer, includes neurons for each class that have been normalized with the Softmax function; the amount of them varies based on the classification category. We used binary and multi-class classifications in this study.

4. Methodology

As seen in Figure 3, the proposed method consists of three main steps. Firstly, we obtain training samples by applying the extraction of patches technique to the dataset. Secondly, stain normalization preprocessing of the dataset is performed to resolve the stain variation in the images. For stain normalization, several methods have been suggested in these studies [49–51]. DL-based approaches for classifying cancer histopathology images employs a training set to detect a wide range of enhancements to distinguish variations within, as well as across, the categories. A wide range of color inconsistencies in the histopathological images may occur due to the color response of the automated scanners, stain supplier materials and processing units or due to various staining procedures in different laboratories. Therefore, stain normalization is a basic step during histopathological image preprocessing. The key benefit of using image patches for each type of training is that it preserves the local characteristic information from the histopathology images, helping the model to learn the local characteristics features. Thirdly, we train our proposed model with these extracted images to classify and differentiate between the benign and malignant tumors. Furthermore, we outline the datasets, image preprocessing, model training, and implementation details below.

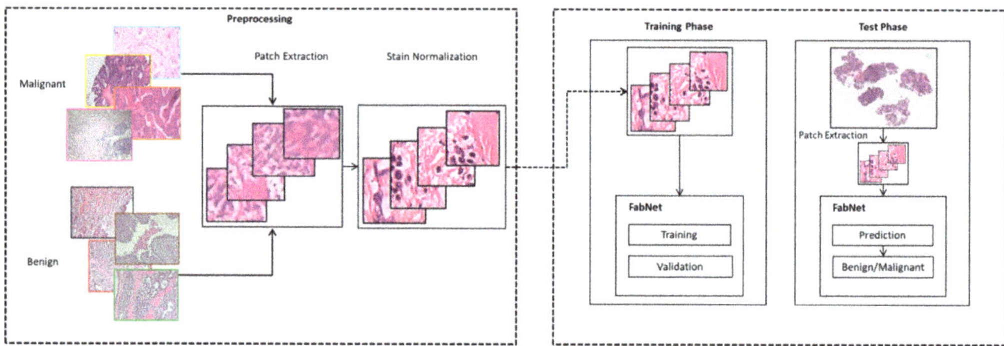

Figure 3. An overview of the proposed methodology to classify the histopathological image.

4.1. Dataset

To evaluate our proposed model, we used the two main, public cancer histology image datasets. Such datasets were considered with three motives: firstly, the diversity of cancer types represented in the histology slides, such as breast cancer and colorectal cancer; secondly, their amount; thirdly, the existence of multiple magnification factors that helped us to carry different tests with the restricted equipment, while modifying different parameters.

4.1.1. BreaHis

In this study, we assessed our model with BreakHis, a publicly available breast-cancer-related histologic dataset [30]. Samples were created using breast tissue biopsy slides that were colored with H&E staining. There are reportedly 7909 histopathological biopsy images of 700×460 pixels in the BreakHis dataset from eighty-two individuals. The dataset consists of two main categories: one of them is benign, and the other one is malignant which are further subdivided into 4 subclasses as per each category. Table 2 shows the statistical specifics of this dataset, and Figure 4, shows a few illustrations of the histological images. For our tests, we randomly divided the entire dataset in into training/testing subgroups at a 70:30 ratio. To assess our model's efficiency in clinical settings, we kept a patient-based distinction between the training and test data. For stain normalization, we adopted the technique suggested in [50]: an innovative composition-preserving color normalization (SPCN) scheme is used in this process.

Table 2. BreakHis dataset categorization at patient level at four magnifications ($40\times$, $100\times$, $200\times$, and $400\times$).

Category	Subtypes	Magnification				Sum	Individuals
		$40\times$	$100\times$	$200\times$	$400\times$		
Benign	Phyllodes Tumor (PHT)	149	150	140	130	569	7
	Fibroadenoma (FID)	253	260	264	237	1014	10
	Adenosis (ADE)	114	113	111	106	444	4
	Tubular Adenona (TUA)	109	121	108	115	453	3
Malignant	Papillary Carcinoma (PAC)	145	142	135	138	560	6
	Ductal Carcinoma (DUC)	864	903	896	788	3451	38
	Lobular Carcinoma (LOC)	156	170	163	137	626	5
	Mucinous Carcinoma (MUC)	205	222	196	169	792	9

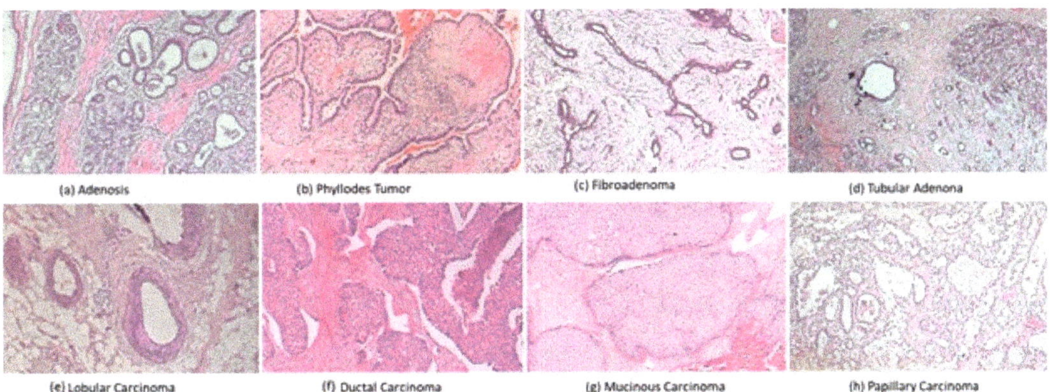

Figure 4. From the BreakHis dataset, the first row depicts benign 4 subclasses, while the second row shows malignant 4 subclasses. These images have a magnification factor of $200\times$.

The illustration of stain normalized images is shown in Figure 5.

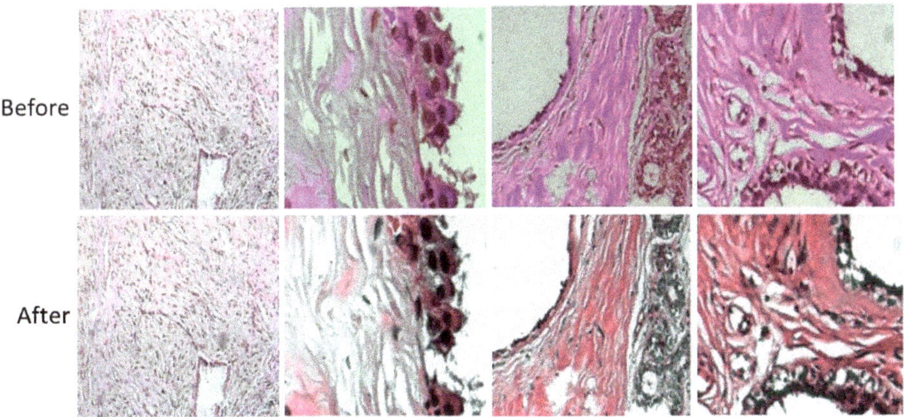

Figure 5. Stain normalized images of 4 different subcategories at a magnification factor of $400\times$.

4.1.2. NCT-CRC-HE-100K

This dataset includes publicly available 100 K images of human colorectal cancer (CRC), as well as normal tissues [52]. To stain normalize this dataset, in which the image size was 224 × 224 pixels, the Macenko approach [53] was used. We used this color normalization technique because the initial images had subtle variations between red and blue tones, resulting in a misleading classification. Figure 6 shows descriptive representations of the sample images. This dataset is divided into nine subclasses, which are adipose tissue (ADI), lymphocytes (LYM), background (BACK), mucus (MUC), smooth muscle (MUS), normal (NORM), debris (DEB), cancer-associated stroma (STR), and tumor (TUM) ones. To improve the variance in this training set, normal tissue samples were obtained primarily from clinical specimens, as well as from gastrectomy samples (such as upper gastrointestinal smooth muscle). The number of distributed training set images in each group was nearly equal, while the test samples contained 7180 images.

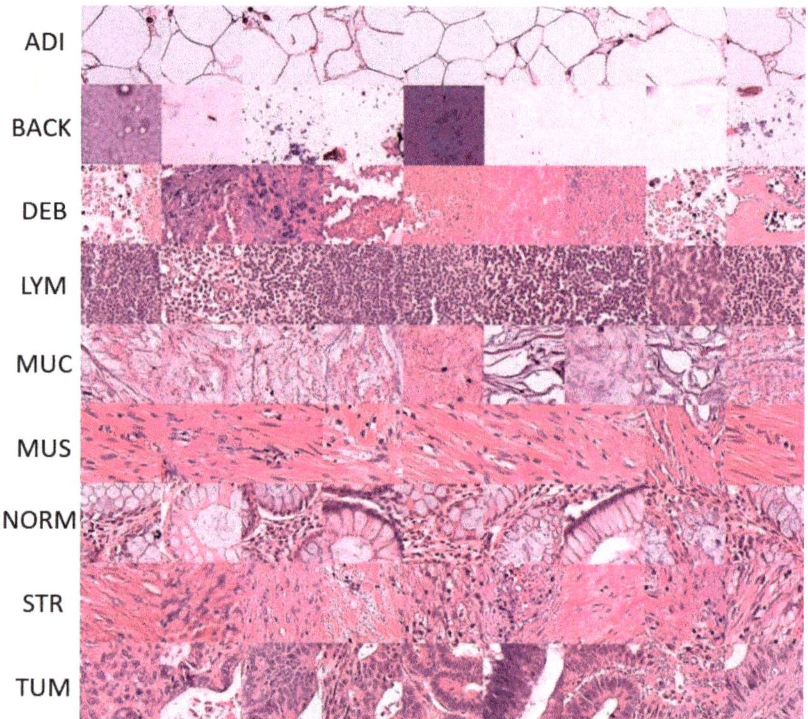

Figure 6. An illustrative image from nine classes of human colorectal cancer datasets.

4.2. Image Representation and Patch Extraction

Table 2 shows that the BreakHis dataset has a data imbalance problem, which was calculated as 0.42 at the case image scale and 0.44 at the patient scale. The data disparity problem can cause a discriminating performance of computer-aided diagnosis (CAD) models against the majority class in classification problems. Equation (2) determines the patch amount obtained from the dataset image of the ith class.

$$N_i = \left\lceil \left(\sum_{i=1}^{n} x_i \middle/ n \middle| x_i \right) \times \beta \right\rceil \tag{2}$$

Equation (2) depicts a mathematical representation of N_i patches derived from the i(th) category, x_i is the i(th) category's number, x_{th} is the i(th) category's number, β is a constant value, and n represents the classes. The fixed parameter (β) was set to 32. After that, each class has nearly the same number of patches. The primary benefit of utilizing patches during training for every individual class is that it preserves the regional distinctive details in the histological image, which enables the model to learn the spatial information [54].

To obtain an image classification, first, we use a patch classifier to compare several distinct magnifications of patches, and afterward, we average the effects for the complete image patches. The extraction and learning of similar features, for instance, the entire tissue composition, nucleus state, and texture features are used to classify the images to the desired categories. We inferred that 224×224, as well as 700×460-pixel patches, would be sufficient to justify the proper cell formation of various tissues. We deduced that 700×460, as well as 224×224 px size for images, would be ample to explain the relevant composition of different tissues.

5. Experimental Results

5.1. Model Training

We assessed the proposed model's efficiency in two areas: (1) sample classification based on binary and multi-class classification, and (2) sample classification based on patient- and image-level classification. We used the datasets discussed in the study. These datasets were subdivided into training validation sets. To find the optimal parameters for our model, we use a five-fold cross-validation scheme. We assess our model with assessment metrics such as accuracy, sensitivity, and precision, and F1 score in the performance assessment. On an NVIDIA GTX 1080Ti, we used the Keras framework to implement the method. The metrics of five successful completed trial experiments are reported. We compared our model's efficiency to that of cutting-edge models such as DenseNet 121 [55], VGG16 [56], and ResNet 50 [57].

5.2. Implementation Details

FabNet model assimilates the fine-to-coarse structural and textural features of multi-scale histopathological images by accretive network architecture that agglomerate hierarchical feature maps to perform significant learning. Our model propagates the features from block to block, and overall, from stage to stage to ensemble the best feature map for learning. We tuned the following hyperparameters in our model, which are a number of convolutional blocks (the internal architecture is defined in Figure 2), epochs, learning rate, optimizer, size of batch, and batch normalization. The epochs were set to 20, 50, 70, and 100, respectively, while 0.01, 0.001, 0.0001, and 10^{-4} learning rates were evaluated. We used a batch size of 16, 32, and 64 due to hardware limitations. We tested the model with different optimizers such as Adadelta, Adamax, SGD, RMSprop, and Nadam, but Adam provided the optimal accuracy. The detailed optimized hypermeters are shown in Table 3.

Table 3. Optimized hyper-parameters for FabNet, Densenet121, DNet, VGG16, and ResNet50.

Dataset	Parameters	FabNet	DenseNet121	VGG16	ResNet50
	Epochs	100	100	100	100
	Learning Rate	10^{-3}	10^{-3}	10^{-3}	10^{-3}
BreakHis	Batch Size	16	16	16	16
	Number of layers	30	121	16	50
	Optimizer	Adam	Adam	Adam	Adam
	Number of parameters	3239 K	7138 K	14,765 K	23,788 K
	Epochs	100	100	100	100
	Learning Rate	10^{-3}	10^{-3}	10^{-3}	10^{-3}
NCT-CRC-HE-100K	Batch Size	64	64	64	64
	Number of layers	30	121	16	50
	Optimizer	Adam	Adam	Adam	Adam
	Number of parameters	3239 K	7138 K	14,765 K	23,788 K

The proposed BreakHis and NCT-CRC-HE-100K datasets intended to serve as a standard for breast and colon cancer CAD systems. Before discussing the results, we define the evaluation matrices, which were used to assess the proposed model. The experimental procedure for evaluating the proposed approach for the BreakHis dataset is similar to that which was used in the previous study [39]. The authors defined two types of accuracies, in which the first one reflects the performance accuracy achieved on the patient scale.

If we suppose N_p represents the images of the patient, while N_c is the patient images that are accurately categorized and N_t are the total patients, the score for an individual patient can be calculated as

$$Patient\ Score = \frac{N_p}{N_c} \tag{3}$$

While the global patient accuracy can be calculated as,

$$Patient\ Level\ Accuracy = \frac{\sum Patient\ Score}{N_t} \tag{4}$$

The second case for the evaluation of classification accuracy is image-level accuracy. If we let N_{tb} be the test image samples for breast cancer and N_{cb} be the images that are classified by CAD system accurately, according to labeled classes, the image level accuracy can be defined as follows,

$$Image\ Level\ accurcy = \frac{N_{tb}}{N_{cb}} \tag{5}$$

The obtained accuracy at the image and patient levels for different magnification levels is shown in Table 4. Largely, a malignant case is considered to be positive during cancer diagnosis, whereas a benign case is considered to be negative. In clinical diagnosis, sensitivity (also known as recall) is more significant for medical professionals. Therefore, in this study, the proposed model is evaluated based on metrics defined below,

$$Precision = \frac{True\ Possitive}{True\ Possitive + False\ possitive} \tag{6}$$

$$Recall = \frac{True\ Possitive}{True\ Possitive + False\ Negative} \tag{7}$$

$$F1\ score = 2 \times \frac{Precision \times Recall}{Precision + Recall} \tag{8}$$

Table 4. Performance comparisons in terms of accuracy for BreakHis dataset.

Accuracy (%)	Method	Magnification Level			
		40×	100×	200×	400×
Patient Level	DenseNet 121 [55]	92.02	90.21	81.94	80.09
	MSI-MFNet [58]	93.04	88.34	92.12	89.19
	Proposed FabNet	99.01	89.26	98.38	96.96
Image Level	DenseNet 121 [55]	94.26	92.71	83.90	82.75
	MSI-MFNet [58]	94.12	89.25	92.45	90.27
	Proposed FabNet	99.03	89.68	98.51	97.10

Table 4 depicts the performance of the proposed model, which outperformed DenseNet 121 and MSI-MFNET in terms of test accuracy at each magnification level using the BreakHis dataset. The model showed superior test accuracy at 40×, 200×, and 400× magnifications. At the 100× magnification level, the model slightly lags behind Dense121, which achieves 90.21% accuracy at the patient level, while it achieves 92.71 for the image-level classifications.

The experiments are performed largely focused on binary and multiclass classification. The patch-wise binary and multi-classification outcomes are shown in Table 5. The results are shown using important metrics such as test accuracy and sensitivity (recall) using the 200× magnified image patches. The results are compared with those of two benchmark models, which are DenseNet121 and MSI-MFNet. The experimental results that are obtained by the proposed FabNet were better than the mentioned models were, with a larger margin in terms of test accuracy for binary classification as well as multi-class classification.

Table 5. Patch wise classification results of FabNet for BreakHis dataset on magnification level $200\times$ in terms of accuracy and sensitivity metrics.

Class	Model	Accuracy	Sensitivity							
			Benign				Malignant			
Binary	DenseNet [55]	0.92	0.75				0.97			
	MSIMFNet [58]	0.92	0.76				0.98			
	FabNet	0.99	0.989				0.990			
			ADE	FIB	PHT	TAD	DUC	LOC	MUC	PAC
Multi	DenseNet121 [55]	0.84	0.60	0.84	0.72	0.84	0.86	0.85	0.97	0.91
	MSIMFNet [58]	0.88	0.60	0.87	0.79	0.89	0.96	0.75	0.98	0.92
	FabNet	0.97	1.00	0.88	1.00	1.00	0.804	0.89	0.784	0.865

In Table 6, the detailed results that are obtained from the proposed model are presented. It is evident that the model exhibited better accuracy for binary classification, as well as multi-classification at contrasting magnifications, for instance, $40\times$, $100\times$, $200\times$, and $400\times$. The model showed better performance for binary classification, for instance, the accuracy at the $40\times$ magnification scale the model achieved 99 percent accuracy. The model showed better performance for many classed as well.

Table 6. Detailed classification results of FabNet on BreakHis dataset based at different magnification levels.

Class	Magnification Level	Accuracy (%)	Precision (%)	Recall (%)	F1 Score (%)
Binary	$40\times$	99.00	98.991	98.986	98.989
	$100\times$	89.26	89.128	89.262	89.195
	$200\times$	99.00	98.352	98.355	98.354
	$400\times$	97.96	97.541	97.521	97.551
Multi	$40\times$	91.26	90.635	89.126	88.289
	$100\times$	97.00	96.531	96.427	95.912
	$200\times$	97.05	85.972	85.526	85.748
	$400\times$	97.20	89.947	89.851	88.899

Table 7 depicts the classification results of the proposed FabNet for the NCT-CRC-HE-100 K dataset. It is evident that the model exhibited an outstanding performance in terms of test accuracy and sensitivity compared to those of the benchmark models such as VGG16, DenseNet 121, and ResNet50.

In Table 8, detailed class-wise scores for important matrices such as precision, sensitivity, and recall are given to elaborate the efficiency using the NCT-CRC-HE-100K dataset.

The ROC curve is a graphical determination of the classification model's results. It is determined by plotting the true positive rate (TPR) against the false positive rate (FPR) at various discriminatory thresholds, where TPR stands for sensitivity or recall, and FPR stands for false positive rate (1-specificity). The ROC curve for a classification algorithm would be a diagonal line from (0,0) to (1,1). Any curve above the diagonal line indicates a decent classification model that randomly outperforms, and any curve below the diagonal line indicates a model that randomly underperforms. The region under the ROC curve, which is often between 0 and 1, is referred to as the AUC. A high AUC means that the classification model is accurate according to the ROC curve concept. The ROC curve graph can be seen for the binary classification of the BreakHis dataset in Figure 6, where class 0 indicates a benign tumor, and class 1 represents a malignant tumor. Figures 7 and 8 depict the ROC curve graph for the multi-classification performance using the BreakHis and NCT-CRC-HE-100K datasets. The confusion matrix for the binary classification of the BreakHis dataset at different magnification scales is shown in Figure 9. As can be seen in the cases of different magnification levels, 40×, 100×, and 200×, our model tends to produce better results for binary classification. Because of the diverse and significant areas in the images, the representation of the confusion matrix results shows that binary scenarios performed better than multi-classification scenarios did. The higher magnification of features gives further structural information to the model, which helps it to acquire a decent depiction of patches with labels.

Table 7. Detailed classification results by FabNet on NCT-CRC-HE-100K dataset concerning benchmark models in terms of accuracy and sensitivity.

Model	Accuracy (%)	Sensitivity								
		ADI	BACK	DEB	LYM	MUC	MUS	NORM	STR	TUM
VGG16 [56]	96.0	0.95	0.93	0.94	0.88	0.96	0.89	0.98	0.91	0.90
ResNet50 [56]	95.9	0.94	0.90	1.00	0.89	0.92	0.88	0.89	0.95	0.98
Dense Net 121 [55]	96.1	0.96	0.70	0.98	0.97	0.92	0.91	0.96	0.93	0.94
FabNet	98.2	0.96	0.98	1.00	1.00	1.00	0.98	0.99	0.94	0.99

Table 8. Class-wise results representation of FabNet in terms of precision, F1 score, and recall using the NCT-CRC-HE-100K dataset.

Class	Precision	F1 Score	Recall
Adipose Tissue	1.00	0.98	0.96
Background	1.00	0.99	0.98
Colorectal Cancer	0.98	0.99	1.00
Debris	1.00	1.00	1.00
Lymphocytes	0.95	0.97	1.00
Mucus	0.94	0.96	0.98
NC Tumor	0.99	0.99	0.99
Colon Mucosa	1.00	0.97	0.94
Cancer Stroma	0.99	0.99	0.99

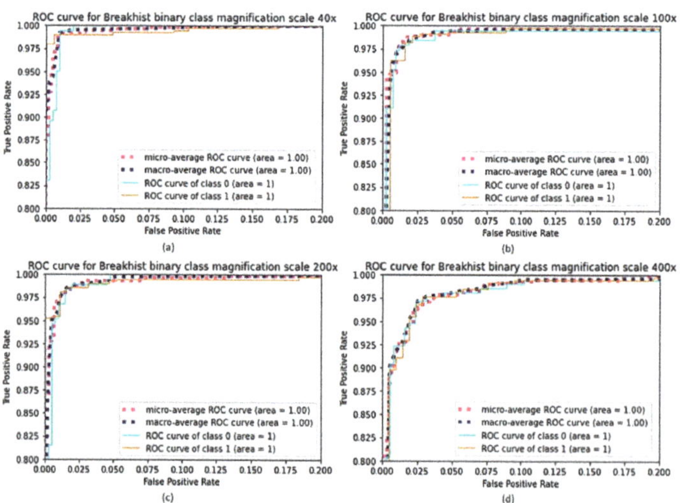

Figure 7. ROC curves of FabNet Model for binary classification, (**a**) 40× magnification, (**b**) BreakHis 100× magnification, (**c**) BreakHis 200× magnification, (**d**) BreakHis 400× magnification.

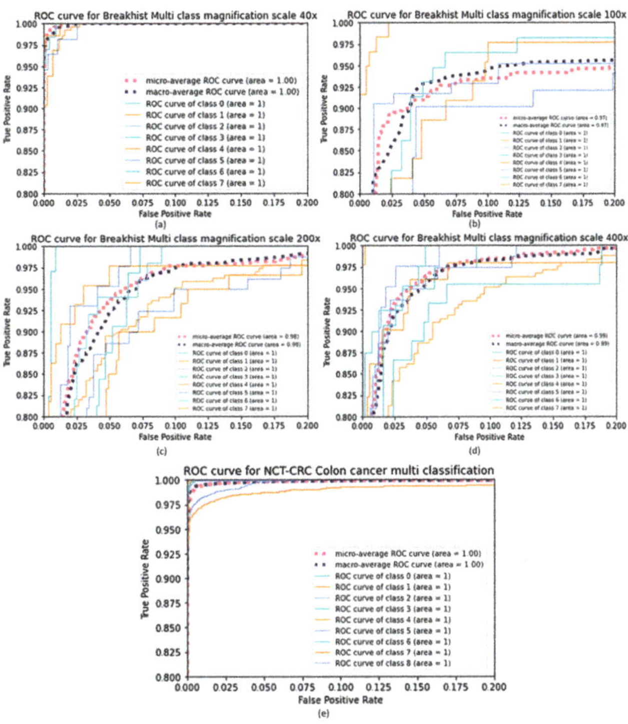

Figure 8. ROC curves of FabNet Model for multi classification, (**a**) 40× magnification, (**b**) BreakHis 100× magnification, (**c**) BreakHis 200× magnification, (**d**) BreakHis 400× magnification (**e**) ROC curves for NCT-CRC colon cancer dataset.

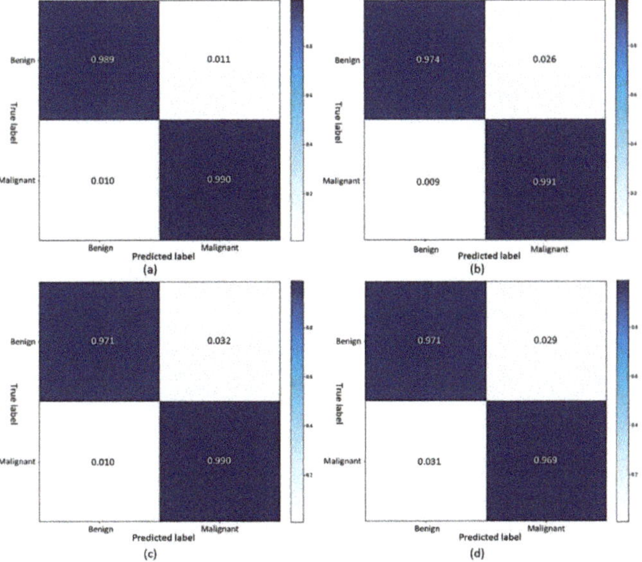

Figure 9. Confusion matrices of FabNet that for BreakHis, (**a**) confusion matrix of 40× magnification, (**b**) confusion matrix of 100× magnification, (**c**) confusion matrix of 200× magnification, (**d**) confusion matrix of 400× magnification.

The confusion matrix results for multi-classification in the case of NCT-CRC color cancer are shown in Figure 10.

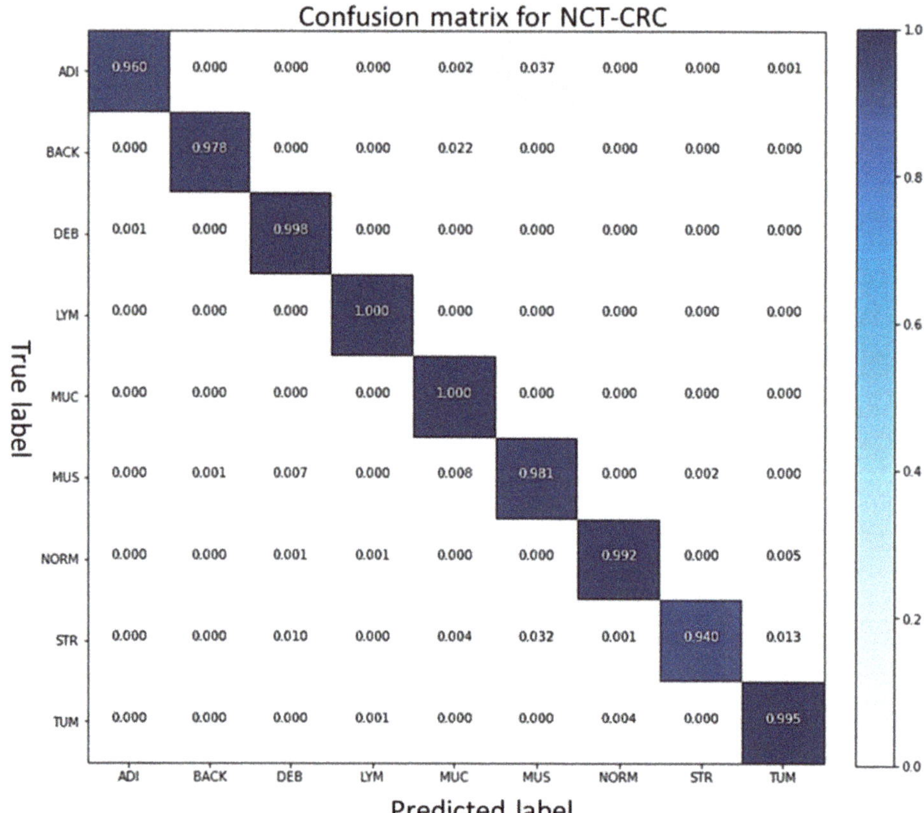

Figure 10. Confusion matrix of FabNet, which shows the best score in the NCT-CRC-HE-100K dataset testing set among 5-fold cross-validation.

Tables 9 and 10 shows the results of proposed model in comparison with benchmarks related to breast and colon histology models.

Table 9 shows the mean and standard deviation of our results by experimenting with satin and without stain normalization to better understand the use of the FabNet model in studying cancer histopathology images.

Table 9. A comparison of FabNet performance with existing studies on BreakHis histology dataset.

Dataset	Author	Year	Preprocessing	Model	Accuracy (%) Magnification Level			
					40×	100×	200×	400×
Break his Dataset	Spanhol et al. [30]	2016	None	PFTAS QDA	83 ± 4.1	82.1 ± 4.9	85.1 ± 3.1	82 ± 3.8
	Spanhol et al. [39]	2016	Image Resize	Pre-Trained AlexNet	88 ± 5.6	84.5 ± 2.4	85.3 ± 3.8	81 ± 4.9
	Spanhol et al. [59]	2017	None	DeCAF Model	84 ± 6.9	83.9 ± 5.9	86.3 ± 3.5	82 ± 2.4
	Kumar et al. [60]	2018	Image Resize	Newly Designed CNN	83 ± 3.2	81.0 ± 4.2	84.2 ± 3.4	81 ± 1.3
	Sudharshan et al. [61]	2019	None	PLTAS NPMIL	92 ± 5.9	89.1 ± 5.2	87.2 ± 4.3	82 ± 3.0
	Gour et al. [62]	2020	Data augmentation	ResHist Model	82 ± 3.3	88.1 ± 2.7	92.5 ± 2.8	87 ± 2.4
	Lingqiao Li et.al [42]	2018	Data Augmentation, Transfer learning	NDCNN	92.8 ± 2.1	93.9 ± 1.9	93.7 ± 2.2	92.9 ± 1.8
	Gandomkar et.al [63]		Data Augmentation,	ResNET152	94.18 ± 2.1	93.2 ± 1.4	94.7 ± 3.6	93.5 ± 2.9
	Proposed	2021	Stain Normalization	FabNet	99 ± 0.2	89.51 ± 1.7	97.41 ± 1.4	96 ± 1.0

Table 10. A comparison of FabNet performance with existing studies on Colorectal histology dataset.

Dataset	Author	Year	Preprocessing	Model	Evaluation Matrices			
					Accuracy	Precision	F1 Score	Sensitivity
Colon (NCT-CRC-HE-100K) dataset	Wang et al. [64]	2017	None	BCNN	92.6	91.2	92.8	90.5
	Sari et al. [65]	2018	None	SSAE/SCAE	93.6	93.4	93.2	92.3
	Kather et al. [66]	2019	Stain Normalization	TL+CNN (VGG)	94.3	92.1	93.5	94.1
	Gosh et al. [67]	2021	None	Ensemble DNN	92.8	92.6	92.2	93.1
	Proposed	2021	None	FabNet	98.3	98.3	98.2	98.2

The model outperformed some of the one in the most current research studies. For example, in [68], they obtained 97.58% and 97.45% accuracy rates with 7.6 million training parameters, whereas we reached a 99.03% accuracy with 3239 K training parameters. Despite having fewer training parameters, our model achieved a higher degree of accuracy. In another study [2], the authors proposed the Inception Recurrent Residual Convolutional Neural Network (IRRCNN) network, which obtained 97.95% accuracy for image classification and 97.65% accuracy for patient classification. Unlike IRRCNN, FabNet obtained a 99.01% patient-level accuracy and 99.03% picture-level accuracy using this dataset. The authors obtained 99.05% accuracy for binary classification and 98.59% accuracy for multiclassification using data augmentation. We obtained comparable outcomes without applying data augmentation. Data augmentation enables a learning model to overcome important training constraints such as overfitting, hence improving its accuracy and generalization capabilities. In the case of our model, we think that its ability for generalization is strengthened despite the absence of data augmentation. A similar accuracy was shown by Rui Man et al. [55] at the $40\times$ magnification level, however our model achieved better results at the $200\times$ and $400\times$ magnification levels. The authors proposed the use of DenseNet121-ino, which has substantially more training parameters than FabNet does.

6. Conclusions

In this paper, we suggested the FabNet model that can learn the fine-to-coarse structural and textural features of multi scale histopathological images by accretive network architecture, which agglomerates hierarchical feature maps to acquire significant classification accuracy. We expanded upon the conventional convolutional neural network architecture by incorporating deeper integration to finely fuse information across layers. This layer expansion had a small impact on the model's depth; however, it made the model more tightly linked with a compact form, ensuring that any piece of detail was transferred to the deeper layers for better learning. Despite having fewer parameters, this lightweight network architecture yielded better classification accuracy than the state-of-the-art models did.

Our model yields improved classification probabilities at both the patch as well as the image levels. The efficiency and reliability of the FabNet were assessed using two public datasets that included breast and colon cancer data based on several experiments for instance, multi- and binary classifications. The suggested FabNet improved upon the existing state-of-the-art models when they were evaluated using both of the public benchmark datasets. The experimental parameters were kept the same for the benchmark models, as well as for the proposed model to precisely conclude the performance. The proposed model achieved 99% accuracy and a 98.9% F1 score in the case of the binary classification of BreakHis at the $40\times$ magnification scale. The model achieved 98.2% test accuracy and a 98.23% F1 score for NCT-CRC-HE-100K colon cancer dataset without employing any data augmentation technique.

We believe that the model can reduce the cancer screening time for pathologists, as well as oncologists. In diverse circumstances, oncologists and researchers working in the field of cancer detection and diagnostics using histological images will benefit from the proposed model's high sensitivity and accuracy. Although the closely coupled architecture tackled the imbalance in the dataset issue, which ultimately resulted in minor effects on the model's performances, since the data imbalance is so prominent in the clinical histology, we intend to look at certain strategies for coping with this problem in the future. We will also look at which feature map combinations which are most significant for classification. The proposed model can be used to perform a variety of tasks related to histological image-based classification in clinical environments.

Author Contributions: Supervision, M.S.A. and H.A. All authors have read and agreed to the published version of the manuscript.

Funding: The authors received no funding.

Institutional Review Board Statement: Not applicable.

Informed Consent Statement: Not applicable.

Data Availability Statement: Publicly available datasets were used in this study. The dataset can be found on www.tubo.tu.ac.kr.

Conflicts of Interest: The authors declare that they have no conflicts of interest to report for this study.

References

1. Rahhal, M.M.A. Breast Cancer Classification in Histopathological Images Using Convolutional Neural Network. *Int. J. Adv. Comput. Sci. Appl. IJACSA* **2018**, *9*. [CrossRef]
2. Alom, M.Z.; Yakopcic, C.; Nasrin, M.S.; Taha, T.M.; Asari, V.K. Breast Cancer Classification from Histopathological Images with Inception Recurrent Residual Convolutional Neural Network. *J. Digit. Imaging* **2019**, *32*, 605–617. [CrossRef]
3. Araújo, T.; Aresta, G.; Castro, E.; Rouco, J.; Aguiar, P.; Eloy, C.; Polónia, A.; Campilho, A. Classification of Breast Cancer Histology Images Using Convolutional Neural Networks. *PloS ONE* **2017**, *12*, e0177544. [CrossRef] [PubMed]
4. Liu, Y.; Chen, C.; Wang, X.; Sun, Y.; Zhang, J.; Chen, J.; Shi, Y. An Epigenetic Role of Mitochondria in Cancer. *Cells* **2022**, *11*, 2518. [CrossRef] [PubMed]
5. Chen, K.; Zhang, J.; Beeraka, N.M.; Tang, C.; Babayeva, Y.V.; Sinelnikov, M.Y.; Zhang, X.; Zhang, J.; Liu, J.; Reshetov, I.V.; et al. Advances in the Prevention and Treatment of Obesity-Driven Effects in Breast Cancers. *Front. Oncol.* **2022**, *12*, 2663. [CrossRef]
6. Chen, K.; Lu, P.; Beeraka, N.M.; Sukocheva, O.A.; Madhunapantula, S.V.; Liu, J.; Sinelnikov, M.Y.; Nikolenko, V.N.; Bulygin, K.V.; Mikhaleva, L.M.; et al. Mitochondrial Mutations and Mitoepigenetics: Focus on Regulation of Oxidative Stress-Induced Responses in Breast Cancers. *Semin. Cancer Biol.* **2022**, *83*, 556–569. [CrossRef] [PubMed]
7. Xie, P.; Ma, Y.; Yu, S.; An, R.; He, J.; Zhang, H. Development of an Immune-Related Prognostic Signature in Breast Cancer. *Front. Genet.* **2020**, *10*, 1390. [CrossRef]
8. Williamson, G.R.; Plowright, H.; Kane, A.; Bunce, J.; Clarke, D.; Jamison, C. Collaborative Learning in Practice: A Systematic Review and Narrative Synthesis of the Research Evidence in Nurse Education. *Nurse Educ. Pract.* **2020**, *43*, 102706. [CrossRef]
9. Bardou, D.; Zhang, K.; Ahmad, S.M. Classification of Breast Cancer Based on Histology Images Using Convolutional Neural Networks. *IEEE Access* **2018**, *6*, 24680–24693. Available online: https://www.google.com/search?q=Bardou%2C+D.%3B+Zhang%2C+K.%3B+Ahmad%2C+S.M.+Classification+of+Breast+Cancer+Based+on+Histology+Images+Using+Convolutional+Neural+Networks.+Ieee+Access+2018%2C+6%2C+24680%E2%80%9324693.&rlz=1C1ONGR_enPK996PK996&oq=Bardou%2C+D.%3B+Zhang%2C+K.%3B+Ahmad%2C+S.M.+Classification+of+Breast+Cancer+Based+on+Histology+Images+Using+Convolutional+Neural+Networks.+Ieee+Access+2018%2C+6%2C+24680%E2%80%9324693.&aqs=chrome..6 9i57.1132j0j4&sourceid=chrome&ie=UTF-8 (accessed on 31 August 2022). [CrossRef]
10. Mccann, M.T.; Ozolek, J.A.; Castro, C.A.; Parvin, B.; Kovacevic, J.; Mccann, M.T.; Member, S.; Ozolek, J.A.; Castro, C.A.; Parvin, B.; et al. Automated Histology Analysis: Opportunities for Signal Processing. *IEEE Signal Process. Mag.* **2014**, *32*, 78–87. [CrossRef]
11. Robertson, S.; Azizpour, H.; Smith, K.; Hartman, J. Digital Image Analysis in Breast Pathology—From Image Processing Techniques to Artificial Intelligence. *Transl. Res.* **2018**, *194*, 19–35. [CrossRef] [PubMed]
12. Ching, T.; Himmelstein, D.S.; Beaulieu-Jones, B.K.; Kalinin, A.A.; Do, B.T.; Way, G.P.; Ferrero, E.; Agapow, P.-M.; Zietz, M.; Hoffman, M.M.; et al. Opportunities and Obstacles for Deep Learning in Biology and Medicine. *J. R. Soc. Interface* **2018**, *15*, 20170387. [CrossRef] [PubMed]
13. Iglovikov, V.; Shvets, A. TernausNet: U-Net with VGG11 Encoder Pre-Trained on ImageNet for Image Segmentation. *arXiv* **2018**, arXiv:1801.05746.
14. Raman, R.; Srinivasan, S.; Virmani, S.; Sivaprasad, S.; Rao, C.; Rajalakshmi, R. Fundus Photograph-Based Deep Learning Algorithms in Detecting Diabetic Retinopathy. *Eye* **2019**, *33*, 97–109. [CrossRef]
15. Tiulpin, A.; Thevenot, J.; Rahtu, E.; Lehenkari, P.; Saarakkala, S. Automatic Knee Osteoarthritis Diagnosis from Plain Radiographs: A Deep Learning-Based Approach. *Sci. Rep.* **2018**, *8*, 1727. [CrossRef] [PubMed]
16. Ma, X.; Niu, Y.; Gu, L.; Wang, Y.; Zhao, Y.; Bailey, J.; Lu, F. Understanding Adversarial Attacks on Deep Learning Based Medical Image Analysis Systems. *Pattern Recognit.* **2021**, *110*, 107332. [CrossRef]
17. Sanyal, R.; Jethanandani, M.; Sarkar, R. DAN: Breast Cancer Classification from High-Resolution Histology Images Using Deep Attention Network. In *Innovations in Computational Intelligence and Computer Vision*; Sharma, M.K., Dhaka, V.S., Perumal, T., Dey, N., Tavares, J.M.R.S., Eds.; Springer: Singapore, 2021; pp. 319–326.
18. Kumar, S.; Sharma, S. Sub-Classification of Invasive and Non-Invasive Cancer from Magnification Independent Histopathological Images Using Hybrid Neural Networks. *Evol. Intell.* **2022**, *15*, 1531–1543. [CrossRef]
19. Dou, J. Clinical Decision System Using Machine Learning and Deep Learning: A Survey. 2022. Available online: https://www.researchgate.net/profile/Jason-Dou/publication/360154101_Clinical_Decision_System_using_Machine_Learning_and_Deep_Learning_a_Survey/links/630b86f5acd814437fe29fe7/Clinical-Decision-System-using-Machine-Learning-and-Deep-Learning-a-Survey.pdf (accessed on 31 August 2022).
20. Amin, M.S.; Ahn, H. Earthquake Disaster Avoidance Learning System Using Deep Learning. *Cogn. Syst. Res.* **2021**, *66*, 221–235. [CrossRef]

21. Amin, M.S.; Yasir, S.M.; Ahn, H. Recognition of Pashto Handwritten Characters Based on Deep Learning. *Sensors* **2020**, *20*, E5884. [CrossRef]

22. Sadiq, A.M.; Ahn, H.; Choi, Y.B. Human Sentiment and Activity Recognition in Disaster Situations Using Social Media Images Based on Deep Learning. *Sensors* **2020**, *20*, 7115. [CrossRef]

23. Lin, M.; Chen, Q.; Yan, S. Network in Network. 2014. Available online: https://arxiv.org/abs/1312.4400 (accessed on 31 August 2022).

24. Szegedy, C.; Liu, W.; Jia, Y.; Sermanet, P.; Reed, S.; Anguelov, D.; Erhan, D.; Vanhoucke, V.; Rabinovich, A. Going Deeper with Convolutions 2014. *arXiv* **2014**, arXiv:1409.4842.

25. He, K.; Zhang, X.; Ren, S.; Sun, J. Identity Mappings in Deep Residual Networks. In Proceedings of the Computer Vision–ECCV 2016: 14th European Conference, Amsterdam, The Netherlands, 11–14 October 2016; Springer International Publishing: Berlin/Heidelberg, Germany, 2016.

26. Srivastava, R.K.; Greff, K.; Schmidhuber, J. Highway Networks. *arXiv* **2015**, arXiv:1505.00387.

27. Yasrab, R. SRNET: A Shallow Skip Connection Based Convolutional Neural Network Design for Resolving Singularities. *J. Comput. Sci. Technol.* **2019**, *34*, 924–938. [CrossRef]

28. Vapnik, V.N. *The Nature of Statistical Learning Theory*; Springer: New York, NY, USA, 2000; ISBN 978-1-4419-3160-3.

29. Kowal, M.; Filipczuk, P.; Obuchowicz, A.; Korbicz, J.; Monczak, R. Computer-Aided Diagnosis of Breast Cancer Based on Fine Needle Biopsy Microscopic Images. *Comput. Biol. Med.* **2013**, *43*, 1563–1572. [CrossRef] [PubMed]

30. Spanhol, F.A.; Oliveira, L.S.; Petitjean, C.; Heutte, L. A Dataset for Breast Cancer Histopathological Image Classification. *IEEE Trans. Biomed. Eng.* **2016**, *63*, 1455–1462. [CrossRef] [PubMed]

31. George, Y.M.; Zayed, H.H.; Roushdy, M.I.; Elbagoury, B.M. Remote Computer-Aided Breast Cancer Detection and Diagnosis System Based on Cytological Images. *IEEE Syst. J.* **2014**, *8*, 949–964. [CrossRef]

32. Breast Cancer Diagnosis from Biopsy Images with Highly Reliable Random Subspace Classifier Ensembles | SpringerLink. Available online: https://link.springer.com/article/10.1007/s00138-012-0459-8 (accessed on 31 August 2022).

33. Robinson, E.; Silverman, B.G.; Keinan-Boker, L. Using Israel's National Cancer Registry Database to Track Progress in the War against Cancer: A Challenge for Health Services. *Isr. Med. Assoc. J. IMAJ* **2017**, *19*, 221–224.

34. Gupta, V.; Bhavsar, A. Breast Cancer Histopathological Image Classification: Is Magnification Important? In Proceedings of the 2017 IEEE Conference on Computer Vision and Pattern Recognition Workshops (CVPRW), Honolulu, HI, USA, 21–26 July 2017; pp. 769–776.

35. Seo, H.; Brand, L.; Barco, L.S.; Wang, H. Scaling Multi-Instance Support Vector Machine to Breast Cancer Detection on the BreaKHis Dataset. *Bioinformatics* **2022**, *38*, i92–i100. [CrossRef]

36. Rashmi, R.; Prasad, K.; Udupa, C.B.K. Breast Histopathological Image Analysis Using Image Processing Techniques for Diagnostic Purposes: A Methodological Review. *J. Med. Syst.* **2021**, *46*, 7. [CrossRef]

37. Gupta, S.; Sinha, N.; Sudha, R.; Babu, C. Breast Cancer Detection Using Image Processing Techniques. In Proceedings of the 2019 Innovations in Power and Advanced Computing Technologies (i-PACT), Piscataway, NJ, USA, 22–23 March 2019; Volume 1, pp. 1–6.

38. Das, A.; Nair, M.S.; Peter, S.D. Computer-Aided Histopathological Image Analysis Techniques for Automated Nuclear Atypia Scoring of Breast Cancer: A Review. *J. Digit. Imaging* **2020**, *33*, 1091–1121. [CrossRef] [PubMed]

39. Spanhol, F.A.; Oliveira, L.S.; Petitjean, C.; Heutte, L. Breast Cancer Histopathological Image Classification Using Convolutional Neural Networks. In Proceedings of the 2016 International Joint Conference on Neural Networks (IJCNN), Vancouver, BC, Canada, 24–29 July 2016; pp. 2560–2567.

40. Krizhevsky, A.; Sutskever, I.; Hinton, G.E. ImageNet Classification with Deep Convolutional Neural Networks. In *Advances in Neural Information Processing Systems*; Curran Associates, Inc.: Red Hook, NY, USA, 2012; Volume 25.

41. Hou, L.; Samaras, D.; Kurc, T.M.; Gao, Y.; Davis, J.E.; Saltz, J.H. Patch-Based Convolutional Neural Network for Whole Slide Tissue Image Classification. In Proceedings of the 2016 IEEE Conference on Computer Vision and Pattern Recognition (CVPR), Las Vegas, NV, USA, 27–30 June 2016.

42. Han, Z.; Wei, B.; Zheng, Y.; Yin, Y.; Li, K.; Li, S. Breast Cancer Multi-Classification from Histopathological Images with Structured Deep Learning Model. *Sci. Rep.* **2017**, *7*, 4172. [CrossRef]

43. Cireşan, D.C.; Giusti, A.; Gambardella, L.M.; Schmidhuber, J. Mitosis Detection in Breast Cancer Histology Images with Deep Neural Networks. In *Medical Image Computing and Computer-Assisted Intervention—MICCAI 2013*; Mori, K., Sakuma, I., Sato, Y., Barillot, C., Navab, N., Eds.; Springer: Berlin, Heidelberg, 2013; pp. 411–418.

44. Wang, H.; Roa, A.C.; Basavanhally, A.N.; Gilmore, H.L.; Shih, N.; Feldman, M.; Tomaszewski, J.; Gonzalez, F.; Madabhushi, A. Mitosis Detection in Breast Cancer Pathology Images by Combining Handcrafted and Convolutional Neural Network Features. *J. Med. Imaging* **2014**, *1*, 034003. [CrossRef] [PubMed]

45. Kashif, M.N.; Raza, S.E.A.; Sirinukunwattana, K.; Arif, M.; Rajpoot, N. Handcrafted Features with Convolutional Neural Networks for Detection of Tumor Cells in Histology Images. In Proceedings of the 2016 IEEE 13th International Symposium on Biomedical Imaging (ISBI), Prague, Czech Republic, 13–16 April 2016; pp. 1029–1032.

46. Tellez, D.; Litjens, G.; Bándi, P.; Bulten, W.; Bokhorst, J.-M.; Ciompi, F.; van der Laak, J. Quantifying the Effects of Data Augmentation and Stain Color Normalization in Convolutional Neural Networks for Computational Pathology. *Med. Image Anal.* **2019**, *58*, 101544. [CrossRef]

7. Bejnordi, B.E.; Zuidhof, G.; Balkenhol, M.; Hermsen, M.; Bult, P.; van Ginneken, B.; Karssemeijer, N.; Litjens, G.; Laak, J. van der Context-Aware Stacked Convolutional Neural Networks for Classification of Breast Carcinomas in Whole-Slide Histopathology Images. *J. Med. Imaging* **2017**, *4*, 044504. [CrossRef] [PubMed]

8. Ehteshami Bejnordi, B.; Mullooly, M.; Pfeiffer, R.M.; Fan, S.; Vacek, P.M.; Weaver, D.L.; Herschorn, S.; Brinton, L.A.; van Ginneken, B.; Karssemeijer, N.; et al. Using Deep Convolutional Neural Networks to Identify and Classify Tumor-Associated Stroma in Diagnostic Breast Biopsies. *Mod. Pathol.* **2018**, *31*, 1502–1512. [CrossRef]

9. Reinhard, E.; Ashikhmin, M.; Gooch, B.; Shirley, P. Color Transfer between Images. *IEEE Comput. Graph. Appl.* **2001**, *21*, 34–41. [CrossRef]

10. Vahadane, A.; Peng, T.; Sethi, A.; Albarqouni, S.; Wang, L.; Baust, M.; Steiger, K.; Schlitter, A.M.; Esposito, I.; Navab, N. Structure-Preserving Color Normalization and Sparse Stain Separation for Histological Images. *IEEE Trans. Med. Imaging* **2016**, *35*, 1962–1971. [CrossRef] [PubMed]

11. Mathews, A.; Simi, I.; Kizhakkethottam, J.J. Efficient Diagnosis of Cancer from Histopathological Images By Eliminating Batch Effects. *Procedia Technol.* **2016**, *24*, 1415–1422. [CrossRef]

12. Kather, J.N.; Halama, N.; Marx, A. 100,000 Histological Images of Human Colorectal Cancer and Healthy Tissue 2018. Available online: https://zenodo.org/record/1214456#.Y98AhK1BxPY (accessed on 30 January 2023).

13. Macenko, M.; Niethammer, M.; Marron, J.S.; Borland, D.; Woosley, J.T.; Guan, X.; Schmitt, C.; Thomas, N.E. A Method for Normalizing Histology Slides for Quantitative Analysis. In Proceedings of the 2009 IEEE International Symposium on Biomedical Imaging: From Nano to Macro, Boston, MA, USA, 28 June–1 July 2009; IEEE: Boston, MA, USA, 2009; pp. 1107–1110.

14. Ono, Y.; Trulls, E.; Fua, P.; Yi, K.M. LF-Net: Learning Local Features from Images. *Adv. Neural Inf. Process. Syst.* **2018**, *31*.

15. Man, R.; Yang, P.; Xu, B. Classification of Breast Cancer Histopathological Images Using Discriminative Patches Screened by Generative Adversarial Networks. *IEEE Access* **2020**, *8*, 155362–155377. [CrossRef]

16. Simonyan, K.; Zisserman, A. Very Deep Convolutional Networks for Large-Scale Image Recognition. *arXiv* **2014**, arXiv:1409.1556.

17. He, K.; Zhang, X.; Ren, S.; Sun, J. Deep Residual Learning for Image Recognition. In Proceedings of the IEEE Conference on Computer Vision and Pattern Recognition, Boston, MA, USA, 7–12 June 2015.

18. Sheikh, T.S.; Lee, Y.; Cho, M. Histopathological Classification of Breast Cancer Images Using a Multi-Scale Input and Multi-Feature Network. *Cancers* **2020**, *12*, 2031. [CrossRef] [PubMed]

19. Spanhol, F.A.; Oliveira, L.S.; Cavalin, P.R.; Petitjean, C.; Heutte, L. Deep Features for Breast Cancer Histopathological Image Classification. In Proceedings of the 2017 IEEE International Conference on Systems, Man, and Cybernetics (SMC), Banff, AB, Canada, 5–8 October 2017; pp. 1868–1873.

20. Kumar, K.; Rao, A.C.S. Breast Cancer Classification of Image Using Convolutional Neural Network. In Proceedings of the 2018 4th International Conference on Recent Advances in Information Technology (RAIT), Dhanbad, India, 15–17 March 2018; pp. 1–6.

21. Sudharshan, P.J.; Petitjean, C.; Spanhol, F.; Oliveira, L.E.; Heutte, L.; Honeine, P. Multiple Instance Learning for Histopathological Breast Cancer Image Classification. *Expert Syst. Appl.* **2019**, *117*, 103–111. [CrossRef]

22. Gour, M.; Jain, S.; Sunil Kumar, T. Residual Learning Based CNN for Breast Cancer Histopathological Image Classification. *Int. J. Imaging Syst. Technol.* **2020**, *30*, 621–635. [CrossRef]

23. Gandomkar, Z.; Brennan, P.C.; Mello-Thoms, C. Computer-Assisted Nuclear Atypia Scoring of Breast Cancer: A Preliminary Study. *J. Digit. Imaging* **2019**, *32*, 702–712. [CrossRef]

24. Wang, C.; Shi, J.; Zhang, Q.; Ying, S. Histopathological Image Classification with Bilinear Convolutional Neural Networks. In Proceedings of the 2017 39th Annual International Conference of the IEEE Engineering in Medicine and Biology Society (EMBC), Jeju Island, Republic of Korea, 11–15 July 2017; pp. 4050–4053.

25. Sari, C.T.; Gunduz-Demir, C. Unsupervised Feature Extraction via Deep Learning for Histopathological Classification of Colon Tissue Images. *IEEE Trans. Med. Imaging* **2018**, *38*, 1139–1149. [CrossRef]

26. Kather, J.N.; Krisam, J.; Charoentong, P.; Luedde, T.; Herpel, E.; Weis, C.-A.; Gaiser, T.; Marx, A.; Valous, N.A.; Ferber, D.; et al. Predicting Survival from Colorectal Cancer Histology Slides Using Deep Learning: A Retrospective Multicenter Study. *PLoS Med.* **2019**, *16*, e1002730. [CrossRef]

27. Ghosh, S.; Bandyopadhyay, A.; Sahay, S.; Ghosh, R.; Kundu, I.; Santosh, K.C. Colorectal Histology Tumor Detection Using Ensemble Deep Neural Network. *Eng. Appl. Artif. Intell.* **2021**, *100*, 104202. [CrossRef]

28. Mewada, H.K.; Patel, A.V.; Hassaballah, M.; Alkinani, M.H.; Mahant, K. Spectral–Spatial Features Integrated Convolution Neural Network for Breast Cancer Classification. *Sensors* **2020**, *20*, 4747. [CrossRef] [PubMed]

Article

Development of a Machine Learning-Based Prediction Model for Chemotherapy-Induced Myelosuppression in Children with Wilms' Tumor

Mujie Li [1,2], Quan Wang [3], Peng Lu [1,2], Deying Zhang [1,2], Yi Hua [1,2], Feng Liu [1,2], Xing Liu [1,2], Tao Lin [1,2], Guanghui Wei [1,2] and Dawei He [1,2,4,*]

1 Department of Urology, Children's Hospital of Chongqing Medical University, Chongqing 400015, China
2 Chongqing Key Laboratory of Children Urogenital Development and Tissue Engineering, Chongqing Key Laboratory of Pediatrics, Ministry of Education Key Laboratory of Child Development and Disorders, National Clinical Research Center for Child Health and Disorders, International Science and Technology Cooperation Base of Child Development and Critical Disorders, Chongqing 400014, China
3 Department of Cardiothoracic Surgery, Children's Hospital of Chongqing Medical University, Chongqing 400015, China
4 Chongqing Higher Institution Engineering Research Center of Children's Medical Big Data Intelligent Application, Chongqing 401331, China
* Correspondence: hedawei@hospital.cqmu.edu.cn

Simple Summary: Wilms' tumor is the most common renal malignant tumor in children, and chemotherapy is an indispensable part of the treatment for most Wilms' tumor patients. Chemotherapy-induced myelosuppression is the most common and serious toxicity of chemotherapy, which can hinder the process of chemotherapy and even endanger life. However, there is a lack of tools to predict chemotherapy-induced myelosuppression. We herein develop a model based on machine learning that can effectively predict the risk of chemotherapy-induced myelosuppression in children with Wilms' tumor, offering the possibility to identify children with high risk of chemotherapy-induced myelosuppression early and take preventive strategies.

Abstract: Purpose: Develop and validate an accessible prediction model using machine learning (ML) to predict the risk of chemotherapy-induced myelosuppression (CIM) in children with Wilms' tumor (WT) before chemotherapy is administered, enabling early preventive management. **Methods:** A total of 1433 chemotherapy cycles in 437 children with WT who received chemotherapy in our hospital from January 2009 to March 2022 were retrospectively analyzed. Demographic data, clinicopathological characteristics, hematology and blood biochemistry baseline results, and medication information were collected. Six ML algorithms were used to construct prediction models, and the predictive efficacy of these models was evaluated to select the best model to predict the risk of grade $\geq$ 2 CIM in children with WT. A series of methods, such as the area under the receiver operating characteristic curve (AUROC), the calibration curve, and the decision curve analysis (DCA) were used to test the model's accuracy, discrimination, and clinical practicability. **Results:** Grade $\geq$ 2 CIM occurred in 58.5% (839/1433) of chemotherapy cycles. Based on the results of the training and validation cohorts, we finally identified that the extreme gradient boosting (XGB) model has the best predictive efficiency and stability, with an AUROC of up to 0.981 in the training set and up to 0.896 in the test set. In addition, the calibration curve and the DCA showed that the XGB model had the best discrimination and clinical practicability. The variables were ranked according to the feature importance, and the five variables contributing the most to the model were hemoglobin (Hgb), white blood cell count (WBC), alkaline phosphatase, coadministration of highly toxic chemotherapy drugs, and albumin. **Conclusions:** The incidence of grade $\geq$ 2 CIM was not low in children with WT, which need attention. The XGB model was developed to predict the risk of grade $\geq$ 2 CIM in children with WT for the first time. The model has good predictive performance and stability and has the potential to be translated into clinical applications. Based on this modeling and application approach, the extension of CIM prediction models to other pediatric malignancies could be expected.

Citation: Li, M.; Wang, Q.; Lu, P.; Zhang, D.; Hua, Y.; Liu, F.; Liu, X.; Lin, T.; Wei, G.; He, D. Development of a Machine Learning-Based Prediction Model for Chemotherapy-Induced Myelosuppression in Children with Wilms' Tumor. *Cancers* **2023**, *15*, 1078. https://doi.org/10.3390/cancers15041078

Academic Editors: Hamid Khayyam, Ali Madani, Rahele Kafieh and Ali Hekmatnia

Received: 13 January 2023
Revised: 3 February 2023
Accepted: 5 February 2023
Published: 8 February 2023

Keywords: chemotherapy-induced myelosuppression; Wilms' tumor; artificial intelligence; machine learning; prediction model

1. Introduction

Wilms' tumor (WT) is the most common renal malignancy in children and has the second highest incidence of pediatric primary abdominal malignancies. Although multidisciplinary treatments have advanced, recurrence occurs in approximately 15% of children with WT, and the survival rate after recurrence is only about 50% [1–3]. As the surgical resection of pediatric tumors is often difficult, chemotherapy is an indispensable part of the treatment for most WT patients.

However, chemotherapy drugs have many toxicities and side effects. Chemotherapy-induced myelosuppression (CIM) is the most common and severe toxicity of chemotherapy for tumors, typically manifesting as anemia, neutropenia, thrombocytopenia, and/or lymphopenia [4–7], leading to an increased risk of life-threatening infection, fatigue, and potential bleeding [8,9]. CIM often forces children to interrupt or postpone their chemotherapy course, severely compromising the effectiveness of treatment and even leading to death due to CIM-related complications. Studies have reported that the mortality rate related to grade 4 CIM can reach 4–12% [10]. Therefore, early identification of children at high risk of CIM and timely implementation of corresponding preventive and therapeutic measures can not only improve the effectiveness of tumor treatment, but also significantly reduce the disease burden caused by the related complications [11].

Studies have shown that risk factors for CIM include age, nutritional status, poor liver and kidney function, low baseline white blood cell count (WBC), chemotherapy cycles, etc. [12–15]. Various mathematical models for predicting CIM or febrile neutropenia (FN) have been proposed [16–18] and successfully applied to predict dynamic changes in neutrophil count [19,20]. However, these studies focused on predicting the risk of FN in adult tumors such as breast cancer, small cell lung cancer, and colorectal cancer [14,21,22].

The predictors of CIM in pediatric malignant solid tumors, especially in WT, have not been reported. In addition, most of the pharmacokinetic mathematical models developed in these studies focus on predicting CIM/FN caused by a single drug, making it difficult to extend to pediatric tumors requiring multidrug combination therapy. Moreover, the application of these models requires repeated and frequent monitoring of changes in hematological parameters and drug concentrations, such invasive tests are often unacceptable to children and parents [20,22], and the relatively backward economic and medical levels in developing countries seem to make the implementation of such monitoring strategies more difficult.

Therefore, CIM or FN prediction models reported in the existing studies are difficult to widely apply to predict CIM in children with WT. It is necessary to develop a CIM prediction model for children with WT that is easy to use and has good prediction efficiency.

At present, artificial intelligence (AI) has been widely applied in the medical field. Machine learning (ML), as a branch of AI, can overcome the shortcomings of traditional logistic regression and mathematical models, and has a strong ability for feature recognition, classification, and prediction [23]. The models established based on machine learning have been successfully used in predicting the prognosis of various tumors or diseases, which presented good predictive ability [24–26]. Shibahara et al. collected pretreatment clinical data of glioma patients treated with nimustine hydrochloride (ACNU), and further successfully established a prediction model of CIM using machine learning, as well as describing the relationship between myelosuppression and hematopoietic stem cells (HSCs) [27]. In our study, various premedication clinical data in each chemotherapy cycle of WT children with a large sample size from the clinical big data platform of our hospital were collected, including blood cells baseline level, liver and kidney function indicators, tumor stage, body weight, body surface area and other variables, and six ML algorithms were used to construct CIM prediction models. Meanwhile, further evaluation of each model was carried out to select the model with the best prediction performance, which can help doctors identify children with WT at high risk of CIM early and develop individualized strategies for prevention, treatment, and follow-up to reduce the disease burden and improve prognosis.

2. Methods

2.1. Patients

The data of patients with WT who received chemotherapy in our hospital from January 2009 to March 2022 were collected from our hospital's clinical big data platform. Inclusion criteria: (1) younger than 18 years old; (2) patients diagnosed with WT; (3) patients having received at least one cycle of chemotherapy; (4) patients having received at least one routine blood test and biochemical blood test before and after chemotherapy. Exclusion criteria: (1) patients with other hematologic diseases or a history of HIV infection or stem cell transplantation; (2) patients with incomplete medical records (missing more than 50% of variables used for analysis); (3) patients with treatment interruption.

2.2. Collection and Definition of Variables

2.2.1. General Variables

Variables such as demographic data, clinicopathological characteristics, the laboratory examination, and medication information after each admission were collected as follows: age, gender, height, weight, tumor stage, COG grade, the routine hematologic index and biochemical index, routine urinalysis, the type of chemotherapy drugs used, chemotherapy cycles, etc.

2.2.2. Outcome Indicators

The occurrence of grade $\geq$ 2 CIM was taken as the outcome indicator. According to the National Cancer Institute Common Terminology Criteria for Adverse Events (CTCAE) version 5.0, if one of the following 4 criteria is met after chemotherapy, it can be defined as grade $\geq$ 2 CIM: (1) WBC < 3.0 × 10^9/L; (2) absolute neutrophil count (ANC) < 1.5 × 10^9/L; (3) hemoglobin level (Hgb) < 100 g/L; (4) platelet count (PLT) < 75 × 10^9/L.

2.2.3. Calculation of Composite Variables

(1) Systemic immune-inflammation index (SII) = PLT × ANC/absolute lymphocyte count (ALC) [28]
(2) Neutrophil to lymphocyte ratio (NLR) = ANC/ALC
(3) Platelet to lymphocyte ratio (PLR) = PLT/ALC
(4) Body surface area (BSA) = 0.035 × body weight + 0.1 (body weight $\leq$ 30 kg)

BSA = 1.05 + (body weight − 30) × 0.02 (body weight > 30 kg)

2.2.4. Derived Variables

Coadministration of highly toxic chemotherapy drugs refers to any high hematologic toxicity chemotherapy drugs used during that chemotherapy cycle.

Chemotherapy drugs are divided into two categories according to the level of hematological toxicity [29,30]: (1) high: cyclophosphamide (CTX), ifosfamide, doxorubicin, epirubicin, actinomycin D, carboplatin, etoposide, topotecan, vindesine; (2) moderate/low: cisplatin, vincristine, bleomycin, fluorouracil.

2.3. Data Preprocessing

2.3.1. Quality Control of Samples

Each chemotherapy cycle of each WT patient was taken as a separate sample. The missing rate of each sample characteristic variable was counted, and 50% was selected as the threshold value according to the distribution of each sample characteristic variable and modeling requirements. If 50% or more of all characteristic variables were missing simultaneously, the sample characteristic variable was considered seriously missing and met the exclusion criteria.

2.3.2. Imputation Methods of Missing Data

For clinical characteristic variables, after the sample size was determined, the missing rate of each characteristic variable was checked, and 20% was selected as the threshold according to the modeling requirements. If the missing rate of the characteristic variable exceeds 20%, the variable will be deleted and not included in the model construction. Other missing categorical variables were imputed with the mode while missing continuous variables were imputed with the median. In addition, chemotherapy drugs with a relative frequency of medication less than 5% were also deleted and not included in the model construction (relative frequency of medication = frequency of drug use/total sample size).

2.4. Model Building

2.4.1. Datasets and Algorithms

Extreme gradient boosting (XGB), logistic regression (LR), random forest (RF), least absolute shrinkage and selection operator (LASSO), support vector machine (SVM), and CatBoost were used to establish the ML model. R version 4.2.0 and Python version 3.7 were used for model construction and statistical analysis. Stratification was performed according to the outcome, and the data set was randomly divided into the training set and the test set at a 7:3 ratio.

2.4.2. Original Variables and Variable Selection

Information value (IV) was used as a correlation indicator, which can be used to measure the difference in the distribution of a variable between the two groups of samples to characterize the predictive ability of the variable on the outcome [31]. The threshold value of IV was set as 0.2, and variables with IV less than 0.2 were deleted. Since the chemotherapy cycle and the type of chemotherapy drugs have been confirmed to be related to the occurrence of CIM, these two variables were included in the model even though their IV were less than 0.2.

For the selected variables related to the outcome, the absolute value of the correlation coefficient was calculated to examine the collinearity, and the threshold was set as 0.8. The variable with the smaller IV was also deleted from the collinear variables exceeding the threshold.

2.4.3. Modelling Procedure

Fivefold cross-validation (CV) was used to divide the CV training set and the CV validation set inside the training set, then the optimal hyperparameter of the model was obtained using Bayesian optimization. According to the optimal hyperparameter, the model was trained again on the entire training set to obtain the final model, and further evaluated the models' prediction performance on the training set and test set.

The area under the curve (AUC), sensitivity (TPR), specificity (TNR), precision (ACC), and precision (PPV) of the receiver operating characteristic curve (ROC) were used to characterize the fitting and accuracy of the model. Population stability index (PSI) was used to measure the stability of the model in the training set and validation set [32]. (PSI < 0.1, the model is stable; PSI: 0.1~0.25, the model is slightly unstable; PSI > 0.25, the model is unstable). Hosmer–Lemeshow test was used to assess the calibration of models. The decision curve analysis (DCA) was used to evaluate the clinical utility of these models. Moreover, Coefficients of weight importance in the final model were provided to rank the feature importance.

2.4.4. Clinical Application of the Model

In order to realize the translation of research results into clinical practice, the model was presented and applied in our hospital information system (HIS) in the form of clinical decision support system (CDSS). After the first hematological examination for each patient, the doctor preliminarily confirms the medication regimen, at which point the system backstage automatically extracts the relevant data from the HIS into the model, then calculate the risk value and present it in the CDSS. "Risk Scoring" is one of the essential modules. A patient's risk score was calculated based on the final model score × 100, where low–medium risk was classified according to negative predictive value (NPV) = 0.8 and medium–high risk was classified according to positive predictive value (PPV) = 0.9. That is, the cutoff value for low–medium risk should ensure a negative prediction rate of >80% for low-risk patients, and the cutoff value for medium–high risk should ensure a positive prediction rate of >90% for high-risk patients.

To further improve the intuitiveness, accessibility, and practicability of the model, a brief description and the scoring basis of the model were presented in the CDSS, and the "Historical Trend" module was added to show the occurrence of CIM in previous admissions. In addition, the system can provide recommendations for possible prevention or intervention strategies based on the model scores.

2.5. Statistical Analysis Methods

Continuous variables were described in the form of the median (lower and upper quantile), and categorical variables were described in the form of frequency and percentage. Wilcoxon rank sum test and chi-square test were used to compare the differences between groups for continuous variables and categorical variables, respectively. $p < 0.05$ was considered statistically different.

The entire modeling procedure is shown in Figure 1.

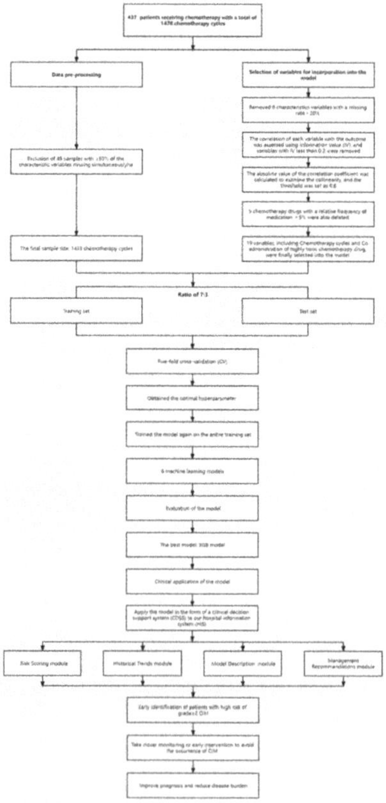

Figure 1. The entire modeling procedure.

3. Result

3.1. Description of Baseline Characteristics

On our hospital's clinical big data platform, 437 cases of WT patients receiving chemotherapy were retrieved, with a total of 1478 chemotherapy cycles. According to the inclusion and exclusion criteria, 45 samples were excluded, resulting in a final sample size of 1433. According to the National Cancer Institute Common Terminology Criteria for Ad-verse Events (CTCAE) version 5.0, grade $\geq$ 2 CIM can be defined if one of the following four criteria is met after chemotherapy: (1) WBC < 3.0 $\times$ 10^9/L; (2) absolute neutrophil count (ANC) < 1.5 $\times$ 10^9/L; (3) hemoglobin level (Hgb) < 100 g/L; (4) platelet count (PLT) < 75 $\times$ 10^9/L. The baseline characteristics of all patients and the comparison of baseline characteristics of patients in different datasets are shown in Table 1, and the comparison of baseline characteristics of patients with and without grade $\geq$ 2 CIM is shown in Table 2.

Table 1. Comparison of baseline characteristics of patients in different data sets.

Variable	ALL (N = 1433)	Training Set (N = 1003)	Test Set (N = 430)	Statistic (Z/χ^2)	*p* Value
Age (days), M (Q1–Q3)	1388 (807–2221)	1391 (810–2245)	1376 (803–2163)	−0.814	0.416
Sex				2.707	0.100
Female	674 (47.0%)	486 (48.5%)	188 (43.7%)		
Male	759 (53.0%)	517 (51.5%)	242 (56.3%)		
Weight (kg), M (Q1–Q3)	14.5 (11.5–19.0)	14.5 (11.5–19.0)	14.0 (11.5–18.0)	−0.844	0.398
BSA (m^2), M (Q1–Q3)	0.61 (0.50–0.77)	0.61 (0.50–0.77)	0.59 (0.50–0.73)	−0.823	0.410
Tumor stage				6.561	0.161
I	116 (8.1%)	84 (8.4%)	32 (7.4%)		
II	224 (15.6%)	149 (14.9%)	75 (17.4%)		
III	495 (34.5%)	360 (35.9%)	135 (31.4%)		
IV	516 (36.0%)	360 (35.9%)	156 (36.3%)		
V	82 (5.7%)	50 (5.0%)	32 (7.4%)		
Risk classification (COG)				0.045	0.831
FH	1022 (71.3%)	717 (71.5%)	305 (70.9%)		
uFH	411 (28.7%)	286 (28.5%)	125 (29.1%)		
Chemotherapy cycles, M (Q1–Q3)	4.0 (2.0–9.0)	4.0 (2.0–8.0)	5.0 (2.0–9.0)	1.408	0.159
Hematologic index, M (Q1–Q3)					
Neutrophil percentage	0.59 (0.48–0.70)	0.59 (0.48–0.70)	0.58 (0.47–0.69)	−1.087	0.277
ANC ($\times 10^9$/L)	3.50 (2.36–4.91)	3.53 (2.35–4.99)	3.49 (2.39–4.83)	−0.031	0.975
Monocyte percentage	0.04 (0.03–0.07)	0.04 (0.03–0.06)	0.04 (0.03–0.07)	−0.659	0.510
AMC ($\times 10^9$/L)	0.28 (0.19–0.39)	0.28 (0.19–0.38)	0.30 (0.19–0.41)	−1.212	0.225
P–LCR (%)	24.2 (19.0–29.8)	24.2 (19.0–29.9)	24.4 (18.9–29.7)	−0.381	0.703
MCV (fL)	82.9 (78.7–87.6)	83.0 (78.6–87.6)	82.9 (78.8–87.5)	−0.283	0.777
MCHC (g/L)	325.0 (315.0–333.0)	325.0 (316.0–333.0)	325.0(315.0–333.0)	−0.100	0.920
MCH (pg)	27.1 (25.4–28.8)	27.0 (25.4–28.9)	27.1 (25.3–28.7)	−0.423	0.673
Lymphocyte percentage (%)	0.30 (0.20–0.43)	0.30 (0.20–0.43)	0.32 (0.21–0.43)	−0.842	0.400
ALC ($\times 10^9$/L)	1.75 (0.95–3.11)	1.71 (0.97–2.99)	1.94 (0.93–3.36)	−1.220	0.223
WBC ($\times 10^9$/L)	6.10 (4.32–8.73)	6.01 (4.27–8.73)	6.28 (4.45–8.74)	−0.809	0.419
RBC ($\times 10^9$/L)	3.96 (3.52–4.36)	3.94 (3.51–4.33)	4.02 (3.53–4.41)	−1.427	0.154
RDW (%)	15.5 (14.0–17.3)	15.5 (14.0–17.3)	15.5 (14.1–17.5)	−0.551	0.582
ARD (fL)	47.0 (41.0–52.0)	47.0 (42.0–52.0)	47.0 (41.0–52.0)	−0.248	0.804
Hematocrit (%)	32.9 (29.9–35.5)	32.8 (29.9–35.3)	33.2 (29.8–35.9)	−1.257	0.209
PDW (fL)	11.0 (9.8–12.4)	11.0 (9.8–12.4)	11.1 (9.8–12.3)	−0.329	0.742
Thrombocytocrit (%)	0.31 (0.24–0.38)	0.31 (0.24–0.38)	0.32 (0.25–0.38)	−0.114	0.909
MPV (fL)	10.0 (9.3–10.7)	9.9 (9.3–10.7)	10.0 (9.3–10.7)	−0.274	0.784
PLT ($\times 10^9$/L)	297.0 (227.0–387.0)	295.0 (223.0–390.0)	304.0 (238.0–378.0)	−0.903	0.366
Hgb (g/L)	107.0 (95.0–116.0)	107.0 (95.0–116.0)	107.0 (96.0–118.0)	−1.100	0.271
SII	575.2 (334.7–951.1)	579.8 (336.0–967.5)	569.4 (333.4–917.1)	−0.398	0.691
NLR	1.97 (1.12–3.37)	2.00 (1.13–3.44)	1.85 (1.09–3.24)	−0.996	0.319
PLR	162.8 (101.4–274.4)	169.7 (102.5–276.5)	149.6 (96.8–268.8)	−1.180	0.238

Table 1. *Cont.*

Variable	ALL (N = 1433)	Training Set (N = 1003)	Test Set (N = 430)	Statistic (Z/χ^2)	p Value
Urinalysis index					
pH	6.52 (6.00–7.00)	6.52 (6.00–7.00)	6.52 (6.00–7.00)	−0.433	0.665
Biochemical index					
LDH (U/L)	286.8 (227.0–418.4)	286.5 (228.0–418.4)	287.0 (225.0–418.4)	−0.246	0.806
UA (μmol/L)	284.8 (242.0–325.0)	284.8 (237.0–325.0)	284.8 (249.5–325.0)	−1.598	0.110
TBIL (μmol/L)	6.80 (4.00–8.10)	6.80 (4.00–8.20)	6.50 (4.00–7.80)	−0.970	0.332
TP (g/L)	63.8 (60.7–67.4)	63.8 (60.9–67.3)	63.8 (60.2–67.8)	−0.313	0.755
Globulin (g/L)	22.2 (19.3–24.4)	22.2 (19.1–24.4)	22.2 (19.6–24.2)	−0.202	0.840
Albumin (g/L)	41.7 (39.6–44.8)	41.7 (39.7–44.9)	41.7 (39.3–44.7)	−0.592	0.554
ALP (U/L)	175.8(133.1–197.3)	175.8 (134.0–199.0)	175.8(132.0–193.5)	−0.637	0.524
Scr (μmol/L)	34.3 (28.0–38.0)	34.3 (28.0–38.0)	34.3 (28.0–38.5)	−1.139	0.255
ALT (U/L)	21.7 (14.4–26.0)	21.1 (14.3–25.6)	22.5 (14.6–27.1)	−1.303	0.192
AST (U/L)	35.9 (28.1–40.0)	35.4 (28.0–39.4)	37.2 (29.0–41.3)	−2.270	0.023
Grade ≥ 2 CIM				0.001	0.977
With	594 (41.5%)	416 (41.5%)	178 (41.4%)		
Without	839 (58.5%)	587 (58.5%)	252 (58.6%)		

AMC: absolute monocyte count; P–LCR: platelet–large cell ratio; MCV: mean corpuscular volume; MCHC: mean corpuscular hemoglobin concentration; MCH: mean corpuscular hemoglobin; ALC: absolute lymphocyte count; WBC: white blood cell count; RBC: red blood cell count; RDW: red blood cell distribution width; ARD: absolute value of RBC distribution; PDW: platelet distribution width; MPV: mean platelet volume; PLT: platelet count; Hgb: hemoglobin; LDH: lactate dehydrogenase; UA: uric acid; TBIL: total bilirubin; TP: total protein; ALP: alkaline phosphatase; Scr: serum creatinine; ALT: alanine transaminase; AST: aspartate transaminase.

Table 2. Comparison of baseline characteristics of patients with and without CIM.

Variable	Grade ≥ 2 CIM		Statistic	p Value
	Without (N = 594)	With (N = 839)		
Age (days), M (Q1–Q3)	1554 (905–2478)	1294 (726–2022)	−4.432	<0.001
Sex			11.368	0.001
Female	248 (41.7%)	426 (50.8%)		
Male	346 (58.3%)	413 (49.2%)		
Weight (kg), M (Q1–Q3)	16.0 (12.0–20.0)	14.0 (11.0–18.0)	−5.388	<0.001
BSA (m^2), M (Q1–Q3)	0.66 (0.52–0.80)	0.59 (0.49–0.73)	−5.385	<0.001
Tumor stage			1.915	0.751
I	51 (8.6%)	65 (7.8%)		
II	92 (15.5%)	132 (15.7%)		
III	211 (35.5%)	284 (33.9%)		
IV	211 (35.5%)	305 (36.4%)		
V	29 (4.9%)	53 (6.3%)		
Risk classification (COG)			3.011	0.083
FH	409 (68.9%)	613 (73.1%)		
uFH	185 (31.1%)	226 (26.9%)		
Chemotherapy cycles	5 (2.0–10.0)	4 (1.0–8.0)	5.574	<0.001
Hematologic index, M (Q1–Q3)				
Neutrophil percentage (%)	0.59 (0.48–0.71)	0.59 (0.47–0.69)	−1.398	0.162
ANC ($\times 10^9$/L)	3.68 (2.65–4.74)	3.38 (2.06–5.10)	−2.631	0.009
Monocyte percentage (%)	0.04 (0.03–0.06)	0.05 (0.03–0.07)	−4.919	<0.001
AMC ($\times 10^9$/L)	0.27 (0.19–0.37)	0.29 (0.20–0.40)	−2.650	0.008
P–LCR (%)	25.2 (20.1–31.3)	23.7 (18.3–28.4)	−3.726	<0.001
MCV (fL)	83.1 (79.4–87.3)	82.9 (78.1–87.7)	−0.853	0.393
MCHC (g/L)	328.0 (319.0–334.0)	322.0 (312.0–332.0)	−7.379	<0.001
MCH (pg)	27.3 (26.0–28.8)	26.9 (24.8–28.8)	−3.545	<0.001

Table 2. *Cont.*

Variable	Grade $\geq$ 2 CIM		Statistic	*p* Value
	Without (N = 594)	**With (N = 839)**	**Statistic**	***p* Value**
Lymphocyte percentage (%)	0.31 (0.19–0.43)	0.30 (0.21–0.43)	−0.562	0.574
ALC($\times 10^9$/L)	1.85 (1.01–3.01)	1.70 (0.92–3.21)	−1.272	0.203
WBC ($\times 10^9$/L)	6.35 (4.80–8.23)	5.98 (3.78–9.05)	−2.253	0.024
RBC ($\times 10^9$/L)	4.21 (3.88–4.54)	3.73 (3.29–4.18)	−13.946	<0.001
RDW (%)	14.8 (13.7–16.1)	16.1 (14.4–18.3)	−9.918	<0.001
ARD (fL)	45.0 (41.0–49.0)	47.7 (42.0–54.0)	−6.889	<0.001
Hematocrit (%)	34.9 (33.0–36.8)	30.8 (27.9–33.6)	−18.756	<0.001
PDW (fL)	11.2 (10.0–12.6)	10.8 (9.7–12.1)	−4.376	<0.001
Thrombocytocrit (%)	0.29 (0.24–0.37)	0.34 (0.25–0.40)	−5.935	<0.001
MPV (fL)	10.0 (9.4–10.9)	9.9 (9.2–10.5)	−4.009	<0.001
PLT ($\times 10^9$/L)	278.0 (218.0–345.0)	316.0 (237.0–413.0)	−5.989	<0.001
Hgb (g/L)	114.0 (107.0–121.0)	98.0 (89.0–110.0)	−19.054	<0.001
SII	516.2 (322.5–909.0)	616.0 (346.0–977.5)	−2.349	0.019
NLR	1.90 (1.13–3.65)	2.00 (1.11–3.22)	−0.650	0.516
PLR	140.0 (94.9–243.2)	180.1 (109.0–301.0)	−4.873	<0.001
Urinalysis index				
pH	6.52 (6.00–7.00)	6.52 (6.00–7.00)	−0.535	0.593
Biochemical index				
LDH (U/L)	275.0 (227.8–418.4)	297.1 (226.8–418.4)	−2.666	0.008
UA (µmol/L)	284.8 (242.0–305.1)	284.8 (241.0–335.0)	−3.076	0.002
TBIL (µmol/L)	6.86 (4.20–8.30)	6.10 (3.80–8.00)	−2.992	0.003
TP (g/L)	63.8 (62.0–68.1)	63.8 (59.9–67.0)	−4.537	<0.001
Globulin (g/L)	22.2 (19.2–23.9)	22.2 (19.5–24.8)	−1.555	0.120
Albumin (g/L)	42.5 (41.6–45.5)	41.7 (38.1–44.1)	−7.927	<0.001
ALP (U/L)	175.8 (160.0–204.8)	159.5 (118.9–188.0)	−8.870	<0.001
Scr (µmol/L)	34.3 (28.0–37.0)	34.3 (27.5–39.0)	−0.525	0.600
ALT(U/L)	21.0 (14.4–24.0)	22.0 (14.2–27.6)	−1.216	0.224
AST (U/L)	35.9 (28.8–38.1)	35.9 (28.0–42.0)	−1.416	0.157

AMC: absolute monocyte count; P–LCR: platelet–large cell ratio; MCV: mean corpuscular volume; MCHC: mean corpuscular hemoglobin concentration; MCH: mean corpuscular hemoglobin; ALC: absolute lymphocyte count; WBC: white blood cell count; RBC: red blood cell count; RDW: red blood cell distribution width; ARD: absolute value of RBC distribution; PDW: platelet distribution width; MPV: mean platelet volume; PLT: platelet count; Hgb: hemoglobin; LDH: lactate dehydrogenase; UA: uric acid; TBIL: total bilirubin; TP: total protein; ALP: alkaline phosphatase; Scr: serum creatinine; ALT: alanine transaminase; AST: aspartate transaminase.

3.2. Selection of Variables during Modeling

Matching the patient's first laboratory examination index after admission, a total of 46 clinically relevant characteristic variables were extracted, of which six characteristic variables (absolute value of basophils, percentage of basophils, cholinesterase, prealbumin, bile acids, and urine pH) had a missing rate of more than 20% and were excluded. Finally, 40 clinical characteristic variables were incorporated into the model for further screening, as shown in Table 3.

3.3. Selection of Chemotherapy Drugs

The relative frequency of the use of each chemotherapy drug is shown in Table 4, among which bleomycin, fluorouracil, topotecan, vindesine, and ifosfamide were excluded because the relative frequency of use was less than 5% and significantly different from that of other drugs. Thus, a total of nine variables including cisplatin, doxorubicin, epirubicin, carboplatin, etoposide, actinomycin D, cyclophosphamide, and vincristine, as well as the coadministration of highly toxic chemotherapy drugs, were incorporated into the final model.

Table 3. 40 clinical characteristic variables to be screened.

Variable	Missing Sample	Miss Rate (%)
Age	0	0.00
Sex	0	0.00
Weight	49	3.42
BSA	49	3.42
Tumor stage	1	0.07
Risk classification (COG)	48	3.35
Chemotherapy cycle	18	1.26
Neutrophil percentage	5	0.35
ANC	18	1.26
Monocyte percentage	8	0.56
AMC	45	3.14
P–LCR	94	6.56
MCV	3	0.21
MCHC	3	0.21
MCH	2	0.14
Lymphocyte percentage	6	0.42
ALC	16	1.12
WBC	2	0.14
RBC	1	0.07
RDW	4	0.28
ARD	102	7.12
Hematocrit	3	0.21
PDW	86	6.00
Thrombocytocrit	114	7.96
MPV	81	5.65
PLT	2	0.14
Hgb	1	0.07
SII	16	1.12
NLR	16	1.12
PLR	16	1.12
LDH	219	15.28
UA	204	14.24
TBIL	220	15.35
TP	219	15.28
Globulin	220	15.35
Albumin	219	15.28
ALP	220	15.35
Scr	205	14.31
ALT	221	15.42
AST	220	15.35

AMC: absolute monocyte count; P–LCR: platelet–large cell ratio; MCV: mean corpuscular volume; MCHC: mean corpuscular hemoglobin concentration; MCH: mean corpuscular hemoglobin; ALC: absolute lymphocyte count; WBC: white blood cell count; RBC: red blood cell count; RDW: red blood cell distribution width; ARD: absolute value of RBC distribution; PDW: platelet distribution width; MPV: mean platelet volume; PLT: platelet count; Hgb: hemoglobin; LDH: lactate dehydrogenase; UA: uric acid; TBIL: total bilirubin; TP: total protein; ALP: alkaline phosphatase; Scr: serum creatinine; ALT: alanine transaminase; AST: aspartate transaminase.

3.4. Variables Finally Selected for the Model

According to the selection criteria of predictive variables, 19 variables finally incorporated into the model are shown in Table 5. In order to improve the interpretability of the final model (XGB), we ranked the feature importance of the incorporated variables. The five variables contributing the most to the model were hemoglobin (Hgb), white blood cell count (WBC), alkaline phosphatase, coadministration of highly toxic chemotherapy drugs, and albumin, as shown in Figure 2.

Table 4. Frequency of use of each chemotherapy drug.

Drug	Relative Frequency	Frequency
Bleomycin	0.001	1
Fluorouracil	0.006	9
Topotecan	0.011	16
Vindesine	0.012	17
Ifosfamide	0.017	23
Cisplatin	0.124	172
Doxorubicin	0.126	176
Epirubicin	0.175	243
Carboplatin	0.254	354
Etoposide	0.342	476
Actinomycin D	0.348	485
Cyclophosphamide	0.504	701
Vincristine	0.703	978

Table 5. Variables finally included in the model.

Variable (n = 19)	IV
Hgb	1.770
RBC	0.708
ALP	0.422
RDW	0.392
WBC	0.372
ANC	0.369
Albumin	0.328
MCHC	0.243
PLT	0.213
Chemotherapy cycles	0.082
Coadministration of highly toxic chemotherapy drug	0.061
Cisplatin	0.028
Vincristine	0.022
Epirubicin	0.013
Carboplatin	0.007
Actinomycin D	0.005
Etoposide	0.001
Cyclophosphamide	0.000
Doxorubicin	0.000

IV: information value; Hgb: hemoglobin; RBC: red blood cell count; ALP: alkaline phosphatase; RDW: red blood cell distribution width; WBC: white blood cell count; MCHC: mean corpuscular hemoglobin concentration; PLT: platelet count.

3.5. Evaluation of the Model

The fitting effect and authenticity evaluation results of each model are shown in Figure 3, Tables 6 and 7, respectively. The results show that the XGB model has the best fitting effect, the largest AUC (training set: 0.981, test set: 0.896), good sensitivity (76.2%), and specificity (93.2%), and better stability. In the XGB model, the feature importance of each variable is shown in Figure 2. The five variables that contribute the most to the model are Hgb, WBC, alkaline phosphatase, coadministration of highly toxic chemotherapy drugs, and albumin. In addition, the XGB model showed the best calibration in the comparison of calibration curves of other models (Figure 4). DCA showed that the XGB model can contribute to clinical decision-making (Figure 5).

Feature Importance (Coefficient)

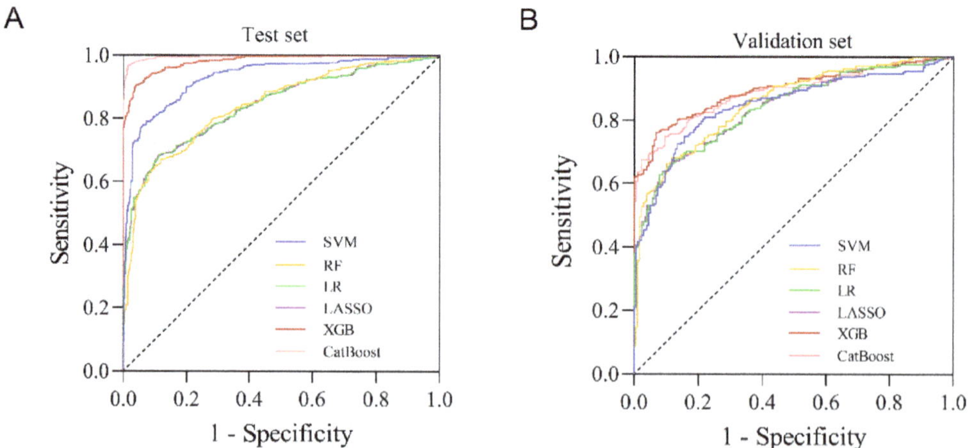

Figure 2. The ranking of feature importance in the XGB model. Briefly, the importance weight of a feature is the sum of the number of its occurrences in all decision trees. In other words, the more a feature is used to build a decision tree in the model, the higher its importance weight will be. Hgb: hemoglobin; WBC: white blood cell count; ALP: alkaline phosphatase; RBC: red blood cell count; MCHC: mean corpuscular hemoglobin concentration; PLT: platelet count; RDW: red blood cell distribution width.

Figure 3. ROC curve of six ML models for predicting grade ≥ 2 CIM. (**A**) In the test set; (**B**) in the validation set. SVM: support vector machine; RF: random forest; LR: logistic regression; LASSO: least absolute shrinkage and selection operator; XGB: extreme gradient boosting.

Table 6. Evaluation of fitting effect of each model.

Model	AUC		PSI
	Training Set	**Test Set**	
XGB	0.981	0.896	0.033
CatBoost	0.996	0.888	0.086
RF	0.842	0.856	0.015
SVM	0.930	0.849	0.066
LR	0.843	0.842	0.007
LASSO	0.843	0.842	0.007

XGB: extreme gradient boosting; LR: logistic regression; RF: random forest; LASSO: least absolute shrinkage and selection operator; SVM: support vector machine; PSI: population stability index.

Table 7. Evaluation of authenticity of each model.

Model	Best Cutoff	TPR	TNR	ACC	PPV
XGB	0.529	76.2%	93.3%	83.3%	94.1%
RF	0.569	68.3%	88.2%	76.5%	89.1%
CatBoost	0.585	75.0%	90.4%	81.4%	91.7%
SVM	0.581	75.0%	84.3%	78.8%	87.1%
LR	0.687	66.3%	88.8%	75.6%	89.3%
LASSO	0.685	66.3%	88.8%	75.6%	89.3%

TPR: sensitivity; TNR: specificity; ACC: precision; PPV: precision; XGB: extreme gradient boosting; LR: logistic regression; RF: random forest; LASSO: least absolute shrinkage and selection operator; SVM: support vector machine.

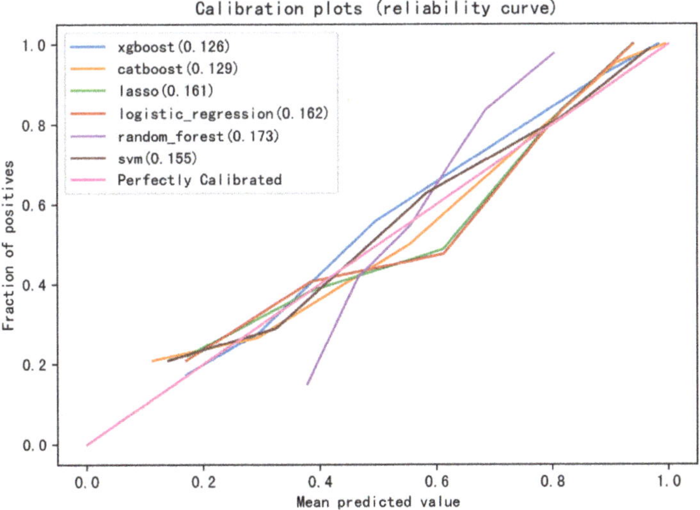

Figure 4. Calibration curves of the six ML models.

3.6. Clinical Application of the Model

Through a series of evaluations of the model, the XGB model with the best predictive efficacy was selected, presented, and applied in our hospital's HIS in the form of CDSS. It includes modules such as the risk scoring and scoring basis of grade ≥ 2 CIM, model description, historical trend of the previous occurrence of CIM, and management recommendations (Figure 6). The predictive model is currently running smoothly in the HIS. Moreover, to better demonstrate how our model works in reality and to further elaborate on the clinical applicability of the model, we ran the model in our hospital HIS to assess the risk of CIM in a particular child (Supplementary Materials).

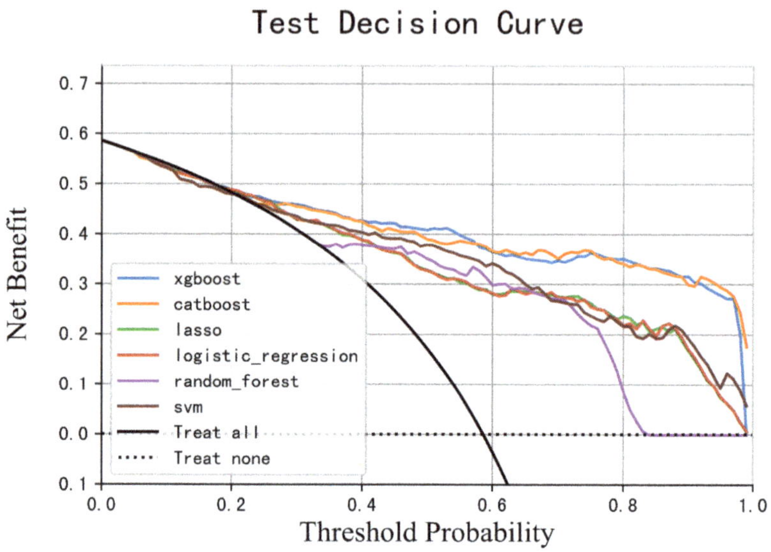

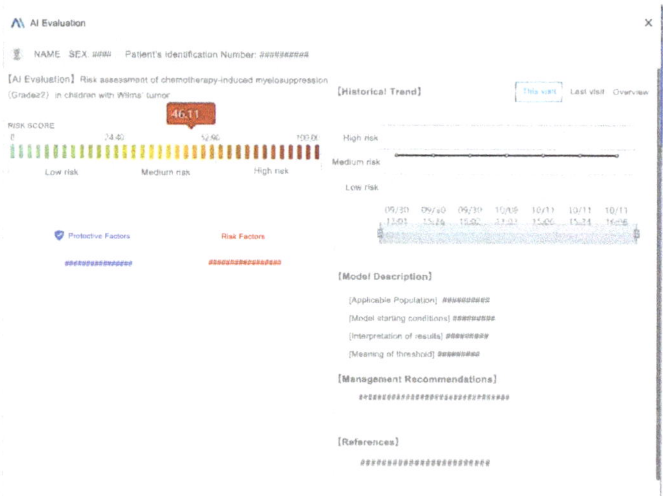

Figure 5. Decision curve analysis of the six ML models.

Figure 6. The interface of the CIM prediction model in the form of CDSS applied in our hospital HIS. AI Evaluation: the "AI Evaluation" module shows the risk scores of patients with grade ≥ 2 CIM calculated by the model, with the corresponding "protective factors" and "risk factors" listed below. **Historical Trend:** the "Historical Trends" module records the occurrence of CIM in previous chemotherapy cycles. **Model Description:** this module provides a detailed description of the applicable conditions and the model results. **Management Recommendations:** according to the prediction results of the model, the management suggestions automatically output by the system backstage are displayed in this module. **References:** this module presents some references.

4. Discussion

4.1. CIM Is Not Rare during the Treatment of Children with WT

Chemotherapy is one of the important means of treating tumors. Currently, most chemotherapy drugs exert their effects through cytotoxicity. Cells with strong proliferative activity may be more sensitive to chemotherapy drugs, making drugs more likely to damage hematopoietic stem cells or blood cell precursors, leading to severe CIM [27,33]. A clinical consensus is that grade $\geq$ 2 CIM requires close monitoring and even timely intervention. Identifying patients with a high risk of grade $\geq$ 2 CIM before administration of chemotherapy drugs can guide doctors to timely administer granulocyte colony-stimulating factor (G-CSF) and other drugs to prevent the occurrence of CIM during the process of closely monitoring the changes in blood cells levels, which avoids the interruption of the chemotherapy course and even the occurrence of more serious complications caused by CIM [34,35]. It is also why we choose the occurrence of grade $\geq$ 2 CIM as the outcome indicator. In this study, grade $\geq$ 2 CIM occurred in 58.5% (839/1433) chemotherapy cycles. Although Castagnola et al. reported that the incidence of FN in children with central nervous system tumors was 27% [36], the outcome of the study was FN rather than CIM, and the different types of tumors studied may also affect the incidence of FN, so our findings cannot be compared with their study. Other studies have reported that the incidence of FN in solid tumors is 13–21%, while FN in hematologic tumors is about 33% [37–39]. Whereas most of the outcome indicators in these studies were FN, and the subjects were adults, which could not be compared with the incidence of CIM in our study. However, this also emphasizes that the incidence of CIM in children with solid tumors is still unknown and more studies are needed to fill in the gaps. In addition, more than half of the chemotherapy cycles in our study presented grade $\geq$ 2 CIM, which fully demonstrates that CIM is not rare in treating pediatric tumors, especially WT, and the development of early prediction models for CIM in children with solid tumors is indeed necessary.

4.2. Contribution of Variables to Model Prediction Results

According to the ranking of IV, 19 variables were finally included in the model. Studies have shown that chemotherapy cycles and regimens can affect the occurrence of CIM, so even if the IV of those relevant variables were less than 0.2, they were still included in our model. Feature importance is an indicator to measure the contribution of each variable to the model's predictive result (Figure 1). In the XGB model, the Hgb level ranked first in the feature importance ranking. This seems to differ from what most studies have reported. More than one study reported that baseline WBC and ANC levels, but not Hgb levels, were the most critical risk factors for CIM or FN [14,40,41]. On the contrary, it has also been reported that a low baseline level of Hgb was associated with CIM in elderly tumor patients [42]. It has been reported that in addition to Hgb, the decrease of alkaline phosphatase, red blood cell count (RBC), and average hemoglobin concentration and the increase of red blood cell distribution width (RDW) can also reflect anemia or hematopoietic abnormalities to some extent [27,33]. Herein, except for RDW, the above five indicators were lower in the CIM group than in the without-CIM group. This may be because most of the children in this study underwent surgery before chemotherapy, and inevitable intraoperative bleeding and the consumption of the tumor on the body led to a lower baseline Hgb or RBC level before chemotherapy. While stimulated by blood loss, the proliferation of bone marrow hematopoietic cells may be more active, thus more likely to be attacked by chemotherapy drugs.

Although Aagaard et al. did not find that low levels of WBC and ANC were associated with the development of bone marrow suppression in their study [43], most studies have shown that low baseline WBC and ANC levels are risk factors for myelosuppression [12–14], and our findings are consistent with them: the low baseline level of WBC and ANC in the XGB model strongly predicts CIM. Due to the short cycle life of granulocytes, it is difficult for haemopoietic stem cells or haemopoietic microenvironment damaged by chemotherapy drugs to generate new granulocytes to replace the consumed granulocytes [27,30]. Hence, a low ANC level is often the earliest manifestation of CIM. Lower baseline WBC or ANC levels mean lower granulocyte reserves, meaning CIM is more likely to occur.

In addition, the low baseline level of albumin may be related to the nutritional status of patients, thus affecting the occurrence of CIM, which is also consistent with the result of another study [44].

Moreover, different patients have different chemotherapy regimens [45–47], and different chemotherapy regimens incorporate chemotherapy drugs with different degrees of hematological toxicity [48,49], so treating each chemotherapy regimen as a variable is unrealistic. As a result, we added the variable "Coadministration of highly toxic chemotherapy drugs" to investigate the effect of highly toxic chemotherapy drugs on the risk of developing CIM. Although its IV was small, its feature importance ranked fourth in the XGB model. It validates that chemotherapy drugs with high hematotoxicity are indeed more likely to cause CIM. Unexpectedly, the ranking of feature importance of chemotherapy drugs in the model seems to be different from our understanding of hematological toxicity of chemotherapy drugs. Low hematologic toxicity drugs such as cisplatin and vincristine ranked even higher than high hematologic toxicity drugs such as doxorubicin and cyclophosphamide. This may be because drugs such as cisplatin and vincristine are more frequently used in chemotherapy regimens for children with WT and are often used in combination with other highly toxic chemotherapeutic drugs. Thus, the ranking of the feature importance of these variables may differ slightly from our general understanding of CIM risk factors. Nevertheless, the XGB model developed in this study still performed surprisingly well in predicting grade ≥ 2 CIM.

4.3. XGB Model Has Good Predictive Performance for Grade ≥ 2 CIM

Since the first mechanism model based on pharmacokinetics and pharmacodynamics was developed, other mathematical models for predicting CIM or investigating the relationship between a chemotherapy drug and changes in blood cell levels have been developed one after another. These mathematical models can simulate hematopoiesis, granulocytopoiesis, myelosuppression, and leukemia cytodynamics. Recently published reviews have provided a comprehensive overview and summary of various models [50,51] and studies have reported associations between the occurrence of CIM and genomic specificity [52–54]. Of these models, the maximum AUC of the model predicting FN or CIM occurrence is only 0.83. Notably, after evaluating the fitting effects of several models used in our study, we found that the XGB model had an AUC of up to 0.981 in the training set and 0.896 in the test set, with satisfactory sensitivity and specificity, as well as good stability. The calibration curve and DCA also suggested that the XGB model had good calibration and could promote clinical decision-making. In addition to good predictive performance, the XGB model we developed has other advantages: the modeling variables we selected were from the baseline data of hematological and biochemical tests before chemotherapy, and the information about the proposed chemotherapy regimen. These variables are readily available prior to drug administration. Children do not need to bear the expensive cost such as genomic marker detection, or the burden and pain caused by frequent laboratory tests.

4.4. Application of CIM Prediction Model in Clinical Practice

Translating clinical research results to clinical applications has been a significant challenge. The clinical decision support system (CDSS) helps doctors improve and enhance the efficiency of decision-making by providing systematic medical knowledge and in depth analysis of medical records through a human–computer interaction model, thereby improving the quality of medical care [55]. CDSS is a vital bridge to facilitate the translation of clinical research into clinical application.

Considering the application scenarios of the CIM prediction model, we present the final model in the form of CDSS in our hospital HIS. Patients undergo hematological and biochemical tests after admission. The doctor then specifies the current chemotherapy regimen, followed by the system backstage immediately extracting the relevant data calculating the CIM risk score through the model and outputting it via CDSS. Doctors can make appropriate treatment plans based on the predicted results. Despite the risk score module, the "Management Recommendations" module and the "Historical Trend" module that records the occurrence of CIM in previous chemotherapy cycles can greatly help doctors make better clinical decisions. To better demonstrate how our model works in reality and to further elaborate on the clinical applicability of the model, we ran the model in our hospital HIS to assess the risk of CIM in a particular child. Please refer to the Supplementary Materials (Figure S1) for sample cases and model results output interface.

By applying this approach, firstly, doctors can identify high-risk patients early and adopt appropriate management plans to improve patients' prognosis. Secondly, the model calculations and results output are carried out automatically by the system backstage, eliminating the inconvenience of other predictive modeling tools requiring manual data input for the corresponding variables. Thirdly, the relevant data of CIM occurrence in each admission will be automatically stored in the system, which will be helpful for other related clinical studies in the future. All of the above fully reflect the practicability, accessibility, and high predictive efficiency of our model in clinical application.

4.5. Limitations and Prospects

However, our study also has some limitations. Firstly, the nature of the retrospective study may inevitably introduce some selection bias; secondly, the risk factors related to CIM, such as prealbumin, BMI, bile acid, bilirubin, etc., which have been reported in other studies [40,56], were not included in the model due to a large amount of missing data. This may be because doctors or patients have insufficient awareness of CIM and do not conduct relevant tests. Thirdly, the dynamic changes in blood cells may be able to predict the specific time when CIM occurs and finding this time point will help doctors develop more accurate prevention strategies for CIM. However, these data were also missing in this study. In addition, our sample size needs to be expanded to make more accurate predictions for different grades of CIM. Furthermore, our model has been successfully piloted in HIS with CDSS, and more data needs to be collected prospectively to further verify the model's accuracy. Finally, different types of tumors may affect the occurrence of CIM, but only children with WT were included in this study. Therefore, the models that can be extended to other pediatric malignant solid tumors need further development. To summarize, a prospective clinical study with large samples and regularly collected data needs to be carried out. We are currently conducting animal experiments related to CIM in order to accurately predict the CIM by finding other more readily available indicators. We intend to validate these indicators in prospective clinical studies and incorporate them into the model for continuous calibration and optimization. Despite these limitations, to our knowledge, this study is the first to use ML algorithms to establish a predictive model for CIM in children with WT, achieving better predictive effects than other pharmacokinetic or mathematical models. Based on the construction method and clinical application approach of this ML model, a CIM prediction model that can be extended to other pediatric malignancies and facilitates widespread clinical applications can be expected.

5. Conclusions

The incidence of grade ≥ 2 CIM was not low in children with WT, which needs more attention. This study developed an ML-based prediction model to predict the risk of grade ≥ 2 CIM in WT children for the first time. The model has good predictive performance and stability and is also convenient for clinical application, which will help doctors identify patients at high risk of CIM earlier, and develop and implement individualized preventive medication strategies, thus reducing the disease burden and economic burden of CIM in children with WT. Based on this modeling and application approach, the extension of CIM prediction models to other pediatric malignancies is expected.

Supplementary Materials: The following supporting information can be downloaded at: https://www.mdpi.com/article/10.3390/cancers15041078/s1. Figure S1. Case example: the interface of CIM prediction model output results. The text in the figure has been translated into English.

Author Contributions: M.L.: Conceptualization, Methodology, Validation, Formal analysis, Investigation Resources, Data Curation, Writing—Original Draft, Writing—Review and Editing, Visualization. Q.W.: Methodology, Validation, Formal analysis, Data Curation, Writing—Review and Editing, Visualization, P.L.: Methodology, Data Curation, Validation, Resources. D.Z.: Resources, Investigation. Y.H.: Resources, Investigation. F.L.: Resources, Investigation. X.L.: Resources, Investigation. T.L.: Resources, Investigation. G.W.: Resources, Supervision. D.H.: Conceptualization, Methodology, Resources, Validation, Writing—Review and Editing, Supervision, Project administration, Funding acquisition. All authors have read and agreed to the published version of the manuscript.

Funding: This study was supported by Contracting Project of Chongqing Talent Program: No. cstc2021ycjh bgzxm0253.

Institutional Review Board Statement: The study was conducted according to the guidelines of the Declaration of Helsinki and approved by the Ethics Committee of the Children's Hospital of Chongqing Medical University (protocol code: 2022-170, approved on 15 April 2022).

Informed Consent Statement: Informed consent was obtained from all subjects involved in this study.

Data Availability Statement: The datasets used and/or analyzed during the current study are available from the corresponding author upon reasonable request.

Acknowledgments: We thank Shanghai Synyi Medical Technology Co., Ltd. for providing the data analysis and statistical platform.

Conflicts of Interest: The authors declare no conflict of interest.

References

1. Kalapurakal, J.A.; Dome, J.S.; Perlman, E.; Malogolowkin, M.; Haase, G.M.; Grundy, P.; Coppes, M.J. Management of Wilms tumour: Current practice and future goals. *Lancet Oncol.* **2004**, *5*, 37–46. [CrossRef] [PubMed]
2. Brok, J.; Treger, T.D.; Gooskens, S.L.; van den Heuvel-Eibrink, M.M.; Pritchard-Jones, K. Biology and treatment of renal tumour in childhood. *Eur. J. Cancer* **2016**, *68*, 179–195. [CrossRef]
3. Malogolowkin, M.; Spreafico, F.; Dome, J.S.; van Tinteren, H.; Pritchard-Jones, K.; van den Heuvel-Eibrink, M.; Bergeron, C.; de Kraker, J.; Graf, N.; On behalf of the COG Renal Tumors Committee and the SIOP Renal Tumor Study Group. Incidence and outcomes of patients with late recurrence of Wilms' tumor. *Pediatr. Blood Cancer* **2013**, *60*, 1612–1615. [CrossRef]
4. Barreto, J.N.; McCullough, K.; Ice, L.L.; Smith, J.A. Antineoplastic Agents and the Associated Myelosuppressive Effects. *J. Pharm. Pr.* **2014**, *27*, 440–446. [CrossRef]
5. Bryer, E.; Henry, D. Chemotherapy-induced anemia: Etiology, pathophysiology, and implications for contemporary practice. *Int. J. Clin. Transfus. Med.* **2018**, *6*, 21–31. [CrossRef]
6. Kuter, D.J. Managing thrombocytopenia associated with cancer chemotherapy. *Oncology* **2015**, *29*, 282–294.
7. Taylor, S.J.; Duyvestyn, J.M.; Dagger, S.A.; Dishington, E.J.; Rinaldi, C.A.; Dovey, O.M.; Vassiliou, G.S.; Grove, C.S.; Langdon, W.Y. Preventing chemotherapy-induced myelosuppression by repurposing the FLT3 inhibitor quizartinib. *Sci. Transl. Med.* **2017**, *9*, eaam8060. [CrossRef]
8. Lennan, E.; Roe, H. Role of nurses in the assessment and management of chemotherapy-related side effects in cancer patients. *Nurs. Res. Rev.* **2014**, *4*, 103. [CrossRef]
9. Epstein, R.S.; Aapro, M.S.; Roy, U.K.B.; Salimi, T.; Krenitsky, J.; Leone-Perkins, M.L.; Girman, C.; Schlusser, C.; Crawford, J. Patient Burden and Real-World Management of Chemotherapy-Induced Myelosuppression: Results from an Online Survey of Patients with Solid Tumors. *Adv. Ther.* **2020**, *37*, 3606–3618. [CrossRef]
10. Talcott, J.A.; Siegel, R.D.; Finberg, R.; Goldman, L. Risk assessment in cancer patients with fever and neutropenia: A prospective two-center validation of a prediction rule. *J. Clin. Oncol.* **1992**, *10*, 316–322. [CrossRef]
11. Epstein, R.S.; Weerasinghe, R.K.; Parrish, A.S.; Krenitsky, J.; Sanborn, R.E.; Salimi, T. Real-world burden of chemotherapy-induced myelosuppression in patients with small cell lung cancer: A retrospective analysis of electronic medical data from community cancer care providers. *J. Med. Econ.* **2022**, *25*, 108–118. [CrossRef] [PubMed]
12. Lyman, G.H.; Morrison, V.A.; Dale, D.C.; Crawford, J.; Delgado, D.J.; Fridman, M. Risk of Febrile Neutropenia among Patients with Intermediate-grade non-Hodgkin's Lymphoma Receiving CHOP Chemotherapy. *Leuk. Lymphoma* **2003**, *44*, 2069–2076. [CrossRef] [PubMed]
13. Lyman, G.H.; Delgado, D.J. Risk and timing of hospitalization for febrile neutropenia in patients receiving CHOP, CHOP-R, or CNOP chemotherapy for intermediate-grade non-Hodgkin lymphoma. *Cancer* **2003**, *98*, 2402–2409. [CrossRef] [PubMed]
14. Lyman, G.H.; Kuderer, N.M.; Crawford, J.; Wolff, D.A.; Culakova, E.; Poniewierski, M.S.; Dale, D.C. Predicting individual risk of neutropenic complications in patients receiving cancer chemotherapy. *Cancer* **2010**, *117*, 1917–1927. [CrossRef]
15. Lyman, G.H. Impact of Chemotherapy Dose Intensity on Cancer Patient Outcomes. *J. Natl. Compr. Cancer Netw.* **2009**, *7*, 99–108. [CrossRef]
16. Friberg, L.E.; Henningsson, A.; Maas, H.; Nguyen, L.; Karlsson, M.O. Model of Chemotherapy-Induced Myelosuppression With Parameter Consistency Across Drugs. *J. Clin. Oncol.* **2002**, *20*, 4713–4721. [CrossRef]

7. Quartino, A.L.; Friberg, L.; Karlsson, M.O. A simultaneous analysis of the time-course of leukocytes and neutrophils following docetaxel administration using a semi-mechanistic myelosuppression model. *Investig. New Drugs* **2010**, *30*, 833–845. [CrossRef]

8. Sanjuan, V.M.; Buil-Bruna, N.; Garrido, M.J.; Soto, E.; Trocóniz, I.F. Semimechanistic Cell-Cycle Type–Based Pharmacokinetic/Pharmacodynamic Model of Chemotherapy-Induced Neutropenic Effects of Diflomotecan under Different Dosing Schedules. *Experiment* **2015**, *354*, 55–64. [CrossRef]

9. Wallin, J.E.; Friberg, L.; Karlsson, M.O. Model-Based Neutrophil-Guided Dose Adaptation in Chemotherapy: Evaluation of Predicted Outcome with Different Types and Amounts of Information. *Basic Clin. Pharmacol. Toxicol.* **2010**, *106*, 234–242. [CrossRef]

20. Netterberg, I.; Nielsen, E.I.; Friberg, L.E.; Karlsson, M.O. Model-based prediction of myelosuppression and recovery based on frequent neutrophil monitoring. *Cancer Chemother. Pharmacol.* **2017**, *80*, 343–353. [CrossRef]

21. Pfeil, A.M.; Vulsteke, C.; Paridaens, R.; Dieudonne, A.-S.; Pettengell, R.; Hatse, S.; Neven, P.; Lambrechts, D.; Szucs, T.D.; Schwenkglenks, M.; et al. Multivariable regression analysis of febrile neutropenia occurrence in early breast cancer patients receiving chemotherapy assessing patient-related, chemotherapy-related and genetic risk factors. *BMC Cancer* **2014**, *14*, 201. [CrossRef] [PubMed]

22. Park, K.; Kim, Y.; Son, M.; Chae, D.; Park, K. A Pharmacometric Model to Predict Chemotherapy-Induced Myelosuppression and Associated Risk Factors in Non-Small Cell Lung Cancer. *Pharmaceutics* **2022**, *14*, 914. [CrossRef] [PubMed]

23. Greener, J.G.; Kandathil, S.M.; Moffat, L.; Jones, D.T. A guide to machine learning for biologists. *Nat. Rev. Mol. Cell Biol.* **2021**, *23*, 40–55. [CrossRef]

24. Gould, M.K.; Huang, B.Z.; Tammemagi, M.C.; Kinar, Y.; Shiff, R. Machine Learning for Early Lung Cancer Identification Using Routine Clinical and Laboratory Data. *Am. J. Respir. Crit. Care Med.* **2021**, *204*, 445–453. [CrossRef]

25. Radakovich, N.; Nagy, M.; Nazha, A. Machine learning in haematological malignancies. *Lancet Haematol.* **2020**, *7*, e541–e550. [CrossRef] [PubMed]

26. Chernbumroong, S.; Johnson, J.; Gupta, N.; Miller, S.; McCormack, F.X.; Garibaldi, J.M.; Johnson, S.R. Machine learning can predict disease manifestations and outcomes in lymphangioleiomyomatosis. *Eur. Respir. J.* **2020**, *57*, 2003036. [CrossRef]

27. Shibahara, T.; Ikuta, S.; Muragaki, Y. Machine-Learning Approach for Modeling Myelosuppression Attributed to Nimustine Hydrochloride. *JCO Clin. Cancer Inform.* **2018**, *2*, 1–21. [CrossRef] [PubMed]

28. Zhu, S.; Cheng, Z.; Hu, Y.; Chen, Z.; Zhang, J.; Ke, C.; Yang, Q.; Lin, F.; Chen, Y.; Wang, J. Prognostic Value of the Systemic Immune-Inflammation Index and Prognostic Nutritional Index in Patients With Medulloblastoma Undergoing Surgical Resection. *Front. Nutr.* **2021**, *8*, 754958. [CrossRef] [PubMed]

29. Mac Manus, M.; Lamborn, K.; Khan, W.; Varghese, A.; Graef, L.; Knox, S. Radiotherapy-associated neutropenia and thrombocytopenia: Analysis of risk factors and development of a predictive model. *Blood* **1997**, *89*, 2303–2310. [CrossRef]

30. Carey, P.J. Drug-Induced Myelosuppression. *Drug Saf.* **2003**, *26*, 691–706. [CrossRef]

31. Gong, Y.-F.; Zhu, L.-Q.; Li, Y.-L.; Zhang, L.-J.; Xue, J.-B.; Xia, S.; Lv, S.; Xu, J.; Li, S.-Z. Identification of the high-risk area for schistosomiasis transmission in China based on information value and machine learning: A newly data-driven modeling attempt. *Infect. Dis. Poverty* **2021**, *10*, 88. [CrossRef]

32. Yurdakul, B.; Naranjo, J. Statistical properties of the population stability index. *J. Risk Model Valid.* **2020**. [CrossRef]

33. Friberg, L.E. Mechanistic Models for Myelosuppression. *Investig. New Drugs* **2003**, *21*, 183–194. [CrossRef]

34. Lyman, G.H.; Kuderer, N.M.; Aapro, M. Improving Outcomes of Chemotherapy: Established and Novel Options for Myeloprotection in the COVID-19 Era. *Front. Oncol.* **2021**, *11*, 697908. [CrossRef]

35. Straka, C.; Oduncu, F.; Hinke, A.; Einsele, H.; Drexler, E.; Schnabel, B.; Arseniev, L.; Walther, J.; König, A.; Emmerich, B. Responsiveness to G-CSF before leukopenia predicts defense to infection in high-dose chemotherapy recipients. *Blood* **2004**, *104*, 1989–1994. [CrossRef]

36. Castagnola, E.; Garrè, M.L.; Bertoluzzo, L.; Pignatelli, S.; Pavanello, M.; Caviglia, I.; Caruso, S.; Bagnasco, F.; Moroni, C.; Tacchella, A.; et al. Epidemiology of Febrile Neutropenia in Children With Central Nervous System Tumor. *J. Pediatr. Hematol.* **2011**, *33*, e310–e315. [CrossRef] [PubMed]

37. Rapoport, B.L.; Aapro, M.; Paesmans, M.; Van Eeden, R.; Smit, T.; Krendyukov, A.; Klastersky, J. Febrile neutropenia (FN) occurrence outside of clinical trials: Occurrence and predictive factors in adult patients treated with chemotherapy and an expected moderate FN risk. Rationale and design of a real-world prospective, observational, multinational study. *BMC Cancer* **2018**, *18*, 917. [CrossRef]

38. Ba, E.A.Y.; Shi, Y.; Jiang, W.; Feng, J.; Cheng, Y.; Xiao, L.; Zhang, Q.; Qiu, W.; Xu, B.; Xu, R.; et al. Current management of chemotherapy-induced neutropenia in adults: Key points and new challenges. *Cancer Biol. Med.* **2020**, *17*, 896–909. [CrossRef] [PubMed]

39. Intragumtornchai, T.; Sutheesophon, J.; Sutcharitchan, P.; Swasdikul, D. A Predictive Model for Life-Threatening Neutropenia and Febrile Neutropenia after the First Course of CHOP Chemotherapy in Patients with Aggressive Non-Hodgkin's Lymphoma. *Leuk. Lymphoma* **2000**, *37*, 351–360. [CrossRef]

40. Razzaghdoust, A.; Mofid, B.; Moghadam, M. Development of a simplified multivariable model to predict neutropenic complications in cancer patients undergoing chemotherapy. *Support. Care Cancer* **2018**, *26*, 3691–3699. [CrossRef]

41. Lyman, G.H.; Abella, E.; Pettengell, R. Risk factors for febrile neutropenia among patients with cancer receiving chemotherapy: A systematic review. *Crit. Rev. Oncol.* **2014**, *90*, 190–199. [CrossRef] [PubMed]

42. Balducci, L. Myelosuppression and its consequences in elderly patients with cancer. *Oncology* **2003**, *17*, 27–32. [PubMed]
43. Aagaard, T.; Roen, A.; Reekie, J.; Daugaard, G.; Brown, P.D.N.; Specht, L.; Sengeløv, H.; Mocroft, A.; Lundgren, J.; Helleberg, M. Development and Validation of a Risk Score for Febrile Neutropenia After Chemotherapy in Patients With Cancer: The FENCE Score. *JNCI Cancer Spectr.* **2018**, *2*, pky053. [CrossRef] [PubMed]
44. Pérez-Pitarch, A.; Guglieri-López, B.; Nacher, A.; Merino, V.; Merino-Sanjuán, M. Impact of Undernutrition on the Pharmacokinetics and Pharmacodynamics of Anticancer Drugs: A Literature Review. *Nutr. Cancer* **2017**, *69*, 555–563. [CrossRef] [PubMed]
45. Metzger, M.L.; Dome, J.S. Current Therapy for Wilms' Tumor. *Oncologist* **2005**, *10*, 815–826. [CrossRef]
46. Fernandez, C.V.; Mullen, E.A.; Chi, Y.-Y.; Ehrlich, P.F.; Perlman, E.; Kalapurakal, J.A.; Khanna, G.; Paulino, A.C.; Hamilton, T.E.; Gow, K.W.; et al. Outcome and Prognostic Factors in Stage III Favorable-Histology Wilms Tumor: A Report From the Children's Oncology Group Study AREN0532. *J. Clin. Oncol.* **2018**, *36*, 254–261. [CrossRef]
47. Dix, D.B.; Seibel, N.L.; Chi, Y.-Y.; Khanna, G.; Gratias, E.; Anderson, J.R.; Mullen, E.A.; Geller, J.I.; Kalapurakal, J.A.; Paulino, A.C.; et al. Treatment of Stage IV Favorable Histology Wilms Tumor With Lung Metastases: A Report From the Children's Oncology Group AREN0533 Study. *J. Clin. Oncol.* **2018**, *36*, 1564–1570. [CrossRef] [PubMed]
48. Moreau, M.; Klastersky, J.; Schwarzbold, A.; Muanza, F.; Georgala, A.; Aoun, M.; Loizidou, A.; Barette, M.; Costantini, S.; Delmelle, M.; et al. A general chemotherapy myelotoxicity score to predict febrile neutropenia in hematological malignancies. *Ann. Oncol.* **2009**, *20*, 513–519. [CrossRef]
49. Pettengell, R.; Bosly, A.; Szucs, T.D.; Jackisch, C.; Leonard, R.; Paridaens, R.; Constenla, M.; Schwenkglenks, M.; for the Impact of Neutropenia in Chemotherapy—European Study Group (INC-EU). Multivariate analysis of febrile neutropenia occurrence in patients with non-Hodgkin lymphoma: Data from the INC-EU Prospective Observational European Neutropenia Study. *Br. J. Haematol.* **2009**, *144*, 677–685. [CrossRef]
50. Craig, M. Towards Quantitative Systems Pharmacology Models of Chemotherapy-Induced Neutropenia. *CPT Pharmacomet. Syst. Pharmacol.* **2017**, *6*, 293–304. [CrossRef]
51. Fornari, C.; O'Connor, L.O.; Yates, J.W.; Cheung, S.A.; Jodrell, D.I.; Mettetal, J.T.; Collins, T.A. Understanding Hematological Toxicities Using Mathematical Modeling. *Clin. Pharmacol. Ther.* **2018**, *104*, 644–654. [CrossRef]
52. Chen, Z.-Y.; Zhu, Y.-H.; Zhou, L.-Y.; Shi, W.-Q.; Qin, Z.; Wu, B.; Yan, Y.; Pei, Y.-W.; Chao, N.-N.; Zhang, R.; et al. Association Between Genetic Polymorphisms of Metabolic Enzymes and Azathioprine-Induced Myelosuppression in 1419 Chinese Patients: A Retrospective Study. *Front. Pharmacol.* **2021**, *12*, 672769. [CrossRef] [PubMed]
53. Björn, N.; Badam, T.V.S.; Spalinskas, R.; Brandén, E.; Koyi, H.; Lewensohn, R.; De Petris, L.; Lubovac-Pilav, Z.; Sahlén, P.; Lundeberg, J.; et al. Whole-genome sequencing and gene network modules predict gemcitabine/carboplatin-induced myelosuppression in non-small cell lung cancer patients. *NPJ Syst. Biol. Appl.* **2020**, *6*, 25. [CrossRef] [PubMed]
54. Ren, W.; Zhou, C.; Liu, Y.; Su, K.; Jia, L.; Chen, L.; Li, M.; Ma, J.; Zhou, W.; Zhang, S.; et al. Genetic associations of docetaxel-based chemotherapy-induced myelosuppression in Chinese Han population. *J. Clin. Pharm. Ther.* **2019**, *45*, 354–364. [CrossRef]
55. Belard, A.; Buchman, T.; Forsberg, J.; Potter, B.K.; Dente, C.J.; Kirk, A.; Elster, E. Precision diagnosis: A view of the clinical decision support systems (CDSS) landscape through the lens of critical care. *J. Clin. Monit. Comput.* **2016**, *31*, 261–271. [CrossRef] [PubMed]
56. Sigurdsson, V.; Haga, Y.; Takei, H.; Mansell, E.; Okamatsu-Haga, C.; Suzuki, M.; Radulovic, V.; van der Garde, M.; Koide, S.; Soboleva, S.; et al. Induction of blood-circulating bile acids supports recovery from myelosuppressive chemotherapy. *Blood Adv.* **2020**, *4*, 1833–1843. [CrossRef] [PubMed]

Article

Development and Evaluation of MR-Based Radiogenomic Models to Differentiate Atypical Lipomatous Tumors from Lipomas

Sarah C. Foreman [1,*,†], Oscar Llorián-Salvador [2,3,4,†], Diana E. David [3], Verena K. N. Rösner [1], Jon F. Rischewski [5], Georg C. Feuerriegel [1], Daniel W. Kramp [1], Ina Luiken [1], Ann-Kathrin Lohse [6], Jurij Kiefer [7], Carolin Mogler [8], Carolin Knebel [9], Matthias Jung [10], Miguel A. Andrade-Navarro [4], Burkhard Rost [3], Stephanie E. Combs [2], Marcus R. Makowski [1], Klaus Woertler [1], Jan C. Peeken [2,11,12,‡] and Alexandra S. Gersing [5,‡]

1 Department of Radiology, Klinikum Rechts der Isar, Technische Universität München, Ismaninger Straße 22, 81675 Munich, Germany
2 Department of Radiation Oncology, Klinikum Rechts der Isar, Technische Universität München, Ismaninger Straße 22, 81675 Munich, Germany
3 Department of Informatics, Bioinformatics and Computational Biology—i12, Technische Universität München, Boltzmannstr. 3, 85748 Munich, Germany
4 Institute of Organismic and Molecular Evolution, Johannes Gutenberg University Mainz, Hanns-Dieter-Hüsch-Weg 15, 55128 Mainz, Germany
5 Department of Diagnostic and Interventional Neuroradiology, University Hospital Munich (LMU), Marchioninistrasse 15, 81377 Munich, Germany
6 Department of Radiology, University Hospital Munich (LMU), Marchioninistrasse 15, 81377 Munich, Germany
7 Department of Plastic Surgery, University Hospital Freiburg, University of Freiburg, Hugstetterstraße 55, 79106 Freiburg im Breisgau, Germany
8 Institute of Pathology, Klinikum Rechts der Isar, Technische Universität München, Ismaninger Straße 22, 81675 Munich, Germany
9 Department of Orthopedics and Sport Orthopedics, Klinikum Rechts der Isar, Technische Universität München, Ismaninger Straße 22, 81675 Munich, Germany
10 Department of Radiology, University Hospital Freiburg, University of Freiburg, Hugstetterstraße 55, 79106 Freiburg im Breisgau, Germany
11 Helmholtz Zentrum München, Deutsches Forschungszentrum für Umwelt und Gesundheit, Institute of Radiation Medicine Neuherberg, 85764 Munich, Germany
12 Deutsches Konsortium für Translationale Krebsforschung (DKTK), Partner Site Munich, 69120 Heidelberg, Germany
* Correspondence: sarah.foreman@tum.de
† These authors contributed equally to this work.
‡ These authors contributed equally to this work.

Citation: Foreman, S.C.; Llorián-Salvador, O.; David, D.E.; Rösner, V.K.N.; Rischewski, J.F.; Feuerriegel, G.C.; Kramp, D.W.; Luiken, I.; Lohse, A.-K.; Kiefer, J.; et al. Development and Evaluation of MR-Based Radiogenomic Models to Differentiate Atypical Lipomatous Tumors from Lipomas. *Cancers* **2023**, *5*, 2150. https://doi.org/10.3390/cancers15072150

Academic Editor: Hamid Khayyam

Received: 17 January 2023
Revised: 10 March 2023
Accepted: 27 March 2023
Published: 5 April 2023

Simple Summary: Differentiating atypical lipomatous tumors from lipomas on MR images is a challenging task due to similar imaging characteristics. Given these challenges, it would be highly beneficial to develop a reliable diagnostic tool, thereby minimizing the need for invasive diagnostic procedures. Therefore, the aim of this study was to develop and validate radiogenomic machine-learning models to predict the MDM2 gene amplification status in order to differentiate between ALTs and lipomas on preoperative MR images. The best machine-learning model was based on radiomic features from multiple MR sequences using a LASSO algorithm and showed a high discriminatory power to predict the MDM2 gene amplification. Due to the varying settings in which patients with lipomatous tumors present, this model may enhance the clinical diagnostic workup.

Abstract: Background: The aim of this study was to develop and validate radiogenomic models to predict the MDM2 gene amplification status and differentiate between ALTs and lipomas on preoperative MR images. Methods: MR images were obtained in 257 patients diagnosed with ALTs ($n = 65$) or lipomas ($n = 192$) using histology and the MDM2 gene analysis as a reference standard. The protocols included T2-, T1-, and fat-suppressed contrast-enhanced T1-weighted sequences. Additionally, 50 patients were obtained from a different hospital for external testing. Radiomic features were selected using mRMR. Using repeated nested cross-validation, the machine-learning models were trained on radiomic features and demographic information. For comparison, the external test set was evaluated by three radiology residents and one attending radiologist. Results: A LASSO classifier trained on radiomic features from all sequences performed best, with an AUC of 0.88, 70% sensitivity, 81% specificity, and 76% accuracy. In comparison, the radiology residents achieved 60–70% accuracy, 55–80% sensitivity, and 63–77% specificity, while the attending radiologist achieved 90% accuracy, 96% sensitivity, and 87% specificity. Conclusion: A radiogenomic model combining features from multiple MR sequences showed the best performance in predicting the MDM2 gene amplification status. The model showed a higher accuracy compared to the radiology residents, though lower compared to the attending radiologist.

Keywords: radiomics; machine learning; soft-tissue sarcomas; radiology; MRI

1. Introduction

Lipomatous tumors are the most common neoplasms encountered by physicians and the most frequent soft-tissue tumors of the extremities [1]. Of these, 40 to 45% are benign adipocytic tumors (lipomas) or atypical lipomatous tumors (ALTs) [2–5]. Lipomas only require treatment if the mass effect causes symptoms such as pain or functional disorders [6]. ALTs may show locally aggressive growth and may dedifferentiate into high-grade sarcomas [7–10]. Therefore, ALTs are typically resected [11]. Histopathological differentiation relies on the detection of atypical hyperchromatic nuclei and the immunohistochemical evaluation of the molecular analysis of the mouse double minute 2 (MDM2) gene [12]. However, the detection of these atypical hyperchromatic cells can be challenging since they are frequently scattered throughout the lesion, and detection is often complicated by fibrous septa, subsequently requiring a careful analysis of the entire tumor [12–14]. Previous studies have shown that the MDM2 amplification status is the most accurate marker to differentiate ALTs and lipomas, and there is a tendency towards sampling errors if the MDM2 status is not determined [12,15–17]. Unfortunately, the majority of MR imaging studies differentiating ALTs from lipomas did not include a molecular analysis, or only performed a molecular analysis in a subset of patients [6,14,18,19].

MR imaging is the standard imaging modality for the assessment of soft-tissue tumors due to its excellent soft-tissue contrast [20–22]. Specific imaging features such as the tumor size, tumor location, presence of thick septa, and amount of contrast uptake can be used to differentiate ALTs from lipomas [6,13,18,19,23]. However, since there is a substantial overlap between these imaging features in both tumor types, differentiating ALTs from lipomas is a challenging task. Moreover, previous studies of systematic radiologic reading have reported relatively low inter-observer reproducibility, with a kappa agreement ranging from 0.17 to 0.42 [13,19,24]. Given these challenges, it would be highly beneficial to develop a reliable diagnostic tool to differentiate ALTs from lipomas on preoperative MR images, thereby minimizing the need for invasive diagnostic procedures.

Machine-learning techniques, including imaging-based radiomics, permit a non-invasive detailed analysis of a tumor phenotype by using a quantitative imaging feature analysis [25,26]. However, one of the main challenges of radiomic models includes reproducibility in different datasets [27,28]. Therefore, the aim of this study was to develop and validate radiogenomic machine-learning models based on multiparametric MR examinations to predict the MDM2 gene amplification status in order to differentiate between ALT and lipomas on preoperative MR images. The models were evaluated using an independent external cohort for testing and were compared to the performance of radiologists.

2. Materials and Methods

The local institutional review boards approved this retrospective multi-center study (ethics committee 666/21 S) The study was performed in accordance with our institutional ethic guidelines and the 1964 Declaration of Helsinki and its later amendments. Written and informed consent was waived for this retrospective anonymized analysis.

2.1. Datasets

We retrospectively reviewed the records of all patients with lipomatous tumors in the upper or lower extremities or trunk that had surgery performed at our sarcoma referral center between 2010 and 2021 ($n = 573$). Of these, 424 patients had a histologically confirmed diagnosis of a lipoma or an ALT. The MDM2 amplification status, determined by fluorescence in situ hybridization (FISH) of the MDM2 gene locus, was available for $n = 257$ patients. Patients without an MDM2 amplification status were excluded. Therefore, in the final dataset, both the histology and the MDM2 gene amplification status were available for all patients. Two senior pathologists specializing in the analysis of soft-tissue tumors provided a final consensus diagnosis based on the MDM2 gene amplification status and histology according to the World Health Organization criteria. The patient selection process is shown in Figure 1.

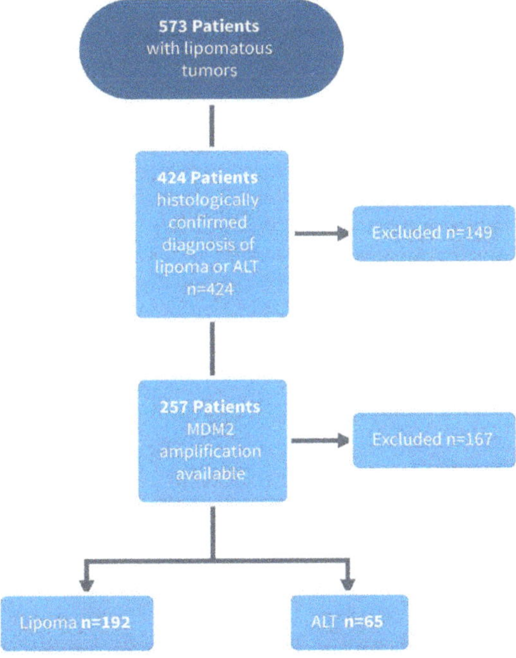

Figure 1. Subject selection flowchart. ALT = atypical lipomatous tumor; MDM2 = murine double minute.

In addition, an external test set was obtained from a further sarcoma referral center, the University Hospital of Freiburg (M1), for final independent testing and geographical validation. The external test set included patients with a diagnosis of a lipoma or an ALT confirmed by their histology and MDM2 amplification status.

2.2. MR Imaging Protocol and Image Segmentation

Pre-operative MR images were acquired using 3 or 1.5 Tesla scanners. Sequences were acquired in at least two planes that were oriented along the short and longitudinal axes of the long articulating bone(s). The protocols included a T2-w turbo spin echo (TSE) sequence (T2w), a T1-w TSE sequence (T1w), and a fat-saturated T1-w TSE sequence after the administration of a contrast agent (T1fsgd). Detailed information on the acquisition parameters is provided in Supplementary Material Table S1.

To define the volumes of interest (VOIs), tumor segmentations were performed manually by two radiology residents (S.C.F. and G.C.F.) using the open-source software 3D Slicer (3D Slicer, Version 4.8, stable release) and extracted as Neuroimaging Informatics Technology Initiative (NIfTI) label maps for further analysis. Multiple delineations were performed by S.C.F. and G.C.F. in 20 randomly selected patients to account for inter-reader variability.

2.3. Radiomic Feature Extraction and Machine-Learning Model Development

All preprocessing steps and radiomic feature extractions were conducted in accordance with the Imaging Biomarker Standardization Initiative guidelines [29] using the Python package PyRadiomics (version 2.2) implemented in Python (3.7), as previously described [30]. Image discretization was conducted using a bin width of 10 to achieve a bin count between 16 and 128, as recommended by the pyradiomics documentation [31]. Image intensity normalization was achieved via redistributing the image at the mean with a standard deviation and a scale of 100. Bspline interpolation was used to perform isotropic resampling to a voxel size of $1 \times 1 \times 1$ mm of the image and VOI mask. A total of 104 features were extracted from the original image of each sequence within the segmented label map (resulting in a total of 312 radiomic features), including first-order features, shape features, and texture features. The latter comprised "gray-level co-occurrence matrix" features, "gray-level size-zone matrix" features, "gray-level run-length matrix" features, "neighboring gray-tone difference matrix" features, and "gray-level dependence matrix" features. No features were extracted from filtered versions of the image due to a missing IBSI consensus. A detailed list of all extracted features is provided in Supplementary Material Table S2. Feature values were transformed to a common scale using min–max normalization in order to conserve their original distribution in the [0,1] range. Data normalization was performed prior to splitting the data into training and testing groups due to the batch harmonization step requirements. Nonparametric ComBatBatch harmonization was applied to account for the variability introduced by different MR scanners, as described previously [30]. Clinical features such as age, sex, and body region of the tumor (torso/head, upper extremity or lower extremity) were also included. Categorical features were encoded into dummy numeric arrays using one hot encoder. All radiomic features susceptible to segmentation variations were excluded using a threshold intraclass correlation coefficient (ICC 3,1) of 0.8. This statistic resulted in 5, 15, and 4 radiomic features that were excluded from the T1w, T2w, and T1fsgd sequences, respectively. ICC 3,1 was chosen, as the raters were not rated as representative of a defined rater group due to their differing extents of training.

An estimate of the number of reduced features to use was calculated using a principal component analysis (PCA) with 95% of data variance: 11 to 13 features for the individual sequences (T1w, T2w, and T1fsgd) and 19 to 21 features for the combined features of all sequences. Each respective number of features was selected using minimum redundancy–maximum relevance (MRMR). Synthetic minority over-sampling and random under-sampling of the majority class were used to counteract the class imbalance. The ratios were tuned to find an optimal balance between data augmentation and data discard with ratios of 0.5–0.6:1 after SMOTE and 0.6–0.8:1 after the random under-sampling of the majority class. The remaining class imbalance was handled by using balanced accuracy as the optimization criteria during hyperparameter optimization. Four machine-learning algorithms were implemented and compared in their performance: the support vector machine (SVM), the random forest classifier (RFC), the least absolute shrinkage and selection operator (LASSO; built from a stochastic gradient descent classifier), and a fully connected feedforward artificial neural network (ANN; multilayer perceptron classifier). A flow chart of the data processing and analysis of the radiomic features can be found in Supplementary Material Figure S1. For each algorithm, models were developed by (i) using demographic information only, (ii) using radiomic features for each individual sequence (T1w, T2w, or T1fsgd), (iii) using the radiomic features of all sequences, and (iv) using a combination of both the radiomic features of all sequences and demographic information. An overview of the radiomic workflow is shown in Figure 2.

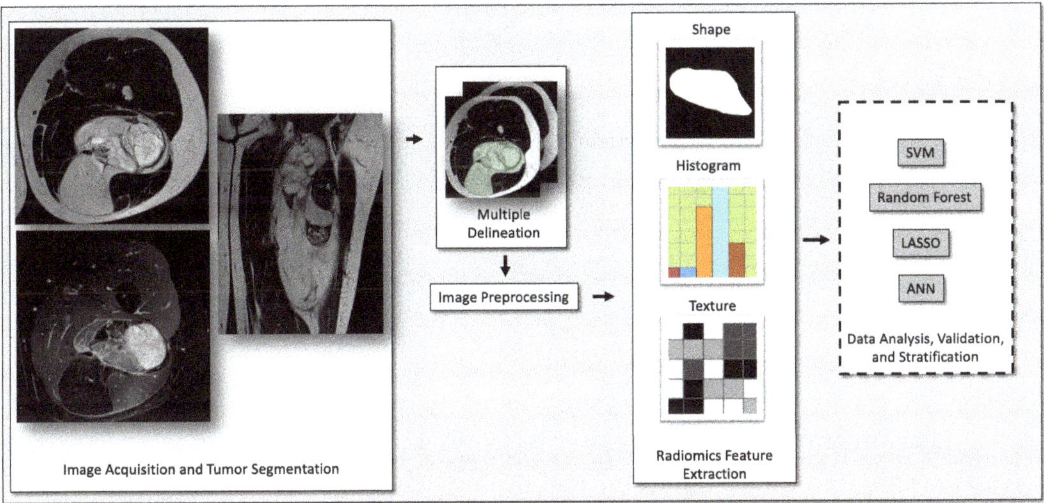

Figure 2. Radiomic workflow. Abbreviations: SVM, support vector machine; LASSO, least absolute shrinkage and selection operator; ANN, artificial neural network.

2.4. Model Optimization, Evaluation, and Statistical Analysis

Training and validation were performed using 3-fold nested cross-validation with 50 repetitions for statistical robustness, for a total of 150 averaged iterations per modeling algorithm and dataset. Hyperparameter optimization was conducted using an exhaustive grid search. This step was performed in the inner fold, after the feature selection step via MRMR, to prevent data leakage. Balanced accuracy was used as the optimization criterion to determine the best set of hyperparameters.

The performance of the models was evaluated with the area under the curve (AUC) obtained from the receiver–operator curve (ROC), plotted after averaging the yielded values. We also included the accuracy, sensitivity, and specificity as the output measures. For an unbiased evaluation, a final cross-validation step was implemented by selecting the best values obtained from the internal dataset before evaluating the performance on the external dataset. Stochastic gradient descent was used to calculate the probability of each class prediction. Calculations of model metrics were performed using scikit-learn (version 1.0.2).

For comparison, MR images of the external test set were rated independently by three radiology residents (I.L., S.C.F., and G.C.F., with 2, 3, and 5 years of experience, respectively) and one musculoskeletal imaging fellowship-trained radiologist (A.S.G., with 10 years of experience) experienced in musculoskeletal tumor imaging. All readers were blinded to all clinical and histopathological findings.

3. Results

3.1. Study Subjects

A total of 257 patients were included in the internal dataset (192 lipomas, 65 ALTs; age, 62.4 ± 14.5 years; 125 (48.6%) women). Fifty patients were included in the external dataset (30 lipomas, 20 ALTs; age, 60.6 ± 12.5 years; 22 (44%) women). All patients had a lipomatous tumor in one of the following six regions: chest, back, neck, leg, arm, hand, or foot. In both datasets, the highest number of patients had a tumor located in the leg (143/257 in the internal dataset and 27/50 in the external dataset), while the fewest number of patients had a tumor located in the foot (two in the internal dataset and none in the external dataset). Table 1 provides an overview of the subject characteristics.

Table 1. Patient characteristics.

Patient Characteristics	Internal Dataset (n = 257)	External Test Set (n = 50)
Age (years) *	62.4 ± 14.5	60.6 ± 12.5
Sex (women)	125	22
Tumor Location (Anatomical Region)		
Chest/Back	19	6
Neck	15	2
Leg	143	27
Arm	75	14
Hand	3	1
Foot	2	0
Lipomas	n = 192	n = 30
Age (years) *	62.3 ± 14.4	57.5 ± 11.1
Sex (women)	88	12
Atypical Lipomatous Tumors (ALT)	n = 65	n = 20
Age (years) *	62.5 ± 15	65.2 ± 13.5
Sex (women)	37	10

* Data are given as mean ± standard deviation.

3.2. Evaluation of the Developed Machine-Learning Models

Table 2 shows the final performance of the developed models on the external test set using demographic information only, radiomic features only (of all sequences combined) and a combination of demographic and radiomic features. The best-performing machine learning model was based on a LASSO algorithm using a combination of all sequences achieving an AUC of 0.88 at 70% sensitivity and 81% specificity with an accuracy of 76% on the external test set. The feature importance table, a confusion matrix, and a boxplot of the prediction probabilities from this model can be found in Supplementary Material Table S5, Supplementary Material Figure S2, and Supplementary Material Figure S3, respectively.

The AUC and accuracy for the individual sequences were lower for most models compared to models based on the radiomic parameters from all sequences combined, with a more imbalanced sensitivity/specificity. For T1w, the LASSO algorithm yielded an AUC of 0.83 at 80% sensitivity and 43% specificity with an accuracy of 58%. For T2w, the AUC was 0.82 at 42% sensitivity and 83% specificity with an accuracy of 69%. The highest AUC (0.84) was yielded for the T1fsgd sequences, though the sensitivity and specificity were highly imbalanced at 6% and 100%, respectively, with an accuracy of 60%. The performance of the developed models for the individual sequences on the external test set is shown in Supplementary Material Table S3.

Table 2. Performance of the machine-learning models on the external test set using demographic information or radiomic features only, as well as combining radiomic features and demographic information for the following model architectures: least absolute shrinkage and selection operator (LASSO), support vector machine (SVM), random forest classifier (RFC), and an artificial neural network (ANN). External performance represents the values yielded when a final cross-validation step considering only the best 150 best hyperparameter sets was implemented to predict the external test set.

Model Architecture	Score	Demographic Features	Combined Sequences	Combined Sequences + Demographic Features
LASSO	AUC *	0.56 (0.540.58) ± 0.07	0.88 (0.85–0.91) ± 0.07	0.72 (0.66–0.78) ± 0.15
	Accuracy	0.58	0.76	0.77
	Sensitivity	0.05	0.70	0.40
	Specificity	0.93	0.81	1.00
SVM	AUC *	0.54 (0.51–0.57) ± 0.12	0.84 (0.80–0.88) ± 0.11	0.85 (0.82–0.88) ± 0.09
	Accuracy	0.56	0.53	0.69
	Sensitivity	0.10	0.90	0.80
	Specificity	0.87	0.31	0.63
RFC	AUC *	0.63 (0.61–0.65) ± 0.06	0.87 (0.85–0.89) ± 0.05	0.87 (0.85–0.89) ± 0.05
	Accuracy	0.50	0.69	0.69
	Sensitivity	0.00	0.50	0.40
	Specificity	0.83	0.81	0.88
ANN	AUC *	0.68 (0.66–0.70) ± 0.08	0.81 (0.77–0.85) ± 0.10	0.81 (0.77–0.85) ± 0.10
	Accuracy	0.60	0.69	0.65
	Sensitivity	0.00	0.70	0.60
	Specificity	1.00	0.69	0.69

* Data are given as mean (95% confidence interval) ± standard deviation.

Interestingly, combining radiomic features and demographic information as the input for the machine-learning models did not improve the performance of the LASSO algorithm to differentiate ALTs from lipomas and resulted in a decrease in the sensitivity from 70% to 40%, though the specificity increased to 100%. The averaged nested cross-validation results of the internal dataset are shown in Supplementary Material Table S4. The training parameters and source code can be found online (https://github.com/deedeedav/alt-lipoma-radiomics (accessed on 9 March 2023)). Figure 3 shows an example of an ALT with typical imaging findings encasing the right gracilis muscle, while Figure 4 shows a typical example of a well-defined intramuscular lipoma in the right posterior thigh. Both cases were identified correctly by the machine-learning model.

3.3. Comparison with Radiologists

The results of the independent radiological readings of the external test are shown in Table 3. The radiology resident with 2 years of experience achieved an accuracy of 60%, a sensitivity of 55%, and a specificity of 63%; the resident with 3 years of experience achieved an accuracy of 70%, a sensitivity of 60%, and a specificity of 77%; and the radiology resident with 5 years of experience achieved an accuracy of 70%, a sensitivity of 80%, and a specificity of 63%. In comparison, the attending radiologist that was experienced in musculoskeletal tumor imaging achieved an accuracy of 90%, a sensitivity of 96%, and a specificity of 87%. Compared to the radiology residents, the model showed a higher accuracy and higher specificity, while the sensitivity was lower compared to the resident with 5 years of experience, but higher compared to the residents with 2 or 3 years of experience. The attending radiologist had a higher accuracy, sensitivity, and specificity. Figure 5 shows an ALT with atypical imaging findings located subcutaneously. The machine-learning model and the attending radiologist classified this tumor as an ALT, while all residents classified this tumor as a lipoma.

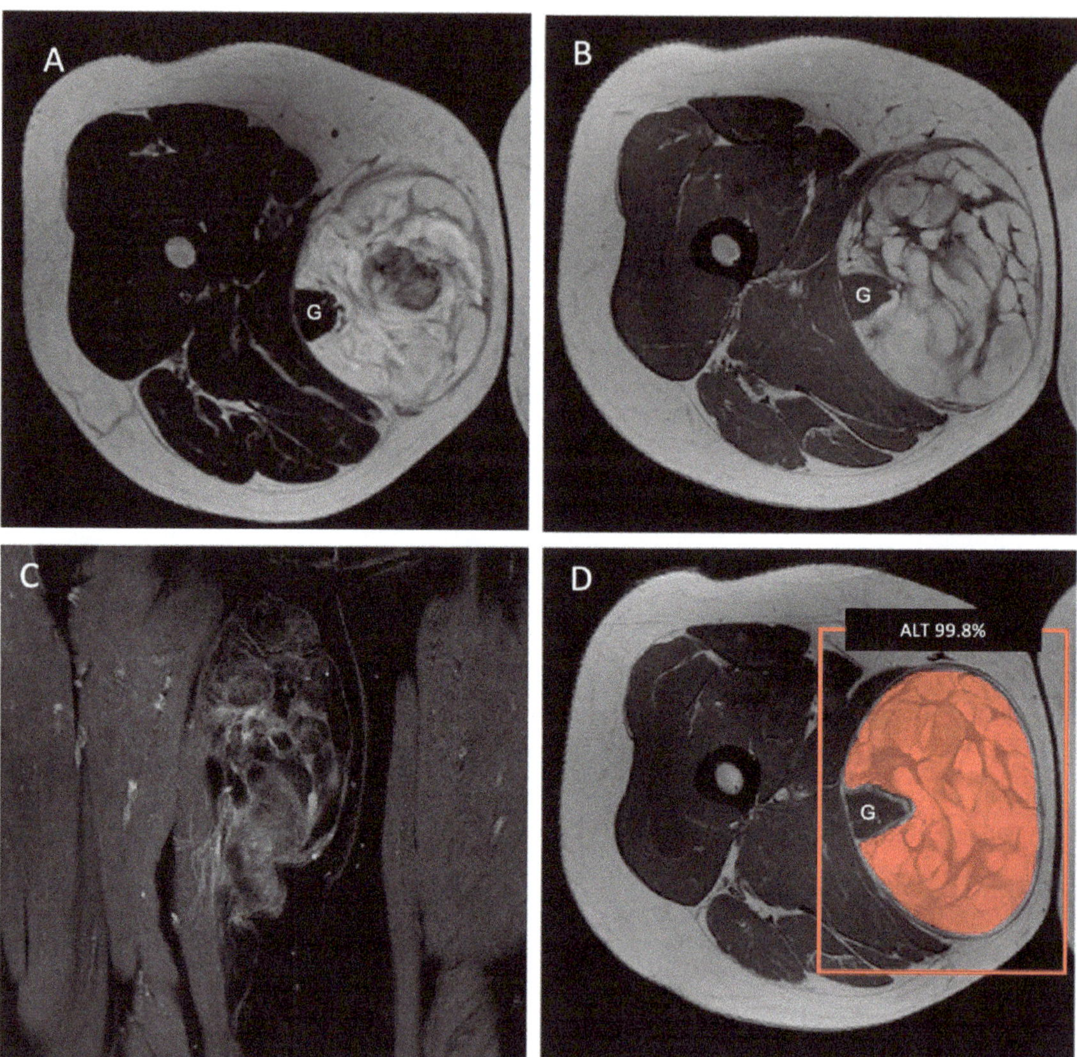

Figure 3. Lipomatous tumor in the medial right thigh, encasing the gracilis muscle (G). (**A**) The axial T2-weighted and (**B**) axial T1-weighted MR images show a large heterogeneous tumor with thick septa. (**C**) Septal contrast enhancement on the coronal T1-weighted images with fat saturation. (**D**) The machine-learning algorithm classified the tumor as an ALT with a probability of 99.8%. This diagnosis was confirmed by pathology and immunohistochemistry after surgical resection.

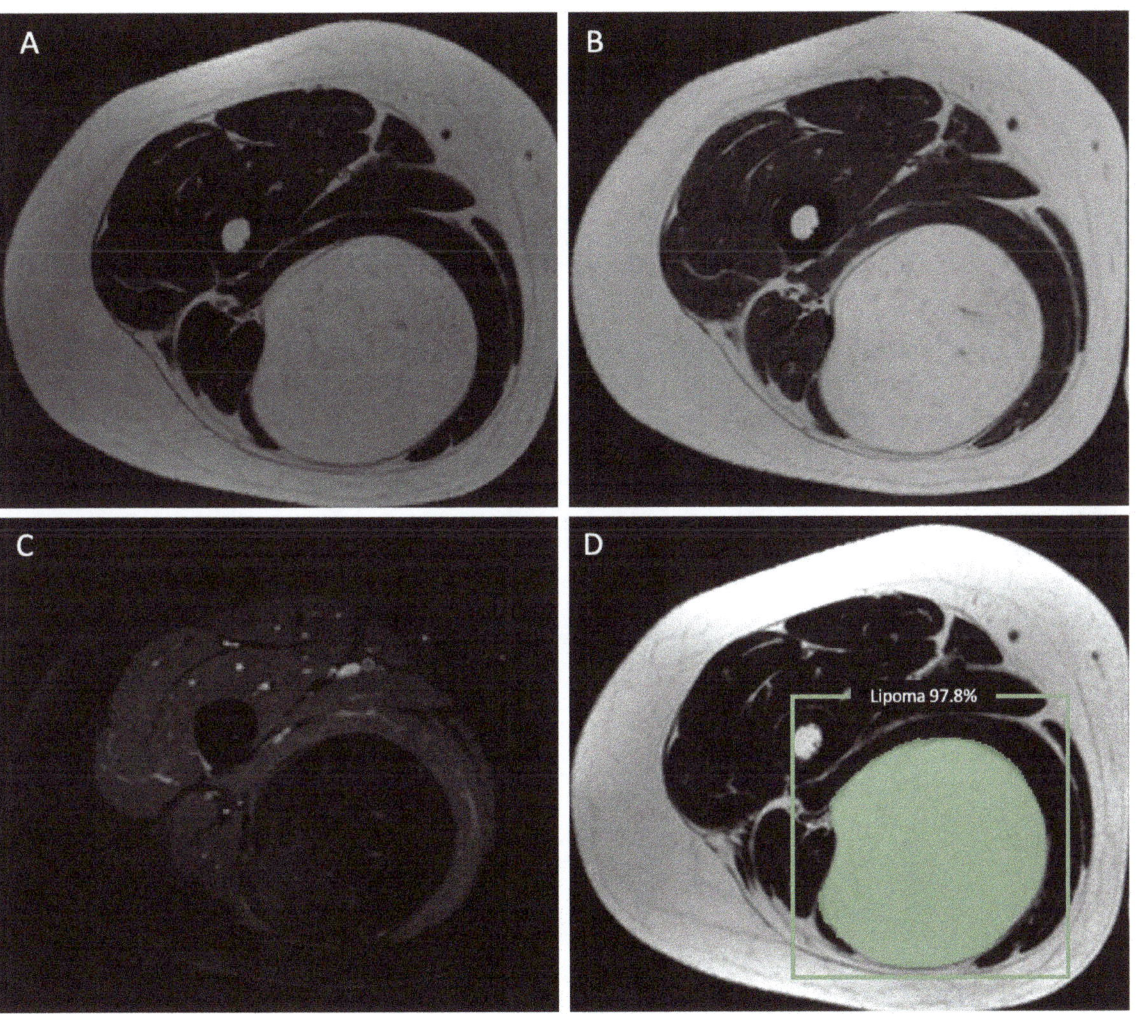

Figure 4. Axial T2-weighted (**A**) and T1-weighted (**B**) MR images showing a well-defined intramuscular lipomatous tumor (lipoma) in the right posterior thigh without significant contrast enhancement on the axial T1-weighted image with fat saturation (**C**). (**D**) The machine-learning model classified this tumor as a lipoma (probability of 97.8%). This was in accordance with the diagnosis made by the radiology residents and the attending radiologist.

Table 3. Performance of the radiology residents with 2, 3, or 5 years of experience and the fellowship-trained radiologist that was experienced in musculoskeletal tumor imaging. Readers were blinded to all clinical and histopathological findings.

Score	Radiology Resident, 2y	Radiology Resident, 3y	Radiology Resident, 5y	Fellowship-Trained Radiologist
Accuracy	0.60 (30/50)	0.70 (35/50)	0.70 (35/50)	0.90 (45/50)
Sensitivity	0.55 (11/20)	0.60 (12/20)	0.80 (16/20)	0.96 (19/20)
Specificity	0.63 (19/30)	0.77 (23/30)	0.63 (19/30)	0.87 (26/30)

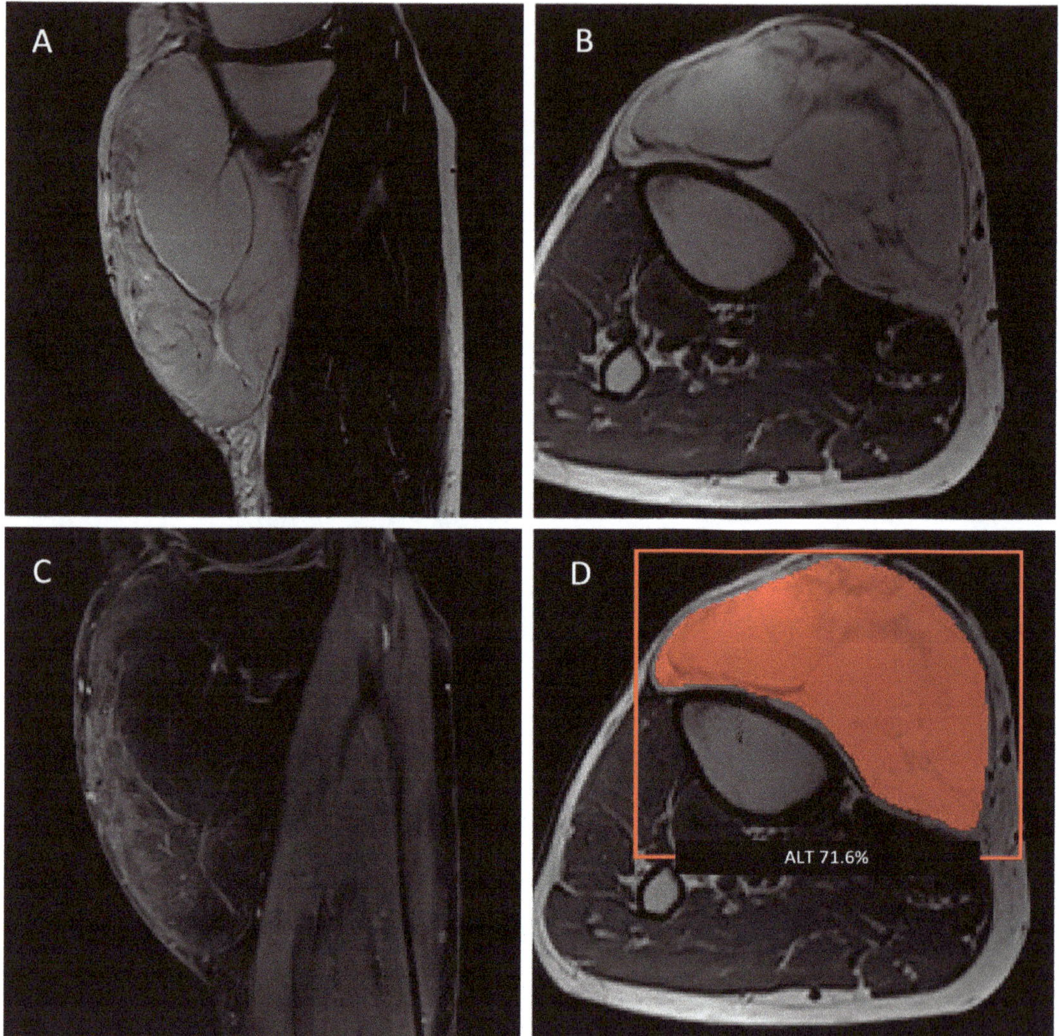

Figure 5. Sagittal T2-weighted (**A**) and axial T1-weighted (**B**) MR images of a lipomatous tumor located subcutaneously, anteromedial to the right proximal tibia. (**C**) A sagittal T1-weighted image with fat saturation shows a moderate septal contrast enhancement. All radiology residents classified this tumor as a lipoma, while the attending radiologist classified this tumor as an ALT. (**D**) The machine-learning algorithm also classified this tumor as an ALT with a probability of 71.6%. The diagnosis of an ALT was confirmed by pathology after surgical resection.

4. Discussion

In this study, machine-learning models were developed and validated to predict the amplification status of the MDM2 gene, to differentiate between atypical lipomatous tumor and lipomas on preoperative MR images, and to compare the results to the performance of radiologists using an external test set. The best-performing model was based on the combination of all MR sequences and achieved an AUC of 0.88 at 70% sensitivity and 81% specificity with an accuracy of 76%. In comparison, the accuracy of the readings by all radiology residents was lower, while the accuracy of the fellowship-trained radiologist was higher. Notably, the performance of the LASSO algorithm for each individual sequence was lower compared to the model that included all sequences (T2w, T1w, and T1fsgd), suggesting that all sequences are required for optimal discrimination.

Radiomic models for differentiating lipomas from ALTs have previously been developed in smaller patient cohorts. Leporq et al. evaluated 2D radiomic models of 40 lipomas and 41 ALTs, including one MR image slice per patient [32]. Their best-performing model achieved an accuracy of 95% at 100% sensitivity and 90% specificity using the histology as the reference standard, though no specific information regarding the MDM2 gene amplification status was included, which may have led to a false classification of ALTs as lipomas [32]. Cay et al. evaluated 45 lipomas and 20 ALTs using histology and MDM2 amplification as the gold standards [33]. They achieved an AUC of 0.987 at 96.8% sensitivity and 93.72% specificity using 1000-fold bootstrapping [33]. However, since there was no separate test set, the algorithm was likely optimized on data used for validation in another bootstrapping iteration; therefore, these results may be inaccurately high [33]. A study by Vos et al. included 116 patients (58 lipomas and 58 ALTs) and used MDM2 amplification as the reference standard [34]. Their model performance was lower compared to our study, yielding an AUC of 0.81 at 66% sensitivity and 84% specificity with an accuracy of 75%. An important limitation of these aforementioned studies is that no external validation on an independent dataset was included. Also notably, the model performance was comparatively high in studies based on smaller patient cohorts ($n < 90$). A possible explanation may be a lack of variation in smaller datasets, which could affect the reproducibility in different datasets. However, this is not clear, since no external testing was included.

Interestingly, combining imaging parameters and clinical data did not improve the performance of most models for differentiating ALTs from lipomas, or only improved the performance marginally. While some demographic differences have been described between patients with ALTs and lipomas [23], it is likely that radiomic MR features are considerably more relevant for differentiating between these tumor types, and including parameters with less predictive power could hinder the capability of the models to identify relevant patterns. It should be noted that only a limited number of clinical features were included (age, sex, and tumor body region). Including additional clinical features may improve the predictive value of the radiomic models. Future studies could also include clinical outcome parameters to detect image-defined high-risk patients, thereby individualizing tumor treatment.

Some limitations are pertinent to this study. Since the cohort included only patients with histopathologically confirmed tumors, this potentially introduced a selection bias. Moreover, our specialized sarcoma center typically only receives larger or atypical lipomas on referral, subsequently increasing the amount of particularly challenging lipoma cases in the dataset. We also used manual segmentations as input for the models, and developing a pipeline that includes automated segmentations would be highly beneficial. In addition, more advanced sequences such as diffusion-weighted imaging or pharmacokinetic dynamic contrast-enhanced imaging were not included in the protocol. Including these sequences could potentially improve the differentiation between ALTs and lipomas. Finally, the developed models only differentiated between ALTs and lipomas, and while this is the most challenging and clinically relevant task, further studies are warranted on the ability to distinguish among all benign and malignant lipomatous tumors.

The advantages of the current study include its multicenter design, which allowed the evaluation of the models on an independent external test set, thereby reducing potential bias introduced by overfitting. Moreover, the dataset used for training was, to the best of our knowledge, the largest MRI dataset of histopathologically confirmed lipomas and ALTs. In addition, a histopathological analysis was conducted by pathologists specialized in the analysis of soft-tissue tumors and included the immunohistochemistry for the assessment of the MDM2 status in all cases. Furthermore, we excluded inter-/intra-reader segmentation-dependent features and included variability features, making the model performance more stable and reliable for other datasets.

5. Conclusions

In conclusion, radiogenomic models were developed that showed a high discriminatory power for predicting the MDM2 gene amplification status to distinguish between atypical lipomatous tumors and lipomas on preoperative MR images. The best-performing model was based on a LASSO algorithm using all MR sequences, with a higher accuracy compared to radiology residents, suggesting that these algorithms would be particularly helpful for radiologists with less experience. Due to the varying settings in which patients with lipomatous tumors present, this model may enhance the clinical diagnostic workup and improve the detection rate for atypical lipomatous tumors.

Supplementary Materials: The following supporting information can be downloaded at: https://www.mdpi.com/article/10.3390/cancers15072150/s1. Supplementary Material Table S1: Magnetic resonance imaging sequence parameters. Supplementary Material Table S2: Extracted radiomic features ($n = 104$). Supplementary Material Table S3: Performance of the machine-learning models on the external test set of each individual sequence (T1w, T2w, and T1fsgd) using the following model architectures: least absolute shrinkage and selection operator (LASSO), support vector machine (SVM), random forest classifier (RFC), and an artificial neural network (ANN). The external performance represents the values yielded when a final cross-validation step considering only the best 150 best hyperparameter sets was implemented. Supplementary Material Table S4: Internal performance representing the averaged values over 150 models resulting from the nested cross-validation using demographic information, radiomic features of each individual sequence (T1w, T2w, and T1fsgd), or radiomic features of all sequences combined, as well as combining radiomic features (of all sequences, and demographic information for the following model architectures: least absolute shrinkage and selection operator (LASSO), support vector machine (SVM), random forest classifier (RFC), and an artificial neural network (ANN). The metrics are given as mean ± standard deviation. Supplementary Material Table S5: Feature importance of the best-performing model (least absolute shrinkage and selection operator (LASSO) trained on features from all radiomic sequences). Supplementary Material Figure S1: Flow chart of the statistical analysis of the extracted radiomic features. Supplementary Material Figure S2: Confusion matrix of the best-performing model, a least absolute shrinkage and selection operator (LASSO) trained on all radiomic sequences. Misclassification rate: 0.23 ((FN + FP)/(N + P)). Supplementary Material Figure S3: Boxplot of the prediction probabilities made by the best-performing model (least absolute shrinkage and selection operator (LASSO) trained on features from all radiomic sequences). The probability cut-off used was 0.5.

Author Contributions: Conceptualization, S.C.F., O.L.-S., M.J., M.A.A.-N., B.R., K.W., C.M., C.K. S.E.C., M.R.M., J.C.P. and A.S.G.; methodology, S.C.F., O.L.-S. and A.S.G.; software, O.L.-S., D.E.D. M.A.A.-N., B.R. and J.C.P.; validation, S.C.F., D.E.D., J.K., M.J., V.K.N.R. and A.-K.L.; formal analysis O.L.-S., D.E.D., G.C.F., C.M., I.L., S.C.F. and A.S.G.; investigation, S.C.F.; resources, J.K., C.M., C.K. M.A.A.-N., B.R., S.E.C., M.R.M., J.C.P. and A.S.G.; data curation, S.C.F., V.K.N.R., D.E.D., O.L.-S. J.F.R. and D.W.K.; writing—original draft preparation, S.C.F. and O.L.-S.; writing—review and editing, S.C.F., O.L.-S., D.E.D., G.C.F., A.-K.L., S.E.C., M.R.M., K.W., J.C.P. and A.S.G.; visualization S.C.F.; supervision, S.C.F., O.L.-S., M.A.A.-N., B.R., S.E.C., M.R.M., K.W., J.C.P. and A.S.G.; project administration, S.C.F., O.L.-S., J.C.P. and A.S.G.; funding acquisition, J.C.P. and A.S.G. All authors have read and agreed to the published version of the manuscript.

Funding: This work was funded by grants from the German Society of Musculoskeletal Radiology (Deutsche Gesellschaft für muskuloskelettale Radiologie; DGMSR), the European Society of Musculoskeletal Radiology (ESSR), the Munich Clinician Scientist Program (MCSP) of the University of Munich (LMU; grant number ACS-10), and the Clinician Scientist Program (KKF) at Technische Universität München (TUM; grant number H-03).

Institutional Review Board Statement: The study was conducted according to the guidelines of the Declaration of Helsinki, and approved by the Institutional Review Board of Klinikum Rechts der Isar Technische Universität München (666/21 S, date of approval 24 November 2021).

Informed Consent Statement: Written informed consent was waived for this retrospective anonymized analysis.

Data Availability Statement: The training parameters and source code can be found online (https://github.com/deedeedav/alt-lipoma-radiomics (accessed on 9 March 2023)).

Conflicts of Interest: The authors declare no conflict of interest.

Abbreviations

atypical lipomatous tumors	ALTs
mouse double minute 2	MDM2
fluorescence in situ hybridization	FISH
turbo spin echo	TSE
volume of interest	VOI
intraclass correlation coefficient	ICC
principal component analysis	PCA
Neuroimaging Informatics Technology Initiative	NIfTI
support vector machine	SVM
random forest classifier	RFC
least absolute shrinkage and selection operator	LASSO
artificial neural network	ANN
area under the curve	AUC
receiver–operator curve	ROC

References

Johnson, C.N.; Ha, A.S.; Chen, E.; Davidson, D. Lipomatous Soft-tissue Tumors. *J. Am. Acad. Orthop. Surg.* **2018**, *26*, 779–788. [CrossRef]

Dalal, K.M.; Antonescu, C.R.; Singer, S. Diagnosis and management of lipomatous tumors. *J. Surg. Oncol.* **2008**, *97*, 298–313. [CrossRef] [PubMed]

Myhre-Jensen, O.; Kaae, S.; Madsen, E.H.; Sneppen, O. Histopathological grading in soft-tissue tumours. Relation to survival in 261 surgically treated patients. *Acta Pathol. Microbiol. Immunol. Scand. A* **1983**, *91*, 145–150.

Rydholm, A.; Berg, N.O. Size, site and clinical incidence of lipoma. Factors in the differential diagnosis of lipoma and sarcoma. *Acta Orthop. Scand.* **1983**, *54*, 929–934. [CrossRef] [PubMed]

Fletcher, C.D. The evolving classification of soft tissue tumours: An update based on the new WHO classification. *Histopathology* **2006**, *48*, 3–12. [CrossRef]

Nagano, S.; Yokouchi, M.; Setoguchi, T.; Ishidou, Y.; Sasaki, H.; Shimada, H.; Komiya, S. Differentiation of lipoma and atypical lipomatous tumor by a scoring system: Implication of increased vascularity on pathogenesis of liposarcoma. *BMC Musculoskelet. Disord.* **2015**, *16*, 36. [CrossRef]

Bassett, M.D.; Schuetze, S.M.; Disteche, C.; Norwood, T.H.; Swisshelm, K.; Chen, X.; Bruckner, J.; Conrad, E.U., 3rd; Rubin, B.P. Deep-seated, well differentiated lipomatous tumors of the chest wall and extremities: The role of cytogenetics in classification and prognostication. *Cancer* **2005**, *103*, 409–416. [CrossRef]

Weiss, S.W.; Rao, V.K. Well-differentiated liposarcoma (atypical lipoma) of deep soft tissue of the extremities, retroperitoneum, and miscellaneous sites. A follow-up study of 92 cases with analysis of the incidence of "dedifferentiation". *Am. J. Surg. Pathol.* **1992**, *16*, 1051–1058. [CrossRef] [PubMed]

Bidault, F.; Vanel, D.; Terrier, P.; Jalaguier, A.; Bonvalot, S.; Pedeutour, F.; Couturier, J.M.; Dromain, C. Liposarcoma or lipoma: Does genetics change classic imaging criteria? *Eur. J. Radiol.* **2009**, *72*, 22–26. [CrossRef]

0. Evans, H.L.; Soule, E.H.; Winkelmann, R.K. Atypical lipoma, atypical intramuscular lipoma, and well differentiated retroperitoneal liposarcoma: A reappraisal of 30 cases formerly classified as well differentiated liposarcoma. *Cancer* **1979**, *43*, 574–584. [CrossRef]

1. Choi, K.Y.; Jost, E.; Mack, L.; Bouchard-Fortier, A. Surgical management of truncal and extremities atypical lipomatous tumors/well-differentiated liposarcoma: A systematic review of the literature. *Am. J. Surg.* **2020**, *219*, 823–827. [CrossRef]

2. Zhang, H.; Erickson-Johnson, M.; Wang, X.; Oliveira, J.L.; Nascimento, A.G.; Sim, F.H.; Wenger, D.E.; Zamolyi, R.Q.; Pannain, V.L.; Oliveira, A.M. Molecular testing for lipomatous tumors: Critical analysis and test recommendations based on the analysis of 405 extremity-based tumors. *Am. J. Surg. Pathol.* **2010**, *34*, 1304–1311. [CrossRef] [PubMed]

3. Brisson, M.; Kashima, T.; Delaney, D.; Tirabosco, R.; Clarke, A.; Cro, S.; Flanagan, A.M.; O'Donnell, P. MRI characteristics of lipoma and atypical lipomatous tumor/well-differentiated liposarcoma: Retrospective comparison with histology and MDM2 gene amplification. *Skelet. Radiol.* **2013**, *42*, 635–647. [CrossRef] [PubMed]

4. Ohguri, T.; Aoki, T.; Hisaoka, M.; Watanabe, H.; Nakamura, K.; Hashimoto, H.; Nakamura, T.; Nakata, H. Differential diagnosis of benign peripheral lipoma from well-differentiated liposarcoma on MR imaging: Is comparison of margins and internal characteristics useful? *AJR Am. J. Roentgenol.* **2003**, *180*, 1689–1694. [CrossRef]

15. Dei Tos, A.P.; Doglioni, C.; Piccinin, S.; Sciot, R.; Furlanetto, A.; Boiocchi, M.; Dal Cin, P.; Maestro, R.; Fletcher, C.D.; Tallini, C Coordinated expression and amplification of the MDM2, CDK4, and HMGI-C genes in atypical lipomatous tumours. *J. Patho* **2000**, *190*, 531–536. [CrossRef]

16. Kulkarni, A.S.; Wojcik, J.B.; Chougule, A.; Arora, K.; Chittampalli, Y.; Kurzawa, P.; Mullen, J.T.; Chebib, I.; Nielsen, G.P.; River M.N.; et al. MDM2 RNA In Situ Hybridization for the Diagnosis of Atypical Lipomatous Tumor: A Study Evaluating DNA, RN/ and Protein Expression. *Am. J. Surg. Pathol.* **2019**, *43*, 446–454. [CrossRef] [PubMed]

17. Kashima, T.; Halai, D.; Ye, H.; Hing, S.N.; Delaney, D.; Pollock, R.; O'Donnell, P.; Tirabosco, R.; Flanagan, A.M. Sensitivity c MDM2 amplification and unexpected multiple faint alphoid 12 (alpha 12 satellite sequences) signals in atypical lipomatous tumc *Mod. Pathol.* **2012**, *25*, 1384–1396. [CrossRef]

18. Kransdorf, M.J.; Bancroft, L.W.; Peterson, J.J.; Murphey, M.D.; Foster, W.C.; Temple, H.T. Imaging of fatty tumors: Distinction c lipoma and well-differentiated liposarcoma. *Radiology* **2002**, *224*, 99–104. [CrossRef] [PubMed]

19. Nardo, L.; Abdelhafez, Y.G.; Acquafredda, F.; Schiro, S.; Wong, A.L.; Sarohia, D.; Maroldi, R.; Darrow, M.A.; Guindani, M.; Lee, S et al. Qualitative evaluation of MRI features of lipoma and atypical lipomatous tumor: Results from a multicenter study. *Skele Radiol.* **2020**, *49*, 1005–1014. [CrossRef]

20. De Schepper, A.M.; De Beuckeleer, L.; Vandevenne, J.; Somville, J. Magnetic resonance imaging of soft tissue tumors. *Eur. Radio* **2000**, *10*, 213–223. [CrossRef] [PubMed]

21. Vilanova, J.C.; Woertler, K.; Narvaez, J.A.; Barcelo, J.; Martinez, S.J.; Villalon, M.; Miro, J. Soft-tissue tumors update: MR imagin features according to the WHO classification. *Eur. Radiol.* **2007**, *17*, 125–138. [CrossRef]

22. Totty, W.G.; Murphy, W.A.; Lee, J.K. Soft-tissue tumors: MR imaging. *Radiology* **1986**, *160*, 135–141. [CrossRef]

23. Knebel, C.; Neumann, J.; Schwaiger, B.J.; Karampinos, D.C.; Pfeiffer, D.; Specht, K.; Lenze, U.; von Eisenhart-Rothe, R.; Rummen E.J.; Woertler, K.; et al. Differentiating atypical lipomatous tumors from lipomas with magnetic resonance imaging: A compariso with MDM2 gene amplification status. *BMC Cancer* **2019**, *19*, 309. [CrossRef] [PubMed]

24. O'Donnell, P.W.; Griffin, A.M.; Eward, W.C.; Sternheim, A.; White, L.M.; Wunder, J.S.; Ferguson, P.C. Can Experienced Observer Differentiate between Lipoma and Well-Differentiated Liposarcoma Using Only MRI? *Sarcoma* **2013**, *2013*, 982784. [CrossRef]

25. Peeken, J.C.; Asadpour, R.; Specht, K.; Chen, E.Y.; Klymenko, O.; Akinkuoroye, V.; Hippe, D.S.; Spraker, M.B.; Schaub, S.K Dapper, H.; et al. MRI-based delta-radiomics predicts pathologic complete response in high-grade soft-tissue sarcoma patient treated with neoadjuvant therapy. *Radiother. Oncol.* **2021**, *164*, 73–82. [CrossRef]

26. Peeken, J.C.; Wiestler, B.; Combs, S.E. Image-Guided Radiooncology: The Potential of Radiomics in Clinical Application. *Recer Results Cancer Res.* **2020**, *216*, 773–794. [CrossRef]

27. Crombe, A.; Fadli, D.; Italiano, A.; Saut, O.; Buy, X.; Kind, M. Systematic review of sarcomas radiomics studies: Bridging the ga between concepts and clinical applications? *Eur. J. Radiol.* **2020**, *132*, 109283. [CrossRef] [PubMed]

28. Gitto, S.; Cuocolo, R.; Albano, D.; Morelli, F.; Pescatori, L.C.; Messina, C.; Imbriaco, M.; Sconfienza, L.M. CT and MRI radiomic of bone and soft-tissue sarcomas: A systematic review of reproducibility and validation strategies. *Insights Imaging* **2021**, *12*, 6 [CrossRef] [PubMed]

29. Zwanenburg, A.; Vallieres, M.; Abdalah, M.A.; Aerts, H.; Andrearczyk, V.; Apte, A.; Ashrafinia, S.; Bakas, S.; Beukinga, R. Boellaard, R.; et al. The Image Biomarker Standardization Initiative: Standardized Quantitative Radiomics for High-Throughpu Image-based Phenotyping. *Radiology* **2020**, *295*, 328–338. [CrossRef] [PubMed]

30. Peeken, J.C.; Neumann, J.; Asadpour, R.; Leonhardt, Y.; Moreira, J.R.; Hippe, D.S.; Klymenko, O.; Foreman, S.C.; von Schack C.E.; Spraker, M.B.; et al. Prognostic Assessment in High-Grade Soft-Tissue Sarcoma Patients: A Comparison of Semantic Imag Analysis and Radiomics. *Cancers* **2021**, *13*, 1929. [CrossRef] [PubMed]

31. Tixier, F.; Le Rest, C.C.; Hatt, M.; Albarghach, N.; Pradier, O.; Metges, J.P.; Corcos, L.; Visvikis, D. Intratumor heterogeneit characterized by textural features on baseline 18F-FDG PET images predicts response to concomitant radiochemotherapy i esophageal cancer. *J. Nucl. Med.* **2011**, *52*, 369–378. [CrossRef] [PubMed]

32. Leporq, B.; Bouhamama, A.; Pilleul, F.; Lame, F.; Bihane, C.; Sdika, M.; Blay, J.Y.; Beuf, O. MRI-based radiomics to predic lipomatous soft tissue tumors malignancy: A pilot study. *Cancer Imaging* **2020**, *20*, 78. [CrossRef] [PubMed]

33. Cay, N.; Mendi, B.A.R.; Batur, H.; Erdogan, F. Discrimination of lipoma from atypical lipomatous tumor/well-differentiatec liposarcoma using magnetic resonance imaging radiomics combined with machine learning. *Jpn. J. Radiol.* **2022**, *40*, 951–96([CrossRef] [PubMed]

34. Vos, M.; Starmans, M.P.A.; Timbergen, M.J.M.; van der Voort, S.R.; Padmos, G.A.; Kessels, W.; Niessen, W.J.; van Leenders, G Grunhagen, D.J.; Sleijfer, S.; et al. Radiomics approach to distinguish between well differentiated liposarcomas and lipomas o MRI. *Br. J. Surg.* **2019**, *106*, 1800–1809. [CrossRef] [PubMed]

Article

CT Radiomics and Clinical Feature Model to Predict Lymph Node Metastases in Early-Stage Testicular Cancer

Catharina Silvia Lisson [1,2,3,*], Sabitha Manoj [1,3,4], Daniel Wolf [1,3,4], Jasper Schrader [5], Stefan Andreas Schmidt [1,2,3], Meinrad Beer [1,2,3,6,7], Michael Goetz [1,3,8,*], Friedemann Zengerling [5,†] and Christoph Gerhard Sebastian Lisson [1,†]

[1] Department of Diagnostic and Interventional Radiology, University Hospital of Ulm, Albert-Einstein-Allee 23, 89081 Ulm, Germany

[2] ZPM—Center for Personalized Medicine, University Hospital of Ulm, Albert-Einstein-Allee 23, 89081 Ulm, Germany

[3] XAIRAD—Artificial Intelligence in Experimental Radiology, University Hospital of Ulm, Albert-Einstein-Allee 23, 89081 Ulm, Germany

[4] Visual Computing Group, Institute of Media Informatics, Ulm University, James-Franck-Ring, 89081 Ulm, Germany

[5] Department of Urology, University Hospital of Ulm, Albert-Einstein-Allee 23, 89081 Ulm, Germany

[6] MoMan—Center for Translational Imaging, Department of Internal Medicine II, University Hospital of Ulm, Albert-Einstein-Allee 23, 89081 Ulm, Germany

[7] i2SouL—Innovative Imaging in Surgical Oncology Ulm, University Hospital of Ulm, Albert-Einstein-Allee 23, 89081 Ulm, Germany

[8] Division Medical Image Computing, DKFZ—German Cancer Research Center, 69120 Heidelberg, Germany

[*] Correspondence: catharina.lisson@uniklinik-ulm.de (C.S.L.); michael.goetz@uni-ulm.de (M.G.); Tel.: +49-731-50061171 (C.S.L.)

[†] These authors contributed equally to this work.

Simple Summary: Testicular germ cell tumour (TGCT) is the most common solid cancer in men below 40. The majority present with disease confined to the testis (stage 1), with its primary treatment being radical orchiectomy. Despite the multiple options for managing stage 1 tumours, optimal management is controversial, with further treatment options including active surveillance, chemotherapy and retroperitoneal lymph node dissection or low dose radiotherapy of the paraaortic region. In this study, the authors incorporated quantitative imaging features and clinical risk factors to stratify patients according to lymph node metastases, thus promoting precision imaging in clinical oncology.

Abstract: Accurate retroperitoneal lymph node metastasis (LNM) prediction in early-stage testicular germ cell tumours (TGCTs) harbours the potential to significantly reduce over- or undertreatment and treatment-related morbidity in this group of young patients as an important survivorship imperative. We investigated the role of computed tomography (CT) radiomics models integrating clinical predictors for the individualised prediction of LNM in early-stage TGCT. Ninety-one patients with surgically proven testicular germ cell tumours and contrast-enhanced CT were included in this retrospective study. Dedicated radiomics software was used to segment 273 retroperitoneal lymph nodes and extract features. After feature selection, radiomics-based machine learning models were developed to predict LN metastasis. The robustness of the procedure was controlled by 10-fold cross-validation. Using multivariable logistic regression modelling, we developed three prediction models: a radiomics-only model, a clinical-only model, and a combined radiomics–clinical model. The models' performances were evaluated using the area under the receiver operating characteristic curve (AUC). Finally, decision curve analysis was performed to estimate the clinical usefulness of the predictive model. The radiomics-only model for predicting lymph node metastasis reached a greater discrimination power than the clinical-only model, with an AUC of 0.87 (±0.04; 95% CI) vs. 0.75 (±0.08; 95% CI) in our study cohort. The combined model integrating clinical risk factors and selected radiomics features outperformed the clinical-only and the radiomics-only prediction models, and showed good discrimination with an area under the curve of 0.89 (±0.03; 95% CI). The decision curve analysis demonstrated the clinical usefulness of our proposed combined model. The

Citation: Lisson, C.S.; Manoj, S.; Wolf, D.; Schrader, J.; Schmidt, S.A.; Beer, M.; Goetz, M.; Zengerling, F.; Lisson, C.G.S. CT Radiomics and Clinical Feature Model to Predict Lymph Node Metastases in Early-Stage Testicular Cancer. *Onco* **2023**, *3*, 65–80. https://doi.org/10.3390/onco3020006

Academic Editor: Fred Saad

Received: 20 February 2023
Revised: 29 March 2023
Accepted: 3 April 2023
Published: 10 April 2023

presented combined CT-based radiomics–clinical model represents an exciting non-invasive tool for individualised LN metastasis prediction in testicular germ cell tumours. Multi-centre validation is required to generate high-quality evidence for its clinical application.

Keywords: radiomics signature; prediction; machine learning; testicular cancer; personalised oncology; precision imaging

1. Introduction

Testicular germ cell tumours (TGCTs) are the most common malignancy among men aged 15–40 [1,2]. Its characteristic patient population and high cure rate make this disease unique, constituting one of the few success stories in cancer care [3,4]. Besides cure reducing the amount of therapy-related acute and long-term toxicity is the goal of care due to the young age of the TGCT patients and the long life expectancy following curative therapy [5–10]. The main risk factors for TGCTs include cryptorchidism, family or personal history of TGCT and contact with organochlorine compounds [11,12]. TGCTs are classified histologically into seminoma and non-seminoma, including pure non-seminoma and mixed germ cell tumours, with seminoma accounting for approximately 55% of all cases with an average age at diagnosis in the fourth decade of life, about eight years later than non seminoma [12]. TGCT are diagnosed by physical examination, testicular ultrasound and specific tumour markers, such as alpha-fetoprotein (AFP), beta-hCG (β-hCG) and lactate dehydrogenase (LDH) [13,14].

Ninety-five percent of all metastases from TGCTs involve the ipsilateral retroperitoneal lymph nodes. Thus the present German guidelines recommend that in early-stage semi noma, patients with certain criteria, such as a tumour with a diameter >4 cm, an adjuvant therapy be applied, consisting of either one to two cycles of carboplatin or radiotherapy of the paraaortic region with 20 Gy [15]. However, retroperitoneal lymph node dissection (RPLND) is the only treatment modality to correctly stage the nodal status of early testicular cancer. Unfortunately, due to the short- and long-term complications, such as retrograde ejaculation, the implementation of adjuvant chemotherapy regimens, and the excellent prognosis with surveillance approaches in stage I disease, RPLND plays a negligible role as the primary treatment of early-stage TGCTs [16]. The most commonly used tumour markers, AFP, β-HCG, and LDH, are not very specific and are present in only about 60% of men with testicular cancer [14,17]. Worse, some conditions lead to false-positive elevation of testicular markers, such as liver disease or genetic reasons [18].

Due to its exceptional spatial resolution, CT imaging is regarded as well suited for identifying pathologically enlarged lymph nodes; in clinical practice, a short axis larger than 7–8 mm is considered pathologic (AUC with a sensitivity and specificity approaching 70%) [19]. Nonetheless, CT cannot distinguish between affected and normal lymph nodes in small lymph nodes [20].

Suboptimal therapeutical management, however, jeopardises the excellent outcomes of TGCT patients, with either over- or undertreatment being equally harmful.

Advanced medical imaging integrating high-resolution image acquisition, powerful computational technologies and artificial intelligence (AI)-based image analysis enabled researchers to develop the field of radiomics [21,22]. This way, data characterisation algorithms can detect specific diagnostic image patterns and convert them into quantitative mineable "big data" [23,24].

In the era of precision medicine, AI-based image analysis addresses the challenges of biopsy with the advantages of being non-invasive, repeatable, and applicable to hard-to reach lesions within the body by analysing texture features of a region of interest (ROI) that reflect tumour physiology and radiologic phenotype according to current data [25,26].

Many studies have evaluated the diagnostic potential of radiomics for classifying lymph nodes in different cancer types, including gastric, rectal, and bladder cancer with promising results [27–30]. AI-based advanced imaging could provide new imaging biomarkers or radiomic signatures to combat the urgent problem of under- or overtreatment of TGCT patients.

Our study is the first to investigate computed tomography (CT) radiomics models integrating clinical risk factors for the individualised prediction of lymph node metastasis in patients with early-stage TGCT, thus promoting precision imaging in clinical oncology.

Based on the findings above, we hypothesised that:

(1) The radiomics features extracted from retroperitoneal lymph nodes might potentially predict TGCT recurrence.
(2) Integrating important clinical factors, including age, histotype, AFP, ß-HCG, and BMI, into a combined clinical-radiomics model might add an incremental value to predict TGCT recurrence.

2. Materials and Methods

2.1. Patients and Imaging Protocol

Ninety-one treatment-naive patients with surgically proven stage I TGCT who underwent contrast-enhanced CT scans at our institution between January 2006 and December 2016 were included in this retrospective study.

Patient demographic, laboratory and clinical data were collected through a careful review of electronic medical records and the radiology information system. Exclusion criteria included incomplete clinical or imaging records and no histologic confirmation after surgery.

The primary endpoint of our study was retroperitoneal LN metastases from TGCT based on subsequent clinical and imaging examinations determined from records in electronic medical records.

Of the 167 patients originally screened, 91 could be included in the final study cohort according to the selection criteria. The patients in the final study cohort were followed up for at least six years after orchiectomy.

A flowchart of the cohort selection is shown in Figure 1.

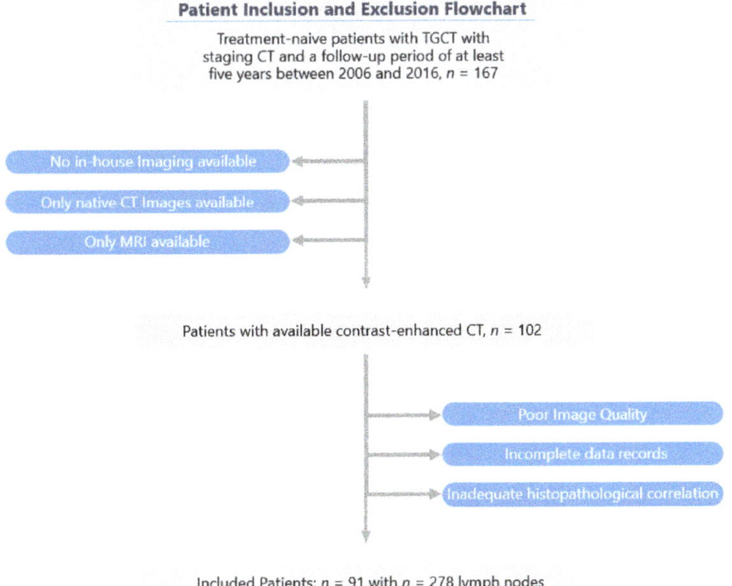

Figure 1. Recruitment pathway of the study.

CT scans were conducted before orchiectomy (+/−2 weeks) (mean time 3 ± 11 days, range 2–24) to determine disease status. Images were obtained as part of the routine staging on the Philips Brilliance CT 16-channel multi-row detector CT or Philips Brilliance CT 64-channel scanner (Philips Healthcare, Cleveland, OH, USA). CT scans were performed using acquisition and reconstruction parameters by the standard protocol after intravenous contrast injection of Ultravist® 370 (Bayer Schering Pharma, Berlin, Germany) at a weight-matched dose with a delay of 70–80 s for the portovenous phase of the chest and abdomen (tube voltage 100 kV–120 kV with automatically calculated tube current, matrix of 512 × 512, in-plane resolution between 0.62 × 0.62 mm and 0.86 × 0.86 mm, section thickness of 2.0–5.0 mm). Using two different CT scanners, a heterogeneous data set was generated to represent a routine clinical scenario as well as possible.

2.2. Segmentation and Radiomic Feature Extraction

First introduced by Haralick et al. in 1973 [31], image feature extraction, such as histogram features or features from the co-occurrence matrix, has demonstrated eminent potential in various questions in different cancers [22,32].

Three-dimensional region-of-interest segmentation, texture analysis, and feature extraction were conducted using mint Lesion™ software (version 3.8.4, mint Medical GmbH, Heidelberg, Germany). Details of the extraction settings are given in Appendix A, Table A1. The schematic diagram for ROI segmentation and feature extraction for model development is shown in Figure 2.

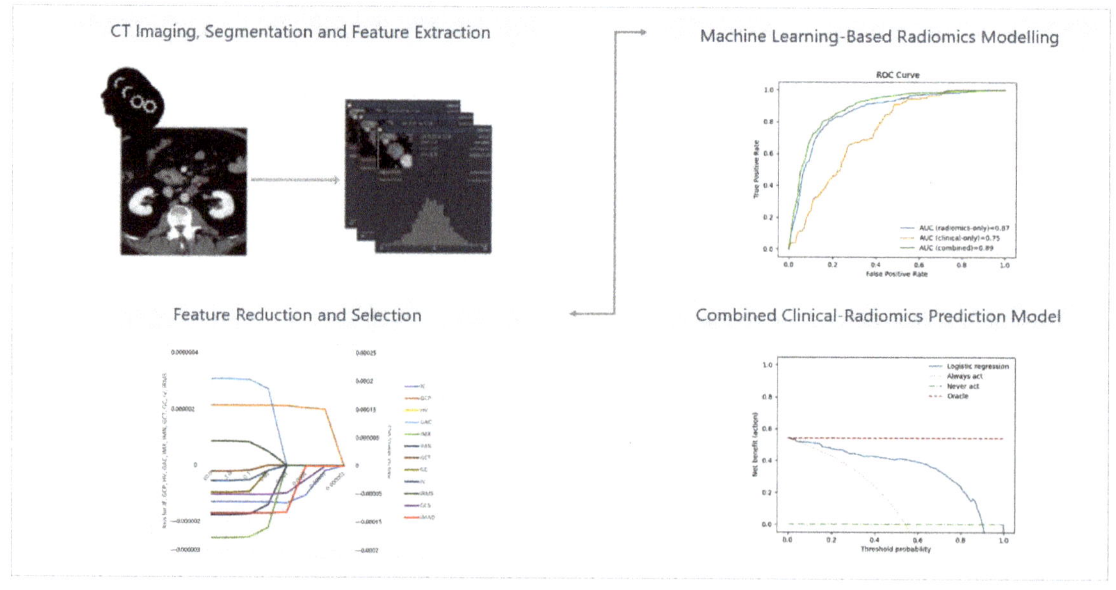

Figure 2. The schematic diagram for ROI segmentation and feature extraction for model development. Details regarding the extraction settings are listed in Appendix A, Table A1.

Two board-certified radiologists, with over 10 years of experience in oncologic imaging and over 8 years' experience in texture analysis, analysed all images.

Three retroperitoneal lymph nodes along the infrarenal aorta were segmented per patient, resulting in 273 eligible samples randomly divided into a training set (n = 191) and a testing set (n = 82) at a ratio of 70:30.

Radiomic features were quantified regarding their distinctive pattern of grey levels within the ROI using texture feature descriptors according to the Image Biomarker Standardization Initiative (IBSI) guidelines [24].

Eighty-five imaging features were extracted from each ROI: features related to the 3D size and shape, first-order statistics characterising the distribution of voxel intensities within the selected region, and features relating to the grey-level co-occurrence matrix (see Tables A2 and A3 in Appendix A).

2.3. Feature Selection and Development of the Predictive Radiomics Model

Analogous to other data mining applications, radiomics extracts many texture features from the regions of interest [33].

For more generalisable, powerful, and faster modelling and reduced overfitting, we selected optimal features using the logistic regression model with the smallest absolute shrinkage and the selection operator (lasso) [34,35]. Each feature had an associated covariate coefficient. With a continuous increase in λ-value, some regression coefficients continuously declined and tended to 0. The remaining variables with non-zero values were chosen as the best-performing predictors. The optimal hyperparameter $\lambda = 0.001$ was found by grid search [36,37].

Multivariable logistic regression developed the most appropriate radiomics model by using the selected radiomic features as the input variables to classify between the binary output variables.

Patients with LN metastases within the 6-year observation period were assigned to the high-risk group, whereas those with complete remission were classified in the low-risk group.

To handle the imbalance between LN metastases (negative vs. positive, 81/10) and avoid bias toward majority class cases to achieve a high classification rate, we applied the synthetic minority over-sampling technique (SMOTE) to the training cohort. SMOTE is an approach in which the minority class is over-sampled by creating "synthetic" examples rather than over-sampling with replacement. Thus, more related minority class samples to learn from are provided, allowing the learner to carve broader decision regions, leading to more coverage of the minority class limitations [38]. For greater generalisability of our results, we performed a stratified 10-fold cross-validation on the under-sampled data in all experiments to train and test the model resulting in a train and test partition of 90% and 10%, respectively, for each fold. We performed patient-specific splits to ensure that each patient's lymph nodes remained together in either the training or test set. We reported the mean and standard deviation of the area under the ROC-curve, accuracy, precision, recall, and F1-sore over the test set results of the ten runs. Furthermore, receiver operating characteristic (ROC) curves were plotted for each cohort. To ensure that our model was more than just a complicated surrogate for volume, we ran our experiments using only Volume and Mean Intensity as input features.

The correlation coefficients and constant of the model were computed (Figure 3, Appendix A, Figure A1). It is worth mentioning that the feature selection and the model construction were all from the date of the training cohort.

Discrimination performance was assessed by the Harrell concordance index (C-index).

The feature selection and the construction of the radiomics signature model were performed using our in-house software programmed with the Python Scikit-learn package (Python version 3.10, Scikit-learn version Scikit-learn 0.23.3, http://scikit-learn.org/) [36,39].

The features IMAD (Intensity Median Absolute Deviation) and GCS (GLCM Cluster Shade) use the secondary axis; all other features use the primary axis.

The following are the abbreviations used for the features:

IE—Intensity Energy

IMAD—Intensity Median Absolute Deviation

GCP—Glcm Cluster Prominence

GCS—Glcm Cluster Shade

HV—Histogram Variance

GAC—Glcm Auto Correlation

IMX—Intensity Maximum

IMN—Intensity Mean
GCT—Glcm Cluster Tendency
GC—Glcm Contrast
IV—Intensity Variance
IRMS—Root Mean Square

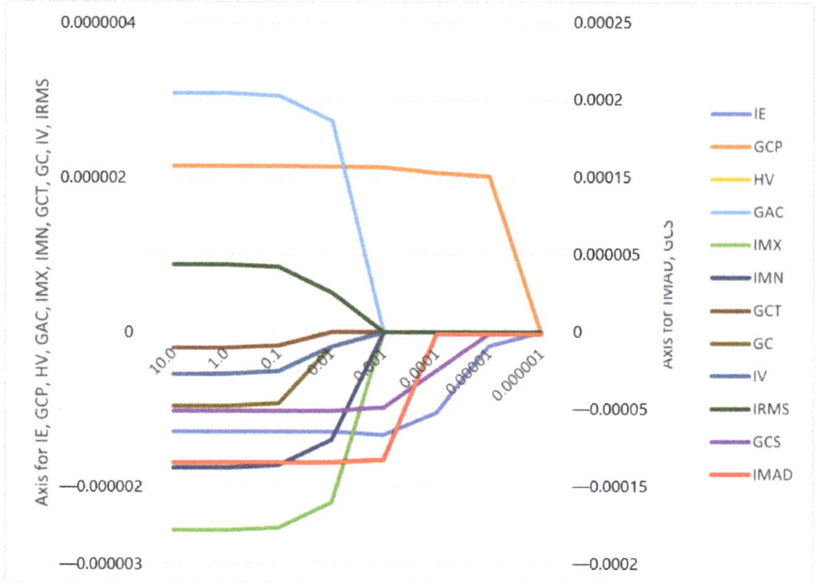

Figure 3. Feature weights generated by the LASSO logistic regression model's coefficients indicate positive or negative correlation with lymph node metastasis.

2.4. Development of the Clinical and the Combined Prediction Models

The clinical factors included in our analysis were age, AFP level, B-HCG level, histotype (seminoma and non-seminoma), and body mass index (BMI). These factors were included as they have all been suggested to be prognostic in TGCT [40–43].

Our study included purely clinical and laboratory chemistry parameters to represent a real-life scenario for the individualised preoperative prediction of LNM at the time of the CT scan.

The selected clinical features and their relationship to lymph node metastasis were assessed with a univariable logistic regression algorithm in the training set. Variables with $p < 0.2$ from the univariable analysis were included for further application in a multivariable logistic regression algorithm using forward stepwise selection. A cutoff value of 0.25 is supported by the literature [44,45].

Then, multivariable logistic regression analysis built three prediction models—a radiomics-only model, a clinical-only model and a combined clinical-radiomics model incorporating the selected radiomics and clinical features.

Their predictive performance for detecting LN metastasis was assessed using the receiver operating characteristic curve (ROC) analysis, in which the areas under the curve (AUC), accuracy, precision, and F1-Score were established.

The clinical utility was demonstrated by decision curve analysis (DCA) to evaluate the net benefits of the prediction models at different threshold probabilities in the training cohort and compare their discriminatory performance.

3. Results

3.1. Clinical Features

The study flowchart is presented in Figure 1.

Ninety-one consecutive patients with histologically-proven TGCT (mean age 35.2 ± 9.4 years, range 18–63) met the criteria for participation in the study. In this cohort, 10 patients (9.1%) relapsed within the six-year observation period (mean 9.8, 35.2 ± 9.4 years, range 18–63); there were no statistically significant differences in clinical characteristics between the LNM-positive group and LNM-negative group. After univariable LR analysis, age, AFP level, B-HCG level, histotype, and body mass index (BMI) were independent predictors in the clinical model.

All patients' baseline clinical characteristics are summarised in Table 1.

Table 1. Baseline demographic and clinical data.

Average age (range)	35.2 ± 9.4 Years (18–63)
Histological type	
Seminoma	60 Patients (66%)
Non-seminoma	31 Patients (34%)
Tumour classification (T)	
T1a	64 (70%)
T1b	27 (30%)
Tumour marker	
AFP positive	21 Patients (19%)
B-HCG positive	40 Patients (44%)
AFP und B-HCG positive	10 Patients (11%)
BMI (range)	25.9 ± 4.6 (19.3–43.9)
Patients' status in 6-year follow up	
Complete remission (CR)	81 (89%)
Relapse of disease (RD) with metastatic lymph nodes	10 (11%)

In total, the dataset consisted of 273 sample instances (three LN ROIs/patient), with 33 instances in the category "relapse of disease" (minority class) and 240 instances in the category "without relapse of disease" (majority class). According to a proportion of 7:3, the 273 sample instances were randomly divided into a training cohort (n = 191) and a test cohort (n = 82).

Due to the class imbalance in the dataset, the under-sampling technique called "Instance Hardness Threshold" was used to balance the data. The balanced data were used for the logistic regression machine learning mode.

3.2. Feature Selection and Performance of the Radiomics Prediction Model

A total of 85 radiomics features were extracted from the venous-phase CT images of the training cohort (Appendix A, Table A2). After screening these features, we chose the 12 radiomics features that had non-zero coefficients using the LASSO logistic regression model as the best-performing predictors for LN metastasis (Figure 3; Appendix A, Table A3).

These features were used as input volume for the machine learning-based radiomics modelling. Traditional measurements of machine learning-based modelling were used, including accuracy, precision, F1-Score, and the area under the ROC curve (AUC), to assess the performance of predicting lymph node metastases.

All tests were two-sided; $p < 0.05$ was considered statistically significant.

In the ROC analysis of the radiomics model, the classification evaluation metrics of the 10-fold cross-validation were AUC 0.84 ± 0.17, accuracy 0.76 ± 0.12, precision 0.80 ± 0.18, recall 0.72 ± 0.23, and F1 score 0.73 ± 0.17 in the training cohort (Table 2).

Table 2. Performance of the radiomics, clinical, and combined models.

Model	AUC (95% CI)	Accuracy	Precision	Recall	F1 Score
Radiomics-only	0.87± 0.04	0.80 ± 0.06	0.81 ± 0.06	0.80 ± 0.08	0.80 ± 0.06
Clinical-only	0.75 ± 0.08	0.68 ± 0.10	0.66 ± 0.11	0.71 ± 0.16	0.68 ± 0.12
Combined clinical-radiomics	0.89 ± 0.03	0.81 ± 0.04	0.80 ± 0.07	0.83 ± 0.06	0.81 ± 0.04

Using only Volume and Intensity Mean as input features led to inferior results with an accuracy of 0.58 ± 0.16, with a precision and recall of 0.11 ± 0.07 and 0.43 ± 0.27 respectively.

3.3. Performance of the Clinical and the Combined Prediction Model

The clinical-only and combined clinical-radiomics models were built by applying multivariable logistic regression analysis.

The predictive performances of the radiomics-only, the clinical-only and the combined clinical-radiomics models on the training cohort are shown in Table 2.

The different models' overall accuracy and F1 score for predicting LN metastases were 77% (range: 65–90%, AUC = 0.60–0.94) and 61% (range: 20–90%).

The combined clinical-radiomics model showed the best prediction accuracy with 90% (AUC 0.94–0.10), indicating that adding radiomics features could improve the predictive performance.

Figure 4 shows the receiver operating characteristic (ROC) curves for the clinical, the radiomics, and the combined clinical-radiomics models on the training cohort.

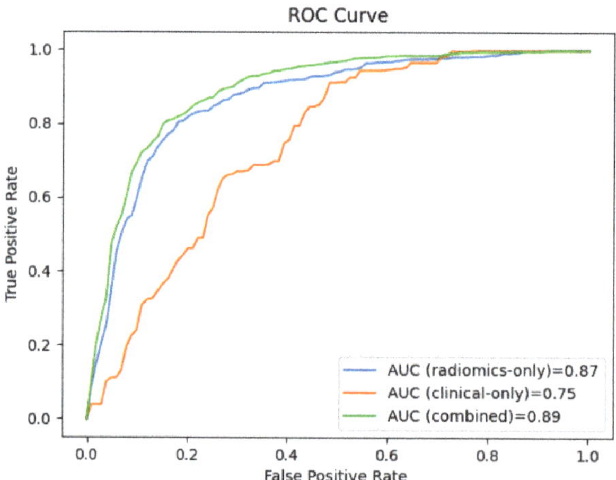

Figure 4. The ROC curves of the radiomics-only, the clinical-only, and the combined clinical-radiomics models show that the combined model outperforms the radiomics and the clinical model in predicting LN metastasis (training cohort 94% vs. 84% and 60%, respectively).

We performed a decision curve analysis to assess the clinical value of the combined clinical-radiomics model. With threshold probability on the x-axis and net benefit on the y-axis, the decision curve analysis graph illustrates the trade-offs between true and false positives (describing benefit and harm) as the threshold probability changes (see Figure 5).

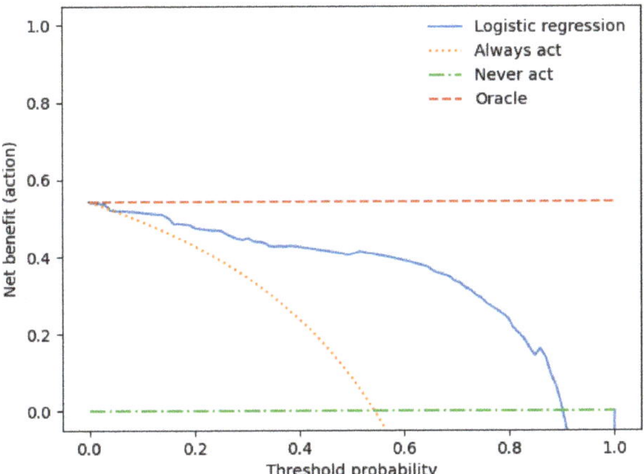

Figure 5. The decision curve analysis for the combined prediction model.

The x-axis represents the threshold probability, the y-axis the net benefit, and the blue line shows the combined prediction model. The green line represents the hypothesis that no patients had LN metastases, and the orange line that all patients had LN metastases. The threshold probability is where the treatment's expected benefit equals the benefit of avoiding treatment. If the possibility of LN metastasis is over the threshold probability, then a therapeutical strategy for LN metastases should be adopted. The DCA of the combined model shows that if the threshold possibility is between 0 and 0.13, then using the combined model to predict LNM adds more benefit than treating either or all patients.

4. Discussion

We developed a clinical-radiomics model for the individualised preoperative prediction of LNM in testicular germ cell tumour (TGCT) patients that consisted of clinical risk factors and radiomics features to identify the stage I (TGCT) patients who required adjuvant therapy and those who did not.

Our main findings can be summarised by the following:

Using multivariable logistic regression analysis, we constructed a radiomics-only model, a clinical-only model, and a combined predictive model integrating clinical and radiomics features. The combined radiomics–clinical model showed the highest accuracy in predicting LNM (AUC = 0.89 ± 0.03; 95% CI); accuracy: 81%, precision 80%, recall 83%, and F1 score 81%.

Most TGCT patients initially present with stage I disease, and >95% of all stage I seminoma or non-seminoma patients are cured regardless of the therapeutical strategy [46–48], resulting in controversies regarding adjuvant chemotherapy, radiotherapy, or retroperitoneal lymph node dissection following orchiectomy due to short- and long-term side effects, such as secondary malignancies, cardiovascular disease, peripheral neuropathy, and loss of antegrade ejaculation [5–7,49].

The serum biomarkers AFP, β-hCG, and LDH are substantial instruments for diagnosing, prognostication, and monitoring testicular cancer, which is reflected in the International Germ Cell Cancer Consensus Group prognostic index [17,50,51]. However, sensitivity is limited; up to 40% of patients with recurrence have "normal" values [52].

Several studies have proposed further prognostic clinical risk factors, including age and BMI, but their roles have not yet been sufficiently clarified, with somewhat controversial discussion [40–43].

To date, neither imaging nor serum tumour markers have been proven to be suitable predictors of the presence of lymph node metastases [53,54]. However, the inherently excellent prognosis can be put at risk by suboptimal treatment, with over- and undertreatment being equally detrimental.

Several studies demonstrate the ability of radiomics based on MR- or CT-imaging to detect lymph node metastasis, including lung, oesophagal, breast, cervical, bladder, and colorectal cancer [28,29,55–58]. Classification accuracy in these studies ranged from 76% to 84%, which is lower than the results of our study.

Until now, few studies have been performed to distinguish between benign and malignant LN in testicular cancer.

In their study, Baessler et al. showed that a machine-learning classifier based on (CT) radiomics could predict the histopathology of lymph nodes after LN dissection following chemotherapy in patients with metastatic non-seminomatous germ cell tumours of the testis [59]. This single-centre retrospective study included eighty patients with a total of 204 lesions classified by a support vector model and achieved 81% classification accuracy.

Nevertheless, in contrast to our study, they did not include clinical variables in their radiomics approach to further increase diagnostic performance.

Furthermore, they split the study cohort, which was altogether of moderate size, into three subgroups, with only 19 patients in the test group and with an overall reduction in statistical significance as a result. To address the moderate dataset, we used a cross validation approach, which involves repeated data splitting to prevent overfitting while obtaining accurate estimates of the model coefficients [60]. Lewin et al. achieved in their retrospective, single-centre study on 77 metastatic TGCT patients with 102 lesions a classification accuracy of only 72% [61].

Lewin et al. used only one single CT scanner. In contrast, our study analysed data from two scanners, thus being more representative of data acquired during routine clinical practice. Like Baessler et al., Lewin et al. did not integrate clinical factors into a combined clinical-radiomics model.

Given our 10-fold cross-validation approach, the a priori inhomogeneity of our dataset and the integration of clinical risk factors, we are convinced that our combined prediction model is more generalisable, and forthcoming investigations should further validate our trained model in prospective studies.

Beyond radiomics-based models, several clinical models exist to predict the occurrence of LNM in TGCT. However, these models yielded conflicting results and could not be included in today's clinical decision-making [53,62–64].

Taken together, identifying and implementing novel biomarkers might be helpful for early diagnosis and monitoring of disease relapse.

Our study is the first to use a combined CT-based radiomics model integrating clinical predictors for the individualised preoperative prediction of LNM in early-stage TGCT to reduce overtreatment in this group of young patients.

However, we acknowledge some limitations in the present study.

As a retrospective study with a modest cohort size, there may be inevitable selection bias. Furthermore, classes were highly unbalanced, in line with the normal distribution, with 80% of all stage 1 TGCT patients showing an excellent outcome. Nevertheless, unlike prior radiomics investigations on LN metastasis that mostly extracted features from the largest cross-sectional area, our study performed whole lesion analysis by considering all available CT slices, thus providing abundant information about tumour heterogeneity.

Second, our case was a single-institution study. Due to our patient population's high cure rate, it is challenging to power studies to examine prognostic and predictive factors adequately. However, prospective and multicenter validation is warranted to obtain higher-quality evidence for clinical use.

Moreover, only one (imaging) modality and the circulating tumour markers β-HCG and AFP were used in this study. Among other prognostic factors, such as lymphovascular or rete testis invasion, tumour size is the most valuable prognostic factor for early-stage seminoma relapse [65,66]. Our study included solely clinical and laboratory parameters that can be collected easily, quickly, and non-invasively so that a preoperative risk assessment of the individual patient can already be made at the time of CT.

In addition to the known serum markers, studies show the potential of non-coding RNAs as biomarkers with stem cell-associated microRNAs (miR-371a-3p and miR-302/367 clusters) outperforming the conventional tumour markers in detecting newly diagnosed TGCT patients [67,68].

If more modalities were combined as a multi-omics approach, the obtained feature pool might increase the ability to predict LNM in patients with testicular cancer.

Our presented CT-based radiomics–clinical model represents an exciting non-invasive prediction tool for individualised prediction of LN metastasis in testicular germ cell tumours to reduce overtreatment in this young group of patients. Multi-centre, retrospective validations and prospective randomised clinical trials should be undertaken to gain high-quality evidence for clinical applications in subsequent studies.

5. Conclusions

In conclusion, our combined clinical-radiomics model applied on preoperative CT imaging represents an exciting new tool for improved prediction of lymph node metastases in early-stage testicular germ cell tumour (TGCT) patients to reduce overtreatment in this group of young patients. The presented approach should be combined with novel clinical biomarkers, such as microRNAs (miR-371a-3p and miR-302/367 cluster) and further validated in larger, prospective clinical trials.

Author Contributions: Conceptualisation, C.S.L., F.Z. and C.G.S.L.; methodology, C.S.L., C.G.S.L. and M.G.; software, S.M., D.W. and M.G.; formal analysis, C.S.L. and C.G.S.L.; investigation, C.S.L., J.S. and C.G.S.L.; resources, C.S.L.; data curation, C.S.L., S.A.S. and C.G.S.L., writing—original draft preparation, C.S.L.; writing—review and editing, F.Z., S.A.S., M.B., D.W. and M.G.; visualisation, C.S.L., S.M., D.W. and C.G.S.L.; supervision, M.B.; project administration, C.S.L. All authors have read and agreed to the published version of the manuscript.

Funding: Radiological Cooperative Network (RACOON).

Institutional Review Board Statement: The study was conducted in accordance with the Declaration of Helsinki and approved by the ethics committee of the medical faculty of the University of Ulm (protocol code 155/18, 25 April 2018).

Informed Consent Statement: Informed consent was obtained from all subjects involved in the study.

Data Availability Statement: Data sharing is not applicable to this article.

Conflicts of Interest: The authors declare no conflict of interest.

Abbreviations

AFP	Alpha-fetoprotein
AUC	Area under the curve
CT	Computed tomography
β-HCG	Beta human chorionic gonadotropin
LNs	Lymph nodes
LNM	Lymph node metastases
LR	Logistic regression
ML	Machine learning
ROC	Receiver operating curve
ROI	Region of interest
TGCT	Testicular germ cell tumour

Appendix A

Table A1. Settings of the radiomics feature extraction.

Setting	Determination
Bin Method	FBN
Bin Amount	32
LoG Filter	0
LoG Sigma	2
Matrix Aggregation Method	3D Average Directions
Resample Filter	1
Resample Spacing X	1
Resample Spacing Y	1
Resample Spacing Z	1
Second-Order Distance	1
Threshold Filter	0

Table A2. Radiomics features extracted for model development.

Radiomics Features of First Order	Radiomics Features of Second Order: Gray Level Co-Occurrence Matrix (GLCM)
Histogram Minimum	Joint Maximum
Histogram Maximum	Joint Average
Histogram Range	Standart Deviation
Histogram Mean	Joint Variance
Histogram Variance	Joint Entropy
Histogram Standart Deviation	Difference Average
Histogram Skewness	Difference Variance
Histogram Kurtosis	Difference Entropy
Histogram Entropy	Sum of Averages
Histogram Uniformity	Sum of Variance
Histogram Mean Absolute Deviation	Sum of Entropy
Histogram Robust Mean Absolute Deviation	Angular Second Moment
Histogram Median Absolute Deviation	Contrast
Histogram Coefficient Variation	Dissimilarity
Histogram Quartile Coefficient Dispersion	Inverse Difference
Histogram Interquartile Range	Inverse Difference Normalised
Histogram P10th	Inverse Difference Moment
Histogram P25th	Inverse Difference Moment Normalised
Histogram P50th	Joint Maximum
Histogram P75th	Joint Average
Histogram P90th	Standart Deviation
Histogram Minimum Histogram Gradient Intensity	Joint Variance
Histogram MaximumHistogram Gradient Intensity	Joint Entropy
Intensity Minimum	Difference Average
Intensity Maximum	Difference Variance
Intensity Range	Difference Entropy
Intensity Mean	Sum of Averages
Intensity Variance	Sum of Variance
Intensity Standart Deviation	Sum of Entropy
Intensity Skewness	Angular Second Moment
Intensity Kurtosis	Contrast
Intensity Energy	Dissimilarity
Intensity P10th	Inverse Variance
Intensity P25th	Correlation
Intensity P50th	Auto Correlation

Table A2. *Cont.*

Radiomics Features of First Order	Radiomics Features of Second Order: Gray Level Co-Occurrence Matrix (GLCM)
Intensity P75th	Cluster Shade
Intensity P90th	Cluster Prominence
Intensity Root Mean Square	Cluster Tendency
Intensity Mean Absolute Deviation	Information Correlation 1
Intensity Robust Mean Absolute Deviation	Information Correlation 2
Intensity Median Absolute Deviation	Inverse Variance 41
Intensity Coefficient Variation	
Intensity Quartile Coefficient Dispersion	
Intensity Interquartile Range 44	

Table A3. Radiomics features selected by LASSO.

Radiomics Features of First Order	Radiomics Features of Second Order: Gray Level Co-Occurrence Matrix (GLCM)
Histogram Variance	Auto Correlation
Intensity Maximum	Cluster Shade
Intensity Mean	Cluster Prominence
Intensity Variance	Cluster Tendency
Intensity Energy	Contrast
Intensity Root Mean Square	
Intensity Median Absolute Deviation	

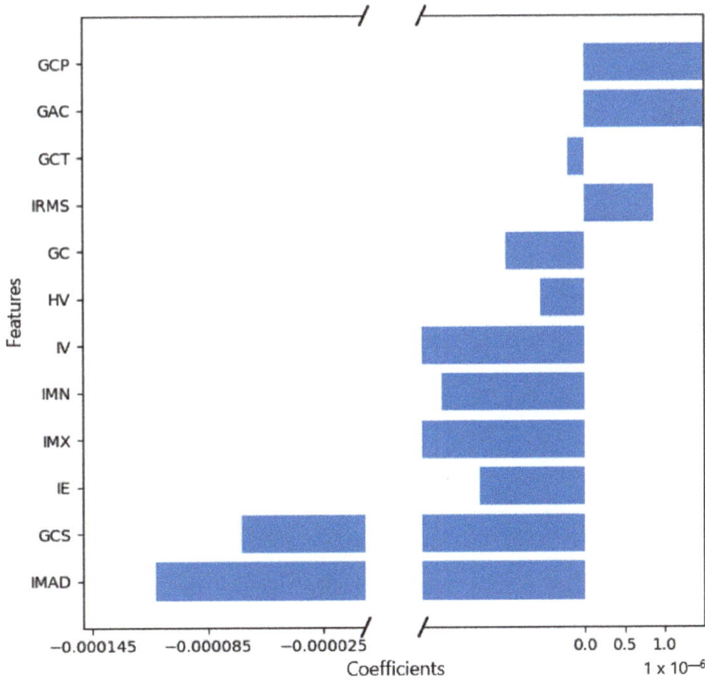

Figure A1. Feature weights generated by the coefficients of the logistic regression model indicating positive or negative correlation with lymph node metastasis.

References

1. Ruf, C.G.; Isbarn, H.; Wagner, W.; Fisch, M.; Matthies, C.; Dieckmann, K.P. *Changes in Epidemiologic Features of Testicular Germ Cell Cancer: Age at Diagnosis and Relative Frequency of Seminoma Are Constantly and Significantly Increasing*; Elsevier: Amsterdam, The Netherlands, 2014; p. 33-e1.
2. Bray, F.; Richiardi, L.; Ekbom, A.; Pukkala, E.; Cuninkova, M.; Møller, H. Trends in testicular cancer incidence and mortality in 22 European countries: Continuing increases in incidence and declines in mortality. *Int. J. Cancer* **2006**, *118*, 3099–3111. [CrossRef] [PubMed]
3. Einhorn, L.H. Treatment of testicular cancer: A new and improved model. *J. Clin. Oncol.* **1990**, *8*, 1777–1781. [CrossRef] [PubMed]
4. Kollmannsberger, C.; Tandstad, T.; Bedard, P.L.; Cohn-Cedermark, G.; Chung, P.W.; Jewett, M.A.; Powles, T.; Warde, P.; Daneshmand, S.; Protheroe, A.; et al. Patterns of relapse in patients with clinical stage I testicular cancer managed with active surveillance. *J. Clin. Oncol.* **2015**, *33*, 51–57. [CrossRef]
5. Fung, C.; Sesso, H.D.; Williams, A.M.; Kerns, S.L.; Monahan, P.; Zaid, M.A.; Feldman, D.; Hamilton, R.J.; Vaughn, D.J.; Beard, C.J.; et al. Multi-institutional assessment of adverse health outcomes among North American testicular cancer survivors after modern cisplatin-based chemotherapy. *J. Clin. Oncol.* **2017**, *35*, 1211. [CrossRef]
6. Huddart, R.; Norman, A.; Shahidi, M.; Horwich, A.; Coward, D.; Nicholls, J.; Dearnaley, D.P. Cardiovascular disease as a long-term complication of treatment for testicular cancer. *J. Clin. Oncol.* **2003**, *21*, 1513–1523. [CrossRef] [PubMed]
7. Travis, L.B.; Ng, A.K.; Allan, J.M.; Pui, C.H.; Kennedy, A.R.; Xu, X.G.; Purdy, J.A.; Applegate, K.; Yahalom, J.; Constine, L.S.; et al. Second malignant neoplasms and cardiovascular disease following radiotherapy. *J. Natl. Cancer Inst.* **2012**, *104*, 357–370. [CrossRef]
8. Kerns, S.L.; Fung, C.; Monahan, P.O.; Ardeshir-Rouhani-Fard, S.; Zaid, M.I.A.; Williams, A.M.; Stump, T.E.; Sesso, H.D.; Feldman, D.R.; Hamilton, R.J.; et al. Cumulative burden of morbidity among testicular cancer survivors after standard cisplatin-based chemotherapy: A multi-institutional study. *J. Clin. Oncol.* **2018**, *36*, 1505. [CrossRef]
9. Agrawal, V.; Dinh, P.C., Jr.; Fung, C.; Monahan, P.O.; Althouse, S.K.; Norton, K.; Cary, C.; Einhorn, L.; Fossa, S.D.; Adra, N.; et al. Adverse health outcomes among US testicular cancer survivors after cisplatin-based chemotherapy vs surgical management. *JNCI Cancer Spectr.* **2020**, *4*, pkz079. [CrossRef]
10. Tandstad, T.; Kollmannsberger, C.K.; Roth, B.J.; Jeldres, C.; Gillessen, S.; Fizazi, K.; Daneshmand, S.; Lowrance, W.T.; Hanna, N.H.; Albany, C.; et al. Practice makes perfect: The rest of the story in testicular cancer as a model curable neoplasm. *J. Clin. Oncol.* **2017**, *35*, 3525. [CrossRef]
11. Rajpert-De Meyts, E.; McGlynn, K.A.; Okamoto, K.; Jewett, M.A.; Bokemeyer, C. Testicular germ cell tumours. *Lancet* **2016**, *387*, 1762–1774. [CrossRef]
12. Cheng, L.; Albers, P.; Berney, D.M.; Feldman, D.R.; Daugaard, G.; Gilligan, T.; Looijenga, L.H. Testicular cancer. *Nat. Rev. Dis. Prim.* **2018**, *4*, 29. [CrossRef]
13. Dieckmann, K.P.; Simonsen-Richter, H.; Kulejewski, M.; Anheuser, P.; Zecha, H.; Isbarn, H.; Pichlmeier, U. Serum tumour markers in testicular germ cell tumours: Frequencies of elevated levels and extents of marker elevation are significantly associated with clinical parameters and with response to treatment. *BioMed Res. Int.* **2019**, *2019*, 5030349. [CrossRef] [PubMed]
14. Gilligan, T.D.; Hayes, D.F.; Seidenfeld, J.; Temin, S. ASCO clinical practice guideline on uses of serum tumor markers in adult males with germ cell tumors. *J. Oncol. Pract.* **2010**, *6*, 199. [CrossRef] [PubMed]
15. Kliesch, S.; Schmidt, S.; Wilborn, D.; Aigner, C.; Albrecht, W.; Bedke, J.; Beintker, M.; Beyersdorff, D.; Bokemeyer, C.; Busch, J.; et al. Management of germ cell tumours of the testis in adult patients. German clinical practice guideline part I: Epidemiology, classification, diagnosis, prognosis, fertility preservation, and treatment recommendations for localized stages. *Urol. Int.* **2021**, *105*, 169–180. [CrossRef]
16. Patel, H.D.; Joice, G.A.; Schwen, Z.R.; Semerjian, A.; Alam, R.; Srivastava, A.; Allaf, M.E.; Pierorazio, P.M. Retroperitoneal lymph node dissection for testicular seminomas: Population-based practice and survival outcomes. *World J. Urol.* **2018**, *36*, 73–78. [CrossRef] [PubMed]
17. Murray, M.J.; Huddart, R.A.; Coleman, N. The present and future of serum diagnostic tests for testicular germ cell tumours. *Nat. Rev. Urol.* **2016**, *13*, 715–725. [CrossRef]
18. Albers, P.; Albrecht, W.; Algaba, F.; Bokemeyer, C.; Cohn-Cedermark, G.; Fizazi, K.; Horwich, A.; Laguna, M.P.; Nicolai, N.; Oldenburg, J. Guidelines on testicular cancer: 2015 update. *Eur. Urol.* **2015**, *68*, 1054–1068. [CrossRef]
19. Hudolin, T.; Kastelan, Z.; Knezevic, N.; Goluza, E.; Tomas, D.; Coric, M. Correlation between retroperitoneal lymph node size and presence of metastases in nonseminomatous germ cell tumors. *Int. J. Surg. Pathol.* **2012**, *20*, 15–18. [CrossRef]
20. Hale, G.R.; Teplitsky, S.; Truong, H.; Gold, S.A.; Bloom, J.B.; Agarwal, P.K. Lymph node imaging in testicular cancer. *Transl. Androl. Urol.* **2018**, *7*, 864. [CrossRef]
21. Obermeyer, Z.; Emanuel, E.J. Predicting the future—Big data, machine learning, and clinical medicine. *N. Engl. J. Med.* **2016**, *375*, 1216. [CrossRef]
22. Lambin, P.; Leijenaar, R.T.; Deist, T.M.; Peerlings, J.; De Jong, E.E.; Van Timmeren, J.; Sanduleanu, S.; Larue, R.T.H.M.; Even, A.J.G.; Jochems, A.; et al. Radiomics: The bridge between medical imaging and personalised medicine. *Nat. Rev. Clin. Oncol.* **2017**, *14*, 749–762. [CrossRef] [PubMed]
23. Gillies, R.J.; Kinahan, P.E.; Hricak, H. Radiomics: Images are more than pictures, they are data. *Radiology* **2016**, *278*, 563. [CrossRef]

4. Zwanenburg, A.; Vallières, M.; Abdalah, M.A.; Aerts, H.J.; Andrearczyk, V.; Apte, A.; Ashrafinia, S.; Bakas, S.; Beukinga, R.J.; Boellaard, R.; et al. The image biomarker standardisation initiative: Standardised quantitative radiomics for high-throughput image-based phenotyping. *Radiology* **2020**, *295*, 328–338. [CrossRef] [PubMed]

5. Sollini, M.; Antunovic, L.; Chiti, A.; Kirienko, M. Towards clinical application of image mining: A systematic review on artificial intelligence and radiomics. *Eur. J. Nucl. Med. Mol. Imaging* **2019**, *46*, 2656–2672. [CrossRef]

6. Ibrahim, A.; Primakov, S.; Beuque, M.; Woodruff, H.; Halilaj, I.; Wu, G.; Refaee, T.; Granzier, R.; Widaatalla, Y.; Hustinx, R.; et al. Radiomics for precision medicine: Current challenges, future prospects, and the proposal of a new framework. *Methods* **2021**, *188*, 20–29. [CrossRef]

7. Dong, D.; Tang, L.; Li, Z.Y.; Fang, M.J.; Gao, J.B.; Shan, X.H.; Ying, X.-J.; Sun, Y.-S.; Fu, J.; Wang, X.-X.; et al. Development and validation of an individualised nomogram to identify occult peritoneal metastasis in patients with advanced gastric cancer. *Ann. Oncol.* **2019**, *30*, 431–438. [CrossRef]

8. Huang, Y.; Liang, C.; He, L.; Tian, J.; Liang, C.; Chen, X.; Ma, Z.-L.; Liu, Z.-Y. Development and validation of a radiomics nomogram for preoperative prediction of lymph node metastasis in colorectal cancer. *J. Clin. Oncol.* **2016**, *34*, 2157–2164. [CrossRef]

9. Wu, S.; Zheng, J.; Li, Y.; Yu, H.; Shi, S.; Xie, W.; Liu, H.; Su, Y.; Huang, J.; Lin, T.; et al. A Radiomics Nomogram for the Preoperative Prediction of Lymph Node Metastasis in Bladder CancerA Radiomics Nomogram for Bladder Cancer. *Clin. Cancer Res.* **2017**, *23*, 6904–6911. [CrossRef]

10. Gao, J.; Han, F.; Jin, Y.; Wang, X.; Zhang, J. A radiomics nomogram for the preoperative prediction of lymph node metastasis in pancreatic ductal adenocarcinoma. *Front. Oncol.* **2020**, *10*, 1654. [CrossRef]

11. Haralick, R.M.; Shanmugam, K.; Dinstein, I.H. Textural features for image classification. *IEEE Trans. Syst. Man Cybern.* **1973**, *SMC-3*, 610–621. [CrossRef]

12. Shen, C.; Liu, Z.; Guan, M.; Song, J.; Lian, Y.; Wang, S.; Tang, Z.; Dong, D.; Kong, L.; Wang, M.; et al. 2D and 3D CT radiomics features prognostic performance comparison in non-small cell lung cancer. *Transl. Oncol.* **2017**, *10*, 886–894. [CrossRef]

13. Duin, R.P.; Pekalska, E. Dissimilarity Representation for Pattern Recognition. In *Foundations And Applications*; World Scientific: Singapore, 2005; Volume 64.

14. Tibshirani, R. The lasso method for variable selection in the Cox model. *Stat. Med.* **1997**, *16*, 385–395. [CrossRef]

15. Tibshirani, R. Regression shrinkage and selection via the lasso: A retrospective. *J. R. Stat. Soc. Ser. B* **2011**, *73*, 273–282. [CrossRef]

16. Pedregosa, F.; Varoquaux, G.; Gramfort, A.; Michel, V.; Thirion, B.; Grisel, O.; Blondel, M.; Prettenhofer, P.; Weiss, R.; Dubourg, V.; et al. Scikit-learn: Machine learning in Python. *J. Mach. Learn. Res.* **2011**, *12*, 2825–2830.

17. Agrawal, T. Hyperparameter Optimization Using Scikit-Learn. In *Hyperparameter Optimization in Machine Learning: Make Your Machine Learning and Deep Learning Models More Efficient [Internet]*; Apress: Berkeley, CA, USA, 2021; pp. 31–51. [CrossRef]

18. Chawla, N.V.; Bowyer, K.W.; Hall, L.O.; Kegelmeyer, W.P. SMOTE: Synthetic minority over-sampling technique. *J. Artif. Intell. Res.* **2002**, *16*, 321–357. [CrossRef]

19. van Rossum, G.; Drake, F.L. *Python/C API Manual-Python 2.6*; CreateSpace: Scotts Valley, CA, USA, 2009.

20. Fosså, S.D.; Cvancarova, M.; Chen, L.; Allan, A.L.; Oldenburg, J.; Peterson, D.R.; Travis, L.B. Adverse prognostic factors for testicular cancer–specific survival: A population-based study of 27,948 patients. *J. Clin. Oncol.* **2011**, *29*, 963–970. [CrossRef]

21. Parker, C.; Milosevic, M.; Panzarella, T.; Banerjee, D.; Jewett, M.; Catton, C.; Tew-George, B.; Gospodarowicz, M.; Warde, P. The prognostic significance of the tumour infiltrating lymphocyte count in stage I testicular seminoma managed by surveillance. *Eur. J. Cancer* **2002**, *38*, 2014–2019. [CrossRef] [PubMed]

22. Lerro, C.; McGlynn, K.; Cook, M. A systematic review and meta-analysis of the relationship between body size and testicular cancer. *Br. J. Cancer* **2010**, *103*, 1467–1474. [CrossRef] [PubMed]

23. Dieckmann, K.P.; Hartmann, J.T.; Classen, J.; Diederichs, M.; Pichlmeier, U. Is increased body mass index associated with the incidence of testicular germ cell cancer? *J. Cancer Res. Clin. Oncol.* **2009**, *135*, 731–738. [CrossRef]

24. Mickey, R.M.; Greenland, S. The impact of confounder selection criteria on effect estimation. *Am. J. Epidemiol.* **1989**, *129*, 125–137. [CrossRef] [PubMed]

25. Bendel, R.B.; Afifi, A.A. Comparison of stopping rules in forward "stepwise" regression. *J. Am. Stat. Assoc.* **1977**, *72*, 46–53.

26. Powles, T.B.; Bhardwa, J.; Shamash, J.; Mandalia, S.; Oliver, T. The changing presentation of germ cell tumours of the testis between 1983 and 2002. *BJU Int.* **2005**, *95*, 1197–1200. [CrossRef] [PubMed]

27. Albers, P.; Siener, R.; Kliesch, S.; Weissbach, L.; Krege, S.; Sparwasser, C.; Schulze, H.; Heidenreich, A.; de Riese, W.; Loy, V.; et al. Risk factors for relapse in clinical stage I nonseminomatous testicular germ cell tumors: Results of the German Testicular Cancer Study Group Trial. *J. Clin. Oncol.* **2003**, *21*, 1505–1512. [CrossRef] [PubMed]

28. Oldenburg, J.; Fosså, S.; Nuver, J.; Heidenreich, A.; Schmoll, H.J.; Bokemeyer, C.; Horwich, A.; Beyer, J.; Kataja, V. Testicular seminoma and non-seminoma: ESMO Clinical Practice Guidelines for diagnosis, treatment and follow-up. *Ann. Oncol.* **2013**, *24*, vi125–vi132. [CrossRef]

29. Kier, M.G.; Hansen, M.K.; Lauritsen, J.; Mortensen, M.S.; Bandak, M.; Agerbaek, M.; Holm, N.V.; Dalton, S.O.; Andersen, K.K.; Johansen, C.; et al. Second malignant neoplasms and cause of death in patients with germ cell cancer: A Danish nationwide cohort study. *JAMA Oncol.* **2016**, *2*, 1624–1627. [CrossRef]

30. von Eyben, F.E. Laboratory markers and germ cell tumors. *Crit. Rev. Clin. Lab. Sci.* **2003**, *40*, 377–427. [CrossRef] [PubMed]

51. Lobo, J.; Leão, R.; Jerónimo, C.; Henrique, R. Liquid biopsies in the clinical management of germ cell tumor patients: State-of-the art and future directions. *Int. J. Mol. Sci.* **2021**, *22*, 2654. [CrossRef]
52. Trigo, J.M.; Tabernero, J.M.; Paz-Ares, L.; García-Llano, J.L.; Mora, J.; Lianes, P.; Esteban, E.; Salazar, R.; López-López, J.; Cortés-Funes, H.; et al. Tumor markers at the time of recurrence in patients with germ cell tumors. *Cancer* **2000**, *88*, 162–16 [CrossRef]
53. Steyerberg, E.; Gerl, A.; Fossa, S.; Sleijfer, D.; de Wit, R.; Kirkels, W.; Schmeller, N.; Clemm, C.; Habbema, J.D.; Keizer, H.J. Validity of predictions of residual retroperitoneal mass histology in nonseminomatous testicular cancer. *J. Clin. Oncol.* **1998**, *16*, 269–27 [CrossRef]
54. Vergouwe, Y.; Steyerberg, E.W.; Foster, R.S.; Habbema, J.D.F.; Donohue, J.P. Validation of a prediction model and its predictors for the histology of residual masses in nonseminomatous testicular cancer. *J. Urol.* **2001**, *165*, 84–88. [CrossRef]
55. Andersen, M.B.; Harders, S.W.; Ganeshan, B.; Thygesen, J.; Torp Madsen, H.H.; Rasmussen, F. CT texture analysis can help differentiate between malignant and benign lymph nodes in the mediastinum in patients suspected for lung cancer. *Acta Radiol.* **2016**, *57*, 669–676. [CrossRef]
56. Tan, X.; Ma, Z.; Yan, L.; Ye, W.; Liu, Z.; Liang, C. Radiomics nomogram outperforms size criteria in discriminating lymph node metastasis in resectable esophageal squamous cell carcinoma. *Eur. Radiol.* **2019**, *29*, 392–400. [CrossRef]
57. Zhang, X.; Yang, Z.; Cui, W.; Zheng, C.; Li, H.; Li, Y.; Lu, L.; Mao, J.; Zeng, W.; Yang, X.; et al. Preoperative prediction of axillary sentinel lymph node burden with multiparametric MRI-based radiomics nomogram in early-stage breast cancer. *Eur. Radiol.* **2021**, *31*, 5924–5939. [CrossRef] [PubMed]
58. Xiao, M.; Ma, F.; Li, Y.; Li, Y.; Li, M.; Zhang, G.; Qiang, J. Multiparametric MRI-Based Radiomics Nomogram for Predicting Lymph Node Metastasis in Early-Stage Cervical Cancer. *J. Magn. Reson. Imaging* **2020**, *52*, 885–896. [CrossRef] [PubMed]
59. Baessler, B.; Nestler, T.; Pinto dos Santos, D.; Paffenholz, P.; Zeuch, V.; Pfister, D.; Maintz, D.; Heidenreich, A. Radiomics allows for detection of benign and malignant histopathology in patients with metastatic testicular germ cell tumors prior to post-chemotherapy retroperitoneal lymph node dissection. *Eur. Radiol.* **2020**, *30*, 2334–2345. [CrossRef] [PubMed]
60. Harrell, F.E., Jr.; Lee, K.L.; Mark, D.B. Multivariable prognostic models: Issues in developing models, evaluating assumptions and adequacy, and measuring and reducing errors. *Stat. Med.* **1996**, *15*, 361–387. [CrossRef]
61. Lewin, J.; Dufort, P.; Halankar, J.; O'Malley, M.; Jewett, M.A.; Hamilton, R.J.; Gupta, A.; Lorenzo, A.; Traubici, J.; Nayan, M.; et al. Applying radiomics to predict pathology of postchemotherapy retroperitoneal nodal masses in germ cell tumors. *JCO Clin. Cancer Inform.* **2018**, *2*, 1–12. [CrossRef]
62. Leão, R.; Nayan, M.; Punjani, N.; Jewett, M.A.S.; Fadaak, K.; Garisto, J.; Lewin, J.; Atenafu, E.; Sweet, J.; Anson-Cartwright, L.; et al. A New Model to Predict Benign Histology in Residual Retroperitoneal Masses After Chemotherapy in Nonseminoma. *Eur. Urol. Focus* **2018**, *4*, 995–1001. [CrossRef]
63. Vergouwe, Y.; Steyerberg, E.W.; Foster, R.S.; Sleijfer, D.T.; Fosså, S.D.; Gerl, A.; de Wit, R.; Roberts, J.T.; Habbema, J.D.F. Predicting Retroperitoneal Histology in Postchemotherapy Testicular Germ Cell Cancer: A Model Update and Multicentre Validation with More Than 1000 Patients. *Eur. Urol.* **2007**, *51*, 424–432. [CrossRef]
64. Peter, A.; Lothar, W.; Susanne, K.; Sabine, K.; Michael, H.; Axel, H.; Walz, P.; Kuczyk, M.; Fimmers, R.; for the German Testicular Cancer Study Group. Prediction of Necrosis After Chemotherapy of Advanced Germ Cell Tumors: Results of a Prospective Multicenter Trial of the German Testicular Cancer Study Group. *J. Urol.* **2004**, *171*, 1835–1838.
65. Zengerling, F.; Kunath, F.; Jensen, K.; Ruf, C.; Schmidt, S.; Spek, A. Prognostic factors for tumor recurrence in patients with clinical stage I seminoma undergoing surveillance—A systematic review. *Urol. Oncol. Semin. Orig. Investig.* **2018**, *36*, 448–458. [CrossRef] [PubMed]
66. Zengerling, F.; Beyersdorff, D.; Busch, J.; Heinzelbecker, J.; Pfister, D.; Ruf, C.; Winter, C.; Albers, P.; Kliesch, S.; Schmidt, S. Prognostic factors in patients with clinical stage I nonseminoma—Beyond lymphovascular invasion: A systematic review. *World J. Urol.* **2022**, *40*, 2879–2887. [CrossRef] [PubMed]
67. Dieckmann, K.P.; Radtke, A.; Spiekermann, M.; Balks, T.; Matthies, C.; Becker, P.; Ruf, C.; Oing, C.; Oechsle, K.; Bokemeyer, C.; et al. Serum Levels of MicroRNA miR-371a-3p: A Sensitive and Specific New Biomarker for Germ Cell Tumours. *Eur. Urol.* **2017**, *71*, 213–220. [CrossRef] [PubMed]
68. Bezan, A.; Gerger, A.; Pichler, M. MicroRNAs in testicular cancer: Implications for pathogenesis, diagnosis, prognosis and therapy. *Anticancer. Res.* **2014**, *34*, 2709–2713. [PubMed]

Current Oncology

Article

Deep Learning Algorithm for Tumor Segmentation and Discrimination of Clinically Significant Cancer in Patients with Prostate Cancer

Sujin Hong [1], Seung Ho Kim [1,*], Byeongcheol Yoo [2] and Joo Yeon Kim [3]

[1] Department of Radiology, Inje University, College of Medicine, Haeundae Paik Hospital, Busan 48108, Republic of Korea
[2] Deepnoid Co., Ltd., Seoul 08376, Republic of Korea
[3] Department of Pathology, Inje University, College of Medicine, Haeundae Paik Hospital, Busan 48108, Republic of Korea
*   Correspondence: radiresi@gmail.com; Tel.: +82-51-797-0382

Abstract: Background: We investigated the feasibility of a deep learning algorithm (DLA) based on apparent diffusion coefficient (ADC) maps for the segmentation and discrimination of clinically significant cancer (CSC, Gleason score $\geq$ 7) from non-CSC in patients with prostate cancer (PCa). Methods: Data from a total of 149 consecutive patients who had undergone 3T-MRI and been pathologically diagnosed with PCa were initially collected. The labelled data (148 images for GS6, 580 images for GS7) were applied for tumor segmentation using a convolutional neural network (CNN). For classification, 93 images for GS6 and 372 images for GS7 were used. For external validation, 22 consecutive patients from five different institutions (25 images for GS6, 70 images for GS7) representing different MR machines were recruited. Results: Regarding segmentation and classification, U-Net and DenseNet were used, respectively. The tumor Dice scores for internal and external validation were 0.822 and 0.7776, respectively. As for classification, the accuracies of internal and external validation were 73 and 75%, respectively. For external validation, diagnostic predictive values for CSC (sensitivity, specificity, positive predictive value and negative predictive value) were 84, 48, 82 and 52%, respectively. Conclusions: Tumor segmentation and discrimination of CSC from non-CSC is feasible using a DLA developed based on ADC maps (b2000) alone.

Keywords: magnetic resonance imaging (MRI); diffusion-weighted imaging (DWI); prostate cancer; Gleason score; deep learning

Citation: Hong, S.; Kim, S.H.; Yoo, B.; Kim, J.Y. Deep Learning Algorithm for Tumor Segmentation and Discrimination of Clinically Significant Cancer in Patients with Prostate Cancer. *Curr. Oncol.* **2023**, *30*, 7275–7285. https://doi.org/10.3390/curroncol30080528

Received: 11 June 2023
Revised: 5 July 2023
Accepted: 25 July 2023
Published: 1 August 2023

1. Introduction

Prostate cancer (PCa) is the second most frequently diagnosed cancer in men worldwide and the fifth most common cause of death [1]. Gleason score (GS) is a classification system based on the structure of PCa and is closely related to tumor aggressiveness. GS7 (particularly 3 + 4, International society of urological pathology (ISUP) grade 2) and above are classified as clinically significant cancers (CSCs) and GS6 (ISUP grade 1) as non-CSC [2].

PCa can be treated individually, depending on the degree of aggressiveness, risk of recurrence, and staging. Non-CSC is associated with relatively lower progression and mortality, suggesting a relatively good prognosis; thus, active surveillance and observation can be followed. However, as CSC is associated with a relatively high probability of adverse outcomes, active treatment, such as radical prostatectomy and/or radiation therapy, is required in general [3]. To date, the National Comprehensive Cancer Network (NCCN) guideline lists active surveillance for patients with favorable intermediate-risk prostate cancer (1 IR factor + Grade 1 or 2 + <50% positive biopsy cores) [4]. Another guideline promotes active surveillance for selected patients with low-volume GS 3 + 4 prostate cancer [5]. Therefore, efforts have been made to determine treatment policies based on risk stratification. However, due to the sampling errors inherent in systemic biopsy [6,7] as well as the possibility of complication associated with invasive approaches [8], interest in evaluating tumor aggressiveness using non-invasive imaging modalities such as magnetic resonance imaging (MRI) has increased.

There have been several promising studies on the usefulness of deep learning algorithms (DLAs), as based on mono-parametric or bi-parametric (bp) MRI for tumor detection of PCa [9–15]. DLA studies based on bp-MRI or mono-parametric MRI for segmentation and classification between CSC and non-CSC are less frequently found in the literature [2,3,9]. One of these studies undertook to distinguish CSC from non-CSC with deep-transfer-learning-based models using combined T2-weighted imaging (T2WI) and diffusion-weighted imaging (DWI) and a corresponding apparent diffusion coefficient (ADC) map, and the study revealed a similar diagnostic performance to that of prostate imaging reporting and data system (PIRADS) v.2.0 [3]. Both of those studies [2,3], however employed sophisticated methods to combine the T2WI and DWI and used a low b value of 800 s/mm^2. PIRADS score, moreover, has inherent limitations, such as a moderate inter-observer agreement and a probability scale by itself [16].

For PIRADS v.2.1, acquisition of high-b-value DWI ($\geq$1400 s/mm^2) is recommended Furthermore, recent studies have shown that DWI b2000 is better than DWI b1000 for the localization of PCa [17,18]. However, to the best of our knowledge, DLA studies based on high-b-value DWI alone are scarce. Thus, we hypothesized that a DLA based on acquired DWI b2000 and corresponding ADC maps as a single input for discriminating CSC from non-CSC might deliver more beneficial results. The purpose of this study was to investigate the feasibility of using a DLA developed based on ADC maps (b2000) alone for tumor segmentation and discrimination of CSC from non-CSC in patients with PCa.

2. Materials and Methods

2.1. Patient Selection Criteria

The pertinent institutional review board approved this retrospective study (IRB number blinded). Informed consent from patients was waived. Between October 2018 and March 2022, the relevant medical records of a total of 157 patients meeting the following inclusion criteria were collected: (i) complete 3T-MRI, including DWI and corresponding ADC maps, (ii) histological diagnosis of PCa and topographic map availability via radical prostatectomy and (iii) GS documentation availability via pathological reports. Among them, 8 patients were excluded based on one of the following exclusion criteria: (i) poor MR image quality due to severe artifacts (n = 1) or (ii) incomplete pathologic topographic map (n = 7). Finally, 149 patients (mean age: 69.2 years, range: 47–84 years) were enrolled for the training and internal validation datasets (80 and 20% of the data, respectively). For external validation, 22 consecutive patients (mean age: 69.6 years, range: 56–80 years), for whom five different MR machines had been employed and different parameters applied, were separately recruited during the same period. The case enrollment process is summarized in Figure 1.

2.2. MRI Technique

All of the MRI examinations for the training and internal validation datasets were performed using a 3.0-T MR machine (Achieva TX; Philips, Best, The Netherlands) with a parallel-array torso coil (SENSE Torso/cardiac coil; USA Instruments, Gainesville, FL, USA).

The scanning protocol was composed of axial, sagittal and coronal T2-weighted turbo spin-echo (TSE) and axial DWI sequences (b values, 0, 100, 1000, 2000 s/mm^2) Corresponding ADC maps were generated for the designated b values, respectively. The detailed scan parameters are summarized in Table 1.

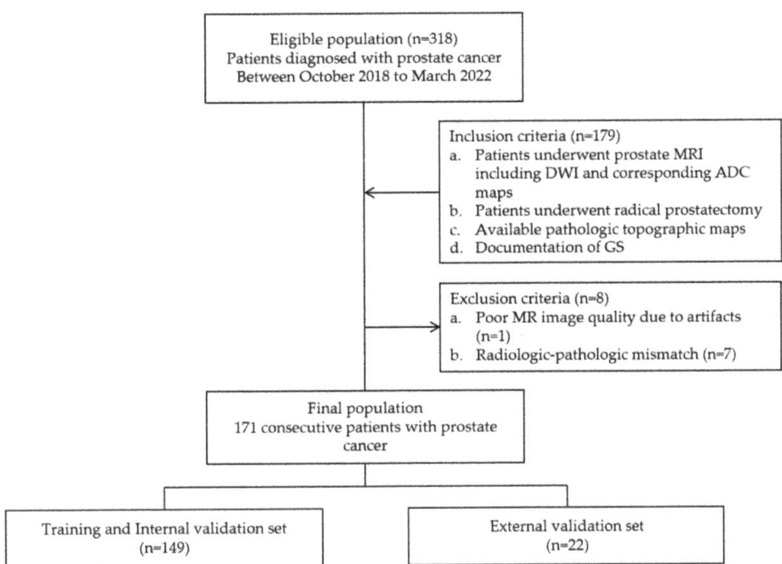

Figure 1. Flowchart of case enrollment process.

Table 1. MRI sequence parameters for training set.

Parameters	T2-Weighted Axial, Sagittal, and Coronal TSE	DWI (b = 0, 100, 1000 and 2000 s/mm^2)
TR (msec)	3370.7	5725
TE (msec)	100	77.8
Slice thickness (mm)	3	3
Slice gap (mm)	0.3	0.3
Matrix size	316 × 272	120 × 118
NEX	1	1
FOV (mm × mm)	220 × 220	240 × 240
Number of slices	30	30

TR, repetition time; TE, echo time; NEX, umber of excitations; FOV, Field of view; TSE, Turbo spin echo. Note that diffusion-weighted imaging (DWI) was performed using the single-shot echo-planar imaging (SS-EPI) technique.

2.3. Data Processing

Two radiologists (with 18 and 3 years of experience, respectively) determined the tumor and whole-gland borders by consensus on axial ADC maps generated from b values of 0 and 2000. For segmentation, they reviewed T2WI in 3 planes and DWI (b = 2000 s/mm^2) after referencing the topographic map as a ground truth. After determination of the tumor and gland borders, the junior radiologist drew the regions of interest (ROIs) along the determined tumor and gland borders on the ADC maps (b = 2000 s/mm^2) using DEEP:LABEL software v.1.0.4 (Deepnoid, Seoul, Republic of Korea). When there were multiple tumors in a patient, the largest one was considered as the index tumor. The reviewers also recorded the PIRADS score for the index tumor based on PIRADS v2.1. The order of patients was random. The reviewers were blinded to the patients' GS.

2.4. DL Architecture for Tumor and Gland Segmentation

As a convolutional neural network (CNN), U-Net was used for tumor and gland segmentation due to its high accuracy at various image sites. This architecture consists of a down-sampling encoder for features learning and an up-sampling decoder for feature production, and it is efficient, even with small datasets [19].

In the gland segmentation, each of the following pre-processing steps was performed for overall segmentation effectiveness. All of the labeled images were cropped with margin of 5 pixels for delineation of the borders of the prostate gland. The Min–Max normalization guaranteed that all features were of the same scale. Finally, all of the images were resized to 128 × 128 pixels for use as inputs to the U-Net architecture for gland segmentation. Several hyper-parameters were tested to train the optimal DLA, for which purpose the Adam optimizer (learning rate: 0.001, decay rate: 0.95) was selected. In the tumor segmentation, the same pre-processing steps were performed, and the Adam optimizer (learning rate: 0.0001, decay rate: 0.95) was again employed for DLA training.

After tumor and gland segmentation, all of the labeled tumor data (148 images for GS6, 580 images for GS7) and gland data (535 images for GS6, 935 images for GS7) were used to evaluate the DLA predictive performance for accuracy, sensitivity, specificity, positive predictive value (PPV), negative predictive value (NPV) and Dice score.

2.5. DL Architecture for Tumor Classification

2.5.1. Training Architecture

For tumor classification, the labeled tumor data were filtered with a cutoff of 25 pixels. Finally, 93 images for GS6 and 372 images for GS7 were used. For balanced training, the GS7 images were randomly allocated into four subsets of 93 images each in order to match the number of GS6 images. Therefore, 186 GS6/7 images were divided into 146 images for use as a training dataset and 40 for use as an internal validation dataset in each session. The Min–Max normalization and resizing steps were performed in the same manner as for the segmentation task.

Several CNNs, such as Inception, ResNet and DenseNet, were trained for tumor classification, and DenseNet 201 was selected for tumor classification due to its superior performance in distinguishing GS6 from GS7 [20–22]. DenseNet connects each layer to every other layer in a feed-forward manner. It also alleviates the vanishing-gradient problem, strengthens feature propagation, encourages feature reuse and substantially reduces the number of parameters [22]. It has shown good performance, even with an insufficient dataset. In the present study, based on four training and internal validation sessions, the DLA with the best diagnostic performance was selected and applied for external validation. All of the data processing as well as DL and training procedures were implemented in DEEPPHI (http://www.deepphi.ai/, accessed on 25 April 2022), a web-based open artificial intelligence platform.

2.5.2. External Validation

For external validation of segmentation and classification, 22 consecutive patients from 5 different institutions (25 images for GS6, 70 images for GS7) representing different MR machines each with different parameters were recruited. The MR machines consisted of 1.5T (n = 1) and 3.0T (n = 21) scanners, and the images with the highest b values of DWI were composed of b800 (n = 1), b1000 (n = 4) and b2000 (n = 17). A total of 95 tumor slices (25 GS6 images, 70 GS7 images) and 180 gland slices were included and analyzed in order to externally validate the DLA that had been developed with the training dataset.

2.6. Reference Standard

Dedicated urologists performed the radical prostatectomies. A dedicated pathologist assessed each pathological slide according to the Gleason grading system [23] and drew up a topographic map that served as the ground truth for tumor segmentation on MRI. For classification of CSC and non-CSC, the GS, as obtained after surgery, was set as the gold standard. CSC was defined as GS $\geq$ 7 and non-CSC as GS6 [24].

2.7. Statistical Analysis

For the categorical data, the chi-square test or Fisher's exact test was used to find any difference between the training and external validation datasets. For the continuous data, the *t*-test was used. The Dice score was used to quantify the performance of image segmentation. A Dice score of 1.0 means perfect overlap, and a score of 0.0 corresponds to no overlap [25]. The diagnostic performance for classification was calculated via receiver operating characteristic (ROC) curve analysis and expressed as the area under the ROC curve (AUC). Diagnostic predictive values, including accuracy, PPV and NPV, were also estimated under the maximal AUC. For all of the statistical calculations, MedCalc software for Windows (MedCalc Software version 20.111, Mariakerke, Belgium) was used. A *p* value of less than 0.05 was considered statistically significant.

3. Results

3.1. Patient Demographics

The age, prostate-specific antigen level, GS, PIRADS score and tumor location were not significantly different between the training and external validation datasets. The average time interval between MRI and surgery was 37.0 days (range, 5–447 days). The average volume of GS 6 tumors was not significantly different from that of GS 7 tumors in both training and internal validation sets (GS 6, 4.1 ± 6.6 cm^3; GS 7, 7.0 ± 7.3 cm^3, *p* = 0.1822) and the external validation set (GS 6, 1.9 ± 1.9 cm^3; GS 7, 6.3 ± 6.1 cm^3, *p* = 0.1348). The patients' demographic data and analysis results are presented in Table 2.

Table 2. Demographic data and analysis results for study population.

Parameter	All	Training and Internal Validation Sets (n = 149)	External Validation Set (n = 22)	*p* Value
Mean Age, years [range]	69.2982 [47–84]	69.2483 [47–84]	69.6364 [56–80]	0.8049
Mean PSA, ng/mL [range]	14.6315 [0.85–149]	14.4478 [0.85–149]	21.1709 [3.0–131]	0.3597
GS, n (%)				
6	46 (27)	40 (27)	6 (27)	0.9307
7	125 (73)	109 (73)	16 (73)	0.9912
3 + 4	89	76	13	
4 + 3	36	33	3	
PIRADS v2.1, n (%)				
3	17 (10)	17 (11)	0 (0)	0.1131
4	55 (32)	49 (33)	6 (27)	0.7006
5	99 (58)	83 (56)	16 (73)	0.3307
Tumor location, n (%)				
Peripheral zone	92 (54)	81 (54)	11 (50)	0.8245
Transitional zone	48 (28)	38 (26)	10 (45)	0.1204
Fibromuscular zone	4 (2)	4 (3)	0 (0)	0.4422
Diffuse	27 (16)	26 (17)	1 (5)	0.1453

GS, Gleason score; PSA, prostate-specific antigen.

3.2. Diagnostic Performance of DLA

In terms of gland segmentation, U-Net had a sensitivity of 95%, a specificity of 96% and a Dice score of 0.951 for internal validation and 92%, 97% and 0.9413, respectively, for external validation (Figure 2). As for tumor segmentation, it had a sensitivity of 82%, a specificity of 96% and a Dice score of 0.822 for internal validation and 77%, 95% and 0.7776, respectively, for external validation (Figure 3) (Table 3).

As for classification, the overall accuracies of internal and external validation were 73 and 75%, respectively. For internal validation, the diagnostic predictive values for CSC (hereafter sensitivity, specificity, PPV and NPV, in order) were calculated as 72, 74, 74 and 72%, respectively. For external validation, the diagnostic predictive values were estimated as 84, 48, 82 and 52%, respectively (Table 4). The DenseNet 201 classifier achieved an AUC of 0.6269. The average precision scores for GS6 and GS7 were 0.4462 and 0.8149, respectively (Figure 4). Out of a total of 95 tumor slices (25 GS6 images, 70 GS7 images), 13 slices of GS6 were over-estimated as GS7 and 11 slices of GS7 were under-estimated as GS6 (Figures 5 and 6).

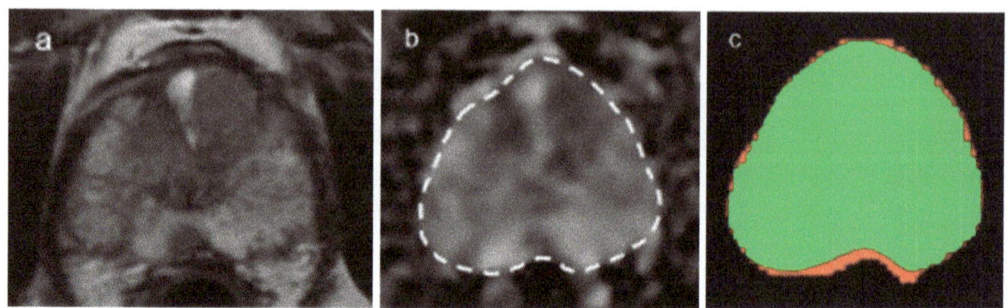

Figure 2. A representative case of gland segmentation. (**a**,**b**) The Dice score for the gland segmentations was 0.94. Axial T2-weighted image (**a**) and corresponding ADC map (**b**) (b = 2000 s/mm^2) with gland segmentation ((**b**), dotted lines). (**c**) Segmentation through the convolutional neural network (CNN, U-Net) shows that the green color represents the matched area and the red color the unmatched area.

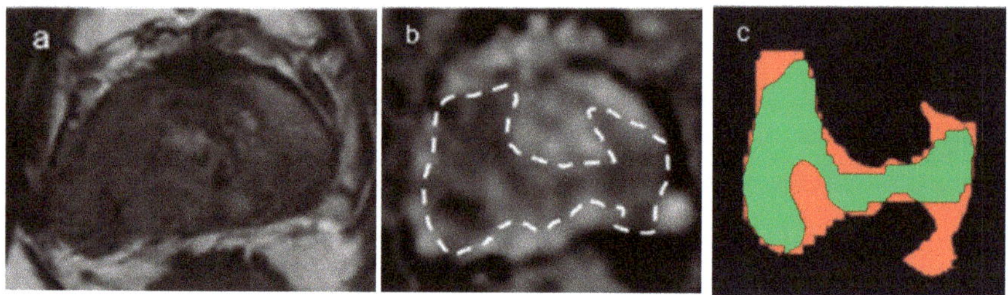

Figure 3. A representative case of tumor segmentation with GS7(4 + 3). (**a**,**b**) The Dice score for the tumor segmentations was 0.78. Axial T2-weighted image (**a**) and corresponding ADC map (**b**) (b = 2000 s/mm^2) with tumor segmentation ((**b**), dotted lines). (**c**) Segmentation through the convolutional neural network (CNN, U-Net) shows that the green color represents the matched area and the red color the unmatched area.

Table 3. Diagnostic predictive values of DLA for segmentation of glands and tumors.

	Accuracy (%)	Sensitivity (%)	Specificity (%)	PPV (%)	NPV (%)	Dice Score
Gland						
Internal validation	96	95	96	95	96	0.951
External validation	95	92	97	96	93	0.9413
Tumor						
Internal validation	93	82	96	83	96	0.822
External validation	92	77	95	79	95	0.7776

U-Net was used for deep learning algorithm (DLA).

Table 4. Diagnostic predictive values of DLA for tumor classification.

	Accuracy (%)	Sensitivity (%)	Specificity (%)	PPV (%)	NPV (%)	AUC
Internal validation set						
CSC	73	72	74	74	72	
External validation set						
CSC	75	84	48	82	52	0.6269

DenseNet 201 was used for the deep learning algorithm (DLA). CSC, clinically significant cancer.

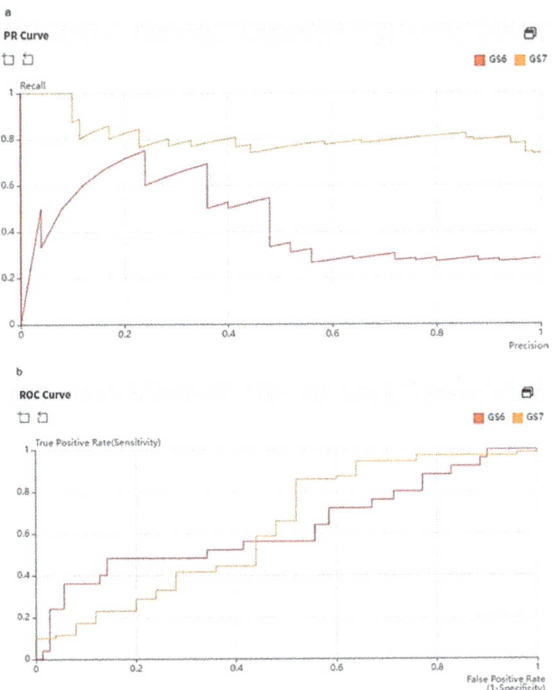

Figure 4. Graphs showing precision recall curve (**a**) and receiver operating characteristic (ROC) curve (**b**) of deep learning algorithm (DLA) for tumor classification as applied to external validation dataset. Average precision for GS6 and GS7 was 0.4462 and 0.8149, respectively. The DenseNet 201 classifier achieved an AUC of 0.6269 for both GS6 and GS7.

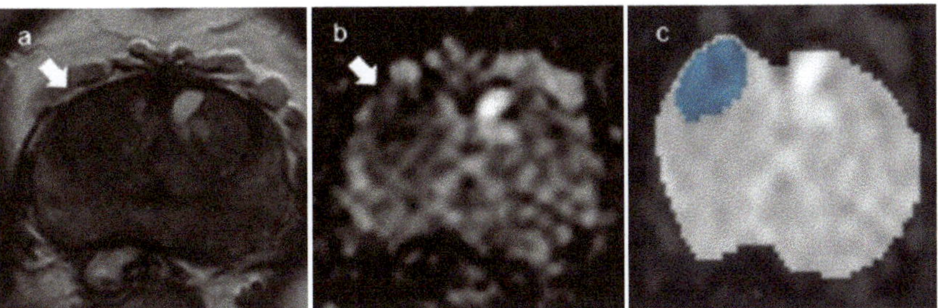

Figure 5. A representative case of misclassification: over-estimation of GS6 as GS7. (**a,b**) Axial T2-weighted image (**a**) and ADC map (**b**) (b = 2000 s/mm^2) show a tumor in the Rt. mid-transitional zone (arrows). (**c**) Segmentation and classification through the convolutional neural network (CNN, DenseNet 201) show the tumor area as blue color.

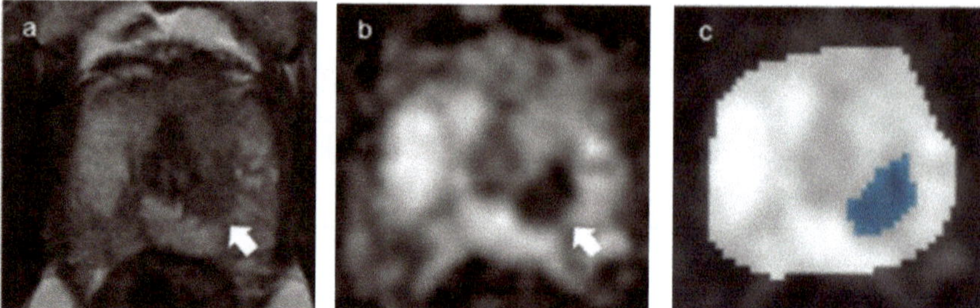

Figure 6. A representative case of misclassification: under-estimation of GS7 as GS6. (**a,b**) Axial T2-weighted image (**a**) and ADC map (**b**) (b = 2000 s/mm^2) show a tumor in the Lt. mid-transitional zone (arrows), respectively. (**c**). Segmentation and classification through the convolutional neural network (CNN, DenseNet 201) show the tumor area as blue color.

4. Discussion

Regarding tumor segmentation, the DLA, which was based on ADC maps (b2000) alone in our study, showed Dice scores of 0.94 and 0.78 for gland and tumor segmentation, respectively. Our observations are similar to those of a previous study on mono-parametric MRI. Alkadi et al. reported that the accuracy of a DLA, which was based on T2WI only for tumor segmentation, was 89% [9]. As for tumor segmentation based on bp-MRI, Schelb et al. reported that the Dice scores for a DLA based on bp-MRI (T2WI + DWI b1500) using U-Net for detection and segmentation of CSCs were 0.35 for tumors and 0.89 for glands [10]. Relative to this latter study, in our opinion, the relatively high Dice score for tumor segmentation in this present study might have been due to the use of DWI b2000. Rosenkrantz et al. revealed that DWI b2000 achieved significantly higher sensitivity for tumor detection than b1000 [17]. Vural et al. found that b2000 showed the best lesion conspicuity and background suppression among b values of 1500, 2000 and 3000 [26]. In addition, Cha et al. reported that the optimal b value of DWI was within a range of 1700–1900 for the detection of a prostatic lesion [27].

In terms of tumor classification, the DLA in the present study showed an accuracy of 75% and an AUC of 0.63 in external validation. Recently, many deep-learning-based computer-aided detection/classification (DL-CADe/CADx) systems have been developed to assist human radiologists. Rampun et al. compared the 11 different CAD systems employed to detect peripheral-zone cancer (GS ≥ 7), only for T2WI on 3T-MRI [12]. The results varied from an AUC of 0.69 (k-Nearest Neighbor classifier) to 0.93 (combined Bayesian Network and Multilayer Perceptron classifiers), according to the applied CNNs. Ishioka et al. reported AUCs ranging from 0.636 to 0.645 for tumor (GS ≥ 6) detection via combined U-Net with ResNet50, as trained on T2WI only with the 1.5T-MRI machine [13]. Although only ADC maps (b2000) were used in our study, the diagnostic performance for tumor classification seems comparable to mono-parametric MRI using T2WI alone.

Beyond mono-parametric MRI, Arif et al. found that a DLA (Keras with TensorFlow) developed based on bp-MRI (T2WI + DWI b800) showed an AUC of 0.89, a sensitivity of 94% and a specificity of 74% for discrimination of CSCs from non-CSCs [2]. In our study, the sensitivity and specificity for GS7 were 84 and 48%, respectively. The relatively low specificity might have been due to the mono-parametric MRI based study, without any other sequences. Zhong et al. compared the diagnostic performance of DLA models trained with T2WI (DLAT2) alone, ADC images (DLAADC) alone and combined T2WI and ADC images (DLAT2 + ADC) in discriminating CSC from non-CSC [3]. All three models showed the same sensitivity of 77%, and the combined T2WI and ADC (b800) information, notably helped to reduce false-positive prediction, thereby improving the specificity from 52 to 64% after adding DLAT2 + ADC to DLAADC.

Considering the previously mentioned merits of DenseNet, including reduction in the vanishing gradient, enhancement of feature propagation, reuse of features, reduction in the number of parameters [22] and its robustness, we think that our DLA, as developed by DenseNet and based on ADC maps, could be a simple and convenient option for the differentiation of CSC from non-CSC.

Our study has several limitations. First, tumor segmentation was conducted not on a three-dimensional (3D)-volume data basis but on a 2D-image basis, due to the inherent technical limitation of the segmentation tool. Therefore, when a classification error occurred in one tumor-bearing slice, there was a tendency that those errors would continue to consecutive slices. As a result, diagnostic performance for tumor classification might have been underestimated. Second, there is a possibility of selection bias, as only GS7 tumors were included in the CSC group. However, GS8-or-higher tumors are frequently advanced cases of metastatic disease, for which systemic chemotherapy would be adopted rather than radical prostatectomy. Considering the purpose of this study, to separate the group capable of surveillance from the group that is not, the study was conducted except for tumors with a score of GS8 or higher that were already inoperable. It would be better to have a larger sample size for GS6 in the external validation set; however, it was difficult to enroll patients with GS6. Patients with GS6 have a relatively good prognosis; thus, active surveillance and observation can be followed instead of radical prostatectomy. Third, the DLA's value added to the human radiologists' performance for tumor classification was not investigated. As for the added value, several previous observations have been reported in the literature [14,15]. Winkel et al. reported that the DL-CAD system increased the diagnostic accuracy in detecting clinically suspicious lesions (PIRADS $\geq$ 4) and reduced both the inter-reader variability and the reading time [14]. However, it was beyond the scope and aim of the present study. To investigate the added value of a DLA to the performance of human radiologists for tumor classification, further studies on DLA efficacy in this regard are warranted.

5. Conclusions

In conclusion, tumor segmentation and classification of PCa through a DLA developed based on ADC maps (b2000) alone are feasible.

Author Contributions: Conceptualization, S.H.K.; methodology, S.H.K.; software, B.Y.; validation, all authors; formal analysis, B.Y.; investigation, all authors; resources, S.H.K.; data curation, all authors; writing—original draft preparation, all authors; writing—review and editing, all authors; visualization, all authors; supervision, S.H.K.; project administration, S.H.K.; funding acquisition, not applicable. All authors have read and agreed to the published version of the manuscript.

Funding: This research received no external funding.

Institutional Review Board Statement: The study was conducted in accordance with the Declaration of Helsinki and approved by the Institutional Review Board (no. 2022-04-032-002, date of approval 2 May 2022).

Informed Consent Statement: Patient consent was waived due to extremely low risk to patients associated with this retrospective study.

Data Availability Statement: The data presented in this study are available on request from the corresponding author. The data are not publicly available due to patients' privacy.

Conflicts of Interest: The authors declare no conflict of interest.

References

1. Sung, H.; Ferlay, J.; Siegel, R.L.; Laversanne, M.; Soerjomataram, I.; Jemal, A.; Bray, F. Global Cancer Statistics 2020: GLOBOCAN Estimates of Incidence and Mortality Worldwide for 36 Cancers in 185 Countries. *CA Cancer J. Clin.* **2021**, *71*, 209–249. [CrossRef] [PubMed]
2. Arif, M.; Schoots, I.G.; Tovar, J.C.; Bangma, C.H.; Krestin, G.P.; Roobol, M.J.; Niessen, W.; Veenland, J.F. Clinically significant prostate cancer detection and segmentation in low-risk patients using a convolutional neural network on multi-parametric MRI. *Eur. Radiol.* **2020**, *30*, 6582–6592. [CrossRef] [PubMed]
3. Zhong, X.; Cao, R.; Shakeri, S.; Scalzo, F.; Lee, Y.; Enzmann, D.R.; Wu, H.H.; Raman, S.S.; Sung, K. Deep transfer learning-based prostate cancer classification using 3 Tesla multi-parametric MRI. *Abdom. Radiol.* **2019**, *44*, 2030–2039. [CrossRef] [PubMed]

4. Schaeffer, E.M.; Srinivas, S.; Adra, N.; An, Y.; Barocas, D.; Bitting, R.; Bryce, A.; Chapin, B.; Cheng, H.H.; D'Amico, A.V.; et al. NCCN Guidelines® Insights: Prostate Cancer, Version 1.2023. *J. Natl. Compr. Cancer Netw.* **2022**, *20*, 1288–1298. [CrossRef]
5. Chen, R.C.; Rumble, R.B.; Loblaw, D.A.; Finelli, A.; Ehdaie, B.R.; Cooperberg, M.R.; Morgan, S.C.; Tyldesley, S.; Haluschak, J.J.; Tan, W.; et al. Active Surveillance for the Management of Localized Prostate Cancer (Cancer Care Ontario Guideline): American Society of Clinical Oncology Clinical Practice Guideline Endorsement. *J. Clin. Oncol.* **2016**, *34*, 2182–2190. [CrossRef] [PubMed]
6. Corcoran, N.M.; Hong, M.K.; Casey, R.G.; Hurtado-Coll, A.; Peters, J.; Harewood, L.; Goldenberg, S.L.; Hovens, C.M.; Costello, A.J.; Gleave, M.E. Upgrade in Gleason score between prostate biopsies and pathology following radical prostatectomy significantly impacts upon the risk of biochemical recurrence. *BJU Int.* **2011**, *108*, E202–E210. [CrossRef] [PubMed]
7. Cohen, M.S.; Hanley, R.S.; Kurteva, T.; Ruthazer, R.; Silverman, M.L.; Sorcini, A.; Hamawy, K.; Roth, R.A.; Tuerk, I.; Libertino, J.A. Comparing the Gleason prostate biopsy and Gleason prostatectomy grading system: The Lahey Clinic Medical Center experience and an international meta-analysis. *Eur. Urol.* **2008**, *54*, 371–381. [CrossRef]
8. Borghesi, M.; Ahmed, H.; Nam, R.; Schaeffer, E.; Schiavina, R.; Taneja, S.; Weidner, W.; Loeb, S. Complications after systematic, random, and image-guided prostate biopsy. *Eur. Urol.* **2017**, *71*, 353–365. [CrossRef]
9. Alkadi, R.; Taher, F.; El-Baz, A.; Werghi, N. A deep learning-based approach for the detection and localization of prostate cancer in T2 magnetic resonance images. *J. Digit. Imaging* **2019**, *32*, 793–807. [CrossRef]
10. Schelb, P.; Kohl, S.; Radtke, J.P.; Wiesenfarth, M.; Kickingereder, P.; Bickelhaupt, S.; Kuder, T.A.; Stenzinger, A.; Hohenfellner, M.; Schlemmer, H.-P.; et al. Classification of cancer at prostate MRI: Deep learning versus clinical PI-RADS assessment. *Radiology* **2019**, *293*, 607–617. [CrossRef]
11. Song, Y.; Zhang, Y.D.; Yan, X.; Liu, H.; Zhou, M.; Hu, B.; Yang, G. Computer-aided diagnosis of prostate cancer using a deep convolutional neural network from multiparametric MRI. *J. Magn. Reson. Imaging* **2018**, *48*, 1570–1577. [CrossRef]
12. Rampun, A.; Zheng, L.; Malcolm, P.; Tiddeman, B.; Zwiggelaar, R. Computer-aided detection of prostate cancer in T2-weighted MRI within the peripheral zone. *Phys. Med. Biol.* **2016**, *61*, 4796–4825. [CrossRef]
13. Ishioka, J.; Matsuoka, Y.; Uehara, S.; Yasuda, Y.; Kijima, T.; Yoshida, S.; Yokoyama, M.; Saito, K.; Kihara, K.; Numao, N.; et al. Computer-aided diagnosis of prostate cancer on magnetic resonance imaging using a convolutional neural network algorithm. *BJU Int.* **2018**, *122*, 411–417. [CrossRef]
14. Winkel, D.J.; Tong, A.; Lou, B.; Kamen, A.; Comaniciu, D.; Disselhorst, J.A.; Rodríguez-Ruiz, A.; Huisman, H.; Szolar, D.; Shabunin, I. A novel deep learning based computer-aided diagnosis system improves the accuracy and efficiency of radiologists in reading biparametric magnetic resonance images of the prostate: Results of a multireader, multicase study. *Investig. Radiol.* **2021**, *56*, 605–613. [CrossRef]
15. Niaf, E.; Lartizien, C.; Bratan, F.; Roche, L.; Rabilloud, M.; Mège-Lechevallier, F.; Rouvière, O. Prostate focal peripheral zone lesions: Characterization at multiparametric MR imaging-influence of a computer-aided diagnosis system. *Radiology* **2014**, *271*, 761–769. [CrossRef]
16. Bhayana, R.; O'Shea, A.; Anderson, M.A.; Bradley, W.R.; Gottumukkala, R.V.; Mojtahed, A.; Pierce, T.T.; Harisinghani, M. PI-RADS versions 2 and 2.1: Interobserver agreement and diagnostic performance in peripheral and transition zone lesions among six radiologists. *Am. J. Roentgenol.* **2021**, *217*, 141–151. [CrossRef]
17. Rosenkrantz, A.B.; Hindman, N.; Lim, R.P.; Das, K.; Babb, J.S.; Mussi, T.C.; Taneja, S.S. Diffusion-weighted imaging of the prostate: Comparison of b1000 and b2000 image sets for index lesion detection. *J. Magn. Reson. Imaging* **2013**, *38*, 694–700. [CrossRef]
18. Tamada, T.; Kanomata, N.; Sone, T.; Jo, Y.; Miyaji, Y.; Higashi, H.; Yamamoto, A.; Ito, K. High b value (2000 s/mm^2) diffusion weighted magnetic resonance imaging in prostate cancer at 3 Tesla: Comparison with 1000 s/mm^2 for tumor conspicuity and discrimination of aggressiveness. *PLoS ONE* **2014**, *9*, e96619. [CrossRef]
19. Zhou, Z.; Siddiquee, M.U. A Nested U-Net Architecture for medical Image Segmentation. In *Deep Learning in Medical Image Analysis and Multimodal Learning for Clinical Decision Support, Proceedings of the 4th International Workshop, DLMIA 2018, and 8th International Workshop, ML-CDS 2018, Held in Conjunction with Miccai 2018, Granada, Spain, 20 September 2018*; Lecture Notes in Computer Science; Springer: Berlin/Heidelberg, Germany, 2018. [CrossRef]
20. Szegedy, C.; Vanhoucke, V.; Ioffe, S.; Shlens, J.; Wojna, Z. Rethinking the inception architecture for computer vision. In Proceedings of the IEEE Conference on Computer Vision and Pattern Recognition, Las Vegas, NV, USA, 26 June–1 July 2016; pp. 2818–2826.
21. He, K.; Zhang, X.; Ren, S.; Sun, J. Deep residual learning for image recognition. In Proceedings of the IEEE Conference on Computer Vision and Pattern Recognition, Las Vegas, NV, USA, 26 June–1 July 2016; pp. 770–778.
22. Huang, G.; Liu, Z.; Van Der Maaten, L.; Weinberger, K. Densely connected convolutional networks. In Proceedings of the IEEE Conference on Computer Vision and Pattern Recognition, Honolulu, HI, USA, 21–26 July 2017.
23. Epstein, J.I.; Egevad, L.; Amin, M.B.; Delahunt, B.; Srigley, J.R.; Humphrey, P.A. The 2014 International Society of Urological Pathology (ISUP) consensus conference on Gleason grading of prostatic carcinoma. *Am. J. Surg. Pathol.* **2016**, *40*, 244–252. [CrossRef]
24. Ploussard, G.; Epstein, J.I.; Montironi, R.; Carroll, P.R.; Wirth, M.; Grimm, M.-O.; Bjartell, A.S.; Montorsi, F.; Freedland, S.J.; Erbersdobler, A. The contemporary concept of significant versus insignificant prostate cancer. *Eur. Urol.* **2011**, *60*, 291–303. [CrossRef]
25. Dice, L.R. Measures of the Amount of Ecologic Association Between Species. *Ecology* **1945**, *26*, 297–302. [CrossRef]

5. Vural, M.; Ertaş, G.; Onay, A.; Acar, Ö.; Esen, T.; Sağlıcan, Y.; Zengingönül, H.P.; Akpek, S. Conspicuity of peripheral zone prostate cancer on computed diffusion-weighted imaging: Comparison of cDWI1500, cDWI2000, and cDWI3000. *BioMed Res. Int.* **2014**, *2014*, 768291. [CrossRef] [PubMed]

7. Cha, S.Y.; Kim, E.; Park, S.Y. Why Is a b-value Range of 1500–2000 s/mm^2 Optimal for Evaluating Prostatic Index Lesions on Synthetic Diffusion-Weighted Imaging? *Korean J. Radiol.* **2021**, *22*, 922. [CrossRef] [PubMed]

Article

Deciphering Machine Learning Decisions to Distinguish between Posterior Fossa Tumor Types Using MRI Features: What Do the Data Tell Us?

Toygar Tanyel [1], Chandran Nadarajan [2], Nguyen Minh Duc [3] and Bilgin Keserci [4,*]

1   Department of Computer Engineering, Yildiz Technical University, Istanbul 34349, Türkiye; toygar.tanyel@std.yildiz.edu.tr
2   Department of Radiology, Gleneagles Hospital Kota Kinabalu, Kota Kinabalu 88100, Sabah, Malaysia; nadarajan.chandran@gleneaglesdr.my
3   Department of Radiology, Pham Ngoc Thach University of Medicine, Ho Chi Minh City 700000, Vietnam; bsnguyenminhduc@pnt.edu.vn
4   Department of Biomedical Engineering, Yildiz Technical University, Istanbul 34349, Türkiye
*   Correspondence: bushido.keserci@gmail.com

Simple Summary: This paper focuses on interpreting machine learning (ML) models' decisions in medical diagnoses, specifically for four types of posterior fossa tumors in pediatric patients. The proposed methodology involves using kernel density estimations with Gaussian distributions to analyze individual MRI features, assess their relationships, and comprehensively study ML model behavior. The study demonstrates that employing a simplified approach in the absence of large datasets can lead to more pronounced and explainable outcomes. Furthermore, the pre-analysis results consistently align with the outputs of ML models and existing clinical findings. By bridging the knowledge gap between ML and medical outcomes, this research contributes to a better understanding of ML-based diagnoses for pediatric brain tumors.

Abstract: Machine learning (ML) models have become capable of making critical decisions on our behalf. Nevertheless, due to complexity of these models, interpreting their decisions can be challenging, and humans cannot always control them. This paper provides explanations of decisions made by ML models in diagnosing four types of posterior fossa tumors: medulloblastoma, ependymoma, pilocytic astrocytoma, and brainstem glioma. The proposed methodology involves data analysis using kernel density estimations with Gaussian distributions to examine individual MRI features, conducting an analysis on the relationships between these features, and performing a comprehensive analysis of ML model behavior. This approach offers a simple yet informative and reliable means of identifying and validating distinguishable MRI features for the diagnosis of pediatric brain tumors. By presenting a comprehensive analysis of the responses of the four pediatric tumor types to each other and to ML models in a single source, this study aims to bridge the knowledge gap in the existing literature concerning the relationship between ML and medical outcomes. The results highlight that employing a simplistic approach in the absence of very large datasets leads to significantly more pronounced and explainable outcomes, as expected. Additionally, the study also demonstrates that the pre-analysis results consistently align with the outputs of the ML models and the clinical findings reported in the existing literature.

Keywords: posterior fossa pediatric brain tumors; magnetic resonance imaging; machine learning; exploratory data analysis; kernel density estimation

Citation: Tanyel, T.; Nadarajan, C.; Duc, N.M.; Keserci, B. Deciphering Machine Learning Decisions to Distinguish between Posterior Fossa Tumor Types Using MRI Features: What Do the Data Tell Us? *Cancers* **2023**, *15*, 4015. https://doi.org/10.3390/cancers15164015

Academic Editor: Sam Payabvash

Received: 25 June 2023
Revised: 22 July 2023
Accepted: 2 August 2023
Published: 8 August 2023

1. Introduction

Brain tumors are the most prevalent type of childhood cancer, comprising over a quarter of all cases. Among these tumors, 60–70% arise in the posterior fossa (PF), with medulloblastoma (MB), ependymoma (EP), pilocytic astrocytoma (PA), and brainstem glioma (BG) being the most common types in children. These tumors can negatively impact mental and physical development.

Clinical information from radiological interpretations and the histopathological features of tumors plays a crucial role in diagnosing, prognosticating, and treating PF tumors in children. Histopathological evaluation, which is necessary for the initial diagnosis, helps to evaluate patient prognosis and direct clinical and therapeutic management. It remains the gold standard in differentiating PF tumors [1,2]. Although biopsies of different PF brain tumors can reveal distinct visual characteristics, they carry significant risks of morbidity and mortality, in addition to being expensive. Recent progress in characterizing tumor subtypes based on cross-sectional diagnostic imaging indicates that it can help to predict differential survival and responses to treatment. This development is particularly promising for future treatment stratification in PF tumors. Hence, developing a novel non-invasive diagnostic tool is essential in classifying tumors based on type and grade and aiding in planning treatment.

Magnetic resonance imaging (MRI) is currently the most preferred non-invasive method. It offers high intrinsic soft-tissue contrast without the risk of ionizing radiation. Conventional MRI protocols, including T1-weighted (T1W), T2-weighted (T2W), and fluid-attenuated inversion recovery (FLAIR) MRI sequences, have shown promising results in differentiating types of PF tumors in children [3–21]. Additionally, diffusion-weighted imaging (DWI) with apparent diffusion coefficient (ADC) maps allows the assessment of physiological features to discriminate between low- and high-grade tumors and their different subtypes [22–37].

While numerous advancements have been made, the diagnosis and prognosis of specific tumor matches still present significant challenges due to the voxel-wise overlap [23,27,38]. The classification process necessitates the inclusion of a tumor's molecular profile as a critical variable to predict the diverse biological behaviors of entities that exhibit histological similarities or even indistinguishability [2]. An extensive exploration of tumor classifications has been conducted using MRI in the literature. Nevertheless, accurately distinguishing between these tumor types remains an active area of research [20,39–42]. The differentiation between MB and EP is of the utmost importance, considering the distinct treatment planning required for each, underscoring the significance of their accurate diagnosis in numerous cases.

Artificial intelligence (AI) applications in pediatric brain tumor research are currently not well documented when compared to the available literature for adults. Challenges arise due to the unique pathology of pediatric cases and limitations in available data, which hinder the development of AI applications specifically tailored to children [43]. While there is growing interest in utilizing AI for pediatric brain tumor classification [44–55], the integration of AI into clinical workflows encounters significant obstacles beyond mere classification. One major challenge is the limited interpretability of many AI methods; creating a "black-box" model raises concerns among clinicians and patients. To address this issue, our research aims to enhance the interpretability of ML models' outcomes, which are frequently either blindly accepted or disregarded due to their black-box nature. To the best of our knowledge, there is a lack of literature specifically focusing on the issue of reasoning and explainability [56].

This study had two main objectives, aiming to bridge the gaps between ML outcomes and medical knowledge. Firstly, it sought to investigate the significance of clinical MRI features in classifying pediatric PF tumors (MB, EP, PA, and BG) through exploratory data analysis (EDA). Secondly, it aimed to offer explanations for the ML outcomes by leveraging the insights gained from the data exploration (Figure 1).

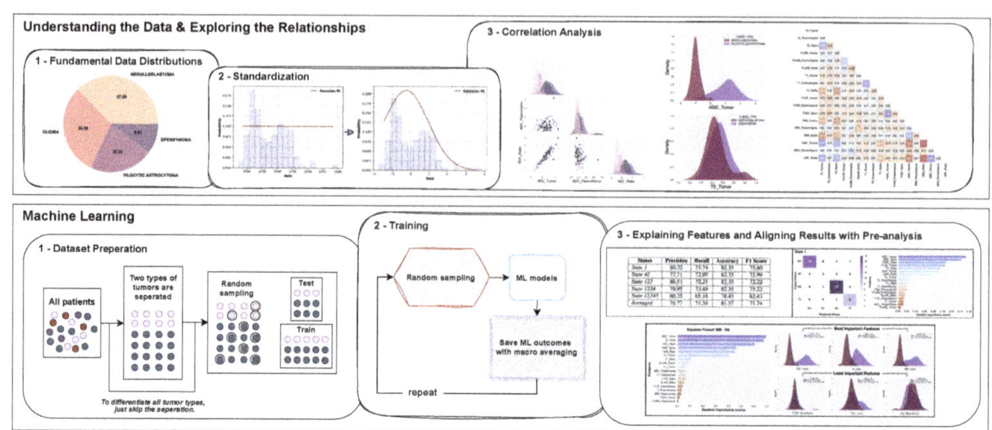

Figure 1. A flowchart depicting the proposed analysis for the classification of pediatric PF tumors, standardization of the dataset, pairwise feature analysis to examine various features of PF tumor types, and aligning interpretations of pre-analysis with ML models' outcomes.

2. Materials and Methods

2.1. Ethics Statement and Patient Characteristics

This prospective study (Ref: 632 QĐ-NĐ2 dated 12 May 2019) was conducted in both Radiology and Neurosurgery departments and approved by the Institutional Review Board in accordance with the 1964 Helsinki declaration. Prior to the MRI procedure, written informed consent was obtained from the authorized guardians of the patients. The study included 112 pediatric patients diagnosed with PF tumors, including 42 with MB, 11 with EP, 25 with PA, and 34 with BG. All BG patients were confirmed based on full agreement between neuroradiologists and neurosurgeons, while the remaining MB, EP and PA patients underwent either surgery or biopsy for histopathological confirmation.

The demographics of the patient population were analyzed to gain insights into their age, gender, and weight distributions. The age statistics revealed a mean age of 6.55 years, with a median age of 6.0 years. The age range varied from a minimum of 0.6 years to a maximum of 15.0 years, reflecting the diversity within our cohort. Regarding gender, we observed a greater representation of males, with a count of 68, compared to females, with a count of 44. The mean weight was calculated to be 22.54 kg, with a median weight of 20.5 kg. The range of weights varied from a minimum of 3 kg to a maximum of 48 kg.

In-depth patient demographics can be found in the accompanying Table 1, which provides a comprehensive overview of the study population. Table 1 includes detailed information on gender, age, and weight for the patients.

Table 1. Patient demographics.

Tumor Type	Gender	Count	Age Mean ± Std	Age Min	Age Max	Weight Mean ± Std	Weight Min	Weight Max
Medulloblastoma	Girl	16	7.16 ± 3.74	0.6	13	20.81 ± 8.53	8	35
	Boy	26	6.77 ± 3.40	1	13	21.19 ± 8.51	9	40
Ependymoma	Girl	3	5.67 ± 0.58	5	6	20.33 ± 4.16	17	25
	Boy	8	4.00 ± 3.42	1	11	18.38 ± 12.35	9	45
Pilocytic Astrocytoma	Girl	11	8.18 ± 3.66	3	14	25.18 ± 12.16	9	48
	Boy	14	5.79 ± 3.24	1	12	24.07 ± 10.22	10	44
Brainstem Glioma	Girl	14	6.43 ± 3.69	1	15	22.86 ± 11.81	3	47
	Boy	20	6.65 ± 2.85	3	14	24.95 ± 7.82	15	48

2.2. Data Acquisition and Assessment of MRI Features

The MRI protocol was performed in the supine position using a 1.5 Tesla MRI scanner (Philips, Best, The Netherlands) and included T1W, T2W, FLAIR, DWI (b values: 0 and 1000) with ADC, and T1 contrast-enhanced (T1CE) sequences with macrocyclic gadolinium-based contrast enhancement (0.1 mL/kg Gadovist, Bayer, Germany or 0.2 mL/kg Dotarem, Guerbet, France).

MR images of all patients were imported into the Medical Imaging Interaction Toolkit, developed by the German Cancer Research Center's Division of Medical Image Computing in Heidelberg, Germany. The radiologists precisely identified the slice in which the largest diameter of the PF tumor was present. For each patient, ROIs corresponding to the posterior fossa tumors and normal-appearing parenchyma were manually delineated on the T1W, T2W, FLAIR, DWI, and ADC images. These delineations were based on the consensus reached by two expert radiologists with over 10 years of experience in interpreting neuro MR images. An example of ROI delineation on a T2W MRI is provided in Figure 2. For additional ROI delineations of other sequences, please refer to Figure S1 in the Supplementary File S1.

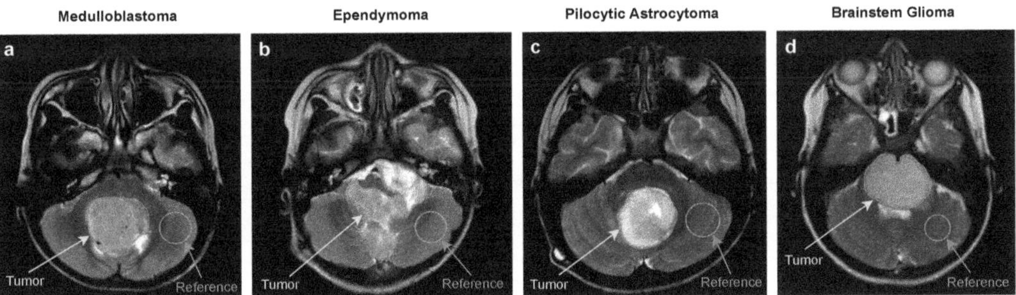

Figure 2. Example of ROI delineation on a T2W MRI. (**a**) MB: 8 years old, boy. (**b**) EP: 3 years old, boy. (**c**) PA: 7 years old, girl. (**d**) BG: 6 years old, girl.

The following MRI features were evaluated: signal intensities (SIs) of T2, T1, FLAIR, T1CE, DWI, and ADC. The ratio of each MRI feature was calculated as the quotient of the tumor's SI and the SI of the normal-appearing parenchyma (Ratio $= \frac{\text{Tumor Feature}}{\text{Parenchyma Feature}}$). Additionally, ADC values were quantified for both the PF tumor and parenchyma regions using the MR Diffusion tool available in the Philips Intellispace Portal, version 11 (Philips, Best, The Netherlands).

2.3. Exploratory Data Analysis

The quality of a dataset has a direct impact on the effectiveness of the trained model. Therefore, EDA plays a crucial role in understanding the data by revealing its inherent structure, identifying anomalies and outliers, extracting significant features, and facilitating the appropriate ML models to establish correlations between MRI feature characteristics and the various types of pediatric PF tumors.

In this study, we performed an exploratory analysis using kernel density estimations (KDE) with Gaussian distributions, focusing on the MRI features. The proposed analysis consisted of three parts: standardization, data analysis involving visualization of the distributions of each MRI feature, as well as exploring relationships between different features, and analyzing the ML models' outcomes through the extracted knowledge from EDA. All figures were generated using the Matplotlib package (version 3.5.2) in Python.

2.4. Standardization

The patient dataset underwent a standardization process known as Z-score normalization. This process was carried out using the Python programming language, specifically Python version 3.9.13, along with the Scikit-Learn library version 1.0.2.

To perform the standardization, the StandardScaler function from Scikit-Learn was utilized. This function ensured that numerical attributes within the patient dataset were transformed into a standardized format. It achieved this by subtracting the mean and scaling the values to have a unit variance.

The StandardScaler function normalizes each feature individually, meaning that each column/feature/variable in the input matrix X will have a mean (μ) of 0 and a standard deviation (σ) of 1. The normalization is accomplished using the formula $z = \frac{x_i - \mu}{\sigma}$, where x_i represents the value of a specific feature for a patient.

2.5. Pairwise Feature Analysis

The pairplot function in the Seaborn Python package (version 0.11.2) enables the visualization of the pairwise relationships between variables in a dataset. Numerical variables are split into a single row on the y-axis and a single column on the x-axis by default. The position of one variable on the vertical or horizontal axis indicates its correlation with another variable in the same row of data. The relationship between the MRI features was further examined through Pearson's correlation coefficients, calculated using the default corr() function in the pandas dataframe.

2.6. Revealing Distribution Differences of Patients between Tumor Types

To effectively illustrate the distinctions among the four PF tumor types, we utilized the kdeplot and pairplot functions from Seaborn as necessary. Additionally, we assigned a hue parameter to represent the tumor type, thereby facilitating a semantic mapping. This assignment transforms the default marginal plot into a layered KDE, which helps to address the challenge of reconstructing the density function f using an independent identically distributed (iid) sample $x_1, x_2, ..., x_n$ from the respective probability distribution. The generalized estimate used in plotting can be expressed as follows:

$$\hat{f}(x) = \frac{1}{nh^d} \sum_{i=1}^{n} k\left(\frac{x - x_i}{h}\right),$$

(1

where h is a bandwidth parameter, and the kernel is commonly a Gaussian,

$$k(z) = \frac{1}{\sqrt{2\pi}} \exp\left(-\frac{1}{2}z^2\right).$$

(2

2.7. Machine Learning

We employed eight ML models, including support vector machine (SVM), linear support vector machine (LSVM), logistic regression (LR), a random forest classifier (RF), a decision tree classifier (DT), a gradient boosting classifier (GBM), a catboost classifier (CB), and an extreme gradient boosting classifier (XGB), to assess the consistency of our interpretations of the raw data with the outcomes. CB and XGB were obtained from their respective libraries (CatBoost version 1.1.1, XGBoost version 1.5.1), while other models were obtained from the Scikit-Learn library.

To ensure methodological consistency, we utilized the default versions of all ML models, as our primary objective was not to maximize the classification scores. It is important to acknowledge that tuning the model parameters could potentially lead to improved results. However, considering the limited data size, the presence of rare tumor types, and the absence of an external dataset from another hospital, such parameter adjustments carried a significant risk of overfitting on our data. In order to mitigate this risk and uphold the credibility of our findings, we chose to adhere to the default model configurations throughout the analysis. This decision safeguarded the integrity of our research and ensured the validity of our conclusions.

Tree-based models, such as RF, DT, GBM, CB, and XGB, are commonly utilized in ML. However, they can be prone to overfitting when the trees are deep and have a large number of features. To address this issue, RF, which is a bagging model, generates a set of decision trees by training on different data samples or subsets of features. XGB, on the other hand, is a sequential model that adopts a different approach to building decision trees. To ensure that our models did not have a bias towards certain features and generalized well, we conducted an analysis of the prioritization and proportional distribution of features used by the RF and XGB models during prediction. This analysis strengthened our explanations of the models' high performance and accuracy in predicting outcomes.

To ensure the reliability of our ML models, particularly with a small dataset, we conducted five runs using stratified random sampling based on tumor type with 55% train and 45% test patients. We used random states to obtain samplings and preserve the train/test distributions for the reproducibility of the experiment. Ultimately, we calculated the averaged outcomes with their standard deviations.

The accuracy metric is not employed in the presentation of our results due to significant class imbalance in our dataset. Utilizing the accuracy metric could have led to misleading results. Instead, we relied on the precision, recall, and F1 score, computed based on the number of true positives (TP), true negatives (TN), false positives (FP), and false negatives (FN), as fundamental evaluation metrics to assess the performance of our classifiers in both binary and multiclass classification tasks. Precision gauges the proportion of correctly predicted positive instances among all positive predictions, highlighting the accuracy of positive classifications. Conversely, recall assesses the proportion of positive instances that were correctly identified by the classifier, emphasizing the completeness of the positive predictions.

While high precision and high recall are typically desirable, we were aware of the potential trade-off between these two metrics in certain scenarios. To gain a comprehensive understanding of our classifier's effectiveness, we utilized the F1 score, which harmoniously considers both precision and recall.

To ensure a precise interpretation of the ML results, we chose not to equalize the labels. Instead, we utilized the macro precision, macro recall, and macro F1 score metrics to ensure that all labels contributed equally to the results. This approach allowed us to assess the classifier's performance while considering the impact of varying patient counts across different labels.

The validation metrics used in ML are as follows:

$$\text{Macro Precision} = \frac{1}{n} \sum_{i=1}^{n} \frac{\text{TP}_i}{\text{TP}_i + \text{FP}_i} \tag{3}$$

$$\text{Macro Recall} = \frac{1}{n} \sum_{i=1}^{n} \frac{\text{TP}_i}{\text{TP}_i + \text{FN}_i} \tag{4}$$

$$\text{Macro F1 Score} = \frac{1}{n} \sum_{i=1}^{n} \frac{2 \times \text{TP}_i}{2 \times \text{TP}_i + \text{FP}_i + \text{FN}_i} \tag{5}$$

where n represents the total number of classes.

2.8. Statistical Analysis

Statistical analysis was performed using the SPSS software (version 25.0, 64-bit edition, IBM Corp., Armonk, NY, USA). A two-sided p-value of <0.05 was considered statistically significant. The statistical summary of the variability of ML outcomes is presented in the mean ± standard deviation format.

2.9. Hardware Requirements for Machine Learning

Designing an ML pipeline with the current number of patients and their tabular data does not require significant computational power. The entire machine learning system was developed utilizing a system equipped with an Apple M1 chip CPU and a memory capacity of 16 GB, namely the Hynix LPDDR4.

3. Results

3.1. Single MRI Feature Analysis

Our feature analysis, which utilized KDE on the standardized distributions presented in Figure 3a–f, yielded several valuable insights.

T2_Tumor features possess distributions that are expected to differentiate PA from EP and MB but cannot differentiate between MB and EP or PA and BG (Figure 3a). Moreover, the T2_Ratio might aid in distinguishing between MB and EP, as well as PA and BG.

The distributions of FLAIR_Tumor and FLAIR_Ratio generate notably different distributions (Figure 3b), even though the Ratio feature is mathematically dependent on the Tumor feature. The FLAIR distributions might be effective in distinguishing between MB and EP, as demonstrated by FLAIR_Tumor, which exhibits a broad EP and a narrow MB distribution. Furthermore, the FLAIR_Ratio exhibits two distinct and narrow Gaussian distributions, which also might aid in distinguishing between MB and EP. In contrast, the other scenarios do not present any discriminative characteristics.

The DWI characteristics (Figure 3c) demonstrate distributions that allow differentiation between MB and PA. Additionally, although to a lesser extent, discrete distributions can be observed in the differentiation between MB and BG, as well as between EP and PA. On the other hand, despite their high distinctive distributions overall, DWI_Ratio features are not expected to be effective in distinguishing between PA and BG due to significant overlap.

ADC (Figure 3d) demonstrates separate distributions in distinguishing each tumor pair, with the highest distinction observed between MB and PA and the least between PA and BG. When considering tumors as a whole rather than in pairs, ADC and DWI present the most distinct distributions for all tumor types. ADC shows highly distinct distributions for each tumor, with DWI following closely behind.

The T1 features, as shown in Figure 3e, do not demonstrate any distinctive distributions that can effectively differentiate between different tumor scenarios. However, the T1_Ratio appears to be a critical factor in distinguishing PA from other types of tumors. In addition, T1CE presents important distinct distributions for all other tumor matches with BG, as depicted in Figure 3f.

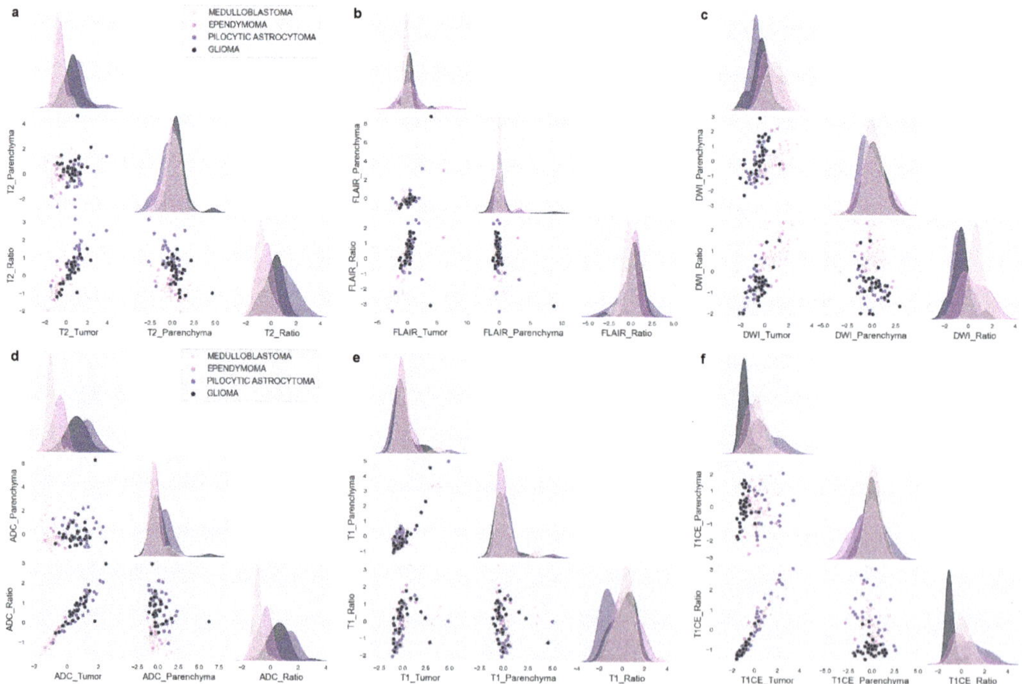

Figure 3. Kernel density estimations with Gaussian distributions of MRI features for PF tumors (**a**) T2, (**b**) FLAIR, (**c**) DWI, (**d**) ADC, (**e**) T1, and (**f**) T1CE.

3.2. Pairwise Analysis of MRI Features

The scatter correlation plots (Figure S2 in the Supplementary File S1) and Pearson's correlation coefficients (Figure 4) illustrate varying degrees of correlation between the MRI features and tumor types. Notably, MB exhibits clustered shapes, while PA appears scattered in most cases, and BG and EP show dispersed and uncertain distributions. Outlier patients with correlated features were identified, and some features exhibited no correlations with the tumor types.

The results of the Pearson's correlation analysis indicated that the T2 and ADC features, with complex distributions compared to other features, exhibited significant positive correlations, particularly T2_Tumor and ADC_Tumor (r = 0.87, $p < 0.0001$), T2_Tumor and ADC_Ratio (r = 0.85, $p < 0.0001$), T2_Ratio and ADC_Tumor (r = 0.78, $p < 0.0001$), and T2_Ratio and ADC_Ratio (r = 0.79, $p < 0.0001$). Conversely, significant negative correlations were observed between the T2 and DWI features, as well as between the DWI and ADC features, namely T2_Tumor and DWI_Tumor (r = -0.46, $p < 0.0001$), T2_Tumor and DWI_Ratio (r = -0.52, $p < 0.0001$), T2_Ratio and DWI_Tumor (r = -0.51, $p < 0.0001$), T2_Ratio and DWI_Ratio (r = -0.44, $p < 0.0001$), ADC_Tumor and DWI_Tumor (r = -0.68, $p < 0.0001$), ADC_Tumor and DWI_Ratio (r = -0.79 $p < 0.0001$), ADC_Ratio and DWI_Tumor (r = -0.66, $p < 0.0001$), and ADC_Ratio and DWI_Ratio (r = -0.78, $p < 0.0001$).

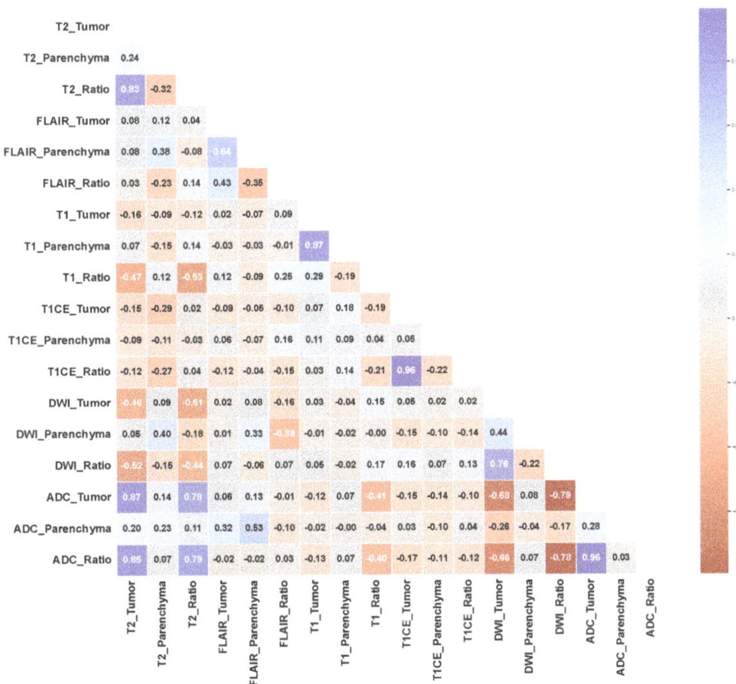

Figure 4. Pearson's correlation coefficients between each MRI feature.

FLAIR_Tumor did not demonstrate a significant correlation with any other features (T2_Tumor (r = 0.08, $p = 0.39$), T1_Tumor (r = 0.02, $p = 0.84$), T1CE_Tumor (r = -0.09, $p = 0.34$), DWI_Tumor (r = 0.02, $p = 0.85$), and ADC_Tumor (r = 0.06, $p = 0.52$)), while FLAIR_Ratio could exhibit correlations in logarithmic or reduced dimensions (T1_Ratio (r = 0.25, $p = 0.008$). Similar patterns to FLAIR were observed for T1_Tumor (T2_Tumor (r = -0.16, $p = 0.08$), T1CE_Tumor (r = 0.07, $p = 0.45$), DWI_Tumor (r = 0.03, $p = 0.76$), and ADC_Tumor (r = -0.12, $p = 0.19$)), and T1_Ratio (T2_Tumor (r = -0.47, $p < 0.0001$), T2_Ratio (r = -0.53, $p < 0.0001$), T1CE_Tumor (r = -0.19, $p = 0.045$), T1CE_Ratio (r = -0.21, $p = 0.03$), ADC_Tumor (r = 0.41, $p < 0.0001$) and ADC_Ratio (r = 0.40, $p < 0.0001$)), emphasizing the importance of using a ratio computed with reference to parenchyma. In contrast T1CE_Tumor and T1CE_Ratio showed dispersed distributions, with non-linear patterns that could be observed for certain tumor types.

3.3. Findings from Machine Learning

The ML procedure involved analyzing feature importance scores, the test scores of eight ML models (Tables S1–S7 in the Supplementary File S2), and confusion matrices to assess the accuracy and reliability of the results. We trained the models on all possible tumor pairs to identify unique features for each case, and the most favorable outcomes are summarized in Table 2. Additionally, we conducted a comprehensive analysis of the feature importance scores for all four tumor types, providing further insights into their distinguishing characteristics.

We focused on the RF and XGB models since RF delivered the best scores for the classification of all tumors, while XGB possesses a distinct tree structure compared to RF allowing us to explore and compare the variations in the ML models' outcomes. Although various ML models could have been employed for this analysis, we specifically chose XGB and RF to illustrate how the methods' structures differ in generating importance scores and to present a clear and concise analysis.

Notably, as shown in Figure 5a, the FLAIR_Ratio was identified as the most discriminating feature in distinguishing between MB and EP in both the RF and XGB models followed by the ADC_Ratio. However, the two models relied on different features for decision-making. Therefore, relying solely on the analysis presented in Figure 5 may not be sufficient for model comparison, as they prioritize different features. The performance evaluation of both models showed that the RF model, which prioritized diffusion features (4 out of top 5, 65.38%), demonstrated greater accuracy in feature selection compared to the XGB model (2 out of top 5, 60.08%). Therefore, the features highlighted by the RF model should be considered more significant in distinguishing between MB and EP.

Table 2. Best test scores for each case from evaluation of 8 different ML models.

	Best Model	**Precision**	**Recall**	**F1 Score**
MB-EP	LR	70.65 ± 4.55	68.21 ± 7.86	67.70 ± 6.19
MB-PA	LSVM	97.46 ± 1.57	97.28 ± 1.70	97.28 ± 1.52
MB-BG	LR	94.94 ± 2.38	94.77 ± 2.50	94.81 ± 2.42
EP-PA	LR	95.90 ± 4.17	96.33 ± 4.11	95.80 ± 3.85
EP-BG	LR	91.46 ± 7.58	87.50 ± 10.94	87.48 ± 9.73
PA-BG	CB	89.95 ± 5.97	88.69 ± 6.86	89.04 ± 6.50
MB-EP-PA-BG	RF	76.77 ± 9.78	71.34 ± 3.53	71.74 ± 5.41

In differentiating between MB and PA, the RF model showed a more dispersed reliance on various features, whereas the XGB model heavily relied on ADC_Tumor (Figure 5b). The study suggests that DWI features play an important role in distinguishing between these tumors, with T2 features being crucial in the decision-making process of the RF model, leading to a higher F1 score (RF: 93.81%, XGB: 90.15%).

To differentiate between MB and BG, our results indicated that the XGB model heavily relied on the ADC_Tumor feature ($\sim$80%), while the RF model utilized a more diverse set of features, such as ADC, DWI, T1CE, and T2 (Figure 5c). Surprisingly, both models performed equally well in this scenario, with an F1 score of 93.62%. Therefore, our results suggest that the ADC_Tumor feature is crucial in distinguishing MB from BG.

In differentiating EP from BG, the ML models were mainly impacted by the ADC_Tumor feature, with a supportive influence of the T1CE_Ratio and T1CE_Tumor features (Figure 5d). The RF model was also influenced by T2_Tumor, which might have led to a slight decrease in the overall F1 score (RF: 69.96%, XGB: 73.34%). As there was significant feature overlap between EP and BG, the performance scores for this classification task were lower compared to other tumor pairs, except MB and EP.

In distinguishing EP from PA, T2 features provided significant discriminative power (Figure 5e). However, the ADC_Tumor and ADC_Ratio features were found to be the leading contributors to the F1 score of 92.18% for the RF model, while the XGB model achieved a score of 81.48% due to its strong dependency on T2.

To differentiate PA from BG, both ML models heavily relied on T1CE features in their decision-making processes, with the T2_Ratio also providing discriminative power (Figure 5f). The XGB model outperformed the RF model, achieving a higher F1 score of 89.01% compared to 87.25%.

Additionally, the LR model achieved the highest F1 score in the MB-EP case and emerged as the dominant model in the MB-BG, EP-PA, and EP-BG cases. Furthermore, the LSVM outperformed other models in distinguishing between MB and PA. For the PA-BG case, the CB model attained high scores.

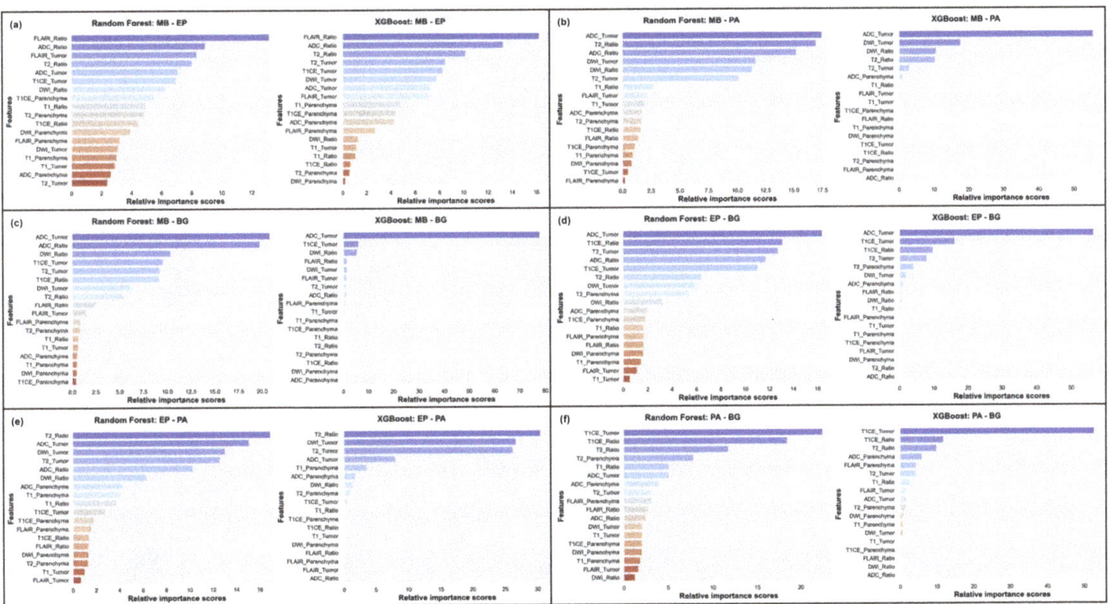

Figure 5. Averaged feature importance scores generated by RF and XGB models for behavior comparison. (**a**) MB-EP; (**b**) MB-PA; (**c**) MB-BG; (**d**) EP-BG; (**e**) EP-PA; (**f**) PA-BG.

The analysis revealed varying degrees of feature importance in differentiating between the four tumor types in the MB, EP, BG, and PA classification task. Among the models, the RF model achieved the highest F1 score of 71.74%, outperforming the other models. Figure 6c illustrates the significant role of ADC features in overall differentiation, followed by the T1CE and T2 features. The DWI and FLAIR features also contributed to the discriminative power, albeit to a lesser extent.

We also identified the most challenging discrimination task, which involved distinguishing between MB and EP (Figure 6a), and the easiest discrimination task, which involved distinguishing between MB and PA (Figure 6b). In the challenging classification problem of MB and EP, the Gaussian distributions of the best distinguishing features were found to overlap significantly, while, in the easiest one, MB and PA, the distributions of the best features did not overlap at all. Conversely, the least important features overlapped completely in every scenario.

The impact of stratified random sampling on the feature selection and performance of the RF model was examined and the findings are as given below (Figure S3 in the Supplementary File S1).

- In State 1, the model misclassified nine patients, including two with a BG tumor, three with an EP tumor, one with an MB tumor, and three with a PA tumor. The most informative features for this classification were ADC_Tumor, ADC_Ratio, and T1CE_Ratio.
- In State 2, the model performed slightly better in predicting BG and could distinguish all MB from other types. However, it misclassified two more PA patients, using ADC_Ratio, ADC_Tumor, and T2_Ratio as the most significant features.
- In State 3, the model could distinguish almost all PA test patients except one. However, it missed one BG, which was previously predicted as EP in State 2.
- In State 4, the model was unable to differentiate three BG, three EP, and three PA test patients from other types.
- In State 5, the model attributed the highest importance to the T1CE_Tumor feature, which led to the misclassification of all EP patients and four BG patients.

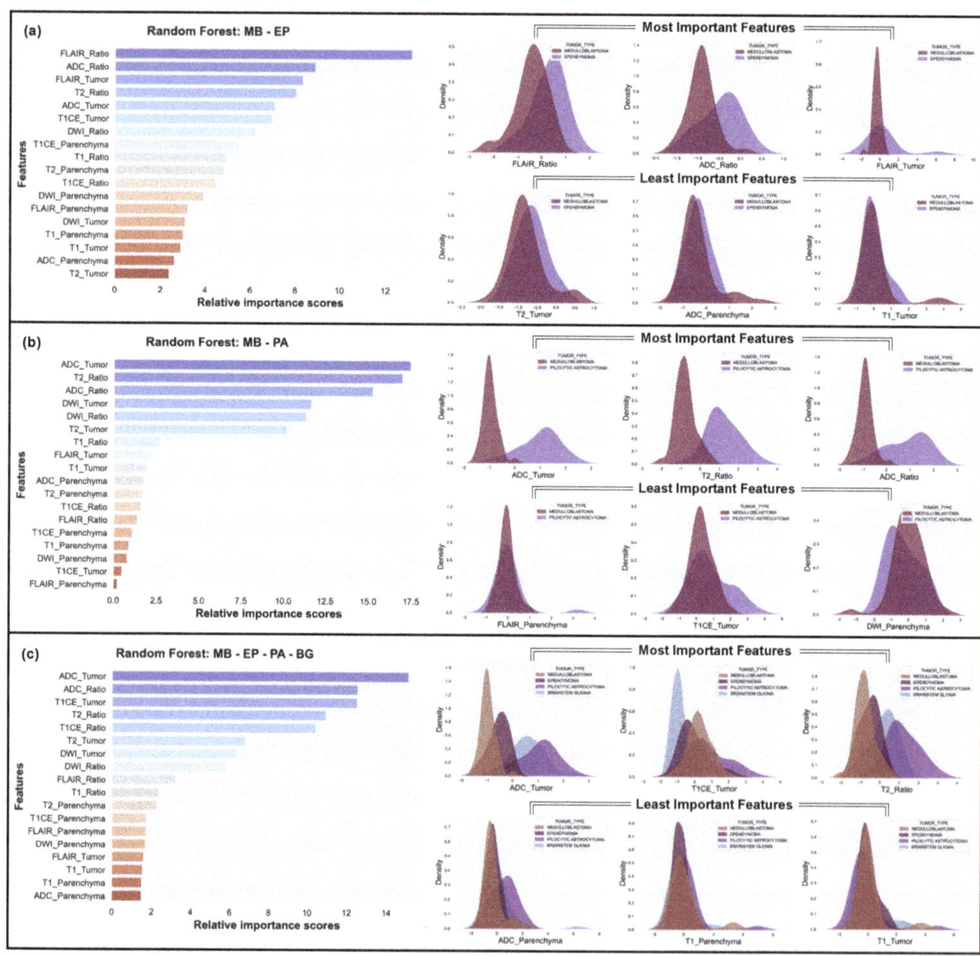

Figure 6. The three most and least effective features for the classifications using the Random Forest (RF) model. (**a**) The hardest case: MB vs. EP; (**b**) The easiest case: MB vs. PA; (**c**) Case of all tumor types.

4. Discussion

Pediatric brain tumors pose a significant clinical challenge due to the substantial degree of spatial heterogeneity in tumor characteristics. Tumors such as those arising from the posterior fossa have a significant imaging feature overlap, leading to difficulty in differentiation, even among experts. The need to differentiate is important due to the different treatment options available for each of them. Thus, precise diagnosis and treatment are crucial in improving outcomes and enhancing quality of life.

Despite the significant advancements observed in AI and medical imaging, the dependability and accuracy of these approaches are profoundly influenced by the quality of the data, meticulous system design, and the comprehensive dissemination of transparent results. Therefore, we conducted a comprehensive and systematic analysis focusing on four distinct tumor types in pediatric brain tumor research. We employed an approach that integrated EDA to interpret ML outcomes, while the ML models provided additional insights into the underlying patterns and relationships among the MRI features.

This study was motivated by the idea that the feature distribution obtained from KDE can provide reliable estimates of ML results prior to the actual model training. The estimation provides insights into which features are the most effective and which features contribute negligibly to the ML models' decisions. To test this hypothesis, we conducted several pre-training analyses without relying on prior clinical knowledge and analyzed the feature distribution plots. In the present research, a thorough investigation of the diverse characteristics of pediatric PF tumor types was carried out through the utilization of Gaussian distributions, which can be observed in Figure 3. Through this analysis, a number of predictions have been drawn, indicating that certain features are likely to be highly effective in distinguishing particular tumor types, while others are deemed to have a limited impact on classification.

The single-feature analysis using Gaussian distributions, shown in Figure 3, revealed that some MRI features are effective in distinguishing specific pediatric PF tumor types, while others have minimal contributions towards classification. ADC and DWI features are the most effective in differentiating between tumor types, with clear differences in the distributions of these features for different tumors, whereas T1 and T1CE features are less effective in distinguishing between tumor types, although there are some differences in the distributions for different tumors. Moreover, T2 and FLAIR features show some differences in the distributions for different tumors, but these are less pronounced than for ADC and DWI.

Our analyses in the single-feature section are consistent with both the clinical and ML results in almost every instance, and we provide corresponding references in this section to validate our findings. Specifically, our analysis indicates that the T2_Tumor feature can effectively differentiate PA from EP and MB (Figure 5b–d), but cannot differentiate between MB and EP or PA and BG (Figure 5a–f). Remarkably, the incorporation of the hand-crafted feature T2_Ratio further enhances the effectiveness of T2 for tumor classification (Figure 5a,b,e,f). This is particularly evident in the differentiation between MB and EP, as well as PA and BG tumors (Figure 5a–f). Our findings also shed light on the potential of FLAIR features in distinguishing between different tumor types (Figures 3b and 5). The distributions of FLAIR_Tumor and FLAIR_Ratio exhibit notable differences, despite the lack of distributional disparities for parenchyma, which serves as a reference point. Specifically, FLAIR_Tumor shows a broad distribution for EP and a narrow distribution for MB, while FLAIR_Ratio displays two distinct and narrow Gaussian distributions. The ML results highlight that FLAIR features are useful in distinguishing between MB and EP tumors (Figure 5a), although the discriminative characteristics are not evident in the rest of the scenarios (Figure 5b–f). These findings are consistent with those of previous studies [10–12,15,17,19,20].

The results of our study also indicate that the DWI characteristics display distinct distributions that enable the differentiation of MB and PA (Figures 3c and 5b). While the distributions are less clear, there are still noticeable differences in the DWI characteristics when distinguishing between MB and BG, as well as between EP and PA (Figures 3c and 5c,e). However, we found that the DWI_Ratio features, despite having highly distinctive distributions overall, were not likely to be useful in distinguishing between PA and BG due to their significant overlap (Figures 3c and 5f). Moreover, our findings revealed that the ADC had distinct distributions for each tumor pair, with the most noticeable distinction between MB and PA and the least between PA and BG (Figures 3d and 5a–f). When considering the distributions of all tumors collectively, rather than in pairs, ADC exhibited the most prominent differences among all tumor types (Figure 6c). The difference in diffusivity between various types of PF tumors is due to their cellular characteristics and arrangement as well as the presence of cystic spaces within the tumor bulk. These results are in line with previous research findings [20,22–24,30–32].

While our study identified several features that can effectively differentiate between different types of PF tumors, not all features are equally informative. Some features may exhibit significant overlap between different tumor types, which can limit their usefulness in certain scenarios. For instance, our study demonstrated that T1_Tumor feature did not exhibit any notably distinctive distributions in distinguishing between different tumor types (Figures 3e and 5a–f). However, it could be seen that T1_Ratio is a crucial factor in differentiating PA from other tumor types (Figure 5b,e,f). Additionally, T1CE displays notable distinctive distributions when differentiating all other tumor types from BG (Figures 3f and 5c,d,f).

Based on the pairwise analysis of the dataset, our findings suggest that MB exhibit a more distinct set of MRI features that are strongly correlated with the tumor type. Conversely, PA appears to be more heterogeneous in terms of its MRI features, and the MRI features associated with BG and EP may not be well defined. Furthermore, the positive correlation between both T2 and ADC features may reflect the diverse nature of these tumor types, with different subtypes exhibiting distinct MRI features. The negative correlation observed between DWI and ADC, as well as DWI and T2 features, may reflect differences in tumor cellularity and tissue microstructure. This finding may have important implications for treatment planning, particularly with regard to therapies that target the tumor microenvironment. The findings of this study offer significant insights into the correlation between tumor types and MRI features.

In an ML classification model, a feature's ability to distinguish between different classes, such as different tumor types, is determined by the degree of separation or overlap between the distributions of the feature values for each class. When the distributions are close together and have significant overlap, the feature is unlikely to provide much discriminative value and will have little impact on the classification decision. Conversely, when the distributions are far apart and have minimal overlap, the feature is more likely to provide discriminative value and will significantly impact the classification decision. Examining the distribution of tumor types across various features can help to identify potential biomarkers that may be useful for diagnostic or prognostic purposes.

Our preliminary analysis in this study agrees with the results obtained from the RF model, thus substantiating its decision-making process. However, there is a possibility of the ML model selecting a non-distinctive feature as the most critical factor, which is irrational. Therefore, it is imperative to provide an explanation for the model's decision. To this end, we have proposed a methodology to elucidate the correlation between the KDE analysis and the averaged feature importance of the ML results, as presented in Figure 6, to bring clarity to this association.

The performance of ML models can be significantly influenced by the distribution of samples in the training and testing sets, even if the samples belong to the same class. To ensure more generalizable results, we utilized stratified random sampling to evaluate the dataset across five different distributions. The feature importance was then computed and averaged over the five distributions, as demonstrated in Figures 5 and 6. However, there was still a considerable degree of variability in the results for each distribution (Figure S3 in the Supplementary File S1). Thus, it is crucial to take into account the sample distribution when assessing ML models and to implement stratified random sampling to ensure robust and generalized outcomes.

We evaluated the performance of eight different ML models and determined that LR is suitable for binary classification tasks. However, in discriminating between all tumor types simultaneously, RF outperformed all other models. To enhance the interpretability of the RF model's results, we aligned its feature importance values with the KDE predictions. To ensure reproducibility, we used five different random seeds and computed the mean of the resulting outputs. Comprehensive analysis of the ML results side by side revealed that no single model outperformed the others, as demonstrated in Table 2. Furthermore, the results differed depending on the models and data structure. To demonstrate this, we compared the RF and XGB models for PF tumor classification, as shown in Figure 5. This approach provided a clear understanding of how the models' behavior influenced the results, which was crucial for the PF tumor classification task.

Analyzing patient distributions can provide insights into subtypes with unusual patterns that ML models may not detect, leading to errors in calculating tumor characteristics. These outliers can reveal unique features that improve the reliability and accuracy of ML models for medical diagnosis and treatment. Clustering and explaining ML models with larger labeled datasets could enhance our understanding of the heterogeneity within patient subtypes in future studies.

In clinical practice, tumors are assessed based on location, the effect exerted by the tumor on the surrounding tissue, and tumor behavior, including the tendency to invade surrounding tissues or the presence of cystic components or calcification. Tumors with classic imaging features in pathognomonic locations can be identified even by novice radiologists with ease. However, distinguishing between tumors with similar characteristics and locations requires a more in-depth analysis of the imaging features, as demonstrated in this study. When two tumors exhibit near-similar characteristics and locations, the ability to differentiate them based on imaging features such as those studied in this work becomes important. AI models trained on MRI sequences can assist in diagnosing similar lesions and aid in management planning. The transparent use of ML methods with pre-analysis and proper testing procedures is crucial for reliable, reproducible, and accurate findings. Ultimately, the primary goal of any analysis is to produce explainable and reproducible results that can be verified by other researchers, improving the diagnostic features and patient outcomes in medical research.

There are some limitations that need to be considered in the present study. First, the dataset used for analysis was limited in scope and size. Although it contained a sufficient number of samples to train ML models, the dataset may not have been representative of all possible scenarios, and the results may not generalize well to other datasets. To address this, we employed stratified random sampling to ensure that each tumor subtype was represented proportionally in the training and testing datasets. This approach helped to minimize bias in the model training and increase the generalizability of our findings. Second, our dataset only included four types of pediatric PF tumors, which may not fully represent the diversity of pediatric brain tumors. Third, the study was limited to the analysis of a single feature and pairwise interactions between features. Other important features or higher-order interactions that were not considered in this analysis may exist, and their inclusion may change the outcome. Future studies with larger sample sizes and additional advanced MRI protocols, such as semiquantitative and quantitative perfusion MRI and MR spectroscopy, could provide more insights into the diagnostic and prognostic value of MRI features for pediatric PF tumors. Additionally, further research is needed to investigate the potential of ML models and EDA to improve the reliability of pediatric PF tumor diagnosis and treatment.

5. Conclusions

The significance of our study lies in its ability to surpass the constraints of prior research in this field. While previous studies have often focused on only one binary differentiation or incorporated numerous exceptions, leading to reduced transparency, our research stands out by offering a comprehensive analysis of four distinct tumor subtypes within a single source. This paper offers a comprehensive and holistic understanding of the subject matter.

Through our analysis, we have uncovered the effectiveness of specific MRI features such as ADC and DWI, in accurately distinguishing between tumor types, while also shedding light on the limited impact of features such as T1 and T1CE. The combination of EDA and ML has provided valuable insights into feature distributions and their importance in classification. Additionally, handcrafted features such as T2_Ratio and T1_Ratio enhanced the effectiveness of T2 and T1 features, respectively, in tumor classification. Overall, we identified RF as a suitable model for tumor classification, while LR emerged as the optimal choice for most binary cases.

In our analysis, we focused on MRI features that demonstrated minimal overlap between tumor types within their KDE distributions, as they offered valuable discriminatory information. Our findings have highlighted the potential of specific features, such as ADC_Ratio and ADC_Tumor, in effectively differentiating between tumor types. This effectiveness can be attributed to the distinct cellular characteristics, arrangement, and presence of cystic spaces within the tumor mass.

We have also demonstrated that in situations where patient data are limited, complex systems may not always be necessary to evaluate feature importance; in fact, they could impair both performance and interpretability. By conducting comprehensive analyses using simpler approaches, we can still extract valuable insights into the significance of specific features. This emphasizes the importance of adaptability and resourcefulness in leveraging available data to make informed decisions in clinical settings.

Supplementary Materials: The following supporting information can be downloaded at: https://www.mdpi.com/article/10.3390/cancers15164015/s1, Supplementary File S1: Additional Figures (Figure S1: Examples of ROI delineations on a ADC, DWI, T1 and T1CE MRI; Figure S2: Pairwise plot with no parenchymas; Figure S3: RF model analysis of MB, EP, PA, and BG with classification metrics, confusion matrices, and feature importance distributions for each stratified random sampling state.) Supplementary File S2: Differentiation scores of machine learning models for all tumor types (Table S1: MB-EP; Table S2: MB-PA; Table S3: MB-BG; Table S4: EP-PA; Table S5: EP-BG; Table S6: PA-BG; Table S7: MB-EP-PA-BG.).

Author Contributions: Conceptualization, T.T.; Methodology, T.T.; Software, T.T.; Validation, T.T.; Formal analysis, T.T., C.N. and N.M.D.; Investigation, N.M.D.; Resources, N.M.D.; Data curation, N.M.D.; Writing—original draft, T.T.; Writing—review and editing, C.N., N.M.D. and B.K.; Visualization, T.T.; Supervision, B.K.; Project administration, B.K.; Funding acquisition, B.K. All authors have read and agreed to the published version of the manuscript.

Funding: This work was supported by the Scientific and Technological Research Council of Türkiye (TUBITAK) [2232- 118C221].

Institutional Review Board Statement: After obtaining approval from the Institutional Review Board of Children Hospital of 02 with approval number [Ref: 632 QĐ-NĐ2 dated 12 May 2019], we conducted the study in both Radiology and Neurosurgery departments in accordance with the 1964 Helsinki declaration.

Informed Consent Statement: Informed consent was obtained from all subjects involved in the study.

Data Availability Statement: The datasets generated and/or analyzed during the current study are not publicly available due to privacy concerns but are available from the corresponding author upon reasonable request. The source codes, EDA outputs, and entire ML evaluation results of the presented study can be accessed at https://github.com/toygarr/save-the-kid.

Conflicts of Interest: The authors declare that they have no known competing financial interests or personal relationships that could have appeared to influence the work reported in this paper.

References

Louis, D.N.; Ohgaki, H.; Wiestler, O.D.; Cavenee, W.K.; Burger, P.C.; Jouvet, A.; Scheithauer, B.W.; Kleihues, P. The 2007 WHO Classification of Tumours of the Central Nervous System. *Acta Neuropathol.* **2007**, *114*, 97–109. [CrossRef] [PubMed]

Louis, D.N.; Perry, A.; Reifenberger, G.; Von Deimling, A.; Figarella-Branger, D.; Cavenee, W.K.; Ohgaki, H.; Wiestler, O.D.; Kleihues, P.; Ellison, D.W. The 2016 World Health Organization classification of tumors of the central nervous system: A summary. *Acta Neuropathol.* **2016**, *131*, 803–820. [CrossRef] [PubMed]

Meyers, S.P.; Kemp, S.S.; Tarr, R.W. MR imaging features of medulloblastomas. *AJR Am. J. Roentgenol.* **1992**, *158*, 859–865. [CrossRef] [PubMed]

Koeller, K.K.; Rushing, E.J. From the archives of the AFIP: Medulloblastoma: A comprehensive review with radiologic-pathologic correlation. *Radiographics* **2003**, *23*, 1613–1637. [CrossRef]

Koeller, K.K.; Rushing, E.J. From the archives of the AFIP: Pilocytic astrocytoma: Radiologic-pathologic correlation. *Radiographics* **2004**, *24*, 1693–1708. [CrossRef]

Koeller, K.K.; Sandberg, G.D. From the archives of the AFIP: Cerebral intraventricular neoplasms: Radiologic-pathologic correlation. *Radiographics* **2002**, *22*, 1473–1505. [CrossRef]

Meyers, S.; Khademian, Z.; Biegel, J.; Chuang, S.; Korones, D.; Zimmerman, R. Primary intracranial atypical teratoid/rhabdoid tumors of infancy and childhood: MRI features and patient outcomes. *Am. J. Neuroradiol.* **2006**, *27*, 962–971.

Arai, K.; Sato, N.; Aoki, J.; Yagi, A.; Taketomi-Takahashi, A.; Morita, H.; Koyama, Y.; Oba, H.; Ishiuchi, S.; Saito, N.; et al. MR signal of the solid portion of pilocytic astrocytoma on T2-weighted images: Is it useful for differentiation from medulloblastoma? *Neuroradiology* **2006**, *48*, 233–237. [CrossRef]

Koral, K.; Gargan, L.; Bowers, D.C.; Gimi, B.; Timmons, C.F.; Weprin, B.; Rollins, N.K. Imaging characteristics of atypical teratoid–rhabdoid tumor in children compared with medulloblastoma. *Am. J. Roentgenol.* **2008**, *190*, 809–814. [CrossRef]

Forbes, J.A.; Chambless, L.B.; Smith, J.G.; Wushensky, C.A.; Lebow, R.L.; Alvarez, J.; Pearson, M.M. Use of T2 signal intensity of cerebellar neoplasms in pediatric patients to guide preoperative staging of the neuraxis. *J. Neurosurg. Pediatr.* **2011**, *7*, 165–174. [CrossRef]

Forbes, J.A.; Reig, A.S.; Smith, J.G.; Jermakowicz, W.; Tomycz, L.; Shay, S.D.; Sun, D.A.; Wushensky, C.A.; Pearson, M.M. Findings on preoperative brain MRI predict histopathology in children with cerebellar neoplasms. *Pediatr. Neurosurg.* **2011**, *47*, 51–59. [CrossRef] [PubMed]

Poretti, A.; Meoded, A.; Huisman, T.A. Neuroimaging of pediatric posterior fossa tumors including review of the literature. *J. Magn. Reson. Imaging* **2012**, *35*, 32–47. [CrossRef]

Rasalkar, D.D.; Chu, W.C.w.; Paunipagar, B.K.; Cheng, F.W.; Li, C. Paediatric intra-axial posterior fossa tumours: Pictorial review. *Postgraduate Med. J.* **2013**, *89*, 39–46. [CrossRef] [PubMed]

Plaza, M.J.; Borja, M.J.; Altman, N.; Saigal, G. Conventional and advanced MRI features of pediatric intracranial tumors: Posterior fossa and suprasellar tumors. *Am. J. Roentgenol.* **2013**, *200*, 1115–1124. [CrossRef] [PubMed]

Porto, L.; Jurcoane, A.; Schwabe, D.; Hattingen, E. Conventional magnetic resonance imaging in the differentiation between high and low-grade brain tumours in paediatric patients. *Eur. J. Paediatric Neurol.* **2014**, *18*, 25–29. [CrossRef] [PubMed]

Koob, M.; Girard, N. Cerebral tumors: Specific features in children. *Diagn. Interv. Imaging* **2014**, *95*, 965–983. [CrossRef] [PubMed]

Orphanidou-Vlachou, E.; Vlachos, N.; Davies, N.P.; Arvanitis, T.N.; Grundy, R.G.; Peet, A.C. Texture analysis of T1-and T2-weighted MR images and use of probabilistic neural network to discriminate posterior fossa tumours in children. *NMR Biomed.* **2014**, *27*, 632–639. [CrossRef]

Moharamzad, Y.; Sanei Taheri, M.; Niaghi, F.; Shobeiri, E. Brainstem glioma: Prediction of histopathologic grade based on conventional MR imaging. *Neuroradiol. J.* **2018**, *31*, 10–17. [CrossRef]

D'Arco, F.; Khan, F.; Mankad, K.; Ganau, M.; Caro-Dominguez, P.; Bisdas, S. Differential diagnosis of posterior fossa tumours in children: New insights. *Pediatr. Radiol.* **2018**, *48*, 1955–1963. [CrossRef]

Duc, N.M.; Huy, H.Q. Magnetic resonance imaging features of common posterior fossa brain tumors in children: A preliminary Vietnamese study. *Open Access Maced. J. Med. Sci.* **2019**, *7*, 2413. [CrossRef]

Duc, N.M.; Huy, H.Q.; Nadarajan, C.; Keserci, B. The role of predictive model based on quantitative basic magnetic resonance imaging in differentiating medulloblastoma from ependymoma. *Anticancer Res.* **2020**, *40*, 2975–2980. [CrossRef] [PubMed]

Rumboldt, Z.; Camacho, D.; Lake, D.; Welsh, C.; Castillo, M. Apparent diffusion coefficients for differentiation of cerebellar tumors in children. *Am. J. Neuroradiol.* **2006**, *27*, 1362–1369. [PubMed]

Jaremko, J.L.; Jans, L.; Coleman, L.T.; Ditchfield, M.R. Value and limitations of diffusion-weighted imaging in grading and diagnosis of pediatric posterior fossa tumors. *Am. J. Neuroradiol.* **2010**, *31*, 1613–1616. [CrossRef]

Gimi, B.; Cederberg, K.; Derinkuyu, B.; Gargan, L.; Koral, K.M.; Bowers, D.C.; Koral, K. Utility of apparent diffusion coefficient ratios in distinguishing common pediatric cerebellar tumors. *Acad. Radiol.* **2012**, *19*, 794–800. [CrossRef]

Bull, J.G.; Saunders, D.E.; Clark, C.A. Discrimination of paediatric brain tumours using apparent diffusion coefficient histograms. *Eur. Radiol.* **2012**, *22*, 447–457. [CrossRef] [PubMed]

Pierce, T.; Kranz, P.G.; Roth, C.; Leong, D.; Wei, P.; Provenzale, J.M. Use of apparent diffusion coefficient values for diagnosis of pediatric posterior fossa tumors. *Neuroradiol. J.* **2014**, *27*, 233–244. [CrossRef] [PubMed]

Porto, L.; Jurcoane, A.; Schwabe, D.; Kieslich, M.; Hattingen, E. Differentiation between high and low grade tumours in paediatric patients by using apparent diffusion coefficients. *Eur. J. Paediatric Neurol.* **2013**, *17*, 302–307. [CrossRef]

28. Poretti, A.; Meoded, A.; Cohen, K.J.; Grotzer, M.A.; Boltshauser, E.; Huisman, T.A. Apparent diffusion coefficient of pediatric cerebellar tumors: A biomarker of tumor grade? *Pediatr. Blood Cancer* **2013**, *60*, 2036–2041. [CrossRef]
29. Gutierrez, D.R.; Awwad, A.; Meijer, L.; Manita, M.; Jaspan, T.; Dineen, R.A.; Grundy, R.G.; Auer, D.P. Metrics and textural features of MRI diffusion to improve classification of pediatric posterior fossa tumors. *Am. J. Neuroradiol.* **2014**, *35*, 1009–1015. [CrossRef]
30. Zitouni, S.; Koc, G.; Doganay, S.; Saracoglu, S.; Gumus, K.Z.; Ciraci, S.; Coskun, A.; Unal, E.; Per, H.; Kurtsoy, A.; et al. Apparent diffusion coefficient in differentiation of pediatric posterior fossa tumors. *Jpn. J. Radiol.* **2017**, *35*, 448–453. [CrossRef]
31. Esa, M.M.M.; Mashaly, E.M.; El-Sawaf, Y.F.; Dawoud, M.M. Diagnostic accuracy of apparent diffusion coefficient ratio in distinguishing common pediatric CNS posterior fossa tumors. *Egypt. J. Radiol. Nucl. Med.* **2020**, *51*, 1–11. [CrossRef]
32. Minh Thong, P.; Minh Duc, N. The role of apparent diffusion coefficient in the differentiation between cerebellar medulloblastoma and brainstem glioma. *Neurol. Int.* **2020**, *12*, 34–40. [CrossRef] [PubMed]
33. Dury, R.J.; Lourdusamy, A.; Macarthur, D.C.; Peet, A.C.; Auer, D.P.; Grundy, R.G.; Dineen, R.A. Meta-Analysis of Apparent Diffusion Coefficient in Pediatric Medulloblastoma, Ependymoma, and Pilocytic Astrocytoma. *J. Magn. Reson. Imaging* **2022**, *56*, 147–157. [CrossRef] [PubMed]
34. Chen, D.; Lin, S.; She, D.; Chen, Q.; Xing, Z.; Zhang, Y.; Cao, D. Apparent Diffusion Coefficient in the Differentiation of Common Pediatric Brain Tumors in the Posterior Fossa: Different Region-of-Interest Selection Methods for Time Efficiency, Measurement Reproducibility, and Diagnostic Utility. *J. Comp. Assist. Tomogr.* **2023**, *47*, 291. [CrossRef] [PubMed]
35. Reddy, N.; Ellison, D.W.; Soares, B.P.; Carson, K.A.; Huisman, T.A.; Patay, Z. Pediatric posterior fossa medulloblastoma: The role of diffusion imaging in identifying molecular groups. *J. Neuroimaging* **2020**, *30*, 503–511. [CrossRef]
36. Gonçalves, F.G.; Zandifar, A.; Ub Kim, J.D.; Tierradentro-García, L.O.; Ghosh, A.; Khrichenko, D.; Andronikou, S.; Vossough, A. Application of Apparent Diffusion Coefficient Histogram Metrics for Differentiation of Pediatric Posterior Fossa Tumors: A Large Retrospective Study and Brief Review of Literature. *Clin. Neuroradiol.* **2022**, *32*, 1097–1108. [CrossRef]
37. Phuttharak, W.; Wannasarnmetha, M.; Waraaswapati, S.; Yuthawong, S. Diffusion MRI in evaluation of pediatric posterior fossa tumors. *Asian Pac. J. Cancer Prev. APJCP* **2021**, *22*, 1129. [CrossRef]
38. Koob, M.; Girard, N.; Ghattas, B.; Fellah, S.; Confort-Gouny, S.; Figarella-Branger, D.; Scavarda, D. The diagnostic accuracy of multiparametric MRI to determine pediatric brain tumor grades and types. *J. Neuro-Oncol.* **2016**, *127*, 345–353. [CrossRef]
39. Deng, J.; Xue, C.; Liu, X.; Li, S.; Zhou, J. Differentiating between adult intracranial medulloblastoma and ependymoma using MRI. *Clin. Radiol.* **2023**, *78*, e288–e293. [CrossRef]
40. Yamaguchi, J.; Ohka, F.; Motomura, K.; Saito, R. Latest classification of ependymoma in the molecular era and advances in its treatment: A review. *Jpn. J. Clin. Oncol.* **2023**, *53*, hyad056. [CrossRef] [PubMed]
41. Wang, S.; Wang, G.; Zhang, W.; He, J.; Sun, W.; Yang, M.; Sun, Y.; Peet, A. MRI-based whole-tumor radiomics to classify the type of pediatric posterior fossa brain tumor. *Neurochirurgie* **2022**, *68*, 601–607. [CrossRef] [PubMed]
42. Yearley, A.G.; Blitz, S.E.; Patel, R.V.; Chan, A.; Baird, L.C.; Friedman, G.K.; Arnaout, O.; Smith, T.R.; Bernstock, J.D. Machine Learning in the Classification of Pediatric Posterior Fossa Tumors: A Systematic Review. *Cancers* **2022**, *14*, 5608. [CrossRef] [PubMed]
43. Huang, J.; Shlobin, N.A.; Lam, S.K.; DeCuypere, M. Artificial intelligence applications in pediatric brain tumor imaging: A systematic review. *World Neurosur.* **2022**, *157*, 99–105. [CrossRef] [PubMed]
44. Li, M.; Shang, Z.; Yang, Z.; Zhang, Y.; Wan, H. Machine learning methods for MRI biomarkers analysis of pediatric posterior fossa tumors. *Biocybern. Biomed. Eng.* **2019**, *39*, 765–774. [CrossRef]
45. Li, M.; Wang, H.; Shang, Z.; Yang, Z.; Zhang, Y.; Wan, H. Ependymoma and pilocytic astrocytoma: Differentiation using radiomics approach based on machine learning. *J. Clin. Neurosci.* **2020**, *78*, 175–180. [CrossRef] [PubMed]
46. Zhou, H.; Hu, R.; Tang, O.; Hu, C.; Tang, L.; Chang, K.; Shen, Q.; Wu, J.; Zou, B.; Xiao, B.; et al. Automatic machine learning to differentiate pediatric posterior fossa tumors on routine MR imaging. *Am. J. Neuroradiol.* **2020**, *41*, 1279–1285. [CrossRef]
47. Dong, J.; Li, L.; Liang, S.; Zhao, S.; Zhang, B.; Meng, Y.; Zhang, Y.; Li, S. Differentiation between ependymoma and medulloblastoma in children with radiomics approach. *Acad. Radiol.* **2021**, *28*, 318–327. [CrossRef]
48. Grist, J.T.; Withey, S.; MacPherson, L.; Oates, A.; Powell, S.; Novak, J.; Abernethy, L.; Pizer, B.; Grundy, R.; Bailey, S.; et al. Distinguishing between paediatric brain tumour types using multi-parametric magnetic resonance imaging and machine learning: A multi-site study. *NeuroImage Clin.* **2020**, *25*, 102172. [CrossRef]
49. Payabvash, S.; Aboian, M.; Tihan, T.; Cha, S. Machine learning decision tree models for differentiation of posterior fossa tumors using diffusion histogram analysis and structural MRI findings. *Front. Oncol.* **2020**, *10*, 71. [CrossRef]
50. Novak, J.; Zarinabad, N.; Rose, H.; Arvanitis, T.; MacPherson, L.; Pinkey, B.; Oates, A.; Hales, P.; Grundy, R.; Auer, D.; et al. Classification of paediatric brain tumours by diffusion weighted imaging and machine learning. *Sci. Rep.* **2021**, *11*, 1–8. [CrossRef]
51. Zhuge, Y.; Ning, H.; Mathen, P.; Cheng, J.Y.; Krauze, A.V.; Camphausen, K.; Miller, R.W. Automated glioma grading on conventional MRI images using deep convolutional neural networks. *Med. Phys.* **2020**, *47*, 3044–3053. [CrossRef]
52. Quon, J.; Bala, W.; Chen, L.; Wright, J.; Kim, L.; Han, M.; Shpanskaya, K.; Lee, E.; Tong, E.; Iv, M.; et al. Deep learning for pediatric posterior fossa tumor detection and classification: A multi-institutional study. *Am. J. Neuroradiol.* **2020**, *41*, 1718–1725. [CrossRef] [PubMed]
53. Saju, A.C.; Chatterjee, A.; Sahu, A.; Gupta, T.; Krishnatry, R.; Mokal, S.; Sahay, A.; Epari, S.; Prasad, M.; Chinnaswamy, G.; et al. Machine-learning approach to predict molecular subgroups of medulloblastoma using multiparametric MRI-based tumor radiomics. *Br. J. Radiol.* **2022**, *95*, 20211359. [CrossRef] [PubMed]

4. Fathi Kazerooni, A.; Arif, S.; Madhogarhia, R.; Khalili, N.; Haldar, D.; Bagheri, S.; Familiar, A.M.; Anderson, H.; Haldar, S.; Tu, W.; et al. Automated tumor segmentation and brain tissue extraction from multiparametric MRI of pediatric brain tumors: A multi-institutional study. *Neuro-Oncol. Adv.* **2023**, *5*, vdad027. [CrossRef] [PubMed]

5. Kashani, H.R.K.; Azhari, S.; Moradi, E.; Samii, F.; Mirahmadi, M.S.; Towfiqi, A. Predictive Value of Blood Markers in Pediatric Brain Tumors Using Machine Learning. *Pediatr. Neurosurg.* **2022**, *57*, 323–332. [CrossRef]

6. Tanyel, T.; Ayvaz, S.; Keserci, B. Beyond Known Reality: Exploiting Counterfactual Explanations for Medical Research. *arXiv* **2023**, arXiv:2307.02131.

Article

The Prognostic Value of *ASPHD1* and *ZBTB12* in Colorectal Cancer: A Machine Learning-Based Integrated Bioinformatics Approach

Alireza Asadnia [1,2,3,†], Elham Nazari [4,†], Ladan Goshayeshi [5,6], Nima Zafari [1], Mehrdad Moetamani-Ahmadi [1,2], Lena Goshayeshi [6], Haneih Azari [1], Ghazaleh Pourali [1], Ghazaleh Khalili-Tanha [1], Mohammad Reza Abbaszadegan [2,3], Fatemeh Khojasteh-Leylakoohi [1,3], MohammadJavad Bazyari [7], Mir Salar Kahaei [2], Elnaz Ghorbani [1], Majid Khazaei [1,3], Seyed Mahdi Hassanian [1,3], Ibrahim Saeed Gataa [8], Mohammad Ali Kiani [3], Godefridus J. Peters [9,10], Gordon A. Ferns [11], Jyotsna Batra [12], Alfred King-yin Lam [13], Elisa Giovannetti [10,14,*] and Amir Avan [1,7,12,*]

[1] Metabolic Syndrome Research Center, Mashhad University of Medical Sciences, Mashhad 91779-48564, Iran; alirezaasadnia@gmail.com (A.A.); nima.zafri@gmail.com (N.Z.); mehrdadahmadi45@yahoo.com (M.M.-A.); azari.hanie@gmail.com (H.A.); ghazalehpourali@gmail.com (G.P.); ghazaleh.khalili24@gmail.com (G.K.-T.); fatemekhjst@gmail.com (F.K.-L.); ghorbanie971@mums.ac.ir (E.G.); khazaeim@mums.ac.ir (M.K.); hasanianmehrm@mums.ac.ir (S.M.H.)
[2] Medical Genetics Research Center, Mashhad University of Medical Sciences, Mashhad 91886-17871, Iran; abbaszadeganmr@mums.ac.ir (M.R.A.); salarkahaei1372@gmail.com (M.S.K.)
[3] Basic Sciences Research Institute, Mashhad University of Medical Sciences, Mashhad 13944-91388, Iran; kianima@mums.ac.ir
[4] Department of Health Information Technology and Management, School of Allied Medical Sciences, Shahid Beheshti University of Medical Sciences, Tehran 19839-69411, Iran; nazarie4001@mums.ac.ir
[5] Department of Gastroenterology and Hepatology, Faculty of Medicine, Mashhad University of Medical Sciences, Mashhad 91779-48564, Iran; ladangoshayeshi@gmail.com
[6] Surgical Oncology Research Center, Mashhad University of Medical Sciences, Mashhad 91779-48954, Iran; goshayeshilena@gmail.com
[7] Department of Medical Biotechnology, Faculty of Medicine, Mashhad University of Medical Sciences, Mashhad 91779-48564, Iran; bazyarimj981@mums.ac.ir
[8] College of Medicine, University of Warith Al-Anbiyaa, Karbala 56001, Iraq; ibraheem@uowa.edu.iq
[9] Department of Biochemistry, Medical University of Gdansk, 80-211 Gdansk, Poland; gj.peters@amsterdamumc.nl
[10] Cancer Center Amsterdam, Amsterdam U.M.C., VU University Medical Center (VUMC), Department of Medical Oncology, 1081 HV Amsterdam, The Netherlands
[11] Brighton & Sussex Medical School, Department of Medical Education, Falmer, Brighton, Sussex BN1 9PH, UK g.ferns@bsms.ac.uk
[12] Faculty of Health, School of Biomedical Sciences, Queensland University of Technology (QUT), Brisbane, QLD 4059, Australia; jyotsna.batra@qut.edu.au
[13] Pathology, School of Medicine and Dentistry, Gold Coast Campus, Griffith University, Gold Coast, QLD 4222, Australia; a.lam@griffith.edu.au
[14] Cancer Pharmacology Lab, AIRC Start Up Unit, Fondazione Pisana per La Scienza, 56017 Pisa, Italy
[*] Correspondence: elisa.giovannetti@gmail.com (E.G.); avana@mums.ac.ir or amir.avan@qut.edu.au (A.A.)
[†] These authors contributed equally to this work.

Citation: Asadnia, A.; Nazari, E.; Goshayeshi, L.; Zafari, N.; Moetamani-Ahmadi, M.; Goshayeshi, L.; Azari, H.; Pourali, G.; Khalili-Tanha, G.; Abbaszadegan, M.R.; et al. The Prognostic Value of *ASPHD1* and *ZBTB12* in Colorectal Cancer: A Machine Learning-Based Integrated Bioinformatics Approach. *Cancers* **2023**, *15*, 4300. https://doi.org/10.3390/cancers15174300

Academic Editor: Dania Cioni

Received: 1 August 2023
Revised: 22 August 2023
Accepted: 24 August 2023
Published: 28 August 2023

Simple Summary: Colorectal cancer (CRC) is among the leading causes of cancer-related deaths. Despite extensive efforts, a limited number of biomarkers and therapeutic targets have been identified. Therefore, novel prognostic and therapeutic targets are needed in the management of patients and to increase the efficacy of current therapy. The majority CRC patients follow the conventional chromosomal instability (CIN), which is started by several mutations such as APC, followed by genetic alterations in KRAS, PIK3CA and SMAD4, as well as the hyperactivation of pathways such as Wnt/TGFβ/PI3K. Although the underlying genetic changes have been well identified, the mutational signature of tumor cells alone does not enable us to subclassify tumor types or to accurately predict patient survival and suppression of those pathways have often not been effective in treatment. Our data showed some new genetic variants in *ASPHD1* and *ZBTB12* genes, which were associated with a poor prognosis of patients.

Abstract: Introduction: Colorectal cancer (CRC) is among the leading causes of cancer-related deaths. Despite extensive efforts, a limited number of biomarkers and therapeutic targets have been identified. Therefore, novel prognostic and therapeutic targets are needed in the management of patients and to increase the efficacy of current therapy. The majority CRC patients follow the conventional chromosomal instability (CIN), which is started by several mutations such as APC, followed by genetic alterations in KRAS, PIK3CA and SMAD4, as well as the hyperactivation of pathways such as Wnt/TGFβ/PI3K. Although the underlying genetic changes have been well identified, the mutational signature of tumor cells alone does not enable us to subclassify tumor types or to accurately predict patient survival and suppression of those pathways have often not been effective in treatment. Our data showed some new genetic variants in *ASPHD1* and *ZBTB12* genes, which were associated with a poor prognosis of patients. Colorectal cancer (CRC) is a common cancer associated with poor outcomes, underscoring a need for the identification of novel prognostic and therapeutic targets to improve outcomes. This study aimed to identify genetic variants and differentially expressed genes (DEGs) using genome-wide DNA and RNA sequencing followed by validation in a large cohort of patients with CRC. **Methods:** Whole genome and gene expression profiling were used to identify DEGs and genetic alterations in 146 patients with CRC. Gene Ontology, Reactom, GSEA, and Human Disease Ontology were employed to study the biological process and pathways involved in CRC. Survival analysis on dysregulated genes in patients with CRC was conducted using Cox regression and Kaplan–Meier analysis. The STRING database was used to construct a protein–protein interaction (PPI) network. Moreover, candidate genes were subjected to ML-based analysis and the Receiver operating characteristic (ROC) curve. Subsequently, the expression of the identified genes was evaluated by Real-time PCR (RT-PCR) in another cohort of 64 patients with CRC. Gene variants affecting the regulation of candidate gene expressions were further validated followed by Whole Exome Sequencing (WES) in 15 patients with CRC. **Results:** A total of 3576 DEGs in the early stages of CRC and 2985 DEGs in the advanced stages of CRC were identified. *ASPHD1* and *ZBTB12* genes were identified as potential prognostic markers. Moreover, the combination of *ASPHD* and *ZBTB12* genes was sensitive, and the two were considered specific markers, with an area under the curve (AUC) of 0.934, 1.00, and 0.986, respectively. The expression levels of these two genes were higher in patients with CRC. Moreover, our data identified two novel genetic variants—the rs925939730 variant in *ASPHD1* and the rs1428982750 variant in *ZBTB1*—as being potentially involved in the regulation of gene expression. **Conclusions:** Our findings provide a proof of concept for the prognostic values of two novel genes—*ASPHD1* and *ZBTB12*—and their associated variants (rs925939730 and rs1428982750) in CRC, supporting further functional analyses to evaluate the value of emerging biomarkers in colorectal cancer.

Keywords: machine learning; colorectal cancer; bioinformatics; biomarker; prognosis

1. Introduction

Colorectal cancer (CRC) is the second most common cause of cancer-related mortality [1], and its incidence is increasing despite the advances in the detection of prognostic and/or therapeutic targets. This is partly due to the limited number of therapeutic agents that have been identified. A high proportion of patients with CRC develop metastatic cancer(s) or become resistant to therapy. Therefore, novel prognostic biomarkers and new therapeutic targets that can help to assess the risk of developing CRC recurrence or increase the efficacy of current therapy are urgently needed.

Integrated analyses of multi-omics data provide useful insight into the pathogenesis of CRC and help to identify novel diagnostic and prognostic biomarkers. With the success of artificial intelligence technologies, machine learning (ML) is being used in healthcare. ML methods provide novel techniques of integration and analyzing omics for the discovery of novel biomarkers [2,3]. Hammad and collaborators [4] identified 105 differential expression genes (DEGs) using datasets from the Gene Expression Omnibus (GEO). Functional enrich-

ment analysis revealed that these genes were enriched in cancer-related biological processe
The protein–protein interaction (PPI) network selected 10 genes, including *IGF1, MYH1*
CLU, FOS, MYL9, CXCL12, LMOD1, CNN1, C3, and *HIST1H2BO*, as hub genes. Suppor
Vector Machine (SVM), Receiving Operating Characteristic (ROC), and survival analyse
demonstrated that these hub genes can be considered potential prognostic biomarker
for CRC.

Maurya et al. [5] used Least Absolute Shrinkage and Selection Operator (LASSO) an
Relief for feature selection from the Cancer Genome Atlas (TCGA) dataset and applied R
K-Nearest Neighbor (KNN), and Artificial Neural Network (ANN) to check the accurac
of the models. The joint set of selected features between LASSO and DEGs was 38 gene
among which *VSTM2A, NR5A2, TMEM236, GDLN*, and *ETFDH* were correlated with th
overall survival (OS) of patients with CRC and could be used as prognostic biomarkers. Fo
example, Liu et al. [6] identified 16 lncRNAs as an immune-related lncRNA signature (IRL
for predicting patients' prognosis of CRC using machine learning-based integrated analysi
They performed further investigations to validate the application of IRLS in practice. Th
efficacy of immune-related lncRNA signature was validated using qRT-PCR on CRC tissue
collected from 232 patients. A prospective cohort study, RECOMMEND (NCT05587452
aimed to assess the accuracy of a novel AI-based integrated analysis screening method fo
CRC and advanced colorectal adenomas using plasma multi-omics data.

Genome-wide association studies (GWAS) have already allowed significant progres
in the understanding of the complex genetics behind the pathogenesis of CRC. There are a
least three major molecular pathways that can lead to CRC, including the chromosoma
instability pathway (characterized by aneuploidy or structural chromosomal abnormalities
chromosomal instability, and mutations (e.g., *APC, KRAS, PIK3CA, SMAD4*, or *TP53*
There is a growing body of evidence on targeting deregulated intracellular pathways, sucl
as the hyperactivation of WNT–β-catenin, PI3K/Akt, or RAS signaling, although it ha
been shown that inhibiting these pathways has often not been effective in the clinica
management of CRC [7–10]. Many patients with CRC had conventional chromosoma
instability (CIN), which is started by several mutations such as APC, followed by geneti
alterations in *KRAS, PIK3CA*, and *SMAD4*, as well as the hyperactivation of pathway
such as Wnt/TGFβ/PI3K. Although the underlying genetic changes have been sufficientl
identified, the mutational signature of tumor cells alone does not enable us to subclassif
tumor types or to accurately predict patients' survival, and the suppression of thos
pathways has often not been effective in treatment [11]. In this study, we attempted t
develop and validate novel prognostic biomarkers based on ML-based integrated analysi
as well as validation of novel candidate genes in two additional cohorts of CRC in DNA an
RNA levels using whole exome sequencing (WES) and reverse transcription polymeras
chain reaction (RT-PCR), respectively.

2. Materials and Methods

2.1. Data Sources and Data Processing

RNA-sequencing (RNA-seq) expression data and clinicopathological information wer
retrieved from The Cancer Genome Atlas (TCGA) database, which included 287 CRC tissu
samples and 41 non-cancers tissue samples. In this study, RNA-seq data were obtaine
from TCGA-colorectal adenocarcinoma. Patients with colorectal cancer were classified int
early-stage and late-stage. Early-stage CRCs were classified into three subgroups based o
microsatellite instability (MSI) status: low MSI (MSI-L), high MSI (MSI-H), and MSI-stabl
(MSI-S). Late-stage CRCs were classified into two subgroups based on the therapeuti
regimens (chemotherapy versus targeted therapy).

2.2. Patient's Samples

Sixty-four CRCs were included in this study based on histological confirmation b
two pathologists. All the eligible patients were chemotherapeutic naive patients treated a

the Omid Hospital of Mashhad University of Medical Sciences. The study was approved by the local Hospital Ethic Committee of Mashhad University of Medical Sciences.

2.3. DNA-Seq and Whole Exome Sequencing

Data from the TCGA database were downloaded and prepared for further analysis in the R programming language. The data were downloaded in the Mutation Annotation Format (MAF). MAF is a standardized format used by TCGA for storing and analyzing various types of somatic mutations in cancer. The patients were divided into two groups: patients in the early stages (I, II) of CRC and patients in the advanced metastatic stage (IV). The first group consisted of 118 patients, while the second group consisted of 28 patients. MAF data belonging to each group is analyzed with the maftools package in R programming.

The genes with a significant *p*-value of less than 0.05 obtained from the survival analysis were combined with the whole exome sequencing data of TCGA for colorectal cancer. Then, the variants of the candidate genes obtained from sequencing data were analyzed using the Maftools package. Then, two candidate genes, ASPHD1 and ZBTB12, were further evaluated for their impact on gene expression using RegulomeDB and 3DSNP. Subsequently, the candidate genes were further confirmed in an additional cohort performed for the Whole Exome Sequencing (WES) data of 15 patients with CRC, as described previously.

2.4. Differential Gene Expression Analysis

Normalization was performed, while the PCA plots, volcano plots, heatmap, and karyoplote were represented by the R packages "ggplot2", "heatmap", and karyoploteR to visualize data. Significance analysis of differentially expressed genes (DEGs) was performed using DESeq2 in R software with the cutoff criteria of | log fold change | $\geq$ 1.5 and an adjusted *p*-value of <0.05.

2.5. Gene Set, Ontology, and Pathway Enrichment Analysis

The significant enrichment analysis of DEGs was assessed based on Gene Ontology (GO), Reactom, GSEA, and Human Disease Ontology (DO). GO analysis (http://www.geneontology.org/) is used for annotating genes and gene products and investigating the biological aspects of high-throughput genome or transcriptome data, including biological processes, cellular components, and molecular function. The Reactom database was used for the analysis of gene functions in biological signaling pathways. We set a *p*-value < 0.05 and a false discovery rate (FDR) < 0.05 as the statistically significant criteria to output. The whole transcriptome was employed for GSEA, and only gene sets with *p*-value < 0.05 and FDR q < 0.05 were set as statistically significant criteria. Statistical significance was set at an adjusted *p*-value of <0.05. Several R packages were utilized to perform enrichment analyses, including ReactomePA, enrichplot, clusterProfiler, and topGO.

2.6. Survival Analysis

The univariate/Cox proportional hazards regression model was used to identify DEGs that were significantly correlated with overall survival and assess the independent prognostic factors. R version 4.2.1 software was used to analyze the data.

2.7. Machine Learning Method

Two machine learning techniques were used, including the decision tree learning and deep learning. Deep learning models were applied to identify the effective factors. The significant variables obtained from the feature selection method (Weight by Correlation) were the final parameters in creating the model. The coefficient of correlation between variables is presented as a correlation matrix. The correlation coefficient is measured from −1 to 1; positive values represent that the variables are in the same direction, and negative correlations show the variables in opposite directions. The lack of correlation was displayed as 0.

2.8. Computational Workflow

Python3.7 was utilized for modeling. Parameters of epochs = 10, activation function = Rectifier, and learning rate = 0.01 were set in deep learning. The standard workflow was utilized as follows: Splitting the source data set into a training set and test set was performed to provide some independent evaluation levels. Subsequently, the model was optimized using the training data and then independently evaluated using the test data. In this study, a 70/30 train/test ratio was determined for the ML models. For each workflow, a model with the fixed optimal hyperparameter values was retrained on data and randomly sampled from the complete dataset. Machine learning method assessment was performed by 5 indicators including accuracy, R2, MSE, and AUC.

$$\text{Accuracy} = (TP + TN)/(TP + TN + FP + FN)$$

where TP is true positive, FP is false positive, TN is true negative, and FN is false negative.

$$\text{MSE (Mean Squared Error)} = (1/n) \times \Sigma \, (\text{actual} - \text{forecast})2$$

where Σ represents a symbol that means "sum", n is the sample size, actual is the actual data value, and the forecast is the predicted data value.

$$\text{R2 (R-Squared)} = 1 - \text{Unexplained Variation/Total Variation}$$

R2 is the coefficient of determination, and it tells you the percentage variation in y explained by x-variables. AUC (Area Under the Curve) represents the degree of separability and illustrates the capability of the model in distinguishing the classes.

2.9. Protein–Protein Interaction (PPI) Network

The STRING database (https://string-db.org/) was checked to find the relationship between the studied proteins obtained from DEG and the proteins that are directly or indirectly involved in the development of cancers. A minimum effective binding score of ≥0.4 was established. Genes with significant interactions were screened.

2.10. Kaplan–Meier Survival Curve

Kaplan–Meier survival curve comparison was conducted to measure the prognostic value of candidate genes in CRC using the log-rank test.

2.11. Receiver Operating Characteristic (ROC) Curve Analysis

Receiver operating characteristic (ROC) curves are a fundamental analytical tool for assessing diagnostic tests and identifying diagnostic biomarkers. ROC curve analysis evaluates the accuracy of a test to differentiate between diseased and healthy cases, thereby measuring the overall diagnostic performance [12]. A ROC curve and the area under the curve (AUC) were employed to determine the specificity, sensitivity, likelihood ratios, positive predictive values, and negative predictive values using the R package (pROC version 1.16.2).

2.12. Quantitative Real-Time-PCR Validation

Total RNAs were extracted from tissues using a total RNA extraction kit according to the manufacturer's protocol (Parstous, Tehran, Iran). RNA quantity and quality were assessed using a Nanodrop 2000 spectrophotometer (BioTek, USA EPOCH), and forty RNAs that passed the quality control were used for the next step. The RNAs were then reverse-transcribed into complementary DNA (cDNA) using a cDNA Synthesis Kit (Parstous, Tehran, Iran) according to the manufacturer's instructions. Primers were designed (Forward Reverse: *ASPHD1*: AGTGGCTCACAATGGCTCC and AAGACAAAGTCGAGGGCCTG and *ZBTB12*: TTGCTCCTCTCCTGCTACACG and AACTGGCTGAGGGCATTCCG), and RT-PCR was performed via the ABI-PRISM StepOne instrument (Applied Biosystems,

Foster City, CA, USA) using the SYBR green master mix (Parstous Co. Tehran, Iran). Gene expression data were standardized to glyceraldehyde 3-phosphate dehydrogenase (GAPDH) using a standard curve of cDNAs obtained from quantitative polymerase chain reaction (qPCR) Human Reference RNA (Stratagene, La Jolla, CA, USA).

3. Results

3.1. Whole Exome Sequencing

The Mutation Annotation Format (MAF) data were divided into two groups: patients in the early stages and advanced metastatic stage, as shown in Figures 1 and 2, containing 118 and 28 patients, respectively. The MAF data were analyzed using the maftools package in R programming. Figures 1 and 2 show different plots, including plot maf Summary, oncoplots, Transition and Transversions reports, Plotting VAF (Variant Allele Frequencies), Somatic Interactions reports, Drug–Gene Interactions, and Oncogenic Signaling Pathways to visualize the MAF distribution in a different group. As shown in Figures 1A and 2A, in the early and late stages, missense mutations were more frequent than other mutations, and they were typically referred to as single-nucleotide polymorphism (SNP) types. Additionally, in both groups, 70–71% of patients had mutations in their APC or TP53 genes. Most of the variants are involved in Wnt/B-catenin _signaling, Genome integrity, and MAPK signaling (Figures 1B and 2B). The clonal status of the most mutated genes can be estimated using the Variant Allele Frequencies plot; clonal genes usually have an average allele frequency of about 50% in pure samples. In the early stages of tumor development, TP53 was observed to have clonal status in the tumor tissue, while SMAD4, RYR4, and TP53 exhibit such a status in the late stages, as shown in Figures 1D and 2D.

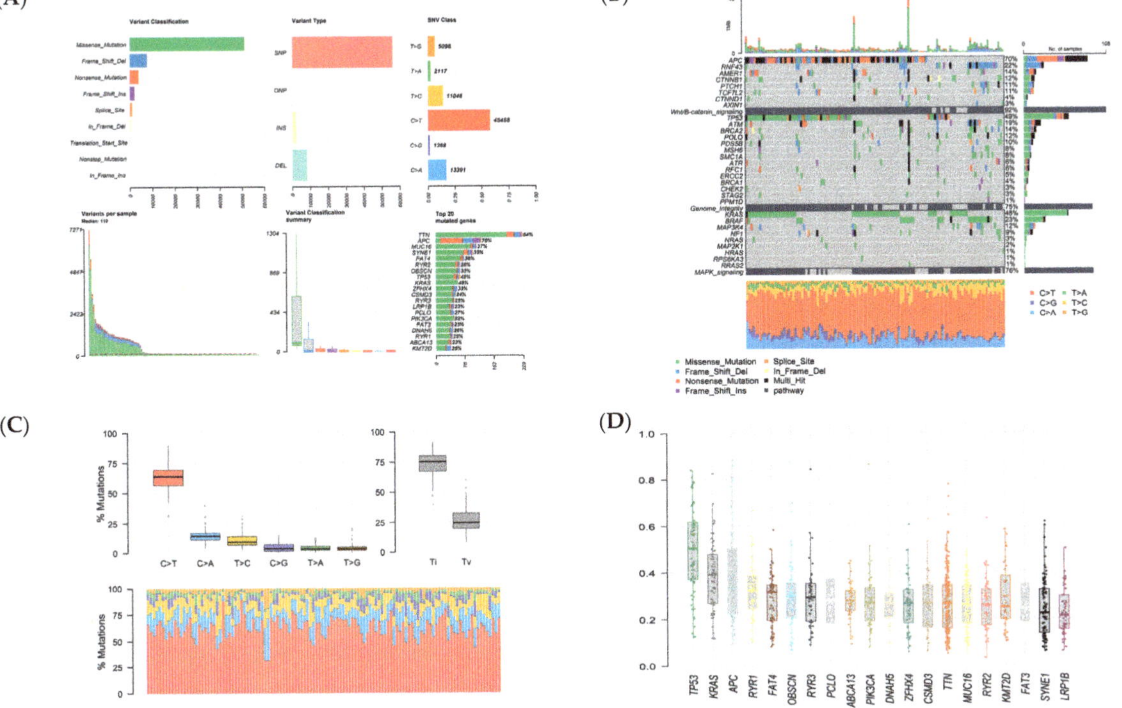

Figure 1. *Cont.*

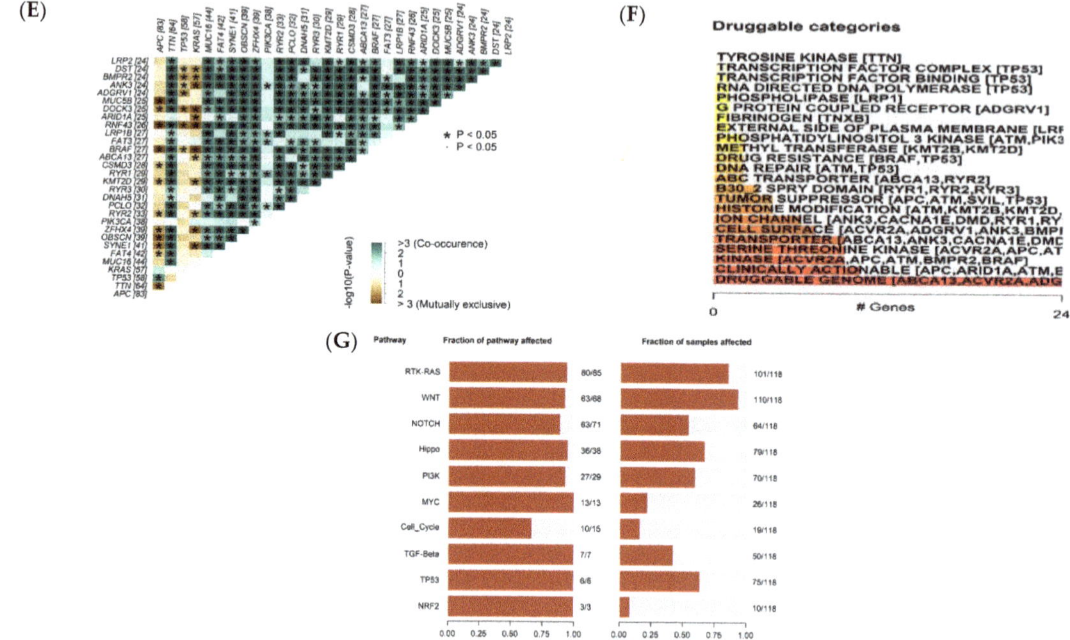

Figure 1. Visualization and summary of the analysis results of MAF data in the early-stage group (I, II stages) with the maftools package. (**A**) Bar and box plots display the frequency of different variants across samples (DEL: Deletion, INS: Insertion, SNP: Single-nucleotide polymorphism, ONP: Oligo-nucleotide polymorphism). (**B**) Oncoplots (note: variants annotated as Multi_Hit are genes that are mutated repeatedly within the same sample). (**C**) Transition and Transversion mutations (Ti: Transition; Tv: Transversions). (**D**) A boxplot of Variant Allele Frequencies. (**E**) Somatic Interactions show results of exclusive/co-occurrence event analysis. (**F**) Drug–gene interaction analysis based on the Drug–Gene Interaction database. (**G**) Oncogenic Signaling Pathways.

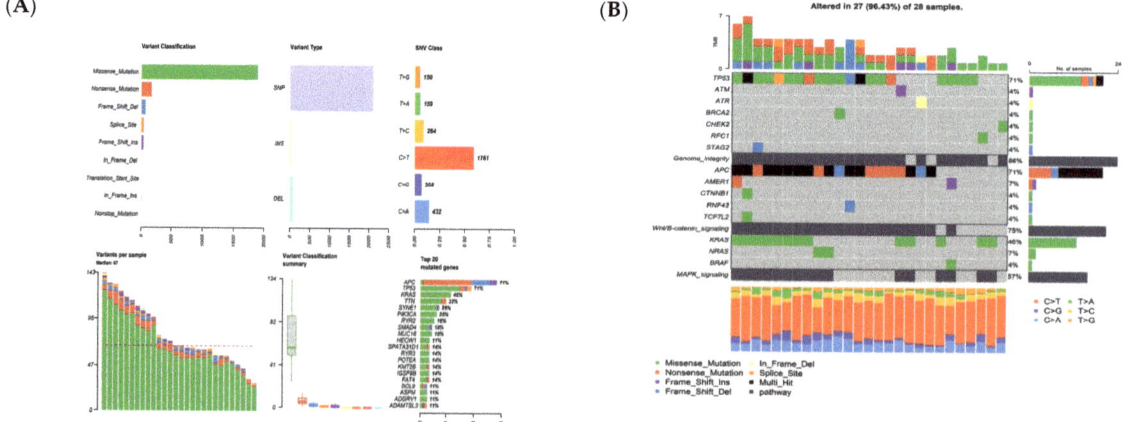

Figure 2. *Cont.*

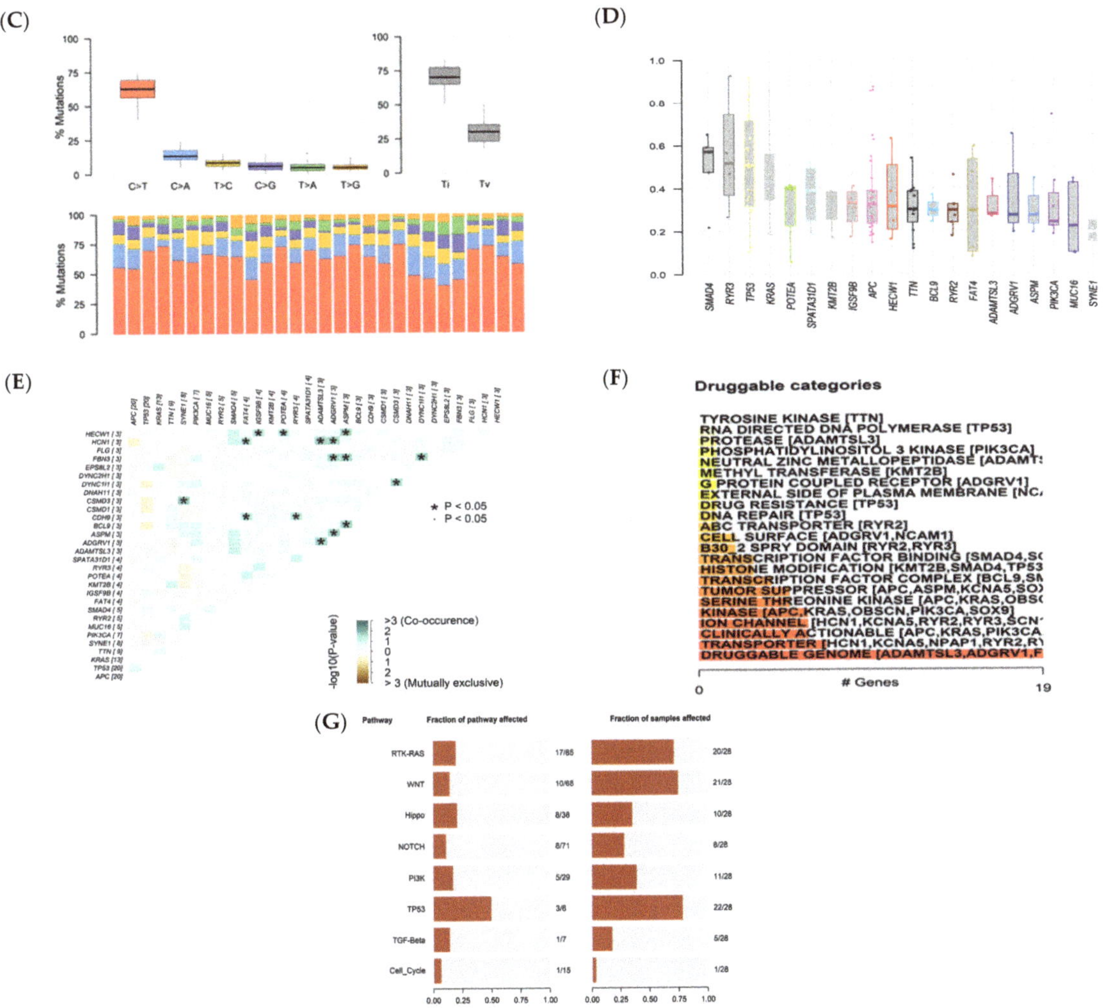

Figure 2. Visualization and summary of the analysis results of MAF data in the advanced-stage group (IV stage) with the maftools package. (**A**) Bar and box plots display the frequency of different variants across samples (DEL: Deletion, INS: Insertion, SNP: Single-nucleotide polymorphism, ONP: Oligo-nucleotide polymorphism). (**B**) Oncoplots (note: variants annotated as Multi_Hit are genes that are mutated repeatedly within the same sample). (**C**) Transition and Transversion mutations (Ti: Transition; Tv: Transversions). (**D**) Boxplot of Variant Allele Frequencies. (**E**) Somatic Interactions show the results of exclusive/co-occurrence event analysis. (**F**) Drug–gene interaction analysis based on the Drug–Gene Interaction database. (**G**) Oncogenic Signaling Pathways.

Somatic Interactions analysis indicated exclusive or co-occurrence (Figures 1E and 2E). Mutually exclusive events happen in cancer when mutations in one gene prevent the occurrence of mutations in another gene. Co-occurring events, on the other hand, arise when mutations in two or more genes occur together more frequently than would be expected by chance. Determining mutually exclusive genes implies that these genes may participate in the same pathway or process, and there might be functional overlap between them. On the other hand, identifying genes that co-occur may indicate that they collaborate to facilitate the growth of tumors, or that their cumulative impact is essential for the development of cancer. The interaction between genes and drugs that target tyrosine kinase, transcription factor complex, DNA repair, and other related processes is illustrated in Figures 1F and 2F. The involvement of mutated genes in colorectal cancer across different oncogenic signaling pathways, including RTK-RAS, Wnt, Hippo, Notch, and others, is demonstrated in Figures 1G and 2G.

3.2. Gene Expression Profiling, Identification of DEGs, and Pathway Enrichment Analysis

We performed gene expression profiling in 287 CRC cases, analyzed by the DESeq2 package, according to the adjusted *p*-value of <0.05 and a $|logFC| \geq 1.5$ (Table S1). The PCA plots, volcano plots, and heat maps of each subgroup are shown in Figures 3 and S1. Moreover, the gene expression of each subgroup, obtained from the DEG analysis was exhibited in the ideogram of chromosomes using the karyoploteR package (Figure 3C). Enrichment analysis showed that DEGs were significantly enriched in biological processes related to cancer progression. Based on GO analysis, the main biological processes involving the DEGs included ion homeostasis, inorganic cation transmembrane transport, and the regulation of hormone levels. In terms of cellular components, the DEGs were mostly enriched in the external encapsulating structure and extracellular matrix (ECM). In terms of molecular functions, the DEGs were linked by cation transmembrane transport activity, receptor regulator activity, signaling receptor activator activity, etc. (Figures 4A and S2–S6).

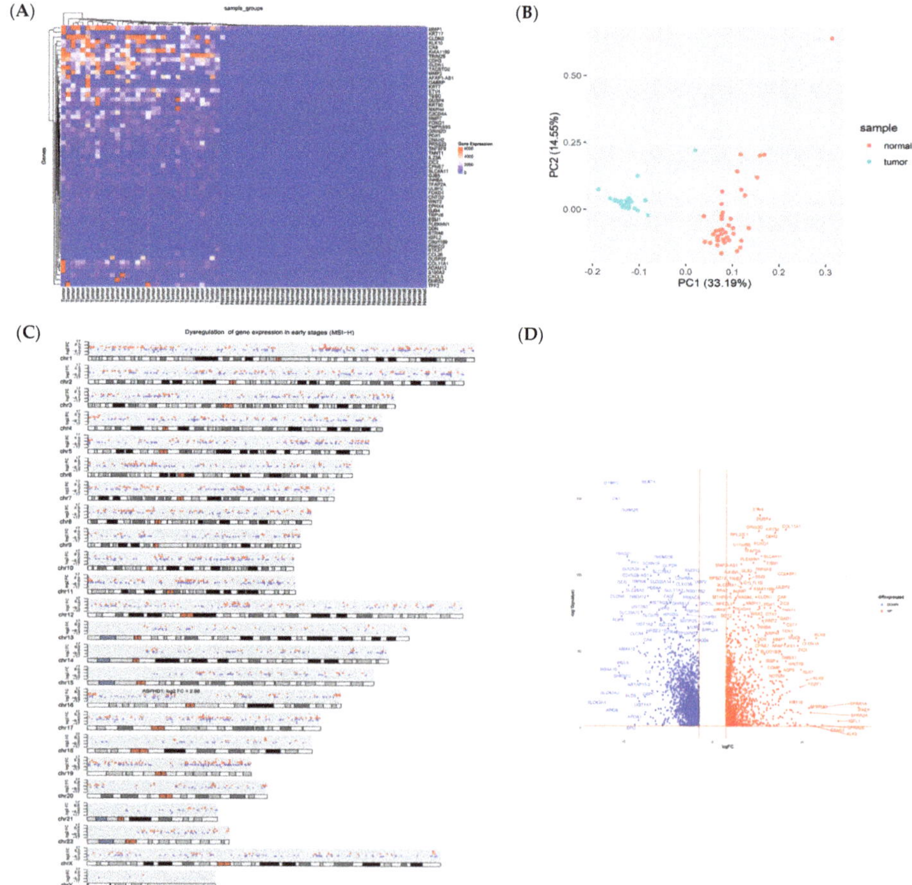

Figure 3. The results of the analysis of differentially expressed genes (DEGs) in colorectal adenocarcinoma (COAD) were generated using R software https://www.r-project.org/. (**A**) The heat map. (**B**) Principal component analysis (PCA). (**C**) karyoplot. (**D**) Volcano plot.

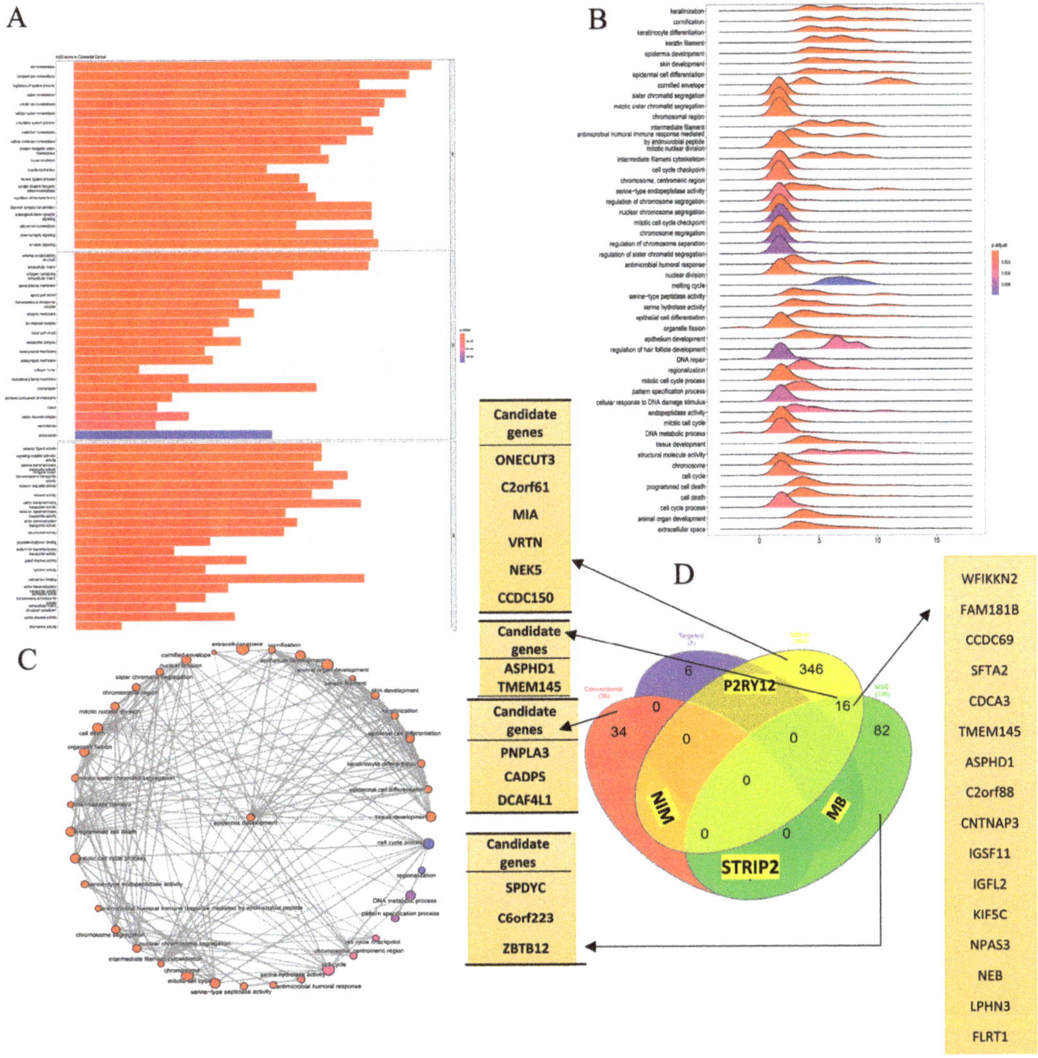

Figure 4. (**A**) Gene Ontology (GO), (**B**) GSEA functional annotation, and (**C**) Reactome functional pathways in colorectal adenocarcinoma (COAD). The *p*-value is less than 0.05 and is shown by the color. (**D**) A Venn diagram indicating the number of survival-related genes and the overlap between the different subgroups.

GSEA analysis showed that there was a relationship between identified DEGs and cell cycle, cell cycle checkpoint, DNA repair, mitotic nuclear division, cellular response to DNA damage stimulus, programmed cell death, epithelial cell differentiation, DNA-binding transcription factor activity, regulation of transcription by RNA polymerase II, Wnt signaling pathway, keratin filaments. According to the Reactom database analysis, DEGs were involved in GPCR signaling and its downstream signaling pathways, the regulation of Insulin-like growth factor (IGF), SLC-mediated transmembrane transport, the degradation of the extracellular matrix (ECM), collagen degradation, biological oxidation, and the activation of matrix metalloproteinases. (Figure 4B,C).

To further explore the prognostic value of emerging DEGs, we performed univariate Cox proportional hazards regression (Table S2).

3.3. Machine Learning Analysis

The results of the ML analysis are shown in Table 1. The deep learning method achieved an accuracy of 97.14%, 97%, 98%, and 92% for predicting CRC in the MSI-H, MSS, chemotherapy, and targeted therapy subgroups, respectively, with AUC values of 1.0, 1.0, 1.0, and 0.88. This model had the best performance in the MSI-H and MSS subgroups. Then, 14 candidate genes were identified as novel genes which were dysregulated in both DNA and RNA levels. Also, the candidate genes and common genes resulting from the survival analysis were then displayed on a Venn diagram (Figure 4D and Table S3). Following the visualization described in the MAF data analysis stage, 232 variants from 14 candidate genes related to survival were analyzed (Figure 5). Then, we confirmed the candidate genes in an additional cohort of our patients, which was detected by whole genome sequencing (WES) in 15 cases. Then, 11 genes emerged between the different cohorts, including ASPHD1, C2orf61, C6orf223, CADPS, CCDC150, DCAF4L1, MIA, NEK5, ONECUT3, PNPLA3, and TMEM145 (Table S4).

Table 1. Results of machine learning analysis.

Subgroups	R2	AUC	MSE	RMSE	Accuracy	Prauc
MSI-H	0.99	1.0	1.95	0.0044	97.14%	1.0
MSI-S	0.99	1.0	0.0023	0.0489	97%	1.0
Receiving chemotherapy	0.95	1.0	0.0076	0.0876	98%	1.0
Receiving targeted therapies	0.64	0.88	0.0554	0.0235	92%	0.95

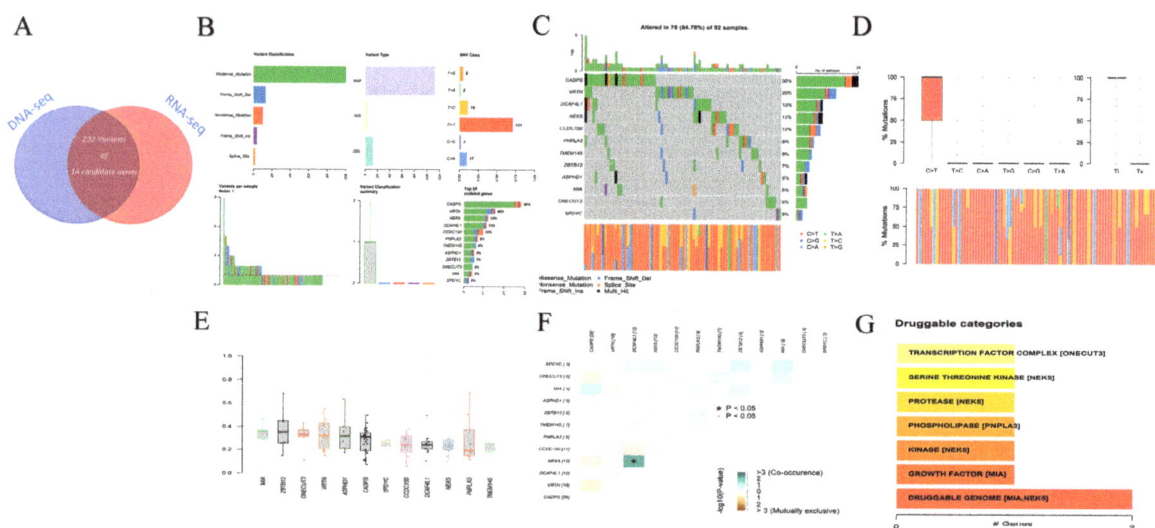

Figure 5. (**A**) A Venn diagram shows the count of variants for 14 candidate genes which are common between DNA-seq and RNA-seq analysis. (**B**) Bar and box plots displaying the frequency of different variants across samples. (**C**) Oncoplots. (**D**) Transition and Transversion mutations. (**E**) Boxplot of Variant Allele Frequencies. (**F**) Somatic Interactions show the results of exclusive/co-occurrence event analysis. (**G**) Drug–gene interaction analysis based on the Drug–Gene Interaction database.

3.4. The Prognostic Value of ZBTB12 and ASPHD1

Of note, RNA-seq data certified the dysregulation of candidate genes identified from Ml and DNA-seq and shortlisted *ZBTB12* and *ASPHD1* as the disease-associated genes (Figure 6). According to the Human Protein Reference Database, *ZBTB12* and *ASPHD1* interact with *HRAS*, *Ras-associated protein 1*, and *HRAS*, *PRRC2A*, *MSL3*, and *PIK3CA* (Figure 6A,B). The results of WES found nine genetic variants in *ASPHD1* and *ZBTB12* (Figure 6C,D). According to the RegulomeDB database and 3DSNP, the rs925939730 variant of the *ASPHD1* and rs1428982750 variant of the *ZBTB1* regulate gene expression and affect chromatin state in the colon and rectum (Tables S5 and S6). Moreover, the rs1428982750 variant was linked to *VARS* and *EHMT2* genes, and the rs925939730 variant was associated with the *MAZ* gene (Tables S7 and S8). The rs1428982750 variant of the *ZBTB12* gene had a score of 0.60906 for its role in gene expression regulation. Also, this variant affected the state of the chromatin transcription activity in the colon and rectum. Chromatin immunoprecipitation coupled with sequencing (CHIP-seq) results showed that the *ZBTB12* gene variant affects the binding site of transcription factors and various regulatory factors. (Figure S7C). The rs925939730 variant of the *ASPHD1* gene had a score of 0.77967 for its role in regulating gene expression. Also, this variant affected the state of the chromatin transcription activity in the colon and rectum. CHIP-seq results showed that the *ASPHD1* gene variant affects the binding site of transcription factors and various regulatory factors. (Figure 6E). The results of the rs1428982750 variant of the *ZBTB12* gene in the 3DSNP database showed that the association of this variant with the regulatory factors of gene expression has a score of 58.4 (Figure S7A). The different positions of this variant. The results of the rs925939730 variant of the *ASPHD1* gene in the 3DSNP database showed that the association of this variant with the regulatory factors of gene expression has a score of 59.7 (Figure S7B).

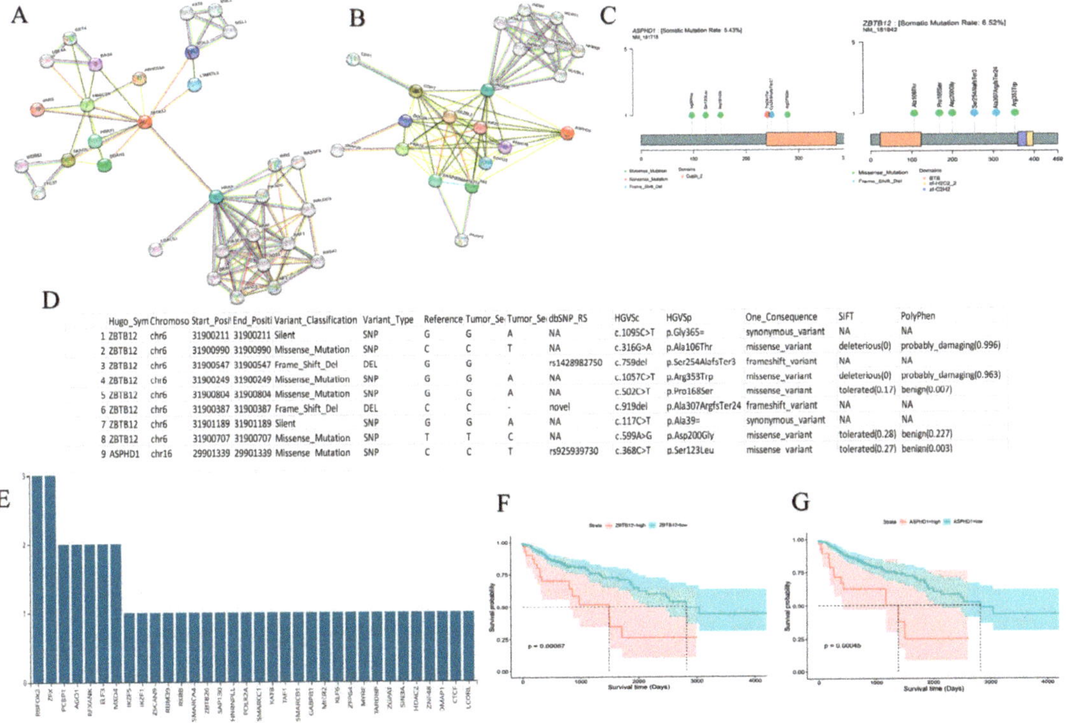

	Hugo_Symbol	Chromoso	Start_Positi	End_Positi	Variant_Classification	Variant_Type	Reference	Tumor_Se	Tumor_Se	dbSNP_RS	HGVSc	HGVSp	One_Consequence	SIFT	PolyPhen
1	ZBTB12	chr6	31900211	31900211	Silent	SNP	G	G	A	NA	c.1095C>T	p.Gly365=	synonymous_variant	NA	NA
2	ZBTB12	chr6	31900990	31900990	Missense_Mutation	SNP	C	C	T	NA	c.316G>A	p.Ala106Thr	missense_variant	deleterious(0)	probably_damaging(0.996)
3	ZBTB12	chr6	31900547	31900547	Frame_Shift_Del	DEL	G	G		rs1428982750	c.759del	p.Ser254AlafsTer3	frameshift_variant	NA	NA
4	ZBTB12	chr6	31900249	31900249	Missense_Mutation	SNP	G	G	A	NA	c.1057C>T	p.Arg353Trp	missense_variant	deleterious(0)	probably_damaging(0.963)
5	ZBTB12	chr6	31900804	31900804	Missense_Mutation	SNP	G	G	A	NA	c.502C>T	p.Pro168Ser	missense_variant	tolerated(0.17)	benign(0.007)
6	ZBTB12	chr6	31900387	31900387	Frame_Shift_Del	DEL	C	C	-	novel	c.919del	p.Ala307ArgfsTer24	frameshift_variant	NA	NA
7	ZBTB12	chr6	31901189	31901189	Silent	SNP	G	G	A	NA	c.117C>T	p.Ala39=	synonymous_variant	NA	NA
8	ZBTB12	chr6	31900707	31900707	Missense_Mutation	SNP	T	T	C	NA	c.599A>G	p.Asp200Gly	missense_variant	tolerated(0.28)	benign(0.227)
9	ASPHD1	chr16	29901339	29901339	Missense_Mutation	SNP	C	C	T	rs925939730	c.368C>T	p.Ser123Leu	missense_variant	tolerated(0.27)	benign(0.003)

Figure 6. *Cont.*

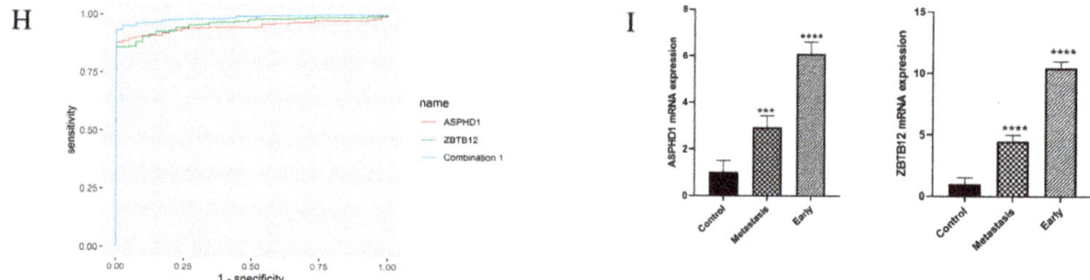

Figure 6. (**A,B**) Protein–protein interaction (PPI) network of the two genes (*ZBTB12, ASPHD1*) identified by survival analysis from STRING. (**C,D**) The different types of *ZBTB12* and *ASPHD1* variants, along with their respective alterations in the amino acid sequence on chromosomes, as well as the rate of somatic mutation. (**E**) CHIP-seq results have shown that variants of two genes (*ZBTB12, ASPHD1*) affect the binding site of transcription factors and various regulatory factors from the Regulome DB Database. (**F,G**) Kaplan–Meier plot of *ZBTB12* and *ASPHD1* with a prognostic value *p*-value < 0.05. (**H**) ROC curve analysis revealed the biomarker potency of *ZBTB12* and *ASPHD1* individually and together using R 4.3.1's combioROC package. (**I**) qRT-PCR results indicate that the expression levels of the two genes (*ZBTB12* and *ASPHD1*) are elevated in tumor tissue compared to non-neoplastic tissue. *** $p > 0.01$; **** $p > 0.001$.

ROC curve data was obtained by plotting the rate of sensitivity versus specificity. Also, Kaplan–Meier revealed that the overall survival of patients with cancer having low *ASPHD1* expression had higher overall survival (OS) than patients with cancer with high *ASPHD1* expression ($p < 0.05$). Similarly, cancers with high *ZBTB12* expression were associated with poor patient survival compared to cancers with low ZBTB12 expression ($p < 0.05$) (Figure 6F,G). As shown in Figure 6H and Tables 2 and 3, *ASPHD1*, *ZBTB12* and their combination were able to discriminate CRC with an area under the curve (AUC) of 0.948, 0.96, and 0.986, respectively. At the cutoff values of 0.863, 0.891, and 0.886, the sensitivities of *ASPHD1*, *ZBTB12*, and their combination were 0.878%, 0.861%, and 0.934% respectively, with specificities of 1. The combination of *ASPHD1* and *ZBTB12* showed higher AUC and sensitivity than each of these candidate genes alone.

Table 2. The area under the curve (AUC) and a cut-off value of ASPHD1, ZBTB12, and their combination in CRC.

Biomarker	AUC	SE	SP	Cutoff	ACC	TN	TP	FN	FP	NPV	PPV
ASPHD1	0.948	0.878	1	0.863	0.893	41	252	35	0	0.539	1
ZBTB12	0.96	0.861	1	0.891	0.878	41	247	40	.	0.506	1
Combination	0.986	0.934	1	0.886	0.942	41	268	19	.	0.683	1

Table 3. Results for the ROC curve for ASPHD1, ZBTB12, and their combination in CRC.

Biomarker	Intercept	Coefficients	Degrees of Freedom	Null Deviance	Residual Deviance	AIC
ASPHD1	−10.37	Log (ASPHD1 + 1):3.032	327	247.2	136.3	140.3
ZBTB12	−22.345	Log (ASPHD1 + 1):5.165	327	247.2	118.3	122.3
Combination 1	−36.814	5,6,2	327	247.2	63.99	69.99

To further verify their values, the expression of these two candidate genes was evaluated in an additional cohort of CRC via qRT-PCR. The data showed a significantly higher expression of ASPHD1 and ZBTB12 in CRC tissues ($p < 0.05$) (Figure 6I).

4. Discussion

Colorectal cancer ranks as the third most common cause of cancer-related mortality [13]. Early diagnosis of this disease leads to more effective treatment, reduced treatment costs, reduced disease progression, and decreased morbidity and mortality. Since cancer is intimately linked to genetic alterations, pinpointing these changes is especially critical for early diagnosis. Implementing the right analyses of gene expression information can promote optimal treatment selection in the early stages of the development of various cancers. Identifying prognostic biomarkers and achieving diagnosis constitute a worthwhile tactic for disease management and care [14,15]. Artificial intelligence (AI) and deep learning (DL) are being widely adopted in medicine to enhance diagnosis, treatment, and research on diagnosing colorectal cancer (CRC) has followed this trend. DL is now integrated across CRC diagnostic approaches such as histopathology, endoscopy, radiology, and biochemical blood tests. By automating complex data analysis, DL allows for more precise CRC detection and characterization. Although AI adoption faces regulatory hurdles, it has the potential to optimize the diagnosis of CRC recurrence and personalized care by synthesizing diverse medical data and uncovering new insights. Overall, AI and DL are transforming the management of patients with CRC through improved diagnostic accuracy [16].

Our previous studies identified prognostic and diagnostic biomarkers in colorectal cancer and gastric cancer using RNA-seq analysis and machine learning [17–19]. In contrast to our previous study, the current study was designed based on an integrated two omics and deep learning approach to identify prognostic and diagnostic biomarkers in colorectal cancer (CRC) patients at different disease stages (early and metastatic). By combining multi-omics data and advanced computational methods, the present study provides novel insights into stratifying CRC patients based on genetic and expression profiles correlated with disease progression and outcomes. To the best of our knowledge, this is the first study showing the potential association of two genetic variants, rs1428982750 in *ZBTB12* and rs925939730 in *ASPHD1* genes, and the prognostic value of these genes in colorectal cancer. Bian Wu et al. used WES and RNA-seq to indicate prognosis prediction in patients with stage IV colorectal cancer. The results showed the following mutations in the genes: *APC*, *TP53*, *KRAS*, *TTN*, *SYNE1*, *SMAD4*, *PIK3CA*, *RYR2*. *BRAF* did not reveal any significant associations between the mutational status of those genes and patient prognosis [20]. Our study revealed that mutations in the genes *ZBTB12* and *ASPHD1* may serve as potential prognostic markers in patients. Specifically, we demonstrated that the mutational status of *ZBTB12* and *ASPHD1* was associated with clinical outcomes in the patient cohort examined. Chen et al. analyzed gene expression data from the GEO and TCGA databases and identified 10 hub genes with high diagnostic values based on ROC curve analysis. A nine-gene prognostic signature was also identified and shown to predict overall survival [21]. Importantly, we validated the expression of *ASPHD1* and *ZBTB12* genes through qPCR and their variants using whole exome sequencing in additional patient cohorts.

Data from the PPI network showed that *ASPHD1* is related to several proteins and genes such as *KIF22*, *INO80E*, *SEZ6L2*, and *DOC2A*, most of which are cancer-related. Kinesin family member 22 (KIF22) is a regulator of cell mitosis and cellular vesicle transport. It is involved in spindle formation and the movement of chromosomes during mitosis. The alteration of *KIF22* is associated with several cancers, including CRC. A previous study indicated that *KIF22* is upregulated in CRC samples and that KIF22 expression is correlated with tumors and the clinical stage of CRC. Moreover, the suppression of KIF22 inhibited cell proliferation and xenograft tumor growth [22].

SEZ6L2 regulates cell fate by involving the transcription of type 1 transmembrane proteins. A study showed that *SEZ6L2* was significantly upregulated in CRC tissues, and this upregulation was associated with poor prognosis in patients with CRC [23]. Lastly, INO80E is involved in transcriptional regulation, DNA replication, and probably DNA repair. Therefore, we hypothesize that *ASPHD1* may play a critical role in the pathogenesis of CRC.

PRRT2 is also related to several kinds of human solid tumors [24]. The results of the Protein–protein interaction network demonstrated that *ZBTB12* is linked to numerous genes, including *HRAS*, *PIK3CA*, *MSL3*, and *PRRC2A*.

Phosphatidylinositol-4,5-bisphosphate 3-kinase (PI3K), an important kinase involved in the PI3K/AKT1/MTOR pathway, plays a crucial role in the growth and proliferation of various solid tumors, and *PIK3CA* is one of the most frequently mutated genes in CRC [25]. Harvey rat sarcoma viral oncogene homolog (HRAS) is involved in the activation of Ras protein signal transduction, and its mutations can be found in bladder and head and neck squamous cell carcinomas [26]. It has been shown that proline-rich coiled-coil2A (PRRC2A) takes part in tumorigenesis and immunoregulation. Recent studies have revealed that PRRC2A impacts pre-mRNA splicing and translation initiation [27]. In this context, several studies have demonstrated that there is a relationship between PRRC2A and several kinds of human cancers, such as hepatocellular carcinoma [28] and non-Hodgkin lymphoma [29].

Collectively, *ASPHD1* and *ZBTB12* are linked to multiple proteins and genes which are associated with cancer initiation and progression. Moreover, our results from WES analysis indicated that the rs925939730 variant of the *ASPHD1* gene and the rs1428982750 variant of the *ZBTB1* gene regulate gene expression and affect the chromatin state in the colon and rectum.

In addition, our findings demonstrated that there was an interaction between the rs1428982750 variant and *VARS* and *EHMT2* genes. Valyl-tRNA synthetase (VARS) was linked with CRC [30], breast cancer [31], and leukemia [30]. Euchromatic histone-lysine N-methyltransferase 2 (EHMT2) methylates histone H3 lysine 9 to generate heterochromatin and inhibit tumor suppressor genes [32]. Furthermore, the rs925939730 variant was associated with the *MAZ* gene. MAZ acts as a transcription factor that can be combined with c-MYC and GA box to regulate the initiation and termination of transcription. The deregulated expression of MYC-associated zinc finger protein (MAZ) is correlated with the progression of tumors such as colorectal adenocarcinoma [33], hepatocellular carcinoma [34], renal cell carcinoma [35], glioblastoma [36], breast carcinoma [37], and prostate adenocarcinoma [38]. Altogether, the *rs925939730* and *rs1428982750* gene variants of *ASPHD1* might be involved in gene expression and epigenetic regulation.

5. Conclusions

Our data show the prognostic value of *ASPHD1* and *ZBTB12* in CRC, warranting further investigations to validate their clinical potential as prognostic markers and predictive markers for colorectal cancer. Our study had some limitations and challenges including the difficulty we experienced obtaining access to more patients for evaluating gene expression, carrying out functional studies, and analyzing other omics data to assess important pathways and biological processes in cancer. Expanding our omics approaches beyond just transcriptomics to also include proteomics, metabolomics, etc., would provide a more comprehensive understanding of the key mechanisms in cancer. Overcoming these limitations will be critical for future efforts to elucidate the complex molecular landscape of cancer and identify novel therapeutic targets or biomarkers.

Supplementary Materials: The following supporting information can be downloaded at: https://www.mdpi.com/article/10.3390/cancers15174300/s1, Figure S1: PCA plots, volcano plots, and heat maps for DEGs in each subgroup of CRC patients from the TCGA database. Figure S2: Pathway enrichment analyses of DEGs in MSI-L CRC patients from the TCGA database. Figure S3: Pathway enrichment analyses of DEGs in MSI-H CRC patients from the TCGA database. Figure S4: Pathway enrichment analyses of DEGs in MSS CRC patients from the TCGA database. Figure S5: Pathway enrichment analyses of DEGs in Receiving targeted therapies CRC patients from the TCGA database. Figure S6: Pathway enrichment analyses of DEGs in CRC patients receiving chemotherapy from the TCGA database. Figure S7: Results of the RegulomeDB and 3DSNP database. Table S1: Number of DEGs in patients with CRC. Table S2: Number of survival-related DEGs. Table S3: Candidate genes based on database search. Table S4: Variants of candidate genes in the Whole Exome Sequencing (WES) data of 15 patients with CRC in the Mashhad population. Table S5: The results of the effect of the rs1428982750 variant of the *ZBTB12* gene on chromatin status in the colon and rectum. Table S6: The results of the effect of the rs925939730 variant of the *ASPHD1* gene on chromatin status in the colon and rectum. Table S7: Results of rs1428982750 variant of *ZBTB12* gene in 3DSNP database. Table S8: Results of rs925939730 variant of *ASPHD1* gene in 3DSNP database.

Author Contributions: Conceptualization, A.A. (Amir Avan) and E.G. (Elisa Giovannetti); Methodology, M.R.A. and M.K.; Software, A.A. (Alireza Asadnia), M.B., E.N. and M.S.K.; Validation, A.A. (Alireza Asadnia), F.K.-L., H.A., G.K.-T. and M.M.-A.; Formal Analysis, A.A. (Alireza Asadnia), E.N. and S.M.H.; Investigation, A.A. (Alireza Asadnia), G.P., L.G. (Ladan Goshayeshi), L.G. (Lena Goshayeshi) and S.M.H.; Resources, M.A.K.; Data Curation, G.P. and A.K.-y.L.; Writing—Original Draft Preparation, A.A. (Alireza Asadnia) and N.Z.; Writing—Review and Editing, A.A. (Alireza Asadnia), A.A. (Amir Avan), J.B., I.S.G., G.J.P., G.A.F. and A.K.-y.L.; Supervision, A.A. (Amir Avan), J.B., G.J.P., M.R.A. and E.G. (Elisa Giovannetti); Project Administration, A.A. (Amir Avan); Funding Acquisition, A.A. (Amir Avan). All authors have read and agreed to the published version of the manuscript.

Funding: This research was funded by National Institute for Medical Research and Development (NIMAD 962782AA); NHMRC—National Health and Medical Research Council (2029788AA), Tour the Cure (AA-RSP-163-2024); AIRC (associazione Italian per la ricerca sul Cancro)—IG-grant 24444; CCA (Cancer Center Amsterdam) Foundation (Elisa Giovannetti); Advance Queensland Industry Research Fellowship (JB).

Institutional Review Board Statement: The study was conducted in accordance with the Declaration of Helsinki and approved by MUMS (IR-MUMS-MEDICAL.REC.1401.404).

Informed Consent Statement: Informed consent was obtained from all patients enrolled in the study.

Data Availability Statement: The datasets are available upon request.

Acknowledgments: The work was supported by Mashhad University of Medical Sciences (Amir Avan) and the National Institute for Medical Research and Development (NIMAD 962782, AA); AIRC (associazione Italian per la ricerca sul Cancro)—IG-grant 24444; CCA (Cancer Center Amsterdam) Foundation (EG); Advance Queensland Industry Research Fellowship (JB).

Conflicts of Interest: The authors declare no conflict of interest.

References

1. Sung, H.; Ferlay, J.; Siegel, R.L.; Laversanne, M.; Soerjomataram, I.; Jemal, A.; Bray, F. Global Cancer Statistics 2020: GLOBOCAN Estimates of Incidence and Mortality Worldwide for 36 Cancers in 185 Countries. *CA Cancer J. Clin.* **2021**, *71*, 209–249. [CrossRef]
2. Yin, Z.; Yao, C.; Zhang, L.; Qi, S. Application of artificial intelligence in diagnosis and treatment of colorectal cancer: A novel Prospect. *Front. Med.* **2023**, *10*, 1128084. [CrossRef]
3. Reel, P.S.; Reel, S.; Pearson, E.; Trucco, E.; Jefferson, E. Using machine learning approaches for multi-omics data analysis: A review. *Biotechnol. Adv.* **2021**, *49*, 107739. [CrossRef]
4. Hammad, A.; Elshaer, M.; Tang, X. Identification of potential biomarkers with colorectal cancer based on bioinformatics analysis and machine learning. *Math. Biosci. Eng.* **2021**, *18*, 8997–9015. [CrossRef]
5. Maurya, N.S.; Kushwaha, S.; Chawade, A.; Mani, A. Transcriptome profiling by combined machine learning and statistical R analysis identifies TMEM236 as a potential novel diagnostic biomarker for colorectal cancer. *Sci. Rep.* **2021**, *11*, 14304. [CrossRef]
6. Liu, Z.; Liu, L.; Weng, S.; Guo, C.; Dang, Q.; Xu, H.; Wang, L.; Lu, T.; Zhang, Y.; Sun, Z.; et al. Machine learning-based integration develops an immune-derived lncRNA signature for improving outcomes in colorectal cancer. *Nat. Commun.* **2022**, *13*, 816. [CrossRef]
7. Tauriello, D.V.F.; Palomo-Ponce, S.; Stork, D.; Berenguer-Llergo, A.; Badia-Ramentol, J.; Iglesias, M.; Sevillano, M.; Ibiza, S.; Cañellas, A.; Hernando-Momblona, X.; et al. TGFβ drives immune evasion in genetically reconstituted colon cancer metastasis. *Nature* **2018**, *554*, 538–543. [CrossRef]
8. Pagès, F.; Mlecnik, B.; Marliot, F.; Bindea, G.; Ou, F.S.; Bifulco, C.; Lugli, A.; Zlobec, I.; Rau, T.T.; Berger, M.D.; et al. International validation of the consensus Immunoscore for the classification of colon cancer: A prognostic and accuracy study. *Lancet* **2018**, *391*, 2128–2139. [CrossRef]
9. Grivennikov, S.I.; Greten, F.R.; Karin, M. Immunity, inflammation, and cancer. *Cell* **2010**, *140*, 883–899. [CrossRef]
10. Calon, A.; Lonardo, E.; Berenguer-Llergo, A.; Espinet, E.; Hernando-Momblona, X.; Iglesias, M.; Sevillano, M.; Palomo-Ponce, S.; Tauriello, D.V.; Byrom, D.; et al. Stromal gene expression defines poor-prognosis subtypes in colorectal cancer. *Nat. Genet.* **2015**, *47*, 320–329. [CrossRef]
11. Isella, C.; Terrasi, A.; Bellomo, S.E.; Petti, C.; Galatola, G.; Muratore, A.; Mellano, A.; Senetta, R.; Cassenti, A.; Sonetto, C.; et al. Stromal contribution to the colorectal cancer transcriptome. *Nat. Genet.* **2015**, *47*, 312–319. [CrossRef]
12. Wixted, J.T.; Mickes, L. ROC analysis measures objective discriminability for any eyewitness identification procedure. *J. Appl. Res. Mem. Cogn.* **2015**, *4*, 329–334. [CrossRef]
13. Siegel, R.L.; Miller, K.D.; Fuchs, H.E.; Jemal, A. Cancer statistics, 2021. *CA Cancer J. Clin.* **2021**, *71*, 7–33. [CrossRef] [PubMed]
14. Ramos, M.; Esteva, M.; Cabeza, E.; Llobera, J.; Ruiz, A. Lack of association between diagnostic and therapeutic delay and stage of colorectal cancer. *Eur. J. Cancer* **2008**, *44*, 510–521. [CrossRef]

15. Gatalica, Z.; Torlakovic, E. Pathology of the hereditary colorectal carcinoma. *Fam. Cancer* **2008**, *7*, 15–26. [CrossRef]
16. Bousis, D.; Verras, G.-I.; Bouchagier, K.; Antzoulas, A.; Panagiotopoulos, I.; Katinioti, A.; Kehagias, D.; Kaplanis, C.; Kotis, K.; Anagnostopoulos, C.-N. The role of deep learning in diagnosing colorectal cancer. *Gastroenterol. Rev./Przegląd Gastroenterol* **2023**, *18*. [CrossRef]
17. Khalili-Tanha, G.; Mohit, R.; Asadnia, A.; Khazaei, M.; Dashtiahangar, M.; Maftooh, M.; Nassiri, M.; Hassanian, S.M.; Ghayour Mobarhan, M.; Kiani, M.A. Identification of ZMYND19 as a novel biomarker of colorectal cancer: RNA-sequencing and machine learning analysis. *J. Cell Commun. Signal.* **2023**, 1–17. [CrossRef]
18. Nazari, E.; Pourali, G.; Khazaei, M.; Asadnia, A.; Dashtiahangar, M.; Mohit, R.; Maftooh, M.; Nassiri, M.; Hassanian, S.M.; Ghayour-Mobarhan, M. Identification of potential biomarkers in stomach adenocarcinoma using machine learning approaches. *Curr. Bioinform.* **2023**, *18*, 320–333. [CrossRef]
19. Azari, H.; Nazari, E.; Mohit, R.; Asadnia, A.; Maftooh, M.; Nassiri, M.; Hassanian, S.M.; Ghayour-Mobarhan, M.; Shahidsales, S.; Khazaei, M. Machine learning algorithms reveal potential miRNAs biomarkers in gastric cancer. *Sci. Rep.* **2023**, *13*, 6147 [CrossRef]
20. Wu, B.; Yang, J.; Qin, Z.; Yang, H.; Shao, J.; Shang, Y. Prognosis prediction of stage IV colorectal cancer patients by mRNA transcriptional profile. *Cancer Med.* **2022**, *11*, 4900–4912. [CrossRef]
21. Chen, L.; Lu, D.; Sun, K.; Xu, Y.; Hu, P.; Li, X.; Xu, F. Identification of biomarkers associated with diagnosis and prognosis of colorectal cancer patients based on integrated bioinformatics analysis. *Gene* **2019**, *692*, 119–125. [CrossRef]
22. Li, B.; Zhu, F.C.; Yu, S.X.; Liu, S.J.; Li, B.Y. Suppression of KIF22 Inhibits Cell Proliferation and Xenograft Tumor Growth in Colon Cancer. *Cancer Biother. Radiopharm.* **2020**, *35*, 50–57. [CrossRef]
23. An, N.; Zhao, Y.; Lan, H.; Zhang, M.; Yin, Y.; Yi, C. SEZ6L2 knockdown impairs tumour growth by promoting caspase-dependent apoptosis in colorectal cancer. *J. Cell Mol. Med.* **2020**, *24*, 4223–4232. [CrossRef]
24. Wang, L.; He, M.; Fu, L.; Jin, Y. Exosomal release of microRNA-454 by breast cancer cells sustains biological properties of cancer stem cells via the PRRT2/Wnt axis in ovarian cancer. *Life Sci.* **2020**, *257*, 118024. [CrossRef] [PubMed]
25. Mei, Z.B.; Duan, C.Y.; Li, C.B.; Cui, L.; Ogino, S. Prognostic role of tumor PIK3CA mutation in colorectal cancer: A systematic review and meta-analysis. *Ann. Oncol.* **2016**, *27*, 1836–1848. [CrossRef] [PubMed]
26. Maffeis, V.; Nicolè, L.; Cappellesso, R. RAS, Cellular Plasticity, and Tumor Budding in Colorectal Cancer. *Front. Oncol.* **2019**, *9*, 1255. [CrossRef] [PubMed]
27. Bohlen, J.; Roiuk, M.; Neff, M.; Teleman, A.A. PRRC2 proteins impact translation initiation by promoting leaky scanning. *Nucleic Acids Res.* **2023**, *51*, 3391–3409. [CrossRef] [PubMed]
28. Liu, X.; Zhang, Y.; Wang, Z.; Liu, L.; Zhang, G.; Li, J.; Ren, Z.; Dong, Z.; Yu, Z. PRRC2A Promotes Hepatocellular Carcinoma Progression and Associates with Immune Infiltration. *J. Hepatocell. Carcinoma* **2021**, *8*, 1495–1511. [CrossRef] [PubMed]
29. Nieters, A.; Conde, L.; Slager, S.L.; Brooks-Wilson, A.; Morton, L.; Skibola, D.R.; Novak, A.J.; Riby, J.; Ansell, S.M.; Halperin, E.; et al. PRRC2A and BCL2L11 gene variants influence risk of non-Hodgkin lymphoma: Results from the InterLymph consortium. *Blood* **2012**, *120*, 4645–4648. [CrossRef] [PubMed]
30. Zhu, Z.; Hou, Q.; Wang, B.; Li, C.; Liu, L.; Gong, W.; Chai, J.; Guo, H. A novel mitochondria-related gene signature for controlling colon cancer cell mitochondrial respiration and proliferation. *Hum. Cell* **2022**, *35*, 1126–1139. [CrossRef]
31. Chae, Y.S.; Lee, S.J.; Moon, J.H.; Kang, B.W.; Kim, J.G.; Sohn, S.K.; Jung, J.H.; Park, H.Y.; Park, J.Y.; Kim, H.J.; et al. VARS2 V552V variant as prognostic marker in patients with early breast cancer. *Med. Oncol.* **2011**, *28*, 1273–1280. [CrossRef] [PubMed]
32. Ryu, T.Y.; Kim, K.; Han, T.S.; Lee, M.O.; Lee, J.; Choi, J.; Jung, K.B.; Jeong, E.J.; An, D.M.; Jung, C.R.; et al. Human gut-microbiome derived propionate coordinates proteasomal degradation via HECTD2 upregulation to target EHMT2 in colorectal cancer. *ISME J.* **2022**, *16*, 1205–1221. [CrossRef] [PubMed]
33. Triner, D.; Castillo, C.; Hakim, J.B.; Xue, X.; Greenson, J.K.; Nuñez, G.; Chen, G.Y.; Colacino, J.A.; Shah, Y.M. Myc-Associated Zinc Finger Protein Regulates the Proinflammatory Response in Colitis and Colon Cancer via STAT3 Signaling. *Mol. Cell Biol.* **2018**, *38*, e00386-18. [CrossRef] [PubMed]
34. Luo, W.; Zhu, X.; Liu, W.; Ren, Y.; Bei, C.; Qin, L.; Miao, X.; Tang, F.; Tang, G.; Tan, S. MYC associated zinc finger protein promotes the invasion and metastasis of hepatocellular carcinoma by inducing epithelial mesenchymal transition. *Oncotarget* **2016**, *7*, 86420–86432. [CrossRef]
35. Ren, L.X.; Qi, J.C.; Zhao, A.N.; Shi, B.; Zhang, H.; Wang, D.D.; Yang, Z. Myc-associated zinc-finger protein promotes clear cell renal cell carcinoma progression through transcriptional activation of the MAP2K2-dependent ERK pathway. *Cancer Cell Int* **2021**, *21*, 323. [CrossRef]
36. Smits, M.; Wurdinger, T.; van het Hof, B.; Drexhage, J.A.; Geerts, D.; Wesseling, P.; Noske, D.P.; Vandertop, W.P.; de Vries, H.E.; Reijerkerk, A. Myc-associated zinc finger protein (MAZ) is regulated by miR-125b and mediates VEGF-induced angiogenesis in glioblastoma. *FASEB J.* **2012**, *26*, 2639–2647. [CrossRef]
37. He, J.; Wang, J.; Li, T.; Chen, K.; Li, S.; Zhang, S. SIPL1, Regulated by MAZ, Promotes Tumor Progression and Predicts Poor Survival in Human Triple-Negative Breast Cancer. *Front. Oncol.* **2021**, *11*, 766790. [CrossRef]
38. Jiao, L.; Li, Y.; Shen, D.; Xu, C.; Wang, L.; Huang, G.; Chen, L.; Yang, Y.; Yang, C.; Yu, Y.; et al. The prostate cancer-up-regulated Myc-associated zinc-finger protein (MAZ) modulates proliferation and metastasis through reciprocal regulation of androgen receptor. *Med. Oncol.* **2013**, *30*, 570. [CrossRef]

Article

GradWise: A Novel Application of a Rank-Based Weighted Hybrid Filter and Embedded Feature Selection Method for Glioma Grading with Clinical and Molecular Characteristics

Erdal Tasci, Sarisha Jagasia, Ying Zhuge, Kevin Camphausen and Andra Valentina Krauze *

Radiation Oncology Branch, Center for Cancer Research, National Cancer Institute, National Institutes of Health, Building 10, Bethesda, MD 20892, USA
* Correspondence: andra.krauze@nih.gov

Simple Summary: Glioma tumor aggressiveness is expressed as tumor grading which is crucial in guiding treatment decisions and clinical trial participation. Accurate and standardized grading systems are essential to optimize care and improve outcomes. However, integrating molecular and clinical information in the grading process has the potential to expose molecular markers that have gained importance in understanding tumor biology as a means of identifying druggable targets. In this study, a novel approach called GradWise is introduced with the goal of enhancing feature selection performance while employing various machine learning models of glioma grading. GradWise combines a rank-based weighted hybrid filter (mRMR) and an embedded feature selection method (LASSO) to select the most relevant features from clinical and molecular predictors and was evaluated using two commonly employed public biomedical datasets, TCGA and CGGA, utilizing two feature selection methods and five supervised models. The findings support existing evidence and provide pioneering results for glioma-specific biomarkers, highlighting the effectiveness of the approach and future directions for biological mechanisms of glioma progression to higher grades.

Citation: Tasci, E.; Jagasia, S.; Zhuge, Y.; Camphausen, K.; Krauze, A.V. GradWise: A Novel Application of a Rank-Based Weighted Hybrid Filter and Embedded Feature Selection Method for Glioma Grading with Clinical and Molecular Characteristics. *Cancers* 2023, 15, 4628. https://doi.org/10.3390/cancers15184628

Academic Editors: Christine Marosi and Bernhard Meyer

Received: 6 July 2023
Revised: 1 September 2023
Accepted: 14 September 2023
Published: 19 September 2023

Abstract: Glioma grading plays a pivotal role in guiding treatment decisions, predicting patient outcomes, facilitating clinical trial participation and research, and tailoring treatment strategies. Current glioma grading in the clinic is based on tissue acquired at the time of resection, with tumor aggressiveness assessed from tumor morphology and molecular features. The increased emphasis on molecular characteristics as a guide for management and prognosis estimation underscores is driven by the need for accurate and standardized grading systems that integrate molecular and clinical information in the grading process and carry the expectation of the exposure of molecular markers that go beyond prognosis to increase understanding of tumor biology as a means of identifying druggable targets. In this study, we introduce a novel application (GradWise) that combines rank-based weighted hybrid filter (i.e., mRMR) and embedded (i.e., LASSO) feature selection methods to enhance the performance of feature selection and machine learning models for glioma grading using both clinical and molecular predictors. We utilized publicly available TCGA from the UCI ML Repository and CGGA datasets to identify the most effective scheme that allows for the selection of the minimum number of features with their names. Two popular feature selection methods with a rank-based weighting procedure were employed to conduct comprehensive experiments with the five supervised models. The computational results demonstrate that our proposed method achieves an accuracy rate of 87.007% with 13 features and an accuracy rate of 80.412% with five features on the TCGA and CGGA datasets, respectively. We also obtained four shared biomarkers for the glioma grading that emerged in both datasets and can be employed with transferable value to other datasets and data-based outcome analyses. These findings are a significant step toward highlighting the effectiveness of our approach by offering pioneering results with novel markers with prospects for understanding and targeting the biologic mechanisms of glioma progression to improve patient outcomes.

Keywords: brain tumor; glioma; grading; feature selection; pattern recognition; classification; data mining; molecular data; oncology

1. Introduction

Tumor grading is the classification of tumor aggressiveness determined via the evaluation of tumor characteristics with additive molecular features under the microscope [1]. Gauging the aggressiveness of a tumor, in this case, glioma represents a surrogate for anticipatory biological behavior. A sense of how the tumor will behave with or without treatment is crucial for decision-making at diagnosis, which connects to managing, monitoring, and treatment planning [2]. Gliomas are the most common primary brain tumor originating from glial cells and can be highly aggressive, progressive, and neurologically devastating [3,4]. Currently, according to the World Health Organization (WHO) guidelines, gliomas are categorized into low-grade (LGGs) and high-grade gliomas (HGGs), with glioblastoma multiforme (GBM) being the most aggressive and invasive. Treatment options and survival rates are highly dependent on tumor grade.

The current approach for treating gliomas is primarily determined by the grade of the tumor and typically involves maximal surgical removal followed by radiation therapy (RT) [2,4,5]. Additionally, patients may receive systemic treatment in the form of chemotherapy using temozolomide (TMZ) administered concurrently or sequentially with either sequential PCV or PC (Procarbazine, CCNU with or without vincristine) as an alternative [2,4,5]. Typically, the diagnosis of glioma is made by obtaining tissue for pathological examination with molecular alterations being increasingly important for CNS tumor classification [6–8]. The isocitrate dehydrogenase (IDH) mutation is now more routinely employed as a molecular marker, given its prognostic value [4,9,10], but is limited by the associated costs and turnaround time of molecular testing, with p.R132H-specific IDH1 immunohistochemistry costing USD 135, single-gene sequencing costing USD 420, and next-generation sequencing costing USD 1800 [9] and the time required for analysis ranging from approximately two days for immunohistochemistry to up to 14 days for next-generation sequencing [9]. IDH mutation vs. IDH wild-type confers superior prognosis, particularly when accompanied by 1p19q co-deletion altering the management of non-GBM gliomas in terms of type and timing of systemic management, while GBMs are treated with standard-of-care concurrent chemo-irradiation irrespective of IDH status. However, despite the superior prognosis conferred by IDH mutation, there is an ongoing lack of clarity as to the mechanism by which IDH mutation connects to the prognosis conferred to patients. There is an ongoing need to identify markers that allow for glioma grading via linkage to biological mechanisms that can be exploited to alter outcomes by modulating tumor resistance and response.

The tumor grading process incorporates clinical features such as age and gender [11], but publicly available datasets lack robust higher-level clinical annotation, which limits the connection between relevant molecular features and clinical data. This gap could be bridged by increasing reimbursement for molecular testing, which could promote more widespread use and benefit value-added care [4]. Therefore, selecting the best molecular and clinical markers that distinguish between tumor grades would not only reduce costs to healthcare systems and patients but also enhance tumor grading performance. This improvement in performance would enable the selection of significant molecular features for future research and testing of novel targeted agents [4]. However, given the partial nature of available molecular information, optimal utilization of such data would require computational analysis. Hence, feature selection plays a vital role in this context.

The feature selection stage is a critical step in machine learning, where the objective is to select a subset of relevant features among all features that improve the accuracy of the model while reducing complexity. Feature selection is generally utilized for data analysis, pattern recognition, data mining, and machine learning tasks. This process aims to improve performance (e.g., tumor grading) and classification accuracy rate and provide computational efficiency by removing irrelevant or redundant features and reducing the dimensionality of data [12–17]. There are various feature selection methods available, such as filter methods, wrapper methods, and embedded methods [12,18], each with its own advantages and limitations. The selected features can be used for further analysis, such as identifying biomarkers, developing predictive models, and gaining insights into the underlying biology of the disease.

Generally, feature selection methods are applied to training-test sets to determine the relevant feature subset, or these methods are aimed at reducing costs and improving classification evaluation results. If there is no training-test set separation, we should use these methods with the cross-validation technique for validation purposes. However identifying molecular feature names is crucial to being able to leverage the identified features biologically and clinically. To this end, feature-weighting, counting, or rank-based approaches can be used [19]. While several markers are being used in the clinic and additional markers are being proposed, given evolving research, there is currently no robust biomarker list or panel that defines glioma grading and employs data from TCGA and CGGA, which are the most commonly available and utilized datasets of clinical and molecular data. In this study, the goal was to identify the most important, discriminative and likely optimal molecular and clinical features for glioma grading by using a novel application of a hybrid rank-based filter and embedded feature selection based-method (GradWise) and five supervised learning models taught with TCGA and CGGA glioma data and to link the results to described mutations in glioma and novel biological applications.

The main contributions of our study are summarized as follows:

Our study proposes the first application and method that employs a rank-based hybrid feature selection method for feature selection and supervised machine learning models to improve glioma grading.

- We combine the advantages of various feature selection methods via a rank-based feature-weighting approach for glioma grading on two commonly used glioma datasets (TCGA and CGGA).
- We utilize feature-weighting to determine which features are significant, enabling validation of this method for glioma grading tasks.
- We conduct a comprehensive computational analysis comparing our feature selection methods, given that these are two commonly employed glioma datasets that share similarities but also exhibit differences.
- Our objective is to determine the optimal combination of feature subsets and learning models during the feature selection stage, aiming to achieve high accuracy with a minimal number of features while accounting for dataset variability in large-scale datasets. This approach seeks to provide accurate results that can be transferred and applied effectively across different scenarios.
- We introduce a TCGA- and CGGA-specific shared feature set and connect identified features for glioma grading with described mutations in glioma and identify potential mechanistic implications for progression to higher grade.

The remaining sections of our study are structured as follows: In Section 2, we provide an overview of the employed methodology and explain the related feature selection, feature-weighting methods, and classification models for glioma grading. In Section 3, we describe the experimental procedures, datasets employed, and evaluation metrics and provide comprehensive experimental results with discussions. Finally, Section 4 encompasses the study's conclusion, a discussion of the results, and potential avenues for future research.

2. Methods

In this section, we present a concise summary of the feature selection and weighting architecture that is being proposed for glioma grading via clinical and molecular characteristics. The subsequent subsections outline the methods used for feature selection, feature-weighting, and classification in this study.

2.1. The Utilized Methodology for Glioma Grading

In this study, we employ a hybrid method for weighting and selecting of features based on ranks [19] which can be used to categorize glioma grades. Our used methodology consists of two main phases: (i) feature selection (FS) and (ii) feature-weighting (FW) [19]. Figure 1 and Table 1 provide a sample algorithmic diagram of our utilized architecture and related processes, showcasing the two feature selection methods used: LASSO and mRMR.

At the outset, all clinical and molecular features are fed into the feature selection (FS) model using a cross-validation technique. For each fold, the feature sets selected by the two FS methods are saved, and their counts are increased based on the corresponding weights assigned by the rank-based approach. Next, the minimum weight-based feature list is evaluated with all weight values. In the final stage, we obtain the final selected feature list by evaluating all weight values and identifying those with the highest accuracy rate.

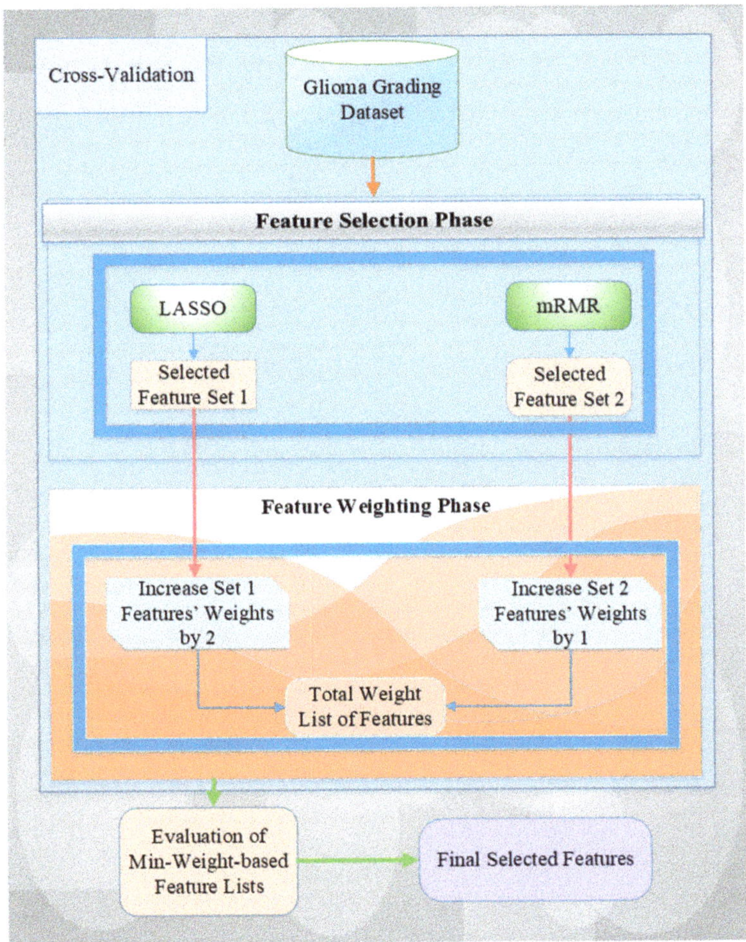

Figure 1. A detailed overview of the proposed methodology.

Table 1. The related textual representation of an employed scheme for the proposed GradWise approach.

1. Input: Clinical and molecular predictors with labels

2. Feature Selection with cross-validation:
For each fold:

        Choose supervised models (e.g., KNN, random forest, SVM, etc.)
- Apply the mRMR (i.e., multivariate filter) FS method
- Apply the LASSO FS (i.e., embedded) FS method
- Select the most relevant features
- Apply feature-weighting (increase the selected features' weights based on their ranks)

3. Try all weights with all machine learning models to determine the maximum accuracy rate with the minimum number of features

4. Evaluation:

        Use TCGA and CGGA datasets
- Calculate performance metrics (e.g., accuracy rate)

5. Results:

        Obtain accuracy rates and feature counts for TCGA and CGGA datasets
- Identify shared biomarkers for glioma grading

6. Conclusion:

        Highlight and discuss the potential advantages of GradWise approach
- Provide pioneering results for glioma-specific biomarker research and conceptualize
- findings given existing evidence for driver mutations and progression to a higher grade in glioma.

In other words, LASSO and mRMR feature selection methods are employed to select features for each fold of the cross-validation of the dataset based on clinical and molecular predictors. After selecting the features, their weights are increased according to their rank-based importance level, which is determined based on their performance results in terms of accuracy (see Tables 2 and 3). Specifically, the weight of a feature selected by LASSO is increased by two, while that of a feature selected by mRMR is increased by one. If the same feature is chosen by both methods for all five folds of cross-validation, its weight is 15, which is the maximum weight value for 5-fold cross-validation. On the other hand, if a feature is not selected by either FS method for a given iteration, we assign it a weight of 0. However, we ensure that the minimum weight value of a selected feature is identified as at least 1 to use all selected features of all cross-validation iterations for the experimental results.

Table 2. The related features and class information for the datasets employed. TCGA has 23 features (3 clinical, 20 molecular), whereas CGGA has 22 features (2 clinical, 20 molecular), given it contains of a Chinese population with race not included in the database.

#	Type	Name	#	Type	Name	#	Type	Name
1	Clinical	Gender	9	Molecular	CIC	17	Molecular	BCOR
2	Clinical	Age	10	Molecular	MUC16	18	Molecular	CSMD3
3	Clinical	Race	11	Molecular	PIK3CA	19	Molecular	SMARCA4
4	Molecular	IDH1	12	Molecular	NF1	20	Molecular	GRIN2A
5	Molecular	TP53	13	Molecular	PIK3R1	21	Molecular	IDH2
6	Molecular	ATRX	14	Molecular	FUBP1	22	Molecular	FAT4
7	Molecular	PTEN	15	Molecular	RB1	23	Molecular	PDGFRA
8	Molecular	EGFR	16	Molecular	NOTCH1	24	Class	Grade

Table 3. The effects of using feature selection methods on the TCGA dataset.

ML-ACC	Without FS	LASSO	mRMR
SVM	86.769	87.007	74.733
LR	86.414	86.414	85.935
KNN	82.837	83.313	82.839
RF	82.841	82.362	81.886
AdaBoost	85.339	85.101	84.621

Following the feature selection and feature-weighting stages, we obtain total weights with the corresponding feature lists and evaluate the minimum weight-based feature lists to select the final features for all weight values. To illustrate this process, consider an example. Suppose we determine the minimum weight as 12 for our study. In this case, we can select the features with weight values of 12, 13, 14, and 15 as the final feature set. We evaluate these weight values based on their performance results in terms of accuracy rate and identify the minimum weight value that achieves the highest accuracy rate with the minimum selected number of features for all values in the dataset.

2.2. Feature Selection and Feature-Weighting

The aim of feature selection methods is to reduce the dimensionality of data space by obtaining a suitable feature subset from all features. This process eliminates redundant, insignificant, or irrelevant features and yields better model interpretation and diagnosis capability, thus accelerating prediction speed and reducing the time requirement of the training stage of the machine learning model [12,20,21]. Additionally, feature selection methods deal with high-dimensional data, computational and storage complexity, data visualization, and high-performance issues for machine learning-related problems in real-world applications [12,22]. Feature selection methods are generally classified into three categories, depending on the evaluation metric of the feature subset: filter, wrapper, or embedded methods [18]. Univariate filter and multivariate filter FS methods are two sub-categories of filter methods that consider relationships between features and/or between features and the target/class or output variable [19]. In this study, we utilized a multivariate filter FS method called minimum redundancy maximum relevance (mRMR) and an embedded-based FS method called LASSO to select the clinical and molecular features of glioma patients' data. We chose to avoid the computational load of the wrapper FS method and the dependence on model-specific features associated with this approach in order to enhance the transferability of our used approach. In the feature-weighting stage, the importance level of each selected feature in discriminating pattern classes is typically represented by a weight value, which can be added or multiplied to feature values [19,23]. In this study, a rank-based feature-weighting approach was adopted. The two-feature selection (FS) methods, LASSO and mRMR, were ranked based on their performance results in terms of the accuracy rate as described in [19].

2.3. Classification

The classification phase is a fundamental task in machine learning that involves assigning predefined labels or categories to input data points based on their features. The goal of classification is to build a predictive model that can accurately classify new instances into their appropriate classes. Classification algorithms learn patterns and relationships from labeled training data, enabling them to make predictions from unlabeled data. Commonly used classification algorithms include k nearest neighbors, logistic regression, support vector machines, random forests, and AdaBoost. We briefly describe these learning models in the subsequent subsections in our previous work [4].

3. Experimental Work

This section describes the experimental processes and environment, relevant parameters, and performance metrics and explains our clinical and molecular dataset for glioma grading. Subsequently, we provide a detailed presentation of our comprehensive computational results, highlighting the impacts of various feature selection methods.

3.1. Experimental Process

To implement the proposed methods in this study, we employed the experimental process previously described [4,19].

We used a preprocessed glioma grading dataset with clinical and molecular features and performed a z-score normalization approach to age feature values before the feature selection and classification phases. We also employed a 5-fold cross-validation technique during the feature selection and weighting processes to ensure robustness. This approach allowed us to evaluate the performance of the utilized learning models and obtain average performance results, enhancing the reliability of our findings. For the evaluation of the learning models in this study, the GBM class was designated the positive class, while the LGG class served as the negative class.

Both the mRMR and LASSO methods were employed at the hybrid feature selection stage. In the mRMR-based feature selection, a heuristic value was utilized by taking the logarithmic value of the total number of features (i.e., $\lceil \log2(\text{Total Number of Features})\rceil$ = round of log2(Total Number of Features(23)) = 5) to determine the number of selected features. For LASSO-based feature selection, a 10-fold cross-validation was performed to determine the optimal alpha parameter value across iterations and identify the number of selected features.

In the feature-weighting stage, a rank-based approach was utilized, where the weights of features were determined based on the performance results of the feature selection methods, specifically the accuracy rate. To identify the final selected feature set, various minimum weight values ranging from 15 to 1 for 5-fold cross-validation were tested to find the subset of features that achieved the highest accuracy rate while using the smallest number of features.

3.2. Dataset

We employed the Cancer Genome Atlas (TCGA) [24] and the Chinese Glioma Genome Atlas (CGGA) [25] databases, which are widely used for analyzing brain tumors (specifically glioma), to assess our employed methodology for rank-based feature-weighting and selection processes. The original TCGA dataset is described in our previous work [4] and Table 2 with the preprocessed TCGA dataset for glioma grading available on the UCI Machine Learning Repository [24]. The CGGA dataset consists of 22 features (one fewer than TCGA) with the same characteristics described in Table 2. The dataset query and storage operations were facilitated via the NIDAP environment [26]. The quantitative description of gene expression (mutated/not mutated frequencies) in TCGA is presented in Supplementary Figure S1 and was described for TCGA by Yan et al. [27]. Quantitative description is available for CGGA at http://www.cgga.org.cn/analyse/WEseq-data-oncoprint-result.jsp and was described by Hu et al. [28].

3.3. Performance Metrics

To assess the performance of the utilized methodology in feature selection and classification, six evaluation metrics were used: classification accuracy (ACC), area under the ROC curve (AUC), F-measure (F1), precision (PRE), recall (REC), and specificity (SPEC) [29] These were described in detail in our previous work [4].

3.4. Computational Results

In this subsection, we present the experimental results showcasing the impact of the feature selection and feature-weighting approaches on the performance analysis of the models investigated in this study. The most optimal results are indicated by bold values # represents the number. The best result for each method is highlighted in bold.

3.4.1. The Effects of Using Feature Selection Methods

During the initial phase, we conducted experiments using five supervised learning models to examine the potential benefits of feature selection (FS) techniques on the glioma grading datasets. Tables 3 and 4 present the average performance results of these models (i.e., 5-fold cross-validation) on TCGA and CGGA datasets in terms of accuracy rate (%).

Table 4. The effects of using feature selection methods on the CGGA dataset.

ML-ACC	Without FS	LASSO	mRMR
SVM	76.564	76.915	73.085
LR	76.570	76.921	76.933
KNN	74.816	76.576	71.670
RF	74.840	72.741	73.442
AdaBoost	74.834	72.033	76.576

The findings in Tables 3 and 4 demonstrate that the supervised learning models with the LASSO method generally achieved better results compared to the mRMR and no feature selection methods. Using the LASSO FS method, we obtained the best accuracy rate value of 87.007 with the SVM model, while without FS, the best accuracy rate value of 86.769 was achieved with the same learning model, according to Table 3 results. The LASSO FS method also provided better results from the CGGA dataset compared to the no FS method in terms of accuracy rate (see Table 3). This is depicted in dark green in Table 4. The LASSO method has three higher performance results than the best accuracy rate value (i.e., 76.570%) of no FS method result on the CGGA dataset. Additionally, the mRMR method yielded accuracy rate values of 85.935 and 76.933 for the TCGA and CGGA datasets, respectively. Following this stage, in which we obtained these results, we proceeded to the next level of the FS process, which involved assigning corresponding ranks to these FS methods, a process known as feature-weighting. We selected the LASSO FS method as the more significant method with respect to the results obtained from Tables 3 and 4 to assign ranks to the corresponding methods (increasing feature weight value by two and one for the LASSO and mRMR FS methods, respectively).

3.4.2. The Effects of Using LASSO and mRMR Feature Selection and Feature-Weighting Methods

After the initial evaluation of features and assigned ranks based on their performance results, the related computational results obtained via the utilization of both LASSO and mRMR-based feature selection (FS) with weighting methods are presented in Tables 5 and 6 and Figures 2 and 3, which denote the mean accuracy rate values obtained using 5-fold cross-validation. k represents the minimum weight value.

As seen in the results given in Table 5, the most optimal outcome was characterized by an accuracy rate of 87.007, a minimum weight value of 10, and the selection of 13 features using the support vector machine model. The selected feature names for the best result from the TCGA dataset are as follows: 'CIC', 'Age', 'IDH1', 'PTEN', 'ATRX', 'PIK3R1', 'NF1', 'IDH2', 'GRIN2A', 'NOTCH1', 'TP53', 'EGFR', 'MUC16'. It is noteworthy that the number of selected features remained constant for both minimum weight values of 10 and 9. Consequently, the selection of the minimum weight value did not affect the results. However, in the scenario where the accuracy rate remains unchanged while different numbers of features for various weight values result, the maximum value could have been assigned as the minimum weight value to facilitate the selection of the minimum number of features. The second-best different result achieved the same accuracy rate of 87.007 with a minimum weight value of 8 and the selection of 18 features using the support vector machine. Figure 2 also depicts the line chart for the comparative illustration of feature-weighting and selection results on the TCGA dataset.

Table 5. Average performance results (i.e., ACC %, CV = 5) obtained utilizing both LASSO and mRMR-based feature selection with feature-weighting methods on the TCGA dataset.

k	# of Features	SVM	LR	KNN	RF	AdaBoost
15	4	85.340	86.054	84.626	80.100	84.264
14	4	85.340	86.054	84.626	80.100	84.264
13	4	85.340	86.054	84.626	80.100	84.264
12	5	85.102	85.816	84.983	80.814	84.502
11	6	85.698	85.816	84.627	81.172	84.859
10	**13**	**87.007**	86.890	84.983	82.481	84.862
9	**13**	**87.007**	86.890	84.983	82.481	84.862
8	18	87.007	86.533	82.599	82.481	85.577
7	18	87.007	86.533	82.599	82.481	85.577
6	20	86.768	86.533	82.479	82.484	85.458
5	20	86.768	86.533	82.479	82.484	85.458
4	22	86.768	86.414	82.718	82.603	85.339
3	22	86.768	86.414	82.718	82.603	85.339
2	23	86.769	86.414	82.837	82.244	85.339
1	23	86.769	86.414	82.837	82.244	85.339

Table 6. Average performance results (ACC %, CV = 5) obtained utilizing both LASSO and mRMR-based feature selection with feature-weighting methods on the CGGA dataset.

k	# of Features	SVM	LR	KNN	RF	AdaBoost
15	4	79.371	78.669	74.477	74.476	77.278
14	4	79.371	78.669	74.477	74.476	77.278
13	4	79.371	78.669	74.477	74.476	77.278
12	**5**	**80.412**	79.014	75.178	75.886	76.225
11	**5**	**80.412**	79.014	75.178	75.886	76.225
10	8	80.073	79.716	76.219	73.799	76.231
9	8	80.073	79.716	76.219	73.799	76.231
8	10	79.722	76.921	75.517	75.535	74.840
7	10	79.722	76.921	75.517	75.535	74.840
6	11	76.219	77.623	75.535	72.396	74.834
5	11	76.219	77.623	75.535	72.396	74.834
4	13	76.915	76.921	75.173	73.799	73.781
3	14	76.915	77.272	74.822	75.892	73.073
2	16	76.915	77.272	75.523	72.752	72.371
1	16	76.915	77.272	75.523	72.752	72.371

As can be seen from Table 6, the best performance results had an accuracy rate value of 80.412 in conjunction with the minimum weight value of 12 and the number of selected features as 5, employing the support vector machine classifier on the CGGA dataset for the glioma grading task. The second-best different result achieved an accuracy rate value of 80.073, with a minimum weight value of 10 and the selection of 8 features, using the support vector machine. The following feature names were selected for the optimal result obtained from the CGGA dataset: **'IDH1'**, **'Age'**, **'PTEN'**, **'PDGFRA'**, and **'NF1'**. Additionally, Figure 3 is a line chart that visually represents and compares the results of feature-weighting and selection from the CGGA dataset.

By employing this approach, it was possible to identify and select shared clinical and molecular predictors from the initial set of 22 or 23 features depending on the dataset used (i.e., TCGA or CGGA) in the glioma grading dataset. These selected and shared four features have the following names: **'IDH1'**, **'Age'**, **'PTEN'**, and **'NF1'**.

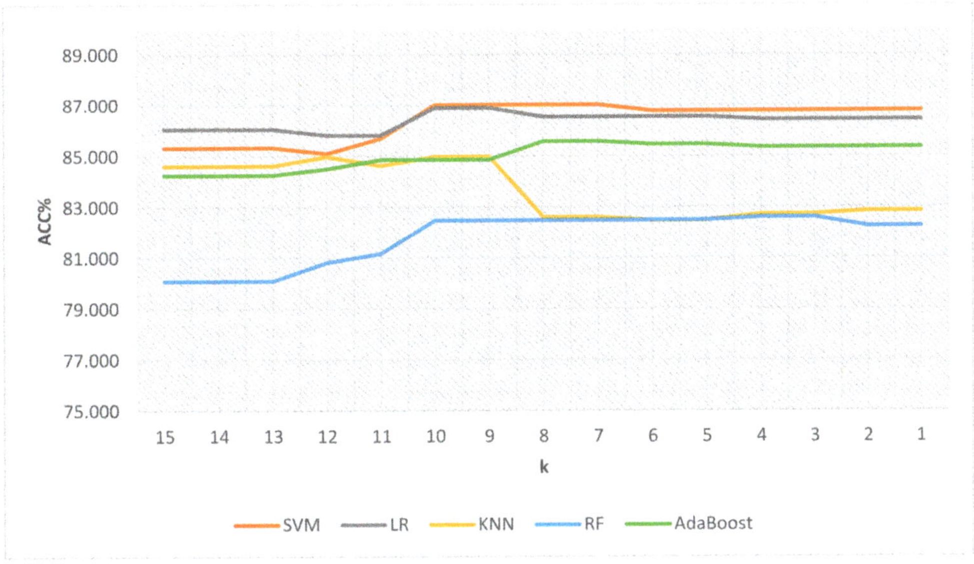

Figure 2. Comparative illustration of feature-weighting and selection results on the TCGA dataset.

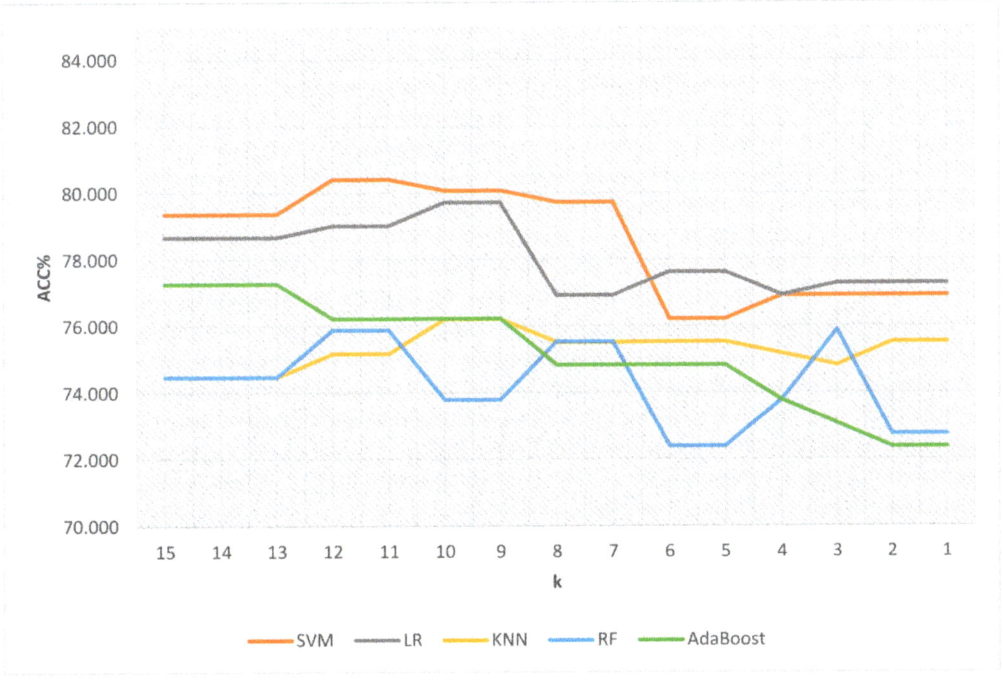

Figure 3. Comparative illustration of feature-weighting and selection results on the CGGA dataset.

3.4.3. Other Performance Results Based on Feature Selection and Weighting Process

We have obtained comprehensive computational results by employing a hybrid feature selection and weighting, taking into consideration six performance metrics: accuracy rate (ACC) %, area under the ROC curve (AUC), F-measure, precision, recall, and specificity. Tables 6 and 7 present the detailed average performance results on the TCGA and CGGA glioma grading datasets with/without feature selection and weighting. The green indicator shows that feature selection and weighting give better results than not applying the feature selection method, while the red indicator represents worse results in the same situation.

Table 7. Comprehensive average performance results (CV = 5) obtained by observing the effects of the feature selection and feature-weighting methods on the TCGA dataset.

	Without FS	With FW and FS		Without FS	With FW and FS	
ML	**ACC%**			**AUC**		
SVM	86.769	87.007		0.904	0.911	
LR	86.414	86.890		0.918	0.923	
KNN	82.837	84.983		0.893	0.906	
RF	82.841	82.481		0.897	0.900	
AdaBoost	85.339	84.862		0.905	0.908	
	Without FS	**With FW and FS**		**Without FS**	**With FW and FS**	
ML	**F1**			**PRE**		
SVM	0.852	0.855		0.801	0.804	
LR	0.847	0.852		0.805	0.808	
KNN	0.802	0.826		0.782	0.802	
RF	0.793	0.792		0.796	0.786	
AdaBoost	0.832	0.829		0.803	0.789	
	Without FS	**With FW and FS**		**Without FS**	**With FW and FS**	
ML	**REC**			**SPEC**		
SVM	0.912	0.915		0.837	0.839	
LR	0.897	0.905		0.843	0.845	
KNN	0.827	0.856		0.832	0.846	
RF	0.796	0.802		0.853	0.842	
AdaBoost	0.869	0.878		0.845	0.830	

As shown in Tables 7 and 8, since we focused on obtaining a high accuracy rate from the best-supervised learning model by using feature selection and weighting approaches in this study, support vector machine with FS and FW provides higher accuracy rate values than no FS method. Regarding Tables 7 and 8, the green color in cells means that the result after FW and FS is higher than the result without FS. Otherwise, the color is orange in cells. The SVM model achieved the highest values on the TCGA dataset, yielding 87.007%, 0.855, and 0.915 for ACC, F-measure, and recall, respectively. The LR model had the highest values for AUC and precision, namely 0.923 and 0.808, respectively, and KNN also yielded the highest value for specificity on the TCGA dataset, namely 0.846. Ash shown by the CGGA dataset-based results (see Table 7), the SVM model on the CGGA dataset achieved the highest values, yielding 80.412%, 0.815, 0.679, 0.807, and 0.913 in terms of ACC, AUC, F-measure, precision, and specificity, respectively. The RF model had the highest value, namely 0.610, in terms of recall on this dataset for this study as well.

Table 8. Comprehensive average performance results (CV = 5) by observing the effects of the feature selection and feature-weighting methods on the CGGA dataset.

	Without FS	With FW and FS		Without FS	With FW and FS
ML	**ACC%**			**AUC**	
SVM	76.564	80.412		0.815	0.798
LR	76.570	79.014		0.792	0.788
KNN	74.816	75.178		0.772	0.753
RF	74.840	75.886		0.758	0.767
AdaBoost	74.834	76.225		0.759	0.749
	Without FS	**With FW and FS**		**Without FS**	**With FW and FS**
ML	**F1**			**PRE**	
SVM	0.609	0.679		0.759	0.807
LR	0.633	0.656		0.706	0.788
KNN	0.555	0.577		0.743	0.706
RF	0.592	0.629		0.659	0.663
AdaBoost	0.625	0.603		0.661	0.717
	Without FS	**With FW and FS**		**Without FS**	**With FW and FS**
ML	**REC**			**SPEC**	
SVM	0.527	0.607		0.901	0.913
LR	0.584	0.582		0.862	0.907
KNN	0.454	0.516		0.908	0.880
RF	0.549	0.610		0.855	0.835
AdaBoost	0.605	0.536		0.829	0.888

3.4.4. Comparison with the Related Methods for Glioma Grading

In this subsection, we compare the performance on the two datasets used of the utilized method of feature selection with another related method from the literature for glioma grading tasks with molecular and clinical characteristics.

In Table 9, the accuracy rates are displayed as percentages along with the number of selected features for each method/dataset combination. Our method surpasses all its competitors by selecting 13 features for the TCGA dataset and 5 features for the CGGA dataset. Taking into account the obtained accuracy rate values, our method also outperforms its competitor [4] with 80.412% from the CGGA dataset.

Table 9. Comparison with the related methods in the literature for glioma grading tasks on the datasets employed.

Dataset	TCGA		CGGA	
Total # of Features	23		22	
Study	**Our Method**	**[4]**	**Our Method**	**[4]**
Selected # of Features	13	14.9	5	17.6
ACC %	87.007	87.606	80.412	79.668
Study	**Our Method**		**[4]**	
Method	mRMR + LASSO		Hierarchical voting-based ensemble scheme	
Advantages	Effective, more realistic, and consistent results, and identified feature names		The method employs an ensemble procedure	

Table 9 gives the significant differences between our approach and those of [4] while noting their respective advantages as well. Since we focused on determining feature names with the minimum number of selected features and maximum accuracy rate by using both feature-weighting and selection strategy in this study, our performance results (i.e., ACC can be lower than the related method's (e.g., [4]) results (e.g., TCGA dataset). However, our methodology presents more robust, effective, realistic results and identified feature names for the related classification task. The utilization of our method, which incorporates feature-weighting and selection approaches, led to a substantial enhancement in identifying the names of the final selected feature set in our biomarker discovery process.

4. Discussion

Based on the empirical findings from our rank-based weighted hybrid filter and embedded feature selection methodology applied to the TCGA and CGGA datasets with molecular and clinical characteristics, several insights emerged. Our utilized feature selection method demonstrated superior performance in terms of the number of selected features as compared to our previous related method [4] when applied to the same datasets. Since this hybrid method harnesses the advantages of two popular and effective feature selection methods, we hypothesize that it generates superior results as compared to the individual selection methods employed in isolation. This is, however, a novel application to this setting (glioma) and these datasets (TCGA/CGGA), and comparison with the results of other studies is yet limited. In a recent benchmarking study of feature selection strategies in multi-omics data, wherein 15 cancer multi-omics datasets were employed to compare four filter methods, two embedded methods, and two wrapper methods with respect to their performance in relation to the prediction of a binary outcome, the authors found that the feature selection methods mRMR, the permutation importance of random forests, and the Lasso method tended to outperform the other methods [30]. Bhadra et al. compared five widely used supervised feature selection methods (mRMR, INMIFS, DFS, SVM-RFE-CBR, and VWMRmR) for multi-omics datasets from a multi-omics study of acute myeloid leukemia (LAML) from TCGA to successfully identify gene signatures in each data subset [31]. Empirical results suggest that our feature selection and weighting methodology with supervised learning models holds promise for glioma grading tasks. By employing two different feature selection methods and five individual learning models using rank-based weighting strategies, we achieved optimal results in this study. Our methodology also generally outperformed the results of using a no feature selection method when tested on two datasets with different model schemes. A pivotal objective of this study was to determine the names of the final features in the set by assigning ranks and weights to the corresponding methods according to their importance level in terms of accuracy rate compared to the no FS methodology. Although the selections by the best supervised learning models and features varied across the datasets, we can conclude that our proposed approach [19] yields more robust and effective results compared to our previous feature selection approach [4]. We anticipate that, as integral sources of shared molecular data, TCGA and CGGA will continue to evolve with increasing number of features, and we hypothesize that providing a method that allows for the selection of the most biologically relevant TCGA and CGGA features, the future cost of molecular characterization may be reduced with increasing prediction performance and potential to examine biological pathways of glioma progression at a higher grade. To address the challenge of having a limited number of cases with high-dimensional features, we employed a 5-fold cross-validation technique to mitigate bias. By combining the strengths of both filter and embedded popular feature selection (FS) methods, we achieved highly effective results, even in the presence of high-dimensional features and the complexity of the problem. It should be noted that the choice of the most suitable feature subset may differ based on the feature selection method(s), parameters, heuristics, data type, and dataset size, as there is no universally optimal method that applies to all situations (as exemplified by the 'no free lunch theorem') [19]. We acquired thirteen and five significant clinical and molecular biomarkers for TCGA and CGGA datasets, respectively. Differences in identified features between TCGA and CGGA can be attributed to the number of patients, the number of features, data distribution, and characteristics in each set. Four features were shared among TCGA and CGGA: **Age, IDH1, PTEN, and NF1.** This indicates that given this input data, these are the currently most robust if not the most informative markers of glioma grading. The shared features align with existing data surrounding the distinction of diagnostic labels in glioma. Age as the sole shared clinical feature between TCGA and CGGA reflects tumor subtype distribution, which is distinctive between lower-grade gliomas occurring more often in younger patients in contrast to high-grade gliomas, including GBM occurring in older patients. IDH status as a co-localizing feature of more favorable biological behavior separates lower grade from higher grade glioma due to its association with LGG as compared to GBM, of which only approximately 10% are IDH mutated [32,33]. PTEN alteration is more associated with aggressive biological behavior and higher-grade glioma [34,35]. The role of NF1 is an equally well-recognized mutation in glioma, albeit not as common [36]. The shared features thus validate the method based on existing literature evidence (see Table 10). The additional features identified in TCGA (**CIC, ATRX, PIK3R1, IDH2, GRIN2A, NOTCH1, TP53, EGFR, MUC16**) and CGGA (**PDGFRA**) provide interesting avenues for analysis of progression to a higher grade in glioma via linkage to known signaling pathways of tumor progression and treatment resistance (Table 1, Figure 4). In particular, MUC16, also known as CA-125, merits additional investigation given its emergence as a distinctive grading feature, since current literature supports this marker in ovarian cancer with clinical use; however, it is identified as mutated in only a relatively small percentage of gliomas [37]. Recent evidence supports its role in tumor grading and prognosis [38], and it carries mechanistic implications via linkage with PDGFRA, a feature also identified in the current study and interestingly identified in CGGA [39]. Connecting PTEN, ATM, and p53, the feature GRIN2A, only described to date in a small minority of GBMs, merits further study as a marker of possible genetic evolution to a higher grade and post-therapeutic adaptation [35]. Similarly, CIC emerged in this method as a distinctive marker. It has been reported in 20% of LGGs [32]; however, CIC protein instability has been associated with tumorigenesis in GBM [40]. The identification of features such as MUC16 (already in use in the clinic albeit not in the glioma setting) and GRIN2A and CIC, both relatively novel, as evidenced by current ingenuity pathway analysis (IPA) (Figure 4) [41], is not currently employed in the clinic but shows promise in analyzing mechanistic progression to higher grade and showcasing the clinical promise of novel applications such as GradWise as potential tools to identify novel biomarker in existing datasets such as TCGA and CGGA. The reality in the clinic is that multiple loci of different molecular subtypes may be present in tumors, complicating diagnosis (Supplementary Figure S2). The method in this study may advance diagnostic capabilities by leveraging the complex feature composites of several markers and molecular subtypes to match them to the most appropriate diagnostic code. This aspect will be further improved by incorporating progression and survival outcomes as well as complex DNA methylation analysis, which is subject to implementation in multidisciplinary pathology discussions and future directions. The limitations of the study include the small scale of data and low quantity of features. TCGA resulted in several more distinctive grading features compared to CGGA, which indicates dataset-dependent limitations grounded in tumor heterogeneity and class imbalance.

Table 10. The 13 features were identified using GradWise. Features that emerged in both TCGA and CGGA are shown in bold.

Feature	Frequency of Mutated Genes in TCGA in GBM [37]	Somatic Genomic Alterations in GBM [33]	% GBM Patients Harboring Specific Oncogenic Mutations in TCGA [42]	Mutation Landscape of LGG [32]	Current Role in Oncology		Mechanistic Connections
					Literature Evidence	Use in Clinic	
Age	n/a	n/a	n/a	n/a	Age-associated with unfavorable neuropathological and radiological features in gliomas [43]	Yes, for clinical decision-making via recursive partitioning criteria	Investigational
IDH1/IDH2	n/a	n/a	3%	77%	IDH mutation in glioma: molecular mechanisms and therapeutic targets [44,45]	Yes, for tumor molecular characterization	HIF-1α
PTEN	34%	31%	19%	n/a	Identification of the Prognostic Signatures of Glioma With Different PTEN Status [34]	Yes, for tumor molecular characterization	TP53, GRIN2A
NF1	11%	11%	9%	n/a	An Update on Neurofibromatosis Type 1-Associated Gliomas [36]	Yes, for clinical decision-making and management discussion	EGFR, PTEN
EGFR	26%	26%	15%	6%	Updated Insights on EGFR Signaling Pathways in Glioma [46]	Yes, for tumor molecular characterization	NOTCH1
TP53	34%	29%	16%	46%	Genetic and histologic spatiotemporal evolution of recurrent, multifocal, multicentric and metastatic glioblastoma [35]	Yes, for tumor molecular characterization	PTEN, GRIN2A
PIK3R1	18%	11%	6%	n/a	Somatic Mutations of PIK3R1 Promote Gliomagenesis [47]	Not currently used in the clinic	PI3K
ATRX	n/a	6%	5%	33%	The Role of ATRX in Glioma Biology [48]	Yes, for tumor molecular characterization	ATM
PDGFRA	n/a	4%	5%	n/a	High frequency of PDGFRA and MUC family gene mutations in diffuse hemispheric glioma, H3 G34-mutant: a glimmer of hope? [39]	Investigational	MUC16
NOTCH1	n/a	n/a	n/a	n/a	Oncogenic and Tumor-Suppressive Functions of NOTCH Signaling in Glioma [49]	Investigational	EGFR
GRIN2A	n/a	n/a	4%	n/a	Somatic mutation of GRIN2A in malignant melanoma results in loss of tumor suppressor activity via aberrant NMDAR complex formation [35]	Investigational	PTEN, TP53
MUC16 (CA-125)	11%	n/a	n/a	n/a	MUC16 mutation is associated with tumor grade, clinical features, and prognosis in glioma patients [38]	Used as a serum biomarker in ovarian cancer with implications for other cancers as well [50]	PDGFRA
CIC	n/a	n/a	n/a	20%	CIC protein instability contributes to tumorigenesis in glioblastoma [40]	Not currently used in clinic	EGFR

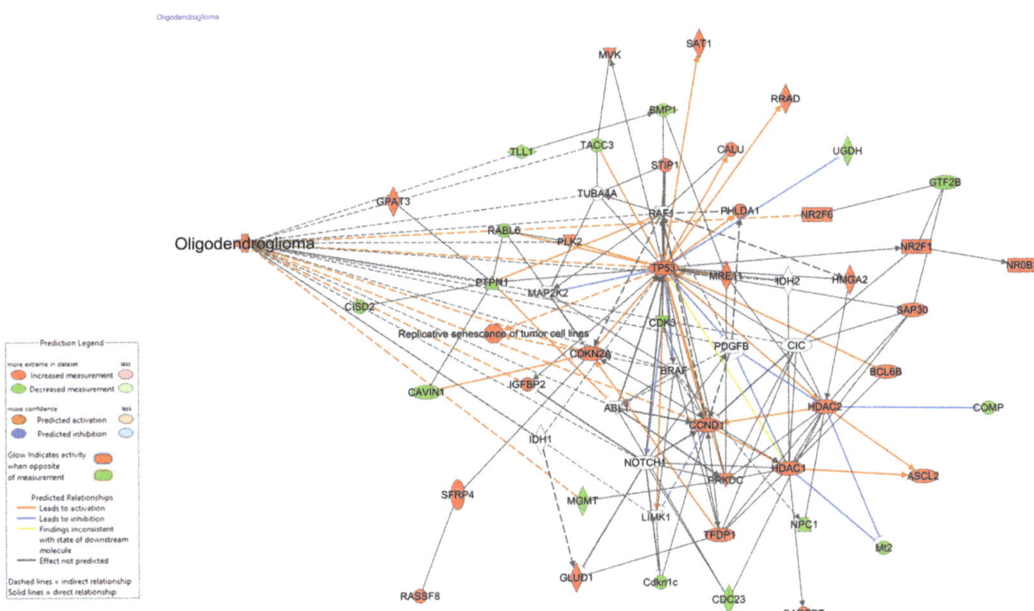

Figure 4. Network output of the 12 GradWise identified molecular features with oligodendroglioma annotation in IPA ((QIAGEN Inc., https://www.qiagenbioinformatics.com/products/ingenuitypathway-analysis) accessed on 6 September 2023) [41]. Several identified features associated with progression to a higher grade and complex biological interplay do not currently exhibit known biological measurement (NOTCH1, PDGFRA, IDH1/2, CIC). GRIN2A and MUC16 did not map in this framework, supporting a more novel role in glioma biology.

5. Conclusions and Future Work

This study introduces GradWise, a novel application of a rank-based weighted hybrid filter and embedded feature selection method employing LASSO and mRMR-based feature selection and weighting methods for glioma grading. The results demonstrate that the method is effective in identifying features representative of tumor grade and is in agreement with existing evidence, and it thus can serve as a framework for feature selection, classification, and pattern recognition towards value-added care, particularly in the context of molecular, clinical, and proteomic markers, while enhancing the predictive performance of models. The exploration of higher-dimensional biomedical datasets, including proteomic or metabolomic data, suggest future directions of this study to further validate this method. Future directions include the aggregation of additional datasets, including clinical, imaging, and omic data, with higher-dimensional features. This will allow us to further leverage and compare the performance and validation of GradWise against other approaches and explore the use of alternatives or combinations of the ensemble machine learning predictors to improve performance results for specific large-scale medical data scenarios.

Supplementary Materials: The following supporting information can be downloaded at: https://www.mdpi.com/article/10.3390/cancers15184628/s1, Figure S1: Expression of mutation features in TCGA dataset; Figure S2: Gene expression profiles in TCGA by primary diagnosis for the 13 features identified (underlined) using GradWise.

Author Contributions: E.T.: Conceptualization, Data Curation, Methodology, Software, Investigation, Writing—original draft preparation, Visualization, Writing—review & editing. S.J.: Review and Editing. Y.Z.: Review and Editing. K.C.: Supervision, Funding acquisition. A.V.K.: Conceptualization, Investigation, Supervision, Project administration, Funding acquisition, Writing—original draft preparation, Visualization, Writing—review & editing. All authors have read and agreed to th published version of the manuscript.

Funding: Funding was provided in part by NCI NIH intramural program (ZID BC 010990).

Institutional Review Board Statement: Not applicable.

Informed Consent Statement: Not applicable.

Data Availability Statement: The data in this paper has been provided from The Cancer Genom Atlas (TCGA) Research Network (https://www.cancer.gov/tcga, accessed on 5 July 2023) and th Chinese Glioma Genome Atlas (CGGA) (http://www.cgga.org.cn/, accessed on 12 November 2022

Acknowledgments: Palantir Foundry was used to integrate, harmonize, and analyze clinical an proteomic data inside the secure NIH Integrated Data Analysis Platform (NIDAP). The results show here are in whole or part based upon data generated by the TCGA Research Network and the Chines Glioma Genome Atlas (CGGA). Palantir Foundry was used in the integration, harmonization, an analysis of TCGA data inside the secure NIH Integrated Data Analysis Platform (NIDAP). We als thank Harpreet Kaur for her contribution to the CGGA dataset curation.

Conflicts of Interest: The authors declare no conflict of interest.

Abbreviations

ACC	Accuracy
AdaBoost	Adaptive Boosting
AUC	Area Under the ROC Curve
CGGA	Chinese Glioma Genome Atlas
CNS	Central Nervous System
F1	F-Measure
GBM	Glioblastoma Multiforme
HGG	High-Grade Glioma
IDH	Isocitrate Dehydrogenase
IPA	Ingenuity Pathway Analysis
KNN	K Nearest Neighbors
LGG	Low-Grade Glioma
LASSO	Least Absolute Shrinkage and Selection Operator
LR	Logistic Regression
mRMR	Minimum Redundancy—Maximum Relevance
NCI	National Cancer Institute
NIDAP	NIH Integrated Data Analysis Platform
NIH	National Institutes of Health
PRE	Precision
REC	Recall
RF	Random Forest
ROC	Receiver Operating Characteristics
RT	Radiation Therapy
SPEC	Specificity
SVM	Support Vector Machine
TCGA	The Cancer Genome Atlas
TMZ	Temozolomide
WHO	World Health Organization

References

1. Marquet, G.; Dameron, O.; Saikali, S.; Mosser, J.; Burgun, A. Grading glioma tumors using OWL-DL and NCI thesauru In Proceedings of the AMIA Annual Symposium Proceedings, Chicago, IL, USA, 10–14 November 2007; American Medica Informatics Association: Washington, DC, USA.
2. Pereira, S.; Meier, R.; Alves, V.; Reyes, M.; Silva, C.A. Automatic brain tumor grading from MRI data using convolutional neura networks and quality assessment. In *Understanding and Interpreting Machine Learning in Medical Image Computing Applications* Springer: Berlin/Heidelberg, Germany, 2018; pp. 106–114.

.   Tasci, E.; Ugur, A.; Camphausen, K.; Zhuge, Y.; Zhao, R.; Krauze, A.V. 3D Multimodal Brain Tumor Segmentation and Grading Scheme based on Machine, Deep, and Transfer Learning Approaches. *Int. J. Bioinfor. Intell. Comput.* **2022**, *1*, 77–95.

.   Tasci, E.; Zhuge, Y.; Kaur, H.; Camphausen, K.; Krauze, A.V. Hierarchical Voting-Based Feature Selection and Ensemble Learning Model Scheme for Glioma Grading with Clinical and Molecular Characteristics. *Int. J. Mol. Sci.* **2022**, *23*, 14155. [CrossRef]

.   Krauze, A. Using Artificial Intelligence and Magnetic Resonance Imaging to Address Limitations in Response Assessment in Glioma. *Oncol. Insights* **2022**, *2022*, 616. [CrossRef]

.   Gaillard, F. WHO Classification of CNS Tumors. Reference Article, Radiopaedia.org. Available online: https://radiopaedia.org/articles/who-classification-of-cns-tumours-1?lang=us (accessed on 2 September 2022).

.   Mirchia, K.; Richardson, T.E. Beyond IDH-mutation: Emerging molecular diagnostic and prognostic features in adult diffuse gliomas. *Cancers* **2020**, *12*, 1817. [CrossRef] [PubMed]

.   Vigneswaran, K.; Neill, S.; Hadjipanayis, C.G. Beyond the World Health Organization grading of infiltrating gliomas: Advances in the molecular genetics of glioma classification. *Ann. Transl. Med.* **2015**, *3*, 95. [PubMed]

.   DeWitt, J.C.; Jordan, J.T.; Frosch, M.P.; Samore, W.R.; Iafrate, A.J.; Louis, D.N.; Lennerz, J.K. Cost-effectiveness of IDH testing in diffuse gliomas according to the 2016 WHO classification of tumors of the central nervous system recommendations. *Neuro-Oncol.* **2017**, *19*, 1640–1650. [CrossRef] [PubMed]

10.   Krauze, A.; Zhuge, Y.; Zhao, R.; Tasci, E.; Camphausen, K. AI-Driven Image Analysis in Central Nervous System Tumors-Traditional Machine Learning, Deep Learning and Hybrid Models. *J. Biotechnol. Biomed.* **2022**, *5*, 1–19.

11.   Diaz Rosario, M.; Kaur, H.; Tasci, E.; Shankavaram, U.; Sproull, M.; Zhuge, Y.; Camphausen, K.; Krauze, A. The Next Frontier in Health Disparities—A Closer Look at Exploring Sex Differences in Glioma Data and Omics Analysis, from Bench to Bedside and Back. *Biomolecules* **2022**, *12*, 1203. [CrossRef]

12.   Guyon, I.; Elisseeff, A. An introduction to variable and feature selection. *J. Mach. Learn. Res.* **2003**, *3*, 1157–1182.

13.   Gokalp, O.; Tasci, E.; Ugur, A. A novel wrapper feature selection algorithm based on iterated greedy metaheuristic for sentiment classification. *Expert Syst. Appl.* **2020**, *146*, 113176. [CrossRef]

14.   Taşci, E.; Gökalp, O.; Uğur, A. Development of a novel feature weighting method using cma-es optimization. In Proceedings of the 2018 26th Signal Processing and Communications Applications Conference (SIU), Izmir, Turkey, 2–5 May 2018.

15.   Taşcı, E.; Uğur, A. Shape and texture based novel features for automated juxtapleural nodule detection in lung CTs. *J. Med. Syst.* **2015**, *39*, 1–13. [CrossRef]

16.   Zanella, L.; Facco, P.; Bezzo, F.; Cimetta, E. Feature Selection and Molecular Classification of Cancer Phenotypes: A Comparative Study. *Int. J. Mol. Sci.* **2022**, *23*, 9087. [CrossRef]

17.   Tasci, E.; Ugur, A. A novel pattern recognition framework based on ensemble of handcrafted features on images. *Multimed. Tools Appl.* **2022**, *81*, 30195–30218. [CrossRef]

18.   Tasci, E.; Zhuge, Y.; Camphausen, K.; Krauze, A.V. Bias and Class Imbalance in Oncologic Data—Towards Inclusive and Transferrable AI in Large Scale Oncology Data Sets. *Cancers* **2022**, *14*, 2897. [CrossRef] [PubMed]

19.   Tasci, E.; Jagasia, S.; Zhuge, Y.; Sproull, M.; Cooley Zgela, T.; Mackey, M.; Camphausen, K.; Krauze, A.V. RadWise: A Rank-Based Hybrid Feature Weighting and Selection Method for Proteomic Categorization of Chemoirradiation in Patients with Glioblastoma. *Cancers* **2023**, *15*, 2672. [CrossRef] [PubMed]

20.   Chandrashekar, G.; Sahin, F. A survey on feature selection methods. *Comput. Electr. Eng.* **2014**, *40*, 16–28. [CrossRef]

21.   Li, J.; Cheng, K.; Wang, S.; Morstatter, F.; Trevino, R.P.; Tang, J.; Liu, H. Feature selection: A data perspective. *ACM Comput. Surv.* **2017**, *50*, 1–45. [CrossRef]

22.   Tang, J.; Alelyani, S.; Liu, H. Feature selection for classification: A review. *Data Classif. Algorithms Appl.* **2014**, *37*, 65–92.

23.   Tahir, M.A.; Bouridane, A.; Kurugollu, F. Simultaneous feature selection and feature weighting using Hybrid Tabu Search/K-nearest neighbor classifier. *Pattern Recognit. Lett.* **2007**, *28*, 438–446. [CrossRef]

24.   Tasci, E.; Zhuge, Y.; Camphausen, K.; Krauze, A.V. *Glioma Grading Clinical and Mutation Features Dataset*; UCI Machine Learning Repository: Irvine, CA, USA, 2022.

25.   Zhao, Z.; Zhang, K.-N.; Wang, Q.; Li, G.; Zeng, F.; Zhang, Y.; Wu, F.; Chai, R.; Wang, Z.; Zhang, C. Chinese Glioma Genome Atlas (CGGA): A comprehensive resource with functional genomic data from Chinese glioma patients. *Genom. Proteom. Bioinform.* **2021**, *19*, 1–12. [CrossRef]

26.   Palantir Foundry—The NIH Integrated Data Analysis Platform (NIDAP); NCI Center for Biomedical Informatics & Information Technology (CBIIT); Software Provided by Palantir Technologies Inc. Available online: https://www.palantir.com (accessed on 5 June 2023).

27.   Yan, Y.; Takayasu, T.; Hines, G.; Dono, A.; Hsu, S.H.; Zhu, J.J.; Riascos-Castaneda, R.F.; Kamali, A.; Bhattacharjee, M.B.; Blanco, A.I.; et al. Landscape of Genomic Alterations in IDH Wild-Type Glioblastoma Identifies PI3K as a Favorable Prognostic Factor. *JCO Precis. Oncol.* **2020**, *4*, 575–584. [CrossRef] [PubMed]

28.   Hu, H.; Mu, Q.; Bao, Z.; Chen, Y.; Liu, Y.; Chen, J.; Wang, K.; Wang, Z.; Nam, Y.; Jiang, B.; et al. Mutational Landscape of Secondary Glioblastoma Guides MET-Targeted Trial in Brain Tumor. *Cell* **2018**, *175*, 1665–1678.e18. [CrossRef]

29.   Fawcett, T. An Introduction to ROC Analysis. *Pattern Recognit. Lett.* **2006**, *27*, 861–874. [CrossRef]

30.   Li, Y.; Mansmann, U.; Du, S.; Hornung, R. Benchmark study of feature selection strategies for multi-omics data. *BMC Bioinform.* **2022**, *23*, 412. [CrossRef] [PubMed]

31. Bhadra, T.; Mallik, S.; Hasan, N.; Zhao, Z. Comparison of five supervised feature selection algorithms leading to top features and gene signatures from multi-omics data in cancer. *BMC Bioinform.* **2022**, *23*, 153. [CrossRef] [PubMed]

32. Lin, W.W.; Ou, G.Y.; Zhao, W.J. Mutational profiling of low-grade gliomas identifies prognosis and immunotherapy-related biomarkers and tumour immune microenvironment characteristics. *J. Cell. Mol. Med.* **2021**, *25*, 10111–10125. [CrossRef]

33. Brennan, C.W.; Verhaak, R.G.; McKenna, A.; Campos, B.; Noushmehr, H.; Salama, S.R.; Zheng, S.; Chakravarty, D.; Sanborn, J.Z.; Berman, S.H.; et al. The somatic genomic landscape of glioblastoma. *Cell* **2013**, *155*, 462–477. [CrossRef] [PubMed]

34. Zhang, P.; Meng, X.; Liu, L.; Li, S.; Li, Y.; Ali, S.; Li, S.; Xiong, J.; Liu, X.; Li, S.; et al. Identification of the Prognostic Signatures of Glioma With Different PTEN Status. *Front. Oncol.* **2021**, *11*, 633357. [CrossRef]

35. Georgescu, M.M.; Olar, A. Genetic and histologic spatiotemporal evolution of recurrent, multifocal, multicentric and metastatic glioblastoma. *Acta Neuropathol. Commun.* **2020**, *8*, 10. [CrossRef]

36. Lobbous, M.; Bernstock, J.D.; Coffee, E.; Friedman, G.K.; Metrock, L.K.; Chagoya, G.; Elsayed, G.; Nakano, I.; Hackney, J.R.; Korf, B.R.; et al. An Update on Neurofibromatosis Type 1-Associated Gliomas. *Cancers* **2020**, *12*, 114. [CrossRef]

37. Sakthikumar, S.; Roy, A.; Haseeb, L.; Pettersson, M.E.; Sundström, E.; Marinescu, V.D.; Lindblad-Toh, K.; Forsberg-Nilsson, K. Whole-genome sequencing of glioblastoma reveals enrichment of non-coding constraint mutations in known and novel genes. *Genome Biol.* **2020**, *21*, 127. [CrossRef] [PubMed]

38. Ferrer, V.P. MUC16 mutation is associated with tumor grade, clinical features, and prognosis in glioma patients. *Cancer Genet.* **2023**, *270–271*, 22–30. [CrossRef]

39. Hu, W.; Duan, H.; Zhong, S.; Zeng, J.; Mou, Y. High frequency of PDGFRA and MUC family gene mutations in diffuse hemispheric glioma, H3 G34-mutant: A glimmer of hope? *J. Transl. Med.* **2022**, *20*, 64. [CrossRef] [PubMed]

40. Bunda, S.; Heir, P.; Metcalf, J.; Li, A.S.C.; Agnihotri, S.; Pusch, S.; Yasin, M.; Li, M.; Burrell, K.; Mansouri, S.; et al. CIC protein instability contributes to tumorigenesis in glioblastoma. *Nat. Commun.* **2019**, *10*, 661. [CrossRef] [PubMed]

41. Krämer, A.; Green, J.; Pollard, J., Jr.; Tugendreich, S. Causal analysis approaches in Ingenuity Pathway Analysis. *Bioinformatics* **2014**, *30*, 523–530. [CrossRef]

42. Zadeh Shirazi, A.; McDonnell, M.D.; Fornaciari, E.; Bagherian, N.S.; Scheer, K.G.; Samuel, M.S.; Yaghoobi, M.; Ormsby, R.J.; Poonnoose, S.; Tumes, D.J.; et al. A deep convolutional neural network for segmentation of whole-slide pathology images identifies novel tumour cell-perivascular niche interactions that are associated with poor survival in glioblastoma. *Br. J. Cancer* **2021**, *125*, 337–350. [CrossRef]

43. Krigers, A.; Demetz, M.; Thomé, C.; Freyschlag, C.F. Age is associated with unfavorable neuropathological and radiological features and poor outcome in patients with WHO grade 2 and 3 gliomas. *Sci. Rep.* **2021**, *11*, 17380. [CrossRef]

44. Han, S.; Liu, Y.; Cai, S.J.; Qian, M.; Ding, J.; Larion, M.; Gilbert, M.R.; Yang, C. IDH mutation in glioma: Molecular mechanisms and potential therapeutic targets. *Br. J. Cancer* **2020**, *122*, 1580–1589. [CrossRef]

45. Yan, H.; Parsons, D.W.; Jin, G.; McLendon, R.; Rasheed, B.A.; Yuan, W.; Kos, I.; Batinic-Haberle, I.; Jones, S.; Riggins, G.J.; et al. IDH1 and IDH2 mutations in gliomas. *N. Engl. J. Med.* **2009**, *360*, 765–773. [CrossRef]

46. Oprita, A.; Baloi, S.C.; Staicu, G.A.; Alexandru, O.; Tache, D.E.; Danoiu, S.; Micu, E.S.; Sevastre, A.S. Updated Insights on EGFR Signaling Pathways in Glioma. *Int. J. Mol. Sci.* **2021**, *22*, 587. [CrossRef]

47. Quayle, S.N.; Lee, J.Y.; Cheung, L.W.; Ding, L.; Wiedemeyer, R.; Dewan, R.W.; Huang-Hobbs, E.; Zhuang, L.; Wilson, R.K.; Ligon, K.L.; et al. Somatic mutations of PIK3R1 promote gliomagenesis. *PLoS ONE* **2012**, *7*, e49466. [CrossRef]

48. Nandakumar, P.; Mansouri, A.; Das, S. The Role of ATRX in Glioma Biology. *Front. Oncol.* **2017**, *7*, 236. [CrossRef] [PubMed]

49. Parmigiani, E.; Taylor, V.; Giachino, C. Oncogenic and Tumor-Suppressive Functions of NOTCH Signaling in Glioma. *Cells* **2020**, *9*, 2304. [CrossRef] [PubMed]

50. Felder, M.; Kapur, A.; Gonzalez-Bosquet, J.; Horibata, S.; Heintz, J.; Albrecht, R.; Fass, L.; Kaur, J.; Hu, K.; Shojaei, H.; et al. MUC16 (CA125): Tumor biomarker to cancer therapy, a work in progress. *Mol. Cancer* **2014**, *13*, 129. [CrossRef] [PubMed]

rticle

Particle Swarm Optimisation Applied to the Direct Aperture Optimisation Problem in Radiation Therapy

Gonzalo Tello-Valenzuela, Mauricio Moyano and Guillermo Cabrera-Guerrero *

Escuela de Ingeniería Informática, Pontificia Universidad Católica de Valparaíso, Av. Brasil 2241, Valparaíso 2362807, Chile; gonzalo.tello.v@mail.pucv.cl (G.T.-V.); mauricio.moyano@pucv.cl (M.M.)
* Correspondence: guillermo.cabrera@pucv.cl

Simple Summary: Intensity Modulated Radiation Therapy (IMRT) is a cancer treatment that targets cancer cells while protecting nearby healthy organs using a linear accelerator. Traditional IMRT planning involves a sequential process: optimizing beam intensities (Fluence Map Optimization) for a set of angles and then sequencing (Multi-Leaf Sequencing). Unfortunately, treatment plans obtained by the sequencing step are severely impaired. One approach that addresses the problem described is the Direct Aperture Optimisation (DAO) approach. The DAO problem aims at simultaneously determining deliverable aperture shapes and a set of radiation intensities. This approach considers physical and delivery time constraints, allowing clinically acceptable treatment plans to be generated. In this work, we adapt the Particle Swarm Optimisation to solve the DAO and introduce a reparation heuristic to enhance treatment plans. We tested our method on prostate cancer patients and found that it delivers radiation more efficiently than the traditional approach, reducing treatment time and improving outcomes.

Abstract: Intensity modulated radiation therapy (IMRT) is one of the most used techniques for cancer treatment. Using a linear accelerator, it delivers radiation directly at the cancerogenic cells in the tumour, reducing the impact of the radiation on the organs surrounding the tumour. The complexity of the IMRT problem forces researchers to subdivide it into three sub-problems that are addressed sequentially. Using this sequential approach, we first need to find a beam angle configuration that will be the set of irradiation points (beam angles) over which the tumour radiation is delivered. This first problem is called the Beam Angle Optimisation (BAO) problem. Then, we must optimise the radiation intensity delivered from each angle to the tumour. This second problem is called the Fluence Map Optimisation (FMO) problem. Finally, we need to generate a set of apertures for each beam angle, making the intensities computed in the previous step deliverable. This third problem is called the Sequencing problem. Solving these three sub-problems sequentially allows clinicians to obtain a treatment plan that can be delivered from a physical point of view. However, the obtained treatment plans generally have too many apertures, resulting in long delivery times. One strategy to avoid this problem is the Direct Aperture Optimisation (DAO) problem. In the DAO problem, the idea is to merge the FMO and the Sequencing problem. Hence, optimising the radiation's intensities considers the physical constraints of the delivery process. The DAO problem is usually modelled as a Mixed-Integer optimisation problem and aims to determine the aperture shapes and their corresponding radiation intensities, considering the physical constraints imposed by the Multi-Leaf Collimator device. In solving the DAO problem, generating clinically acceptable treatments without additional sequencing steps to deliver to the patients is possible. In this work, we propose to solve the DAO problem using the well-known Particle Swarm Optimisation (PSO) algorithm. Our approach integrates the use of mathematical programming to optimise the intensities and utilizes PSO to optimise the aperture shapes. Additionally, we introduce a reparation heuristic to enhance aperture shapes with minimal impact on the treatment plan. We apply our proposed algorithm to prostate cancer cases and compare our results with those obtained in the sequential approach. Results show that the PSO obtains competitive results compared to the sequential approach, receiving less radiation time (beam on time) and using the available apertures with major efficiency.

Citation: Tello-Valenzuela, G.; Moyano, M.; Cabrera-Guerrero, G. Particle Swarm Optimisation Applied to the Direct Aperture Optimisation Problem in Radiation Therapy. *Cancers* **2023**, *15*, 4868. https://doi.org/10.3390/cancers15194868

Academic Editor: Franz Rödel

Received: 15 September 2023
Revised: 29 September 2023
Accepted: 2 October 2023
Published: 6 October 2023

Keywords: direct aperture optimisation; intensity-modulated radiation therapy; particle swarm optimisation

1. Introduction

Cancer is a type of disease that causes abnormal growth of cells in the body, leading to the formation of carcinomas, which can eventually turn into malignant tumours. In 2020 the International Agency for Research on Cancer reported 19.3 million new cancer cases and nearly 10 million cancer-related deaths [1]. There are various methods for treating cancer, and the treatment choice largely depends on the specific type of cancer and its impact on the patient's health.

Radiotherapy is a commonly used cancer treatment technique involving exposing patients to ionising radiation to target cancerous cells. There are various forms of radio therapy, such as Volumetric Modulated Arc Therapy (VMAT), Stereotactic Body Radiation Therapy (SBRT), and Intensity Modulated Radiation Therapy (IMRT), among others. IMRT is one of the most widely used methods of radiation therapy, and is delivered using a linear accelerator (linac) machine [2] (Figure 1). IMRT aims to effectively deliver the prescribed radiation dose to the cancerous cells while minimising the exposure of healthy structures [3]. This is achieved by modulating the radiation passing through the linac using a multi-leaf collimator (MLC) device.

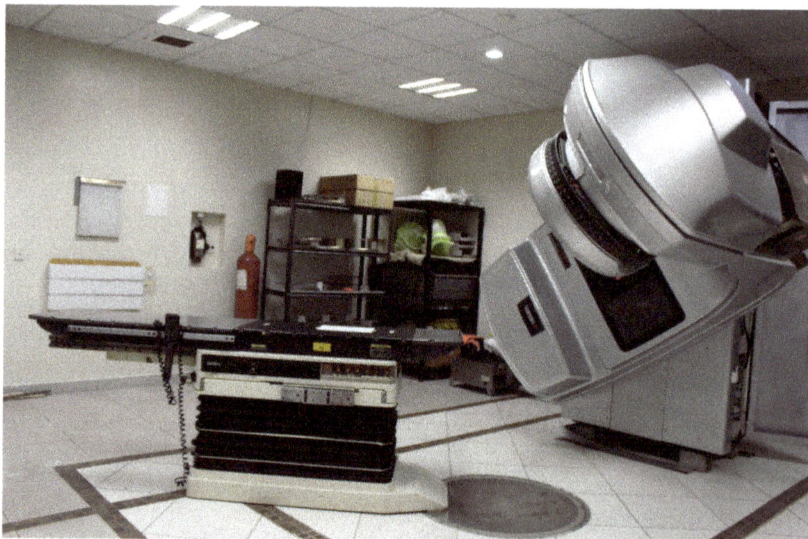

Figure 1. Linear accelerator from the Centro Oncologico Hondureño in Honduras.

The IMRT technique enables the delivery of an optimal radiation dose to the tumour while minimising exposure to surrounding healthy organs [4]. However, finding a treatment plan that balances the desired dose to the tumour and minimal side effects on surrounding organs is highly complex. To address this, the IMRT planning process is typically split into three sequential sub-problems: beam angle optimisation (BAO), fluence map optimisation (FMO), and multi-leaf collimator sequencing [5]. First, the BAO problem aims to identify the best possible combination of beam angles from which the radiation should be delivered, also known as the beam angle configuration (BAC). Once a BAC has been selected, the optimal intensities for that BAC must be found (Fluence Map Optimisation problem, FMO). Finally, in the MLC sequencing problem, we compute a set of deliverable aperture shapes and their corresponding intensities.

This sequential approach ends with a treatment plan consisting of a large set of aperture shapes (with corresponding intensity values). Unfortunately, having too many apertures and larger intensity values per aperture means longer treatment time. The total delivery time of a treatment plan is calculated considering both the *beam-on* time and the decomposition times. The beam-on time (BoT) is the total time a patient is exposed to radiation. The decomposition time is the time the linear accelerator needs to move from one bean angle in a BAC to the next one and the time needed by the MLC to move from one aperture shape to the other [6,7].

As a general rule, prolonged treatment time is something we want to avoid, as it increases the attention time per patient and, thus, reduces the number of patients treated per day [8]. Further, longer treatment plans are more likely to suffer from inaccuracies produced, for instance, by patient's movements.

One strategy commonly used to minimise the total delivery time of treatment plans generated using the sequential approach described before is to reduce the number of apertures. This can be made by "rounding" the intensity values computed during the FMO phase. Unfortunately, such strategies can severely impair the final treatment plan quality.

One alternative to the sequential approach that does not require any "rounding" process is the direct aperture optimisation problem (DAO). The main idea in DAO is to solve the FMO problem considering a limited number of deliverable aperture shapes and the physical constraints associated with the MLC sequencing.

To solve the DAO problem, we must find a set of aperture shapes and their associated intensity values [9]. Usually, aperture shapes are optimised using heuristic strategies [10,11] or looking for the best possible combination of aperture shapes from a pre-defined set of apertures [12]. To optimise intensity values, gradient-based optimisation methods are usually implemented. Compared to the sequential approach, the treatment plan obtained using DAO is not only deliverable, but also better regarding the objective function value [13].

In this paper, we implement a particle swarm optimisation algorithm (PSO) combined with a mathematical programming technique to solve the DAO problem. PSO is recognised for effectively solving large-scale nonlinear optimisation problems through a good balance between exploitation (local search) and exploration (global search) [14,15]. While the PSO algorithm finds the best aperture shapes at each beam angle for a given BAC, the mathematical programming algorithm optimises each aperture's intensity value. Also, we present a reparation heuristic for those aperture shapes that have a negligible effect on the treatment plan. To analyse our algorithm results, we use a set of clinical cases of prostate cancer and compare the treatment plans obtained by our algorithm to those obtained by the traditional sequential approach. The results show that our algorithm can find deliverable treatment plans using fewer apertures and significantly reduce the beam-on time compared to the traditional sequential approach. Compared to deliverable treatment plans with a similar number of apertures, our algorithm outperforms them regarding objective function values.

The remainder of this paper is organised as follows: Section 2 introduces the general concepts of IMRT and DAO and the mathematical models we will consider in this study. In Section 3, the algorithms we implement in this paper are presented. Section 4 presents the results obtained by our algorithm applied to a prostate case. A discussion of these results is also included in this section. Finally, in Section 5, we draw the main conclusions of our work and outline future work.

2. IMRT and the DAO Problem

In this section, we first discuss the main features of the IMRT problem and how to model it. Then, we introduce the DAO problem and present a brief literature review, focusing on the algorithms that have been previously proposed to solve the DAO problem.

2.1. Intensity Modulated Radiation Therapy

To mathematically model the IMRT problem, we first need to discretise each beam angle into *beamlets*, and each region (tissues and tumour) into a set of small sub-volumes called *voxels* [16]. See Figure 2 for a graphical representation of these concepts.

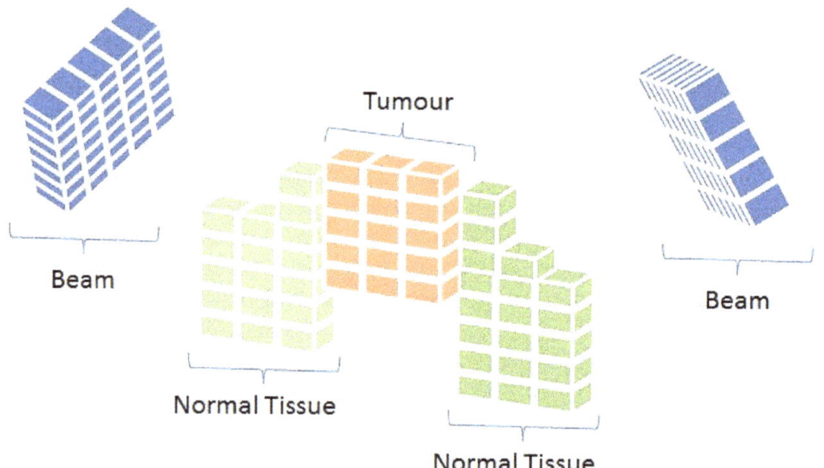

Figure 2. Representation of beam angles and organs discretised into beamlets and voxels, respectively (Cabrera-Guerrero et al. [17]).

Thus, the IMRT problem can be modelled using the representation depicted in Figure 2 [12,17–21]. First, we model the dose distribution deposited in the voxels that compose a region. As mentioned above, beam angles are divided into a set of n beamlets being n, the total number of beamlets summed over all the possible beam angles. Let $\mathscr{A}$ be a BAC and $x \in \mathbb{R}^n_{\geq 0}$ be an intensity vector or fluence map solution for $\mathscr{A}$. Each vector component x_b represents the length of time the patient is exposed to the radiation of the b-th beamlet. The radiation dose deposited into each voxel v of region r by fluence map x is computed by the expression [16,20]

$$d_v^r(x) = \sum_{b=1}^{n} \left(D_{v1i}^r x_b \right) \quad \forall v = 1, 2, \ldots, m^r, \tag{1}$$

where m^r is the total number of voxels in the region r, $r \in R = \{O_1, \ldots, O_Q, T\}$ is an element of the index set of regions, with the tumour indexed by $r = T$ and the organs at risk and normal tissue indexed by $r = O_q$ with $q = 1, \ldots, Q$. $D^r \in \mathbb{R}^{m^r \times n}$ is the dose deposition matrix related to region r, where $D_{vb}^r \geq 0$ defines the rate at which the radiation dose along beamlet b is deposited into voxel v of region r (As shown Figure 3). The set $\mathcal{X}(\mathscr{A}) \subseteq \mathbb{R}^n$ is the set of all feasible fluence maps when the BAC $\mathscr{A}$ is considered. Note that searching for an optimal fluence map x over the $\mathcal{X}(\mathscr{A})$ space implies solving the FMO problem.

Based on the dose distribution in Equation (1), physical and biological models have been proposed in the literature (see Ehrgott et al. [16] for a survey). This study uses the convex nonlinear penalty function in [22,23]. In this model, each voxel is penalised according to the squared difference between the actual and the prescribed doses. This formulation yields a quadratic programming problem with only linear non-negativity constraints on the fluence values [22]. This model is as follows:

$$\min_x z(x) = \sum_{r \in R} \left[\frac{1}{m^r} \sum_{i=1}^{m^r} \left[\underline{\lambda}_r (Y_r - d_v^r(x))_+^2 + \overline{\lambda}_r (d_v^r(x) - Y_r)_+^2 \right] \right] \tag{2}$$

where parameter m^r is, again, the number of voxels of the region r and Y_r is the desired dose for the voxels of the region r. The function $(\cdot)_+$ is the maximum between 0 and $(\cdot)$. $d_v^r(x)$ gives the dose delivered by fluence map x to voxel v of the region r (see Equation (1)). and $\underline{\lambda}_r$ and $\overline{\lambda}_r$ are the penalty weights parameter of under-dose and overdose related to region r, respectively. Since the Equation (2) is convex, the optimal fluence maps can be obtained using mathematical programming techniques.

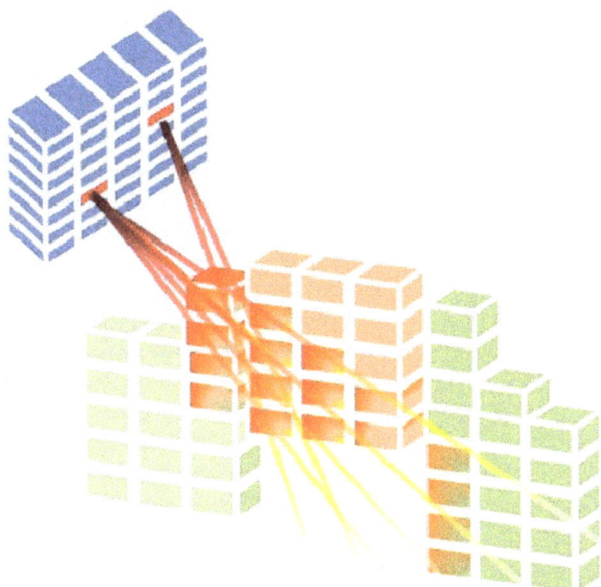

Figure 3. Radiation is delivered from a subset of beamlets, and it irradiates voxels at both tumour and organs at risk (Cabrera-Guerrero et al. [17]).

2.2. Direct Aperture Optimisation

The Direct Aperture Optimisation [10] merges the FMO and MLC problems, optimising the fluence map considering the constraints imposed by the MLC device. This means that the decision variables we focus on are not the beamlet intensities (as we did in the FMO problem), but the beamlet apertures and their corresponding aperture intensities. One consequence of this change is that the model becomes a mixed integer nonlinear problem as the beamlet apertures are binary variables (open/closed). Having binary variables makes the problem too hard to be solved by mathematical programming techniques, as we used to do with the FMO problem.

Let us consider a BAC $\mathscr{A} = \{A_1, \ldots, A_U\}$, where $U \in \mathbb{N}_{>0}$ represents the number of beams that are part of the BAC $\mathscr{A}$. Consider that we represent a DAO solution as the set $\mathbb{H} = \{(P^1, I^1), \ldots, (P^N, I^N)\}$, where the (P^c, I^c) tuples correspond to a set of Θ^c aperture and intensity values for some beam angle c. We define each aperture shape $S_i^c \in P^c$ as a matrix of binary variables. Figure 4 gives an example of a tuple (P^c, I^c) for a beam angle c.

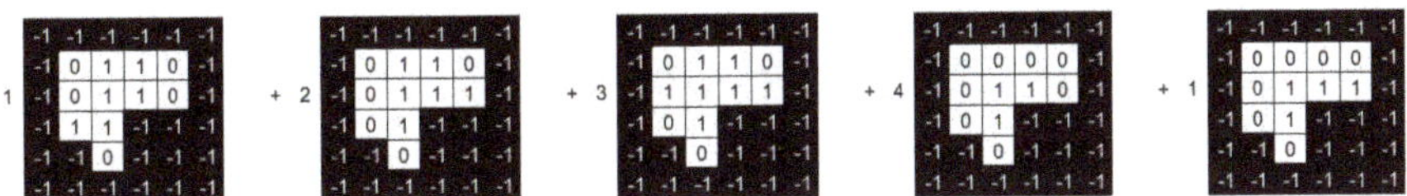

Figure 4. Set of aperture shapes and intensity values associated with a beam angle.

As we can see, the value of an element in the matrix is 1 if the radiation passes through the associated beamlet and 0 otherwise. The elements with value -1 are not considered, as the associated beamlets do not hit any voxel from the tumour. Also note that because of MLC physical constraints, the matrix S_i^c is a consecutive 1's matrix (C1), that is, for each row, 1 values must be consecutive, with no 0 value in between them.

To evaluate $z(x)$, it is necessary to obtain the fluence map x, used in Equation (1), from the DAO solution. To this end, we first need to compute an aggregated matrix for each tuple in H. This aggregated matrix can be obtained through a positive linear combination of the aperture shapes S_i^c and their corresponding intensities I_i^c for angle $\mathcal{A}$:

$$A_c = \sum_{i=1}^{\Theta^c} S_i^c \cdot I_i^c \tag{3}$$

Then, we need to convert the aggregated matrix A_c obtained in Equation (3) to a fluence map x vector. We perform this by mapping the position of each beamlet in the aggregated matrix of beam angle $\mathcal{A}$ to its corresponding position b in the fluence map solution x of beam angle $\mathcal{A}$. Figure 5 shows how to do this.

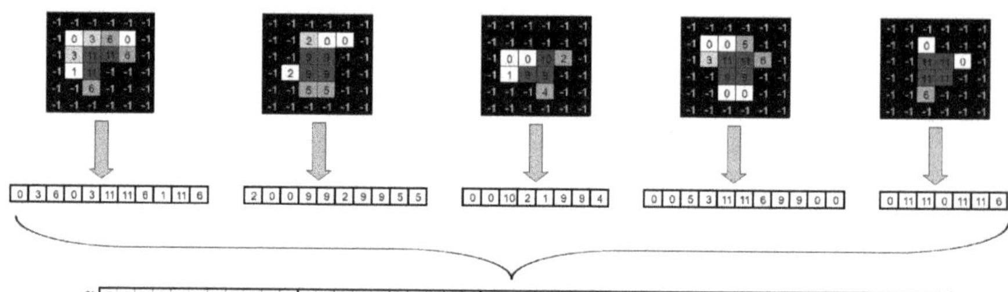

Figure 5. Generation of a fluence map from an angle's apertures and associated intensities.

Direct Aperture Optimisation Related Work

The DAO problem was first introduced by Shepard et al. [10]. In their paper, the authors identify as input of the problem the beam angles, the beam energies, and the number of apertures per beam angle. At the same time, the decision variables are the aperture shapes and their intensities. Currently, several different techniques have been used to solve the DAO problem. Some of these techniques are classified as stochastic search methods. These methods apply small changes in the leaf position of the apertures. When a change in the leaf position improves the objective function, it is accepted. It is important to remark that the changes in this method are stochastic [3,10–12,21,24–27].

Other methods for solving the DAO problem are based on gradient leaf refinement. In these methods, the leaf position is used as the optimisation variable. The relationship between the objective function and the leaf position is established, and the first derivative is given. Such algorithms have been applied to various commercial therapeutic systems, including the direct machine parameter optimisation model used in Pinnacle and RayStation systems [28,29]. Column generation methods have also been proposed in the literature [9,30–33]. In these methods, the initial apertures are not set at the beginning of an iteration; instead, deliverable apertures are individually added to the treatment plan. The iteration process involves two steps. First, the price problem is solved to generate the deliverable aperture that can improve the objective function, which is added to the treatment plan. Then, the new set of aperture weights is optimised in the master problem.

Unfortunately, the methods above also suffer from some issues. For instance, column generation approaches usually converge very fast; however, they do not allow for a hard limit on the number of apertures, which may translate to unreasonably long total treatment times and negligibly small apertures [34]. A relevant issue in stochastic search and gradient based leaf refinement techniques is generating the initial solution. The quality of the initial solution influences the quality of the given final solution, as seen in [12,24].

All in all, solving the DAO problem using a limited number of apertures and obtaining good objective quality function values is an open problem that is worth to be studied.

3. Solution Method

This section introduces our hybrid PSO algorithm to solve the DAO problem. The main goal of our algorithm is to obtain a high-quality treatment plan for IMRT that consists of a set of deliverable set of aperture shapes and their corresponding intensity values.

In Section 3.1, we explain the original PSO algorithm proposed in [35] and how we adapt it to the DAO problem. Then, in Section 3.2, we define a reparation heuristic that uses a mathematical programming algorithm to improve the solution found by our PSO algorithm.

3.1. Particle Swarm Optimisation

The PSO is a nature-inspired population-based metaheuristic algorithm that imitates the social behaviour of birds in nature. This swarm consists of particles that search the objective space intending to find different high-quality solutions. Each particle is, in turn, composed of two fitness-related elements. The first element is the current fitness value of the i-th particle, and the second element is the fitness value of the best position the i-th particle has ever found during the algorithm execution, $pbest_i$. Finally, the algorithm also keeps track of the best fitness value found so far, $gbest$.

The PSO starts with an initial population of particles whose positions have been randomly assigned. The i-th particle's position at iteration t is represented by x_i^t. The direction of particles in each iteration is determined by a velocity variable denoted by v_i^t that obtains its value from Equation (4).

$$v_i^{t+1} = cf * (wv_i^t + c_1 r_1 (pbest_i - x_i^t) + c_2 r_2 (gbest - x_i^t)), \tag{4}$$

where t is the current iteration, $pbest_i$ is the best position the i-th particle has achieved, and $gbest$ is the best position any particle in the swarm has achieved. Parameter cf is the constriction factor used to adjust the velocity of each particle and obtain a balance between exploration and exploitation. The parameter w is the algorithm's inertia and controls the last velocity contribution. Parameter c_1 and c_2 are learning factors for managing the impact of $pbest_i$ and $gbest$. Parameters r_1 and r_2 are random numbers between 0 and 1. The new position of each particle is updated by adding the current velocity to the function of the position of the particle, as shown in Equation (5).

$$x_i^{t+1} = x_i^t + v_i^{t+1} \tag{5}$$

In the proposed algorithm, we represented the particles as shown in Figure 6. As we can see, the particle is composed of three attributes, namely the current fitness value (a real-valued attribute), its best singular position (a treatment plan), and its current position (a treatment plan). Naturally, the current fitness value results from evaluating the current particle's position in the objective function considered by the algorithm.

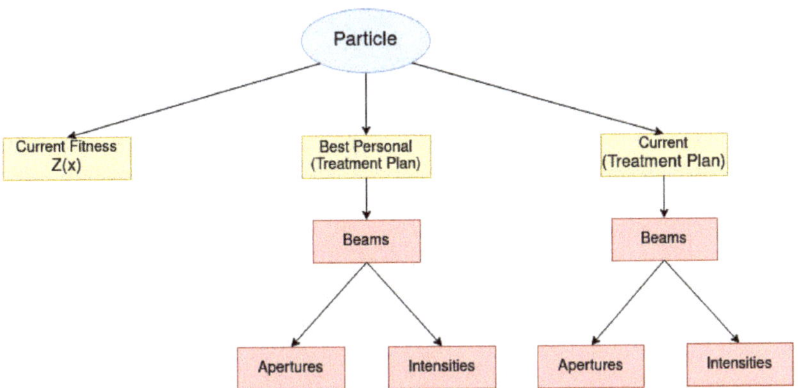

Figure 6. DAO solution on particle representation.

As in any other heuristic algorithm, solutions generated by the PSO algorithms are not (necessarily) optimal. One drawback of the PSO implemented here is that there is no relation between the intensity values associated with an aperture and the aperture itself. Unfortunately, as mentioned before, the aperture shape optimisation problem is an NP-hard problem that mathematical programming solvers cannot solve in a reasonable time. Unlike this, the apertures' intensity optimisation problem (also known as aperture weight optimisation [9] or segmentation weight optimisation) is a convex continuous problem that can quickly be solved for solvers such as Gurobi (see, for instance, [24–26]). Then, we propose to implement a hybrid PSO with a mathematical programming algorithm to solve the DAO problem. We use the PSO algorithm to find a set of aperture shapes and their corresponding intensities, which the linear solver will then optimise.

To better understand the algorithm's behaviour, we can see in Figure 7 how an aperture shape and the associated intensities change in each step. Considering the representation of the treatment plan mentioned in Section 2.2, we can represent the aperture shape and the intensities obtained by the PSO algorithm like a tuple (P^c, I^c). The intensities I^c are optimised by the solver at the end of each iteration of the PSO algorithm. As a result, we obtain a new tuple (P^c, I'^c) where, as mentioned before, some intensities in I'^c are set to zero by the solver. To improve the aperture shapes that resulted in (near) zero intensity value after the solver optimisation, we use a reparation heuristic. This heuristic only modifies P^c leading to a new tuple (P'^c, I'^c). Finally, the reparation heuristic passes on the Solver the tuple (P'^c, I'^c) so we can obtain the optimal intensity values for the new set of apertures P'^c, generating the tuple (P'^c, I''^c). Finally, the treatment plan defined by the tuple (P'^c, I''^c) is passed onto the PSO algorithm for the next iteration. This process is repeated until the PSO algorithm meets some termination criterion (e.g., it reaches a predetermined number of iterations).

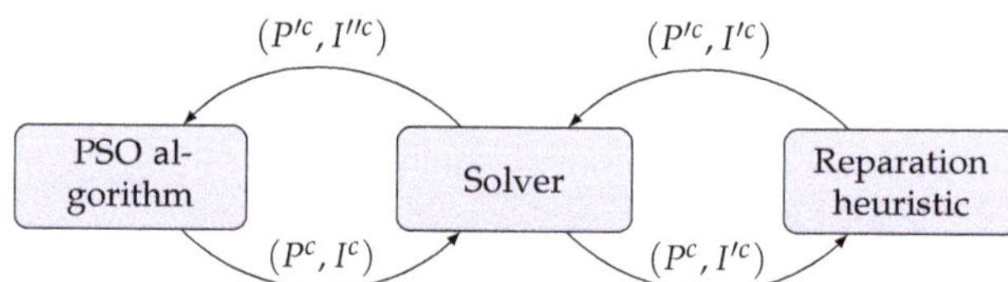

Figure 7. Interaction between PSO algorithm, linear solver and reparation heuristic.

3.2. Reparation Heuristic

As mentioned in the previous paragraph, as a result of the solver usage, we obtain the optimal intensities for each aperture at each beam angle. Since the optimisation solver is conditioned to the aperture shapes obtained at each iteration by the PSO algorithm, it is not unusual that some of the intensities end up in the optimisation process with values close to zero.

In practice, apertures with associated intensities near to zero value are equivalent to having no aperture at all, i.e., an insignificant (or null) impact on the treatment plan. Further, improving the shapes of those apertures with intensity values close to zero is complex. To address this issue, we propose a reparation heuristic that allows us to avoid (as much as possible) those apertures with a negligible effect on the treatment plan.

Figure 8 shows a numerical example of the intensities optimisation process. On top of the image, we can see four aperture shapes with their associated intensities. We can see that all the intensities are modified on the bottom part of the same image.

The main idea of the reparation heuristic proposed here is to replace those apertures with intensity values closer to zero with apertures that (hopefully) can help after running the solver. Particularly, we aim to irradiate those parts of the aperture shape that are not irradiated from any other aperture of the beam angle.

To this end, we generate a new aperture that results from overlapping the apertures with an intensity value greater than 1. We call this new aperture the "overlapped aperture", and the ones with intensity values greater than one "the original apertures". Figure 9 shows an example of the apertures overlapping process. Fields with a value of 1 correspond to beamlets radiation passes through. Fields with zero value correspond to the beamlets closed in the original apertures. Finally, −1 corresponds to the inactive beamlets (those that do not hit the tumour).

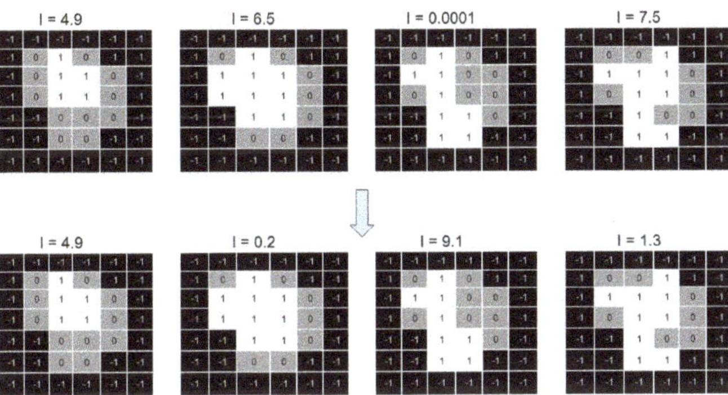

Figure 8. Representation of the change in the intensities of a set of apertures using the solver.

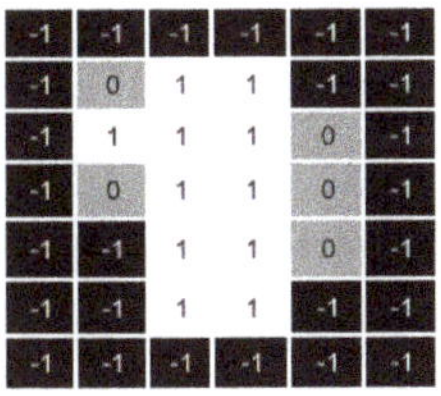

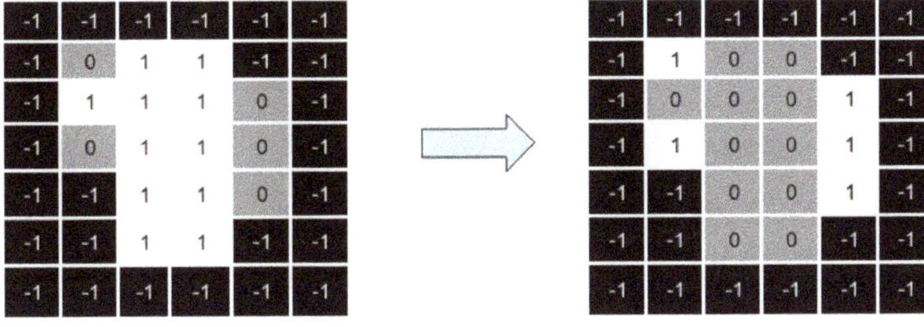

Figure 9. Overlapping matrix from the apertures with intensities over one.

As shown in Figure 9, the overlapped aperture corresponds to the original apertures' aggregation, i.e., the overlapped aperture keeps open beamlets that are open in at least one original aperture and sets closed those beamlets that are closed in all the original apertures.

As a result of this aggregation process, we have a matrix showing all the beamlets currently open in at least one original aperture. As mentioned above, we want to diversify our search, and thus, we want to irradiate from those fields that are not currently in use.

To this end, the reparation heuristic generates the complementary matrix of the overlapped matrix, as shown in Figure 10.

Figure 10. Complementary matrix generated from the overlapping matrix.

It is important to keep in mind some considerations about the application of our reparation heuristic. First, in some cases, the shape of the complementary matrix does not satisfy the MLC physical constraints and can not directly replace the original aperture. In that case, we can select part of the aperture that is actually deliverable and remove those parts that do not satisfy MLC physical constraints. As shown in Figure 11, we divide the complementary matrix into two different apertures that satisfy the MLC physical constraints.

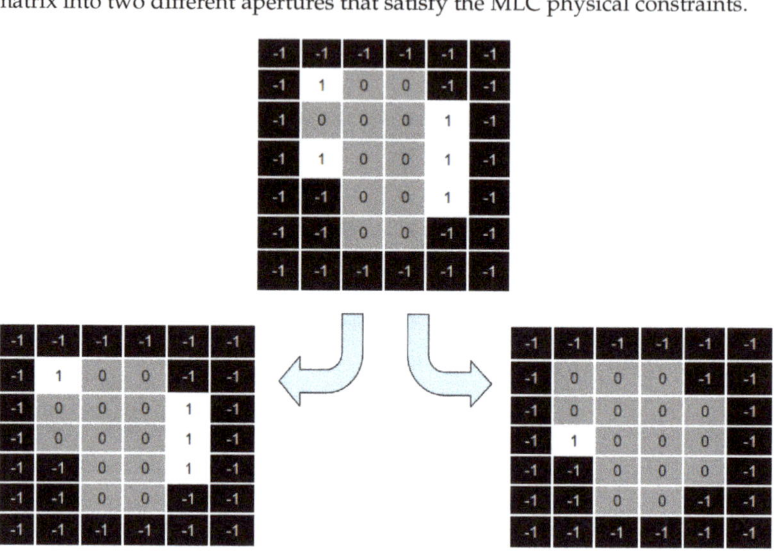

Figure 11. Dividing the complementary matrix so we can obtain deliverable aperture shapes.

Second, suppose the number of original apertures with an intensity value close to zero is more than one. In that case, we must divide the complimentary matrix to generate as many new apertures as needed. Note that this situation can help us to solve our first consideration (undeliverable aperture shapes), as we can divide the complementary matrix in such a way that all (or most of) the open beamlets in the complementary matrix can be added to the new apertures (see, for instance, Figure 11).

Finally, the reparation heuristic replaces those apertures with (near) zero intensity values by the aperture shapes obtained in the previous step. We need to note that, in some cases, one or more apertures still with (near) zero intensity values as the number of deliverable aperture shapes produced by the reparation heuristic is less than the number of apertures with (near) zero intensity values. Once we obtained the repaired aperture shapes, we optimised the intensities values, as shown in Figure 12.

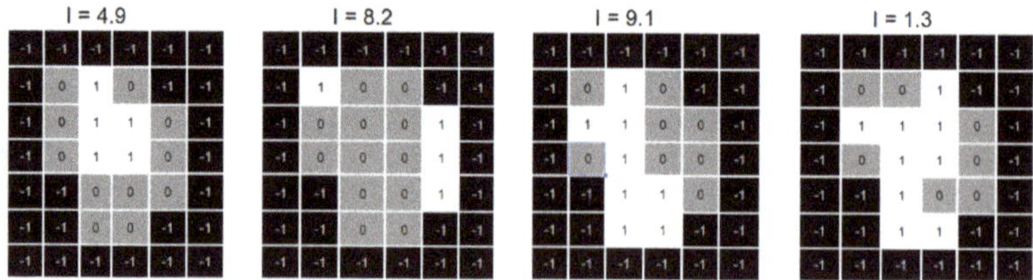

Figure 12. Representation of the apertures obtained after the reparation process.

4. Computational Experiments

This Section introduces the experiments performed by our algorithm and analyses the obtained results. The Section is divided into three subsections. In Section 4.1, we introduce the set of instances considered in our study and the parameters used by the PSO. In Section 4.2, we obtain the best parameters for the PSO algorithm using the framework Irace [36]. Finally, in Section 4.3, we compare our PSO to two algorithms used in the literature. Comparison is made regarding the obtained objective function values, the required number of aperture shapes, and their beam-on time.

4.1. Experimental Setup

In this work, we perform a set of initial experiments on the prostate case instance from *CERR package* [37] and also examine a prostate case acquired from Clinica Alemana de Santiago, Chile. This particular patient is denoted as TRT001 [19]. We use this prostate case to evaluate the performance of the PSO algorithm introduced in Section 3.1. For the CERR and TRT001 cases, we consider three organs: the prostate, where the tumour is located, the bladder, and the rectum (see Figure 13). We label the rectum and the bladder as organs at risk (OARs) and the prostate as planning target volume (PTV).

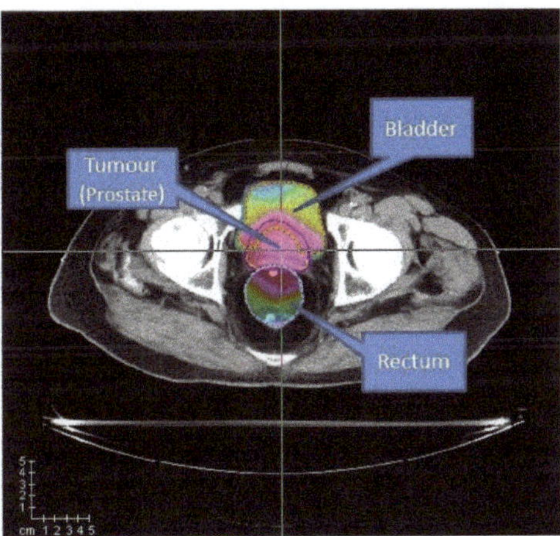

Figure 13. Prostate case from CERR. Two OARs (bladder and rectum) are considered.

The number of voxels per region in the CERR case is 15,172 for the prostate, 22,936 for the bladder and 18,128 for the rectum. We consider 72 beam angles, all of which are on the same plane. Similarly, in the TRT001 case, the prostate comprises 13,081 voxels, the bladder holds 19,762 voxels, and the rectum encompasses 8500 voxels.

Like other works in the problem we consider a set of 14 equidistant BACs [12,17,18,21,24,25]. Each BAC consists of five beam angles for the CERR and TRT01 instances, as shown in Table 1.

Table 1. Equidistant BACs and their corresponding number of beamlets for the CERR and TRT01 cases.

BAC	Beam Angles					# Beamleats CERR	# Beamleats TRT
	θ_1	θ_2	θ_3	θ_4	θ_5		
1	0	70	140	210	280	336	327
2	5	75	145	215	285	336	329
3	10	80	150	220	290	333	328
4	15	85	155	225	295	333	330
5	20	90	160	230	300	329	334
6	25	95	165	235	305	328	334
7	30	100	170	240	310	333	333
8	35	105	175	245	315	336	330
9	40	110	180	250	320	337	329
10	45	115	185	255	325	335	329
11	50	120	190	260	330	331	335
12	55	125	195	265	335	329	335
13	60	130	200	270	340	329	332
14	65	135	205	275	345	328	328

Table 2 details the prescribed doses, Y_r, considered per each organ at all the instances and the weights for both under-dose $\underline{\lambda}_r$ and overdose $\overline{\lambda}_r$.

Table 2. Value of T_i, $\underline{\lambda}_i$ and $\overline{\lambda}_i$ for function $z(x)$.

Organ	Y_r	$\underline{\lambda}_r$	$\overline{\lambda}_r$
PTV	76 Gy	5	5
Rectum	65 Gy	0	1
Bladder	65 Gy	0	1

4.2. Irace Parameter

To optimise the parameter used in the PSO algorithm implemented, we tried the package Irace [36]. This package is an extension of the iterative F-race algorithm (I/F race) [38,39]. The principal use of this method is for the automatic configuration of optimisation algorithms. This is performed by finding the most appropriate configuration of parameters from a set of instances executed in the algorithm. This package has also been used for the parameters optimisation of the algorithm proposed by Caceres et al. [21]. That said, using IRace aims to find suitable parameters for our PSO implementation. The parameters to optimise within the IRace package are shown in Table 3.

Table 3. Parameters of PSO used in Irace.

Parameter	Description	Range
$Npop$	Number of population	$Npop \in [100, 600]$
$c1_a$	Local Learning factor on Apertures	$c1_a \in [0, 2]$
$c2_a$	Global Learning factor on Apertures	$c2_a \in [0, 2]$
w_a	Inertia weight on Apertures	$w_a \in [0, 2]$
cf_a	Constriction factor on Aperture	$cf_a \in [0, 2]$
$c1_i$	Local Learning factor on Intensities	$c1_i \in [0, 2]$
$c2_i$	Global Learning factor on Intensities	$c2_i \in [0, 2]$
w_i	Inertia weight on Intensities	$w_i \in [0, 2]$
cf_i	Constriction factor on Intensities	$cf_i \in [0, 2]$

Table 4 shows the results provided by the IRace package:

Table 4. Best parameters' values obtained by IRace.

Parameter	Value
$Npop$	418
$c1_a$	1.8751
$c2_a$	0.2134
w_a	0.5774
cf_a	1.6641
$c1_i$	0.3158
$c2_i$	1.7017
w_i	0.5331
cf_i	1.2389

The number of iterations used by our algorithm is given by Equation (6), where we set the evaluation to 40,000 (number obtained testing the algorithm) and an $Npop$ of 518 (given in Table 4) doing a total of 95 iterations, like limits for the algorithm.

$$Iterations = \frac{evaluation}{Npop}. \tag{6}$$

4.3. Experiments on Test Instances

In our experiments, we measure the performance of the proposed PSO using the best-found parameter configuration, described in Section 4.2. Note that we run our algorithm 30 times per BAC, as 30 is a widely accepted value for statistical analysis [40].

Tables 5–8 report the results obtained by both the sequential and the PSO approaches when applied to the CERR and TRT001 cases. As mentioned in the previous section, the IMRT sequential approach obtains a fluence map, optimising the dose-volume model of the FMO problem. Next, the MLC sequencing problem is solved for the resulting fluence maps by using a well-known algorithm from [7], which finds a set of apertures that minimise the BoT. In Tables 5 and 6, column $z(x*)$ corresponds to the cost of the optimal fluence map using the function in Equation (2). Columns $z(r(x*))$, $z(r_2(x*))$ and $z(r_4(x*))$ correspond to the cost of the fluence maps with intensities rounded to the nearest integer, the nearest multiple of 2, and the nearest multiple of 4, respectively. For each rounding, we also report the number of apertures generated by the MLC sequencing algorithm (#ap) and the BoT.

Table 5. Results reported by the traditional two-step approach in the CERR dataset.

BAC	$z(x*)$	$z(r(x*))$	# ap	BoT	$z(r_2(x*))$	# ap	BoT	$z(r_4(x*))$	# ap	BoT
1	42.98	44.84	140	196	49.29	87	192	61.54	51	204
2	43.40	43.40	140	215	48.76	84	212	61.72	52	224
3	43.70	44.98	144	203	48.83	87	202	72.87	49	208
4	43.53	45.06	145	206	51.77	89	208	66.48	50	212
5	43.23	44.55	142	200	47.40	89	202	67.48	51	204
6	43.05	44.47	149	212	49.23	90	208	66.05	50	208
7	42.86	44.48	152	212	48.05	96	214	62.96	49	212
8	43.06	44.70	146	197	48.00	88	196	61.75	48	196
9	43.66	45.03	141	186	50.62	83	190	70.76	46	192
10	44.14	45.71	144	200	51.21	89	204	59.64	47	200
11	43.83	45.02	138	190	51.97	86	190	68.84	47	200
12	43.31	44.35	144	214	47.38	94	212	64.03	55	228
13	42.84	44.98	157	229	49.05	98	226	82.49	56	232
14	42.85	44.24	142	217	48.57	92	214	68.45	51	220
Average	43.32	44.71	144	205	49.30	89	205	66.80	50	210

Table 6. Results reported by the traditional two-step approach in patient TRT001 in the CAS dataset.

BAC	$z(x*)$	$z(r(x*))$	# ap	BoT	$z(r_2(x*))$	# ap	BoT	$z(r_4(x*))$	# ap	BoT
1	55.78	56.92	146	220	63.05	92	222	89.97	52	224
2	56.35	58.38	138	212	63.66	89	212	84.22	49	212
3	56.72	58.39	141	211	63.55	85	210	77.01	49	216
4	56.55	57.99	132	210	63.22	88	220	74.48	51	224
5	55.98	57.97	138	210	64.31	90	214	81.44	49	220
6	55.19	56.37	139	208	59.81	87	204	80.24	47	196
7	55.21	56.54	129	192	59.55	78	192	78.24	44	196
8	56.14	57.26	131	187	62.71	84	188	82.21	46	188
9	56.62	58.13	136	218	62.58	88	210	76.24	52	216
10	56.94	58.30	140	207	63.59	85	206	92.40	50	212
11	56.74	58.27	152	231	61.47	100	234	84.80	56	232
12	56.17	57.99	144	218	61.18	96	218	79.20	54	216
13	55.32	57.54	134	204	59.75	87	204	76.08	49	204
14	55.46	56.85	142	212	59.99	95	214	82.88	41	212
Average	56.08	57.64	138	210	62.03	88	210	81.39	49	212

Tables 7 and 8 report the results obtained by our PSO algorithm. Due to its stochastic nature, the strategy was run 30 times on each instance. We report the mean over the 14 instances of each set, the best value for each set, the mean number of apertures with intensity different to zero, and the mean BoT. We need to point out that apertures for which the intensity is set to zero by the mathematical programming solver in the last iteration are considered closed.

Table 7. Results reported by the PSO algorithm in the CERR dataset.

BAC	$z(x*)$	# ap	BoT
1	56.34	11.70	63.47
2	57.49	13.07	64.13
3	57.39	12.63	61.31
4	57.33	12.67	61.60
5	56.24	12.13	65.38
6	54.76	12.59	60.72
7	54.38	12.60	62.46
8	54.57	12.57	66.49
9	57.53	12.29	61.28
10	57.36	11.67	64.57
11	56.18	12.75	68.62
12	54.96	12.35	61.54
13	55.85	12.25	60.35
14	54.41	12.60	62.49
Average	56.06	12.42	63.17

Table 8. Results reported by the PSO algorithm in the TRT001 dataset.

BAC	$z(x*)$	# ap	BoT
1	71.33	12.40	62.44
2	72.39	12.70	62.13
3	73.79	12.00	62.96
4	73.85	12.20	63.07
5	73.80	11.10	58.99
6	73.75	11.40	63.91
7	72.52	10.90	58.94
8	72.65	10.20	60.30
9	74.27	12.10	66.04
10	77.58	11.40	63.30
11	74.63	12.60	62.51
12	72.00	12.10	60.13
13	69.75	11.30	60.32
14	71.31	10.50	59.74
Average	73.18	11.64	61.77

When comparing the objective function value reported by the PSO and the optimal (but not deliverable) fluence map, the difference is 29.41% and 30.49% for CERR and TRT001, respectively, with the PSO algorithm being the one with the higher objective value. This difference in the objective function value is reduced when the rounding process is applied to the optimal fluence map. For instance, when the optimal fluence map is rounded to the nearest multiple of $1(z(r(x^*)))$ and $2(z(r_2(x^*)))$, the difference is 25.39% and 13.71% for the CERR case and 26.98% and 17.98% for the TRT001 case, respectively. Further, rounding to the nearest multiple of $4(z(r_4(x^*)))$ leads to an impairment in the quality of the rounded treatment plan that makes solutions provided by our PSO algorithm become better in all cases. Further, even though our algorithm is not better than the $r_1(x^*)$ and $r_2(x^*)$ treatment plans (with respect to the objective function value), the number of aperture shapes our solutions need is always smaller than the apertures needed by the solutions obtained by the sequential approach. Also, it is interesting to note that even though our approach is not directly focused on reducing the beam on time value, our approach reports better values in all cases compared to the sequential approach. This is mainly because of the fact that we use far fewer aperture shapes in our final treatment plans.

In addition, we report the dose-volume histogram (DVH) for the CERR and TRT001 in Figures 14 and 15, respectively. DVH curves specify the received dose level by different volumes of structures. In the case of CERR, we can see that our algorithm obtains treatments that do not overdose the voxels in the PTV. Unlike this, the solutions obtained by the optimal fluence map overdose above 30% of PTV voxels. When observing the OARs, our algorithm overdoses more voxels than the optimal fluence map. However, the max overdose received for the voxels is less than the received by the optimal fluence map. In the case of TRT001, the PSO and the optimal fluence map have a similar curve, where both do not overdose the PTV. When observing the OARs, our algorithm overdoses more voxels than the optimal fluence map. It is necessary to remember that the optima fluence map is not a deliverable treatment and needs to pass for the MLC sequencing problem.

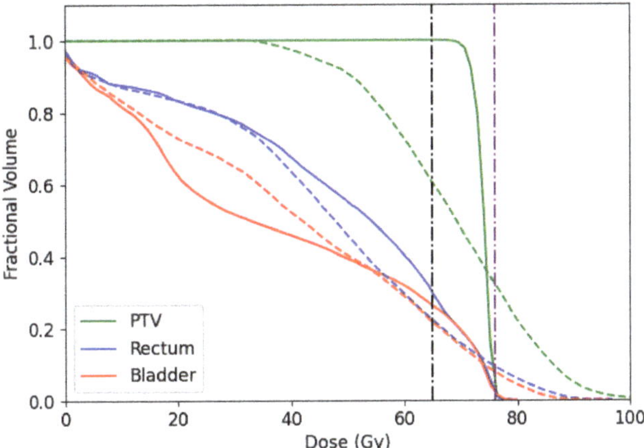

Figure 14. Dose-volume histogram comparing dose obtained by PSO algorithm (solid line) and optimal fluence map obtained by FMO (dashed line) for a prescribed dose of 76 Gy to PTV, and 65 Gy to the rectum and bladder (purple and black horizontal dashed-point line, respectively) with BAC 1 in CERR instance.

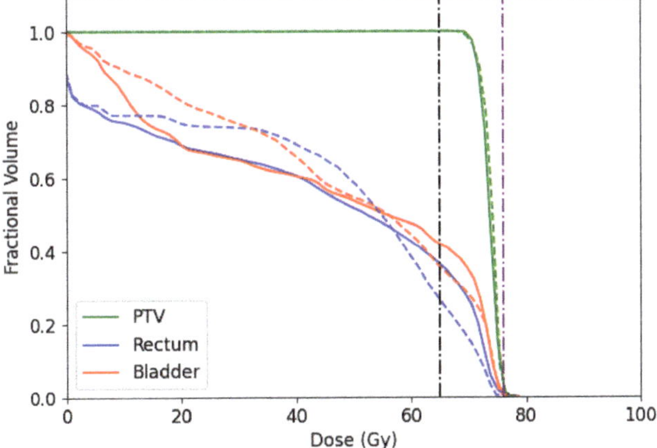

Figure 15. Dose-volume histogram comparing dose obtained by PSO algorithm (solid line) and optimal fluence map obtained by FMO (dashed line) for a prescribed dose of 76 Gy to PTV, and 65 Gy to the rectum and bladder (purple and black horizontal dashed-point line, respectively) with BAC 1 in TRT001 instance.

5. Conclusions

This paper introduces a hybrid heuristic based on PSO and mathematical programming to solve the DAO problem in radiation therapy for cancer treatment. The proposed PSO heuristic finds a set of deliverable aperture shapes and their corresponding intensities for each beam angle within a clinically acceptable time. Further, even though our heuristic algorithm was allowed to use only five aperture shapes per beam angle, they could find very competitive treatment plans.

Comparing our algorithm with the traditional sequential approach shows that the proposed algorithm can obtain competitive results regarding the objective function value. However, the difference with the optimal solution generated by the FMO is still significant. On the opposite, when evaluating the number of apertures generated by our algorithm, we can observe a substantial reduction compared to the traditional approach. This is very important as fewer aperture shapes mean, in general, shorter treatment times, which is something desirable from a clinical point of view.

In future work, we can see different research lines to improve the obtained results. First, we believe that improving the reparation heuristic to activate apertures that have intensities close to zero would allow us to find better-quality treatment plans. This is because the more apertures are used, the better the treatment plan quality. Note that, as mentioned before in the paper, this would be at the cost of longer treatment times. In addition, we seek to extend our single-objective PSO algorithm to a multi-objective one. This is because IMRT is an inherently multi-objective problem, since there is a compromise between tumour irradiation and avoiding damage to the organs at risk. Extending our approach to a multi-objective one is a challenging task from both computational and clinical points of view. However, we are sure that addressing the problem as a multi-objective one will help us better understand the underlying trade-offs between tumour control and OARs sparing.

Author Contributions: Conceptualization, G.T.-V., M.M. and G.C.-G.; methodology, M.M. and G.C.-G.; software, G.T.-V., M.M. and G.C.-G.; validation, M.M.; formal analysis, M.M.; investigation, G.T.-V. and M.M.; resources, G.C.-G.; data curation, M.M. and G.C.-G.; writing—original draft preparation, G.T.-V., M.M. and G.C.-G.; writing—review and editing, G.T.-V., M.M. and G.C.-G.; visualization, G.T.-V. and M.M.; supervision, G.T.-V., M.M. and G.C.-G. All authors have read and agreed to the published version of the manuscript.

Funding: This research was partly funded by the "FONDECYT Regular" project, number 1211129.

Institutional Review Board Statement: Not applicable.

Informed Consent Statement: Not applicable.

Data Availability Statement: The data can be shared upon request.

Acknowledgments: The authors gratefully thank Centro Oncologico Hondureño for giving access to pictures from its linear accelerator.

Conflicts of Interest: The authors declare no conflict of interest.

Abbreviations

The following abbreviations are used in this manuscript:

ASO	Aperture Segmentation Optimisation
AWO	Aperture Weight Optimisation
BAC	Beam angle configuration
BAO	Beam angle optimisation
BoT	Beam-on-Time
CERR	Computational Environment for Radiological Research
DAO	Direct Aperture Optimisation
DVH	Dose-Volume histogram
FMO	Fluence map optimisation
IMRT	Intensity modulated radiotherapy treatment
MLC	Multi-leaf Collimator
Npop	Population of solution

OAR   Organ at risk
PTV   Planning target volume
PSO   Particle Swarm Optimisation
SBRT   Stereotactic Body Radiation Therapy
VMAT   Volumetric Modulated Arc Therapy

References

1. Sung, H.; Ferlay, J.; Siegel, R.L.; Laversanne, M.; Soerjomataram, I.; Jemal, A.; Bray, F. Global Cancer Statistics 2020: GLOBOCAN Estimates of Incidence and Mortality Worldwide for 36 Cancers in 185 Countries. *CA Cancer J. Clin.* **2021**, *71*, 209–249. [CrossRef]
2. Cho, B. Intensity-modulated radiation therapy: A review with a physics perspective. *Radiat. Oncol. J.* **2018**, *36*, 1. [CrossRef] [PubMed]
3. Fallahi, A.; Mahnam, M.; Niaki, S. *Two Metaheuristic Algorithms for Direct Aperture Optimization in Intensity Modulated Radiation Therapy: Real-World Case Study for Liver Cancer*; IEEE: Piscataway, NJ, USA, 2021.
4. Hong, T.; Ritter, M.; Tome, W.; Harari, P. Intensity-modulated radiation therapy: Emerging cancer treatment technology. *Br. J. Cancer* **2005**, *92*, 1819–1824. [CrossRef] [PubMed]
5. Cao, D.; Afghan, M.K.N.; Ye, J.; Chen, F.; Shepard, D.M. A generalized inverse planning tool for volumetric-modulated arc therapy. *Phys. Med. Biol.* **2009**, *54*, 6725–6738. [CrossRef]
6. Ahuja, R.K.; Hamacher, H.W. A network flow algorithm to minimize beam-on time for unconstrained multileaf collimator problems in cancer radiation therapy. *Networks* **2005**, *45*, 36–41. [CrossRef]
7. Baatar, D.; Hamacher, H.; Ehrgott, M.; Woeginger, G. Decomposition of integer matrices and multileaf collimator sequencing. *Dis. Appl. Math.* **2005**, *152*, 6–34. [CrossRef]
8. Dzierma, Y.; Nuesken, F.G.; Fleckenstein, J.; Melchior, P.; Licht, N.P.; Rübe, C. Comparative Planning of Flattening-Filter-Free and Flat Beam IMRT for Hypopharynx Cancer as a Function of Beam and Segment Number. *PLoS ONE* **2014**, *9*, e94371. [CrossRef]
9. Romeijn, H.E.; Ahuja, R.K.; Dempsey, J.F.; Kumar, A. A Column Generation Approach to Radiation Therapy Treatment Planning Using Aperture Modulation. *SIAM J. Optim.* **2005**, *15*, 838–862. [CrossRef]
10. Shepard, D.M.; Earl, M.A.; Li, X.; Naqvi, S.; Yu, C.X. Direct aperture optimization: A turnkey solution for step-and-shoot IMRT. *Med. Phys.* **2002**, *29*, 1007–1018. [CrossRef]
11. Cotrutz, C.; Xing, L. Segment-based dose optimization using a genetic algorithm. *Phys. Med. Biol.* **2003**, *48*, 2987–2998. [CrossRef]
12. Pérez Cáceres, L.; Araya, I.; Soto, D.; Cabrera-Guerrero, G. Stochastic Local Search Algorithms for the Direct Aperture Optimisation Problem in IMRT. In *Hybrid Metaheuristics*; Springer International Publishing: Berlin, Germany, 2019; pp. 108–123.
13. Ludlum, E.; Xia, P. Comparison of IMRT planning with two-step and one-step optimization: A way to simplify IMRT. *Phys. Med. Biol.* **2009**, *53*, 807–821. [CrossRef] [PubMed]
14. del Valle, Y.; Venayagamoorthy, G.K.; Mohagheghi, S.; Hernandez, J.C.; Harley, R.G. Particle Swarm Optimization: Basic Concepts, Variants and Applications in Power Systems. *IEEE Trans. Evol. Comput.* **2008**, *12*, 171–195. [CrossRef]
15. Binkley, K.; Hagiwara, M. Balancing Exploitation and Exploration in Particle Swarm Optimization: Velocity-based Reinitialization. *Trans. Jpn. Soc. Artif. Intell.* **2008**, *23*, 27–35. [CrossRef]
16. Ehrgott, M.; Güler, Ç.; Hamacher, H.W.; Shao, L. Mathematical optimization in intensity modulated radiation therapy. *Ann. Oper. Res.* **2010**, *175*, 309–365. [CrossRef]
17. Cabrera-Guerrero, G.; Rodríguez, N.; Lagos, C.; Cabrera, E.; Johnson, F. Local Search Algorithms for the Beam Angles Selection Problem in Radiotherapy. *Math. Probl. Eng.* **2018**, *2018*, 4978703. [CrossRef]
18. Cabrera-Guerrero, G.; Lagos, C.; Cabrera, E.; Johnson, F.; Rubio, J.M.; Paredes, F. Comparing local search algorithms for the beam angles selection in radiotherapy. *IEEE Access* **2018**, *6*, 23701–23710. [CrossRef]
19. Cabrera-Guerrero, G.; Mason, A.J.; Raith, A.; Ehrgott, M. Pareto local search algorithms for the multi-objective beam angle optimisation problem. *J. Heuristics* **2018**, *24*, 205–238. [CrossRef]
20. Cabrera, G.; Ehrgott, M.; Mason, A.J.; Raith, A. A matheuristic approach to solve the multiobjective beam angle optimization problem in intensity-modulated radiation therapy. *Int. Trans. Oper. Res.* **2018**, *25*, 243–268. [CrossRef]
21. Cáceres, L.P.; Araya, I.; Cabrera-Guerrero, G. Stochastic local search for the Direct Aperture Optimisation Problem. *Exp. Syst. Appl.* **2021**, *182*, 115206. [CrossRef]
22. Romeijn, H.E.; Ahuja, R.K.; Dempsey, J.F.; Kumar, A.; Li, J.G. A novel linear programming approach to fluence map optimization for intensity modulated radiation therapy treatment planning. *Phys. Med. Biol.* **2003**, *48*, 3521–3542. [CrossRef]
23. Aleman, D.M.; Kumar, A.; Ahuja, R.K.; Romeijn, H.E.; Dempsey, J.F. Neighborhood search approaches to beam orientation optimization in intensity modulated radiation therapy treatment planning. *J. Glob. Optim.* **2008**, *42*, 587–607. [CrossRef]
24. Moyano, M.; Cabrera-Guerrero, G. Local Search for the Direct Aperture Optimisation in IMRT. In Proceedings of the 2020 39th International Conference of the Chilean Computer Science Society (SCCC), Coquimbo, Chile, 16–20 November 2020; pp. 1–6.
25. Moyano, M.; Cabrera-Guerrero, G.; Tello-Valenzuela, G.; Lagos, C. An Hybrid Local Search for the Direct Aperture Optimisation Problem. *IEEE Trans. Emerg. Top. Comput. Intell.* **2023**, 1–15. [CrossRef]
26. Cao, R.; Pei, X.; Zheng, H.; Hu, L.; Wu, Y. Direct aperture optimization based on genetic algorithm and conjugate gradient in intensity modulated radiation therapy. *Chin. Med. J.* **2014**, *127*, 4152–4153. [PubMed]

27. Li, Y.; Yao, J.; Yao, D. Genetic algorithm based deliverable segments optimization for static intensity-modulated radiotherap *Phys. Med. Biol.* **2003**, *48*, 3353–3374. [CrossRef] [PubMed]
28. Hardemark, B.; Liander, A.; Rehbinder, H.; Löf, J. *Direct Machine Parameter Optimization with RayMachine in Pinnacle*; RaySearc White Paper; RaySearch Laboratories: Stockholm, Sweden, 2003.
29. Worthy, D.; Wu, Q. Parameter optimization in HN-IMRT for Elekta linacs. *J. Appl. Clin. Med. Phys.* **2009**, *10*, 43–61. [CrossRef]
30. Preciado-Walters, F.; Langer, M.P.; Rardin, R.L.; Thai, V. Column generation for IMRT cancer therapy optimization with implementabl segments. *Ann. Oper. Res.* **2006**, *148*, 65–79. [CrossRef]
31. Carlsson, F. Combining segment generation with direct step-and-shoot optimization in intensity-modulated radiation therap *Med. Phys.* **2008**, *35*, 3828–3838. [CrossRef]
32. Zhang, L.; Gui, Z.; Yang, J.; Zhang, P. A Column Generation Approach Based on Region Growth. *IEEE Access* **2019**, *7*, 31123–3113 [CrossRef]
33. Salari, E.; Unkelbach, J. A column-generation-based method for multi-criteria direct aperture optimization. *Phys. Med. Biol.* **201** *58*, 621–639. [CrossRef]
34. Ripsman, D.; Purdie, T.; Chan, T.; Mahmoudzadeh, H. Robust Direct Aperture Optimization for Radiation Therapy Treatmen Planning. *INFORMS J. Comput.* **2022**, *34*, 2017–2038. [CrossRef]
35. Kennedy, J.; Eberhart, R. Particle swarm optimization. In Proceedings of the ICNN'95—International Conference on Neural Network Perth, WA, Australia, 27 November–1 December 1995; Volume 4, pp. 1942–1948.
36. López-Ibáñez, M.; Dubois-Lacoste, J.; Pérez Cáceres, L.; Birattari, M.; Stützle, T. The irace package: Iterated racing for automati algorithm configuration. *Oper. Res. Perspect.* **2016**, *3*, 43–58. [CrossRef]
37. Deasy, J.O.; Blanco, A.I.; Clark, V.H. CERR: A computational environment for radiotherapy research. *Med. Phys.* **2003**, *30*, 979–98 [CrossRef] [PubMed]
38. Balaprakash, P.; Birattari, M.; Stützle, T. Improvement Strategies for the F-Race Algorithm: Sampling Design and Iterative Refinemen In *Hybrid Metaheuristics*; Bartz-Beielstein, T., Blesa Aguilera, M.J., Blum, C., Naujoks, B., Roli, A., Rudolph, G., Sampels, M., Eds Springer: Berlin/Heidelberg, Germany, 2007; pp. 108–122.
39. Birattari, M.; Yuan, Z.; Balaprakash, P.; Stützle, T. F-Race and Iterated F-Race: An Overview. In *Experimental Methods for the Analysis c Optimization Algorithms*; Bartz-Beielstein, T., Chiarandini, M., Paquete, L., Preuss, M., Eds.; Springer: Berlin/Heidelberg, German 2010; pp. 311–336.
40. Hays, W.L.; Winkler, R.L. *Statistics: Probability, Inference, and Decision*; Technical Report; Houghton Mifflin Harcourt School: Bostor MA, USA, 1970.

Article

Artificial Intelligence Reveals Distinct Prognostic Subgroups of Muscle-Invasive Bladder Cancer on Histology Images

Okyaz Eminaga [1,*], Sami-Ramzi Leyh-Bannurah [2], Shahrokh F. Shariat [3], Laura-Maria Krabbe [4], Hubert Lau [5,6], Lei Xing [7] and Mahmoud Abbas [8,*]

1   AI Vobis, Palo Alto, CA 95054, USA
2   Department of Urology, Pediatric Urology and Uro-Oncology, Prostate Center Northwest, St. Antonius-Hospital, 33705 Gronau, Germany
3   Department of Urology, Comprehensive Cancer Center, Medical University of Vienna, 1090 Vienna, Austria; sfshariat@gmail.com
4   Department of Urology, University Hospital of Muenster, 48419 Muenster, Germany
5   Department of Pathology, School of Medicine, Stanford University, Stanford, CA 94305, USA; hlau@stanford.edu
6   Department of Pathology, Veterans Affairs Palo Alto Health Care System, Palo Alto, CA 94304, USA
7   Department of Radiation Oncology, School of Medicine, Stanford University, Stanford, CA 94305, USA
8   Department of Pathology, University Hospital of Muenster, 48419 Muenster, Germany
*   Correspondence: okyaz.eminaga@aivobis.com (O.E.); mahmoud.abbas@ukmuenster.de (M.A.)

Simple Summary: This study developed an interpretable scoring system using artificial intelligence and bladder tissue images. It identified two distinct risk groups with different outcomes in high-grade bladder cancer. The scoring system was associated with various molecular features and gene mutations. This system can save shared clinical decision making and cost by identifying patients who need further molecular testing.

Abstract: Muscle-invasive bladder cancer (MIBC) is a highly heterogeneous and costly disease with significant morbidity and mortality. Understanding tumor histopathology leads to tailored therapies and improved outcomes. In this study, we employed a weakly supervised learning and neural architecture search to develop a data-driven scoring system. This system aimed to capture prognostic histopathological patterns observed in H&E-stained whole-slide images. We constructed and externally validated our scoring system using multi-institutional datasets with 653 whole-slide images. Additionally, we explored the association between our scoring system, seven histopathological features, and 126 molecular signatures. Through our analysis, we identified two distinct risk groups with varying prognoses, reflecting inherent differences in histopathological and molecular subtypes. The adjusted hazard ratio for overall mortality was 1.46 (95% CI 1.05–2.02; z: 2.23; $p = 0.03$), thus identifying two prognostic subgroups in high-grade MIBC. Furthermore, we observed an association between our novel digital biomarker and the squamous phenotype, subtypes of miRNA, mRNA, long non-coding RNA, DNA hypomethylation, and several gene mutations, including FGFR3 in MIBC. Our findings underscore the risk of confounding bias when reducing the complex biological and clinical behavior of tumors to a single mutation. Histopathological changes can only be fully captured through comprehensive multi-omics profiles. The introduction of our scoring system has the potential to enhance daily clinical decision making for MIBC. It facilitates shared decision making by offering comprehensive and precise risk stratification, treatment planning, and cost-effective preselection for expensive molecular characterization.

Keywords: deep learning; digital biomarker; bladder cancer; FGFR3; WSI; risk stratification; histology images

Citation: Eminaga, O.; Leyh-Bannurah, S.-R.; Shariat, S.F.; Krabbe, L.-M.; Lau, H.; Xing, L.; Abbas, M. Artificial Intelligence Reveals Distinct Prognostic Subgroups of Muscle-Invasive Bladder Cancer on Histology Images. *Cancers* **2023**, *15*, 4998. https://doi.org/10.3390/cancers15204998

Academic Editor: Christine Decaestecker

Received: 7 August 2023
Revised: 12 September 2023
Accepted: 3 October 2023
Published: 16 October 2023

1. Introduction

Bladder Cancer (BC) is the tenth most common cancer in the United States, mostly affecting people older than 55 years. Bladder cancer (BC) exhibits a gender disparity affecting men approximately four times more frequently than women [1]. Furthermore, BC encompasses a broad spectrum of disease behavior, ranging from a slow-growing non-muscle-invasive form (NMIBC) to a highly aggressive muscle-invasive variant (MIBC). Although most BC patients are diagnosed with NMIBC, up to 25% of BC are identified as MIBC with substantial risk for mortality [2]. BC cases with stage I or II show a 5-year relative survival rate of 96% or 70%, respectively, whereas 38 of 100 cases with stage III will survive 5 years; cases with stage IV have the poorest survival outcome with 6% of a 5-year relative survival rate. Moreover, BC reveals distinct multilevel molecular subtype profiles associated with prognosis and treatment responses [3]. However, determining multilevel molecular subtype profiles (i.e., protein expression, gene mutation, mRNA, DNA methylation, and miRNA) requires a complex and expensive infrastructure likely unavailable in most cancer centers worldwide. Therefore, a cost-effective solution could ideally help to manage the patient selection according to their risk of having progressive cancers or to identify cases likely to benefit from certain treatment regimens.

Recent studies revealed the potential of deep learning (DL) to predict a new generation of digital biomarkers for detection, prognosis, molecular signature, and treatment response in different cancers, including bladder cancer [4–6]. For instance, Woerl et al. reported the potential of DL to forecast the molecular subtypes of MIBC by analyzing hematoxylin and eosin (H&E) slides [7]. As a proof-of-concept, Mundhada et al. have shown the DL capability to distinguish low-grade from high-grade histology [8]. Zheng et al. purposed a DL framework to predict survival from histology images with BC [9]. While deep learning holds immense potential, addressing certain tendencies that have arisen within its application is essential. Specifically, a majority of prior research treated confidence scores as equivalent to probability scores, disregarding the well-recognized problem of overconfidence in deep learning models [10,11]. Furthermore, these studies have not provided a feasible means of interpreting whether the feature distributions in the latent feature spaces reflect alterations in histological patterns that contribute to the prediction scores.

Given the limitations of previous studies, our hypothesis posits that morphometrical patterns observed at the histological level are indicative of prognostic confidence scores, which are then associated with omics signatures specific to advanced bladder cancers. Our primary objective is to identify prognostic subgroups that reveal associations with molecular subtypes, utilizing histology images, including bladder cancers and weakly supervised learning. The major contribution of the current work is to provide a novel strategy that facilitates the development of interpretable prognostic scores derived from a collection of mixed histology patterns associated with molecular subtypes and potential treatment options for bladder cancers.

2. Methods

2.1. Survival Modeling

2.1.1. Data

Complete data were available for 113 patients diagnosed with urothelial carcinoma of the bladder (BC) from the Prostate, Lung, Colon, and Ovarian Cancer Screening (PLCO) trial. PLCO is a randomized controlled trial aimed to determine whether certain screening exams reduce mortality from prostate, lung, colorectal, and ovarian cancer (NCT00339495) [12,13]. Although this trial did not screen for BC, it tracked diagnoses of BC during the trial period. Briefly, 154,900 participants from the general population aged 55 through 74 years were enrolled between 1993 and 2001 [14]. Only subjects without a history of prostate, lung, colorectal, or ovarian cancer were enrolled. Cancer diagnoses were confirmed by retrieving results and information from medical records and the cancer registry system. This study used a linkage with the National Death Index to extend mortality follow-up to a maximum of 19 years after randomization [15]. During the study follow-up period, 1430 cases of BC were diagnosed, from which the PLCO study organizer randomly selected 285 cases to scan representative whole slides with samples containing BC. All samples were originally obtained through transurethral resection of bladder tumors.

After excluding the slide images of cases with missing follow-up information, a total of 196 H&E-stained slides of the bladder cancer cohort were available from nine U.S. centers and digitally scanned at 40× objective magnification (one pixel corresponds to ~0.2532 μm) using a Leica Biosystems device (Wetzlar, Germany) and stored in SVS format.

We split these images, as the development set, into a training set, optimization set, and validation set by institutions to prevent overlapping between these sets; cases of a center having the largest portion in our cohort were selected for the training set, the center with the smallest portion was considered for optimization set and the remaining centers for the validation set. Figure 1 summarizes our framework for developing the digital biomarker for mortality.

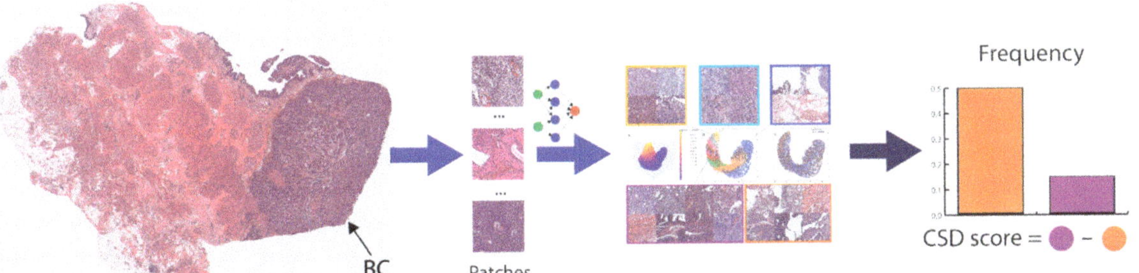

Figure 1. Illustrates the abstract AI framework for our approach. First, the tissue area is masked and tiled into small patches labeled with the cancer-specific survival status (weakly labeling). We trained the model to predict the cancer-specific survival status, and we then explored the distribution of the latent features and histology patterns stratified by the prediction deciles to develop the cancer-specific score system consisting of two prediction deciles (orange and lilac colors) reflecting distinguishable histology patterns.

2.1.2. Image Preprocessing

The rectangle boundary of the tissue area was estimated after thresholding the gray color version of the thumbnail image (1× magnification) for each image and up-scaled to correspond to 40× magnification. After that, the tissue area was divided into 2048 × 2048 pixels (px) tiles, and tiles mostly (>50% of the tile pixels) matching the white background colors were excluded. The resulting tiles were downsized to 512 × 512 px (~10× objective magnification). Each tile originated from the same patient and was labeled for the binary cancer-specific death status (CSD) on the death certificate.

2.1.3. Model Development

The current study applied the neural architecture search (NAS) algorithm for Plexus-Net [16] and the training set to determine the optimal model architecture for CSD prediction. Here, we used the grid search and an abstract search space covering the type of block (i.e., attention block, ResNet, or inception block), depth (i.e., how often to repeat the blocks), and the branching factor (i.e., number of multi branches in the network) of the convolutional neural network and the transformer inclusion, resulting in the examination of 1296 models with different architecture configurations (Table 1). In addition, we applied the widely accepted optimization algorithm "ADAM" with the standard hyperparameter configuration and the cross-entropy loss function to train each model for one epoch. For the NAS, the batch size was set to 64 patches and the learning rate to 1×10^{-3}. To optimize the computational efficiency of NAS, we employed a downsizing technique from our previous work, reducing the patches to a 32×32 pixel dimension [16]. This approach allows us to focus computational resources on smaller patches, reducing complexity while extracting meaningful information. The downsized patches balance between computational efficacy and the ability to explore diverse architectural designs, streamlining the NAS process for large-scale experiments and real-world applications [16]; the two-fold cross-validation was applied to train and evaluate each model for balanced classification accuracy. Finally, the final model architecture with the highest average performance on two-fold cross-validation was selected.

Table 1. The search space for the neural architecture search.

Parameter	Options
Block architecture (microarchitecture design)	Inception block (Inception) Residual block (ResNet) Conventional block (VGG) Attention block (soft_att)
Width	2, 4, 6
Depth	3, 4, 5
Length (pathways)	2, 3, 4
Junctions (interconnection between pathways)	1, 2, 3
Global pooling	Average vs. Maximum
Addition of transformer	Yes vs. No

The resulting model was then trained on the whole training set with 512×512 px patches until convergence. During model training, we set an early stopping algorithm (stop training when the loss values on the optimization setting are not improved for ten epochs) to mitigate the model overfitting; Adam with weight decay was applied as instructed by the authors for model training while the learning rate was set to 1×10^{-4}. The binary patch label was randomly smoothed with $+/- 0.25$ to moderate the model overconfidence in addition to model overfitting and to improve the model calibration. The image augmentation was applied and included random rotation, flipping, clipping, and color space augmentations, as described previously in the image preprocessing section. For each epoch, we validated the model performance on the optimization set at the patient level. Here, we measured the average confidence scores for CSD on all patches for each patient and the discriminative accuracy for CSD prediction using a time-dependent area under the receiver operating characteristic curves (AUROC) and c-index at the case level.

We applied the validation set to validate the case-level model accuracy at the patient level and to visualize the feature space of the last convolutional layer (not the global pooling) using t-SNE (t-distributed stochastic neighbor embedding). We then clustered the feature spaces according to the deciles of CSD prediction to visualize the correspondence between CSD prediction and feature space. After that, we determined two deciles based on the feature clusters. The first decile cluster (reference decile, r) shows a feature space dominant for negative patches, whereas the second decile cluster corresponds to the median decile (m).

After finding deciles r and m, we developed an algorithm to estimate the CSD score for each case as follows:

(1) We first calculated the patch frequency for 10 bins with equal width (histogram) at the case level. The bin width was calculated for each case using Equation (1).

$$bin\ width = (s_{max} - s_{min})/\sqrt{10} \tag{1}$$

where s_{max} is the maximum CDS score, and s_{min} is the minimum CSD score for each case.

(2) Secondly, we applied the maximum normalization to the patch frequencies, including D_r and D_m, to achieve a value range between 0 and 1 for all bins. Third, we estimated the unadjusted CSD score (S_u) using the following equation:

$$S_u = D_m - D_r \tag{2}$$

(3) Since out-of-distribution data may have a different frequency distribution than the development set, we introduced the following algorithm to adjust the CSD score estimation without having the ground truth:

    (a)   Calculate the mean μ of S_u;

    (b)   Calculate the median $\mu_{1/2}$ of $(S_u - \mu)$;

    (c)   Adjust the scores by μ and $\mu_{1/2}$ according to the equation:

$$S_{CSD} = -(\mu + \mu_{1/2}) + D_m - D_r \tag{3}$$

We also applied thresholding to S_{CSD} to define a binarized risk category. The threshold (T) was determined using the following equation:

$$T = \mu_0 + 1.05\,\sigma_0 + \alpha \tag{4}$$

where σ_0 is the standard deviation of S_u, σ_0 and μ_0 were calculated on development set, and α is the correction factor that counts the difference between μ_0 and the mean of out-of-distribution cohort (μ_c) and can be expressed as

$$\alpha = \mu_0 - \mu_c \tag{5}$$

Since the bin range differs from case to case by the CSD score range, we asserted that the median for the case-wise midrange of CSD scores for D_r (MR = 0.17; interquartile range, IQR: 0.16–0.18) and D_m (MR = 0.41; IQR: 0.37–0.42) was comparable between the development and out-of-the distribution cohort (external validation) to ensure the generalization of binning with equal width.

2.2. Evaluation

2.2.1. Data

We obtained 457 H&E-stained whole slide images from The Cancer Genome Atlas (TCGA)—Urothelial Bladder Carcinoma cohort [17], from which 412 images included survival information. This TCGA cohort contains genetic, demographic, and clinical outcome data for various cancers, and this data is made publicly available through their online platform (NCI Genomic Data Commons). The TCGA study for bladder cancer has received contributions from 36 institutions worldwide. The sources of bladder cancer tissue specimens were radical cystectomy (RC) specimens. The slides with bladder cancer tissue specimens were digitally scanned at $40\times$ objective magnification (one pixel corresponds to ~0.2532 μm on average) using a Leica Biosystems device (Wetzlar, Germany) and stored in SVS format. Clinicopathological and follow-up information was available at the case level. We also applied the same image preprocessing strategy and scoring system described earlier to this cohort. All images with available molecular profiles and clinicopathological and follow-up information were considered. Each case corresponded to a single whole-slide image.

2.2.2. Prognosis

We assessed the prognostic value of our novel risk group using the univariate and multivariate Cox proportional hazards models. In multivariate analysis, cancer-stage grouping and age at diagnosis were added to adjust the hazard ratio for the novel risk group. The outcome was the overall survival (OS) from the diagnosis, as the TCGA dataset is highly qualitative and widely used for overall survival analyses in cancer research [18]. Patients lost to follow-up were censored at the date of the last contact.

2.2.3. Association with Familiar Molecular Signatures of Bladder Cancer

We evaluated seven histopathologic (e.g., squamous phenotype) and 126 molecular signatures (e.g., the mutation in FGFR3 and molecular subtypes) investigated by the TCGA study [17] in bladder cancers (see the signature list in the Supplementary Materials File S1) for their association with the categorized risk score groups. In addition, for any significant signatures with more than two categories, we performed post hoc comparison analyses to determine which categories significantly differ between the novel two risk groups.

2.3. Metrics, Statistics and Software

We applied the time-dependent AUROC at the fifth follow-up year [19] and univariate and multivariate Cox regression analyses to assess our novel scoring system on the development set before the external validation.

The classification and accuracy of prognosis were quantified with AUROC and Harrel's c-index [20,21]. The goodness of fit was measured according to the Akaike information criterion (AIC) and Bayesian information criterion (BIC), where the lower the value, the better the model fit [22–24]. Finally, Kaplan–Meier survival estimates were applied to approximate the survival probability for our novel risk classification.

The chi-square tests were performed to determine whether there is an association between categorical variables (n $\times$ m contingency tables). In contrast, the Fisher test was applied to estimate the odd ratios and assess 2 $\times$ 2 contingency tables. Finally, we used the Wilcoxon Rank Sum Test to assess the differences in a numerical variable between the novel risk groups.

The comparison analyses for categorical signatures include repeating the Fisher test for each signature category as one-versus-other and the significance determination for each comparison test according to the Benjamini-Hochberg (B-H) procedure [25]. Here, the critical value was calculated for each comparison test after the p-values of comparison tests were ranked from low to high. The following equation was used to estimate the critical value at a false discovery rate (FDR) of 0.20:

$$\text{Critical value} = \text{rank}/(\text{number of comparisons}) \times 0.20 \tag{6}$$

A comparison test is deemed significant according to the last p-value lower than its critical value. The Pearson Correlation coefficient estimated the correlation between two numerical variables, while the Kendall rank correlation coefficient (τ) was estimated to measure the ordinal association between one numerical variable and one categorical variable or between two categorical variables [26,27]. VIF (variance inflation factor) was used to assess the multicollinearity in the COX regression model [28].

Model development and analyses were performed using Keras 2.6 [29], TensorFlow 2.10 [30], Python™ 3.8, and the R statistical package system (R Foundation for Statistical Computing, Vienna, Austria). All statistical tests were two-sided, and statistical significance was set at $p \leq 0.05$ for prognosis or $p \leq 0.10$ to consider molecular or histopathologic signatures for comparative analyses.

3. Results

3.1. Survival Modeling

Table 2 summarizes the cohort description of the development set. We found no significant difference in the cohort characteristics between the subsets (i.e., training, optimization, and validation sets). We considered the diverse BLC pathologies (not limited to muscle-invasive bladder cancer) to increase the likelihood of capturing differential histopathological patterns by our model for prognosis. In alignment with the literature, 77% of training set cases were non-muscle invasive BC and representative of the population. The optimization set was utilized to fine-tune the model, enabling it to distinguish between non-lethal and lethal patches while considering the various WHO Grades that exhibit heterogeneous patterns. Using a small sample size for optimization allowed the domain expert to manually review the predicted patch classes and streamline performance optimization accordingly. The validation set had a balanced distribution of NMIBC and MIBC cases, and G1/2 and G3 cases, thereby minimizing the effect of sampling bias.

Table 2. The cohort description of the development set. [+]Given the study's history and design, the previous grade was available.

Characteristic	Training Set N = 26	Optimization Set N = 6	Validation Set N = 81	*p*-Value [1]
Age at diagnosis in year, median (IQR)	65.0 (62.2–68.8)	69.0 (65.8–70.0)	65.0 (61.0–68.0)	0.50
Sex, n (%)				0.13
Male	25 (96%)	5 (83%)	64 (79%)	
Female	1 (3.8%)	1 (17%)	17 (21%)	
WHO Grade 1973, n (%)[+]				0.26
G1	5 (19%)	1 (17%)	15 (19%)	
G2	10 (38%)	3 (50%)	15 (19%)	
G3	11 (42%)	2 (33%)	47 (58%)	
Unknown	0 (0%)	0 (0%)	4 (4.9%)	
AJCC tumor staging				
T stage, n (%)				0.35
Ta	4 (15%)	0 (0%)	1 (1.2%)	
Tis	11 (42%)	5 (83%)	31 (38%)	
T1	5 (19%)	1 (17%)	19 (23%)	
T2	3 (12%)	0 (0%)	22 (27%)	
T3	2 (7.7%)	0 (0%)	4 (4.9%)	
T4	0 (0%)	0 (0%)	1 (1.2%)	
Unknown	1 (3.8%)	0 (0%)	3 (3.7%)	
N stage, n (%)				0.46
Nx/N0	25 (96%)	6 (100%)	74 (91%)	
N1	0 (0%)	0 (0%)	4 (4.9%)	
Unknown	1 (3.8%)	0 (0%)	3 (3.7%)	
M stage, n (%)				0.47
Mx/M0	25 (96%)	6 (100%)	75 (93%)	
M1	0 (0%)	0 (0%)	3 (3.7%)	
Unknown	1 (3.8%)	0 (0%)	3 (3.7%)	
Follow-up duration in months, median (IQR)	172 (130–201)	151 (87–192)	168 (130–197)	0.80
Cancer-specific death, n (%)	6 (23%)	1 (17%)	25 (31%)	0.60
Whole-slide images, n (%)	46 (23.5%)	8 (4.1%)	142 (72.4%)	-
Patches, n (%)	26,949 (16.5%)	7574 (4.6%)	129,122 (78.9%)	-

[1] Kruskal–Wallis rank sum test; Pearson's Chi-squared test.

The Neural Architecture Search (NAS) examined 1296 PlexusNET architecture configurations (duration: ~12 h) [16] and suggested a shallow model (model configuration: VGG D6L2J1F2 + transformer and global average pooling; these parameters regulate the design of the model architecture and the model scaling) having only 23,783 parameters and 20 fully connected representation features as the best model configuration for cancer-specific death (CSD) prediction. The Levene test indicated a significant difference in 18 out of the 20 two-dimensional feature maps between the patches derived from patients who died due to bladder cancer and those who survived. In other words, the feature maps were found to be unequal or dissimilar between the two groups, indicating the extraction of significant feature representation for CSD from histology images (Figure 2).

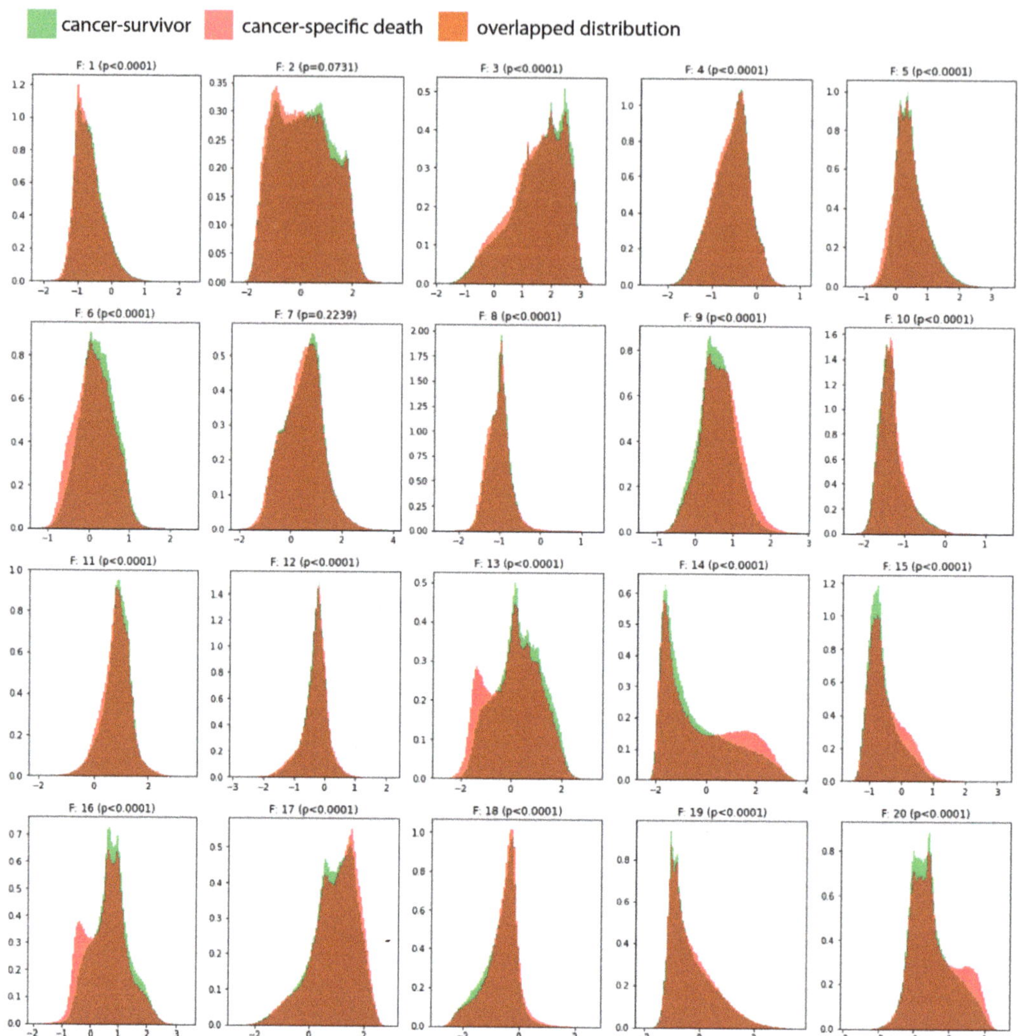

Figure 2. Lists the density histograms for the last 8×8 two-dimensional feature maps (pixels) according to the cancer-specific death status at the pixel level (i.e., pixel values) on 25,000 patches from the validation set. A size of 1×1 pixel on a feature map corresponds to an area with 64×64 pixels on the corresponding patch image (512×512 pixels). The Levene test was applied to assess the equality of variance between cancer survivors and cancer-specific death patches. We identified that some features (e.g., F13, F14, F15, F16, and F20) revealed histogram ranges for pixel values of specific feature maps more common in cancer-specific death patches (red areas).

Following the instructions provided in the Section 2 to derive a risk score from histology images, we visualized the feature space and determined the feature subspaces using the prediction deciles. The t-SNE visualization of the feature space showed that the prediction deciles sorted feature points, and the evaluation of the corresponding patch images confirmed the differences in histopathology appearance according to the deciles (Figure 3). Therefore, based on the t-SNE feature visualization and the assessment of the histopathology appearance, the second decile (D2) and the fifth decile (D5) met the selection criteria described in the Section 2.

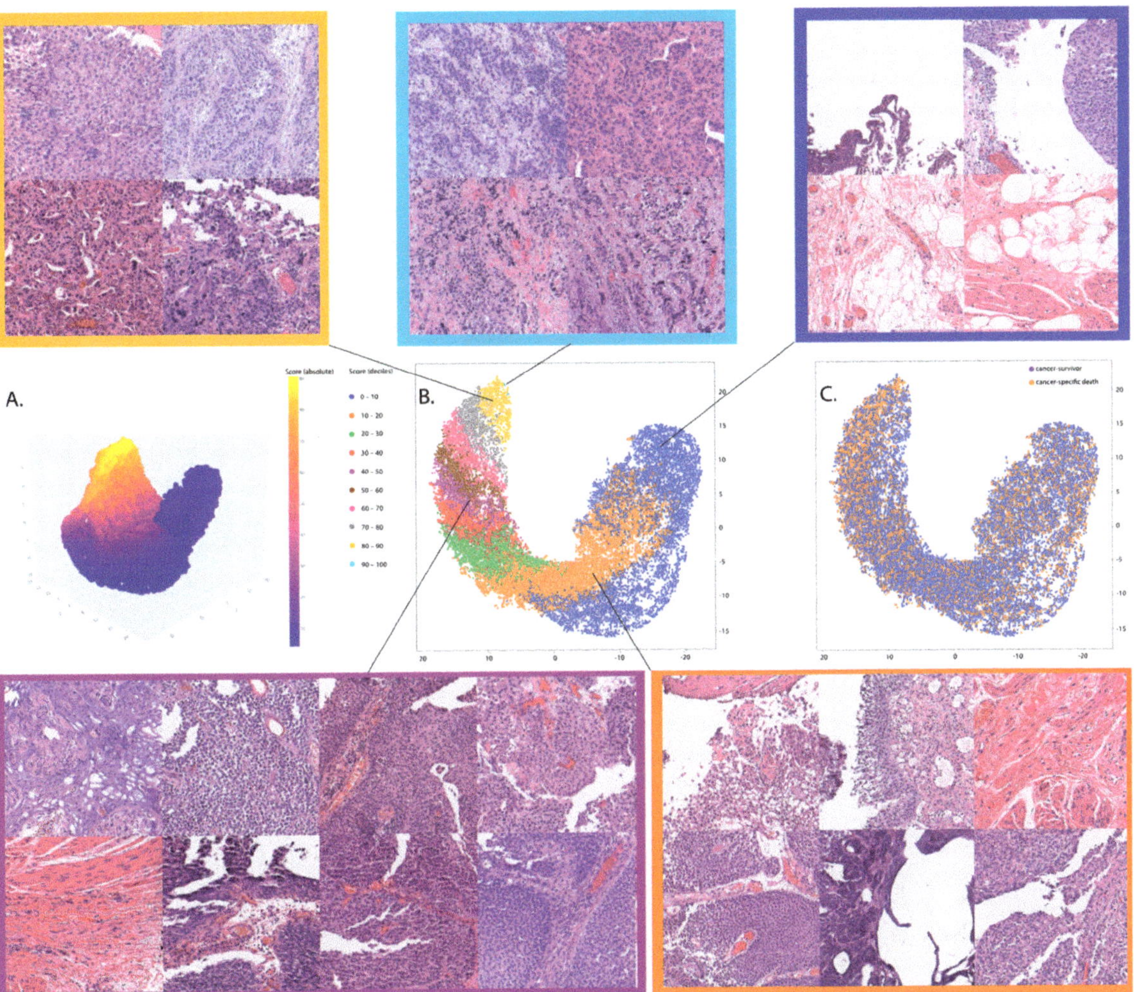

Figure 3. Summarizes the t-SNE visualization of penultimate features intuitively sorted by the deciles of the model inference scores (predications aka confidence) on representative 25,000 patch images randomly selected from the validation set. These patches represent the entire cases (n = 81) of the validation set. The corresponding patches were evaluated and identified to be altered by the prediction deciles. Based on the data evaluation and the domain knowledge, we selected the second decile and fifth decile; the second decile (orange color) was predominantly associated with negative patches (>50%), including bladder cancer, while the fifth decile (lilac color) was the center decile between the first and the ninth decile (the tenth decile was not considered due to its negligible sample size). (**A**) The 3D feature visualization; (**B**) the 2D visualization of features stratified by prediction deciles; (**C**) the 2D visualization of features stratified by the cancer-specific death status.

At the patient level, the risk score was prognostic for cancer-specific mortality (HR: 8.0; 95% CI: 1.4–46.1; z: 2.332; $p = 0.0197$). The 5-year AUC was 0.772 ± 0.04. The multivariate Cox regression analysis further strengthened the independent prognostic significance of our novel risk score, even after adjusting for age at diagnosis and tumor grade. Including our novel scoring system in the analysis offered an alternative approach for assessing histopathological characteristics that is distinct from tumor grade (Table 3).

Table 3. The multivariate Cox regression analysis for cancer-specific mortality. HR: Hazard ratio, CI: Confidence Interval. Grading on the PLCO validation set. Due to the PLCO study design, only the WHO 1973 grading was available. Nonetheless, it is important to emphasize that WHO grading is a well-established prognostic parameter, lending significance to its inclusion in our analysis.

Variable	HR	95% CI	z	p
Age at diagnosis	1.03	(0.96–1.11)	0.87	0.39
Grading (WHO 1973)				
G1 (ref.)	–	–	–	–
G2	2.21	(0.20–24.48)	0.64	0.52
G3	11.99	(1.61–89.21)	2.43	0.02
Unknown	11.72	(1.03–133.02)	1.99	0.05
Risk score	8.39	(1.53–46.12)	2.45	0.01

The Kaplan–Meier Curve revealed that the risk score (categorized) delivered two distinctive risk groups ($p = 0.014$), as shown in Figure 4. The median survival for the high-risk group was achieved between 204 (17 years) and 216 months (18 years) after the initial diagnosis.

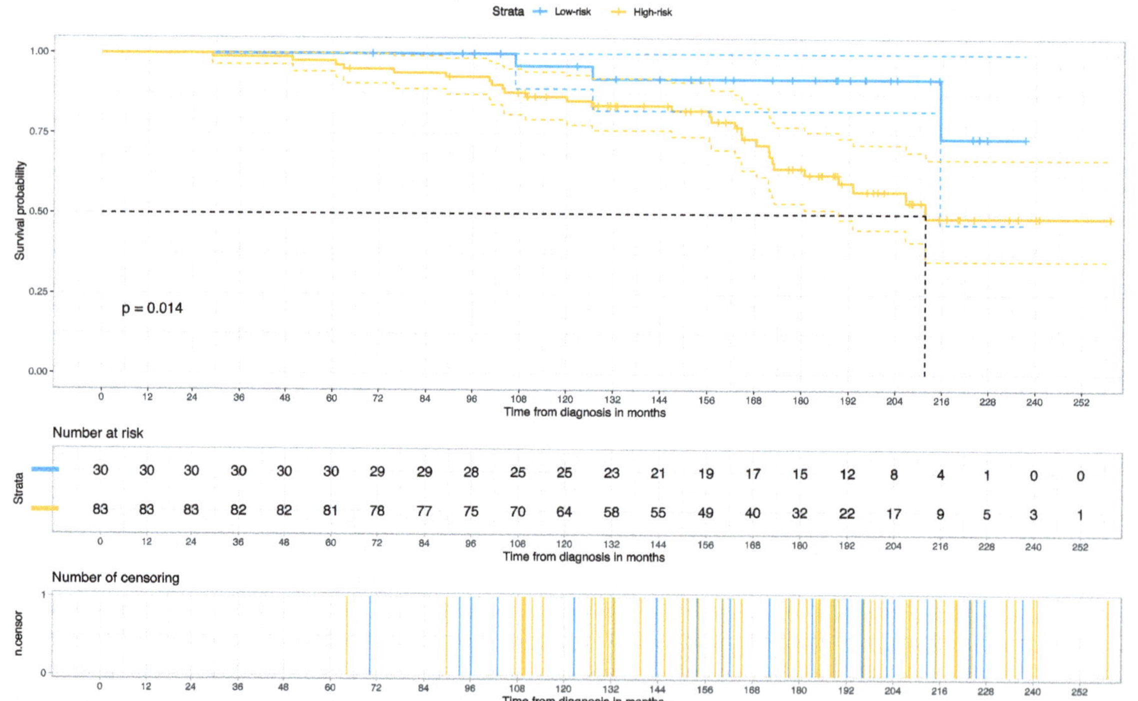

Figure 4. The Kaplan–Meier curve for cancer-specific survival stratified by the categorized risk scores (Low-risk vs. High-risk) on the PLCO validation cohort. The dot line reveals the median survival (the time it takes to reach 50% survival) between 210 and 216 months.

3.2. Prognosis for Muscle-Invasive Bladder Cancer

Table 4 summarizes the cohort description of the external validation set. The vast majority of cases included high-grade MIBC. The distribution of the risk scores around the cohort-specific threshold (T = 0) is shown in Figure 5. The categorization of the risk score was driven by the dominance of either D2 or D5 in each case, and D2 and D5 were associated with distinct histopathologic patterns of bladder cancers in the TCGA cohort (Figure 6).

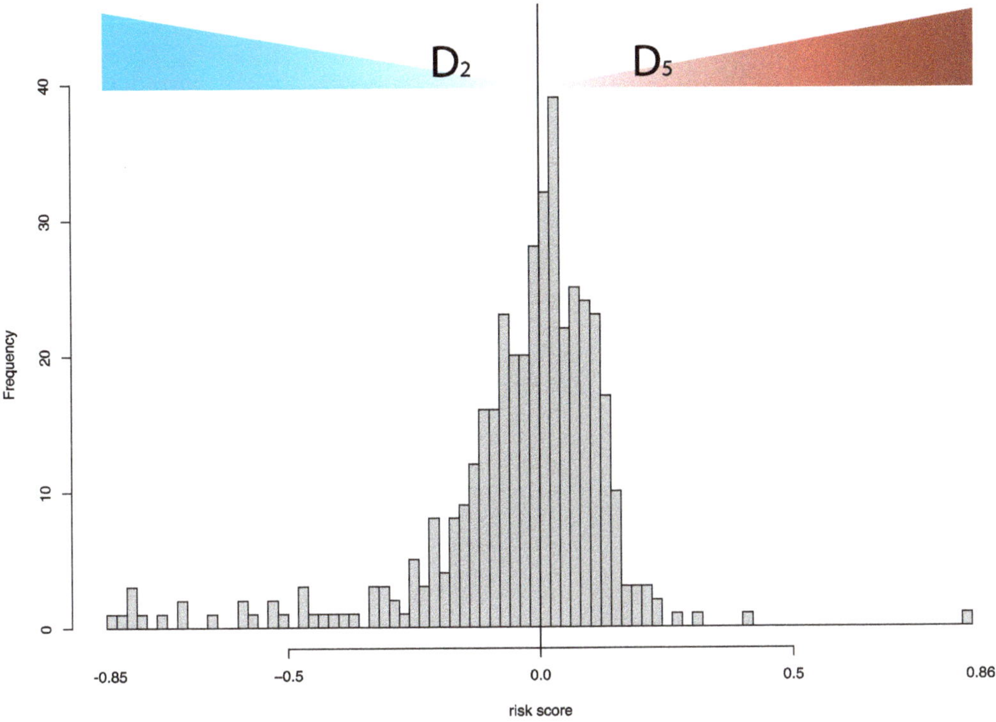

Figure 5. The distribution of the case risk scores is determined by D2 and D5, which are the relative frequencies of patches for the second and fifth deciles of the prediction for each case. The frequency corresponds to the case number. The cohort-specific threshold was estimated to be 0 for the TCGA dataset. Thresholding the risk scores results in two risk groups, where D2 and the high-risk group by D5 dominate the low-risk group. Figure 6 illustrates the histopathologic patterns associated with D2 and D5.

Table 4. The cohort description of the external validation set.

Characteristic	N = 412
Age at diagnosis in years, median (IQR)	68 (60–76)
Sex, n (%)	
Female	107 (26%)
Male	305 (74%)
pM	
M0/x	398 (97%)
M1	11 (2.7%)
Unknown	3 (0.7%)
pN	
N0x	282 (68%)
M1	123 (30%)
Unknown	7 (1.7%)
pT	
T1	2 (0.5%)
T2	112 (27%)
T3	190 (46%)
T4	54 (13%)
Unknown	54 (13%)

Table 4. *Cont.*

Characteristic	N = 412
Grade, n (%)	
Unknown	1 (0.2%)
High grade	390 (95%)
Low grade	21 (5.1%)
History of non-muscle invasive bladder cancer, n (%)	
Unknown	127 (31%)
NO	227 (55%)
YES	58 (14%)
Bladder cancer pathologic stage, n (%)	
I–II	151 (36.7%)
III	130 (31.6%)
IV	130 (31.6%)
Unknown	1 (0.2%)
Death, n (%)	185 (45%)
Follow-up duration in month, median (IQR)	19 (12–33)

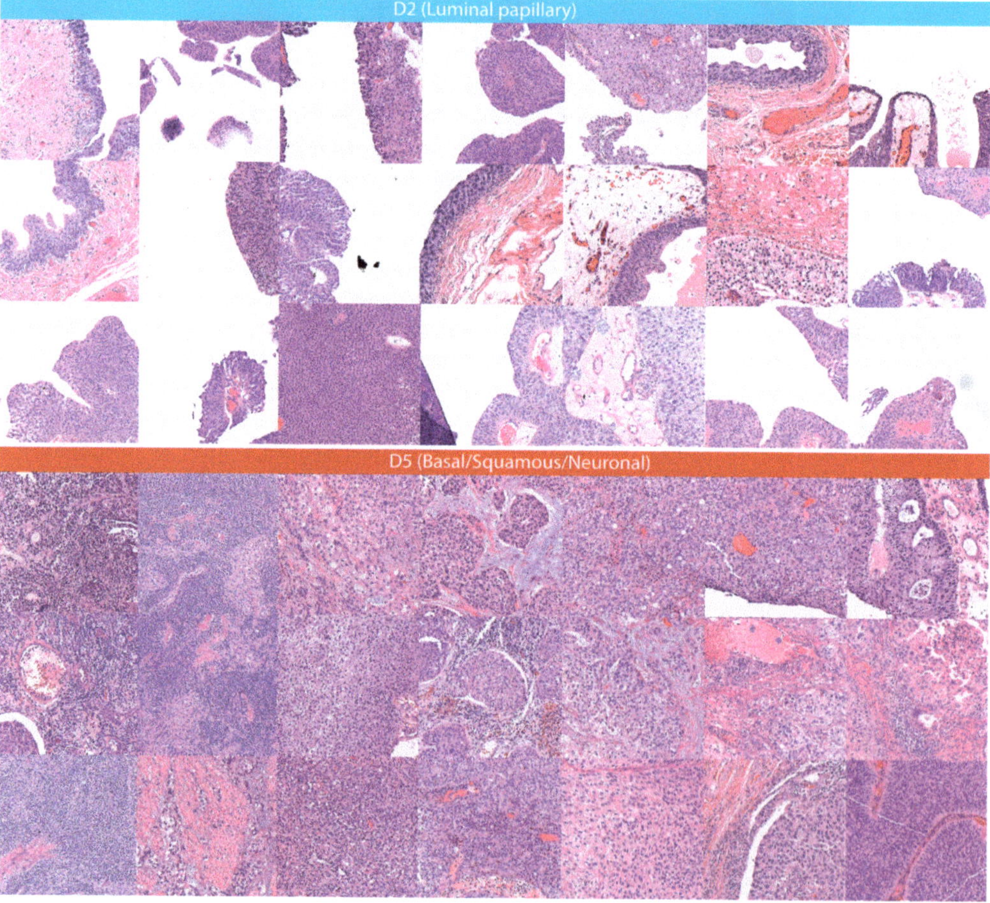

Figure 6. Exemplifies the distinct histopathologic patterns for D2 and D5 on the TCGA cohort. The absolute difference in the proportions between D2 and D5 in histology images determines whether the case is assigned to a low- or high-risk group (Figure 5). A negligible small fraction of patches in D2 solely included arteria vessels as luminal structures.

We found that the risk groups are prognostic for overall survival on the external validation set (HR: 1.46; 95% CI: 1.05–2.02; z: 2.23; p = 0.03). The multivariate Cox regression analysis showed that risk groups are, in addition to the pathologic stage and age at diagnosis, independent prognosticators for overall survival as well (Table 5). The multicollinearity for these covariates was negligibly small (VIFs < 2).

Table 5. Multivariate Cox regression analysis for overall mortality. HR: Hazard ratio, CI: Confidence Interval. The AJCC pathologic tumor stage is a result of combining the subcategories of the TNM classification. We excluded the tumor grade as the muscle invasive bladder cancers are typically high-grade, and 95% of tumor grades in our cohort has high-grade BC.

Variable	HR	95% CI	z	p
High- vs. Low-risk group	1.35	(1.01–1.80)	1.99	0.0462
Age at diagnosis	1.02	(1.00–1.03)	2.32	0.0201
AJCC pathologic tumor stage				
I/II (ref.)	–	–	–	–
III	1.51	(1.03–2.21)	2.10	0.0357
IV	2.21	(1.54–3.18)	4.30	<0.0001

The Kaplan–Meier curve and the log-rank test indicate that the risk groups were statistically distinct (p = 0.037), as shown in Figure 7. Both risk groups reached the median overall survival, but at different time points (~30 months for high-risk vs. ~60 months for low-risk); the high-risk group reached the median survival ~2.5 years earlier than the low-risk group for muscle-invasive bladder cancers. Figure 8 provides the Kaplan–Meier curve for the stages of bladder cancer for comparison.

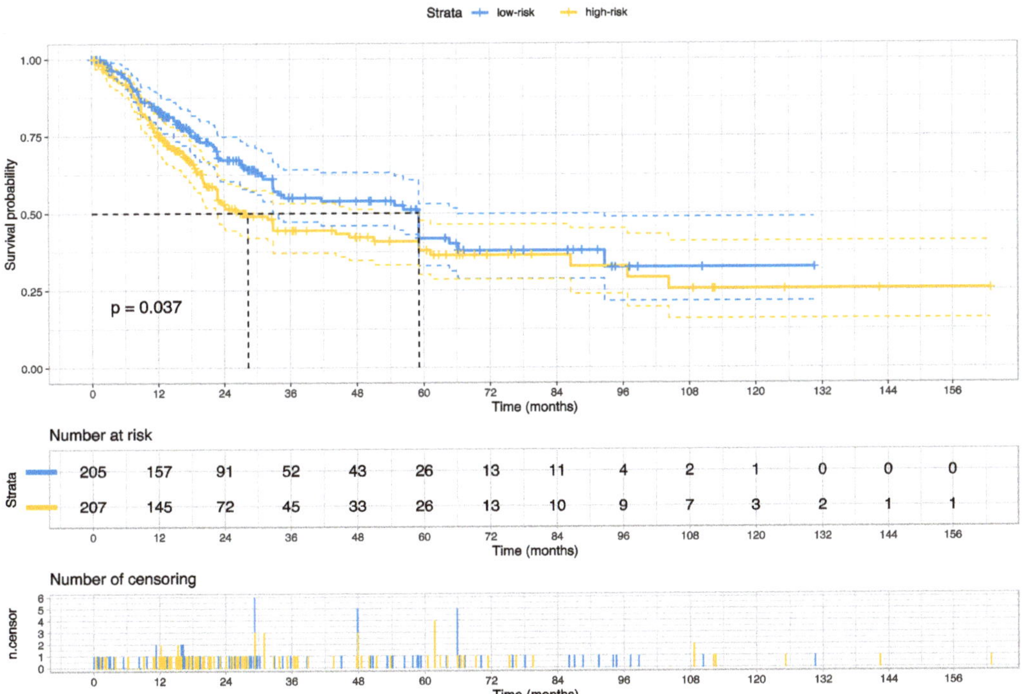

Figure 7. The Kaplan–Meier curve for overall survival stratified by the categorized risk scores (low-risk vs. high-risk) on the external validation set (TCGA cohort). p value was estimated using the log Rank test. The dot lines reveal the median survival for each risk group.

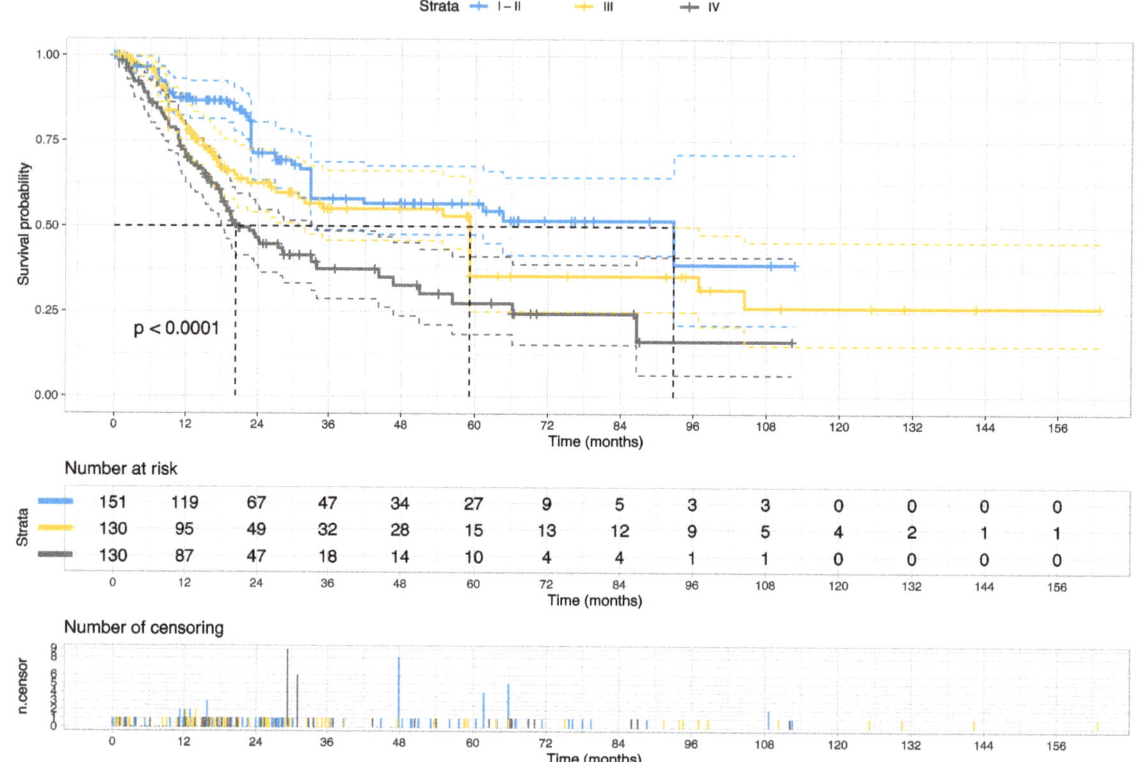

Figure 8. The Kaplan–Meier curve for overall survival stratified by the AJCC pathologic stages of bladder cancer on the external validation set (the TCGA Cohort). This staging system combines the subcategories of the TNM classification. *p* value was estimated using the log Rank test. The single case with unknown stage information was not visualized. The dot lines reveal the median survival.

3.3. Association with Molecular Signatures of Bladder Cancer

We identified molecular and pathologic signatures significantly associated with the risk groups at case level, as shown in Table 6. Specifically, the TCGA clusters for miRNA, mRNA, lncRNA, and DNA methylation were associated with our novel risk groups. In addition, multiple mutations, including TSC1, FGFR3, and ERBB3, occurred differently between the novel risk groups.

The luminal papillary cluster was associated with the low-risk group, whereas the basal/squamous cluster and the neuronal cluster were associated with the high-risk group (Table 7). Moreover, cluster 2 for DNA hypomethylation is associated with the high-risk group; in contrast, cluster 4, with lesser DNA hypomethylation than cluster 2, was associated with the low-risk group. At the long non-coding RNA level, cluster 3 was frequently seen in the low-risk group and cluster 4 in the high-risk group. At the miRNA level, cluster 3 was more frequent in the low-risk group, and cluster 4 was common in the high-risk group.

The low-risk group included 72% of the TSC1 mutation (28 of 39 TSC1 mutations) or 67% of the ERBB3 mutation (30 of 45 ERBB3 mutations) in bladder cancer (Tables 8 and 9). The odd ratio of TSC1 mutation was 0.36 (95% CI: 0.15–0.76; *p* = 0.004), and the odd ratio of ERBB3 was 0.46 (95% CI: 0.22–0.91; *p* = 0.0179) for high-risk groups.

The true positive rate of our low-risk group was 65% for FGFR3 mutations (Table 10) with an AUC of 0.593 (95% CI: 0.55–0.69). The odd ratio for FGFR3 mutation in the high-risk group was 0.49 (95% CI, 0.27–0.87; *p* = 0.0102). The high-risk group included 63.3%

of the squamous pathology. The supplementary section provides different results for significant signatures.

Table 6. Analysis summary of signatures and features associated with the risk group. *p*-values for a signature were estimated using Chi-Squared tests.

Signature	*p* Value	Features Associated with (↓) Low-Risk or (↑) High-Risk Group
microRNA cluster	0.003998001	(↓) Cluster 3 (↑) Cluster 1
mutation in TSC1	0.006496752	(↓) TSC1 mutation
mRNA cluster	0.009995002	(↓) Luminal papillary (↑) Basal/Squamous (↑) Neuronal
mutation in FGFR3	0.010994503	(↓) FGFR3 mutation
lncRNA cluster	0.012493753	(↓) Cluster 3 (↑) Cluster 4
mutation in ERBB3	0.016991504	(↓) ERBB3 mutation
mutation in FAT1	0.023488256	(↓) FAT1 mutation
mutation in PIK3CA	0.028485757	(↓) PIK3CA mutation
mutation in KANSL1	0.033983008	(↑) KANSL1 mutation
mutation in TMCO4	0.038480760	(↓) TMCO4 mutation
mutation in KDM6A	0.044977511	(↓) KDM6A mutation
mutation in METTL3	0.057971014	(↓) METL3 mutation
Squamous pathology	0.066466767	(↑) Squamous histopathology
mutation in PSIP1	0.075462269	(↓) PSIP1 mutation
mutation in ZNF773	0.092453773	(↓) ZNF773 mutation
Hypomethylation cluster	0.092953523	(↓) Cluster 4 (↑) Cluster 2
mutation in GNA13	0.093953023	(↓) GNA13 mutation

Table 7. The distribution of molecular clusters for mRNA, lncRNA, miRNA, and DNA hypomethylation.

Risk Groups	Molecular Signatures	
	mRNA	
	Luminal papillary	Basal/Squamous/Neuronal
Low-risk	85 (59%)	56 (36%)
High-risk	58 (41%)	101 (64%)
	lncRNA	
	Cluster 3	Cluster 4
Low-risk	47 (64%)	61 (41%)
High-risk	26 (36%)	87 (59%)

Table 7. *Cont.*

Risk Groups	Molecular Signatures	
	miRNA	
	Cluster 3	Cluster 1
Low-risk	77 (62%)	30 (39%)
High-risk	47 (38%)	47 (61%)
	DNA hypomethylation	
	Cluster 4	Cluster 2
Low-risk	23 (68%)	27 (39%)
High-risk	11 (32%)	42 (61%)

Table 8. The distribution of TSC1 mutation between the risk groups.

	TSC1 Gene	
Risk groups	wild-type	mutated
Low-risk	177 (47%)	28 (72%)
High-risk	196 (53%)	11 (28%)

Table 9. The distribution of ERBB3 mutation between the risk groups.

	ERBB3 Gene	
Risk groups	wild-type	mutated
Low-risk	175 (48%)	30 (67%)
High-risk	192 (52%)	15 (33%)

Table 10. The distribution of FGFR mutation between the risk groups.

	FGFR3 Gene	
Risk groups	wild-type	mutated
Low-risk	163 (47%)	42 (65%)
High-risk	184 (53%)	23 (35%)

4. Discussion

In this study, we developed and externally validated an AI-based algorithm that stratifies muscle-invasive bladder cancer by mortality risk directly from histology images. Moreover, our novel risk groups can reveal which histopathological pattern is dominant in tissues with bladder cancers. Our approach is feasible thanks to the intuitively well-sorted feature space generated by weakly supervised learning. This property has made it possible to discretize the feature space into ten small segments organized decile-wise, allowing us to evaluate the histopathological patterns for each prediction decile.

Earlier studies in bladder cancer applied deep learning to infer staging [31], grade [32,33], recurrence risk [34], FGFR3 mutation [35], and specific molecular subtypes [7] from histology images. Although some previous studies examined the prediction of molecular targets, the current study found that prognostic histopathological patterns for bladder cancer are rather associated with multi-omics profiles (i.e., transcriptomic, genomics, and epigenomics); these multi-omics profiles are already covering the specific molecular subtypes and the FGFR3 mutations investigated earlier, and we have shown that the accuracy of our risk groups for FGFR3 mutation is similar to the previous report, signifying the impact of multi-omics profiles as confounding factors on the results of earlier studies. In support of our findings, the BLCA-TCGA study (molecular characterization of bladder cancer) revealed that the molecular subtypes and signatures are linked with each other and distinct histopathologic patterns (e.g., papillary, basal/squamous) were connected with omics profiles that are prognostic and have different therapeutic targets [3,17]. A comparable study in Lung cancer reported that omics features are predictive of histology patterns as well [36].

Although multiple studies identified the detection potential of single mutations or specific molecular subtypes from histology images [37–41], the histopathological appearance is mainly driven by a collection of multifaceted molecular modulations and reflects the cancer malignancy and survival. Subsequently, establishing a direct association between a single molecular signature and histology images must be inadequate, given other confounders for bladder cancers.

Our novel risk groups are linked with therapeutic targets like FGFR3 (erdafitinib) [42], ERBB3 (afatinib) [43], PI(3)K (LY294002, other mTOR inhibitors) [44,45], and TSC1 (nab-sirolimus, study no.: NCT05103358) [3] as well as with female gender-biased gene mutations like KDM6A mutation (a histone lysine demethylase) [46]. Accordingly, our novel risk group holds a potential clinical utility in pre-screening for mono- and combinational target therapies (Figure 9). This potential will be more evident once prospective randomized studies to validate the clinical utility of our approach for patient selection in the real-world clinical setting are available.

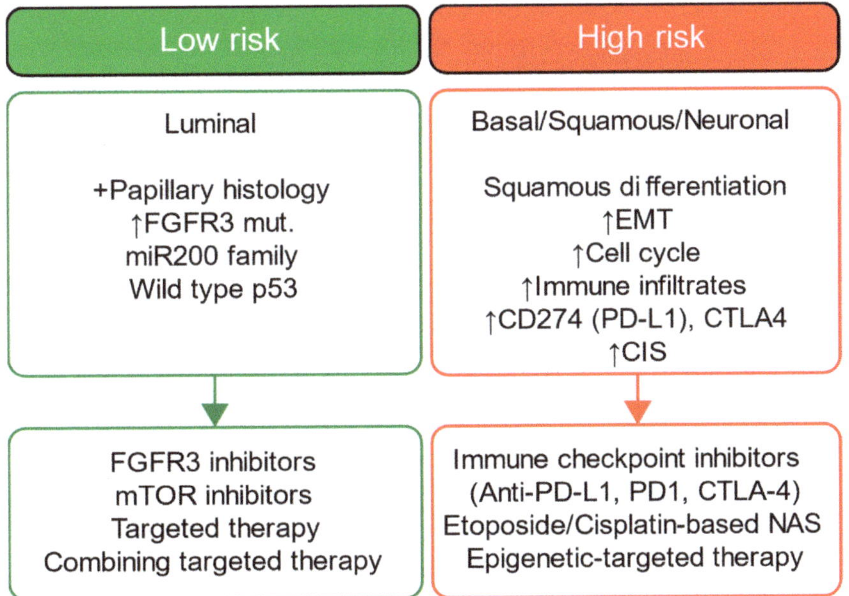

Figure 9. Overview of each risk group's molecular characteristics and proposed treatment options. CIS: Carcinoma in situ; NAS: neoadjuvant chemotherapy; EMT: Epithelial–Mesenchymal Transition. The information is based on the TCGA-BLCA studies that investigated the treatment responses of main molecular subtypes (i.e., luminal, basal, squamous, and neural subtypes). We emphasize that this overview is abstract and not comprehensive and aims to generate hypotheses for potential treatment options for each risk group. The overview covers only the common main molecular subtypes (i.e., luminal and basal) for each risk group. The molecular features for these subtypes are already investigated by TCGA-BLCA studies.

A detailed examination of the multi-omics profiles associated with our risk groups reveals unique molecular regulatory profiles at the microRNA, lncRNA, and DNA methylation levels. We found that the low-risk group is linked with molecular subtypes with good survival for coding and non-coding RNAs or DNA methylation. These multi-omics sub types are associated with papillary tumors, high FGFR3 mutations and miR-200 levels, and low Epithelial–Mesenchymal Transition (EMT) scores, CD274 (PD-L1) and PDCD1 (PD-1) level [17]. In contrast, the high-risk group is linked with molecular subtypes with poor survival for coding and non-coding RNA, which are further associated with lymphocyte infiltration, the high expression of CIS (carcinoma in situ) signature genes, CD274 (PD-L1) and PDCD1 (PD-1) levels, high TP53 mutations and EMT scores [17]. The high-risk group is additionally linked to cluster 2 for DNA hypomethylation, which has more DNA hyperme thylation signals (more gene inactivation) than cluster 4, which is linked with the low-risk group [17]. Our data further facilitates deriving a hypothesis that the low-risk group, with favorable multi-omics profiles, is likely more responsive to different targeted therapies than the high-risk group, and the high-risk group may benefit from immune checkpoint inhibitors (i.e., anti-PD-1 or PD-L1); our data also suggest that epigenetic therapy could be a potential therapeutic option for our high-risk group. Figure 9 summarizes each risk group's molecular characteristics and potential treatment options.

Comparable studies utilized activation maps or tiles with top scores to interpret the model inference. However, the trustworthiness of activation maps could be more questionable as deep neural network classifiers have an opportunistic nature, and the existing saliency methods inherit a high risk for misinterpretation, limited reproducibility and sparse visualization [47,48]. Moreover, it should be considered that tiles with top scores ignore the variance in histology patterns between two categories after thresholding predictions, as evident by our data on the correlation between histology patterns and prediction deciles.

We applied the neural architecture search to achieve a data-driven architecture design with a better trade-off between accuracy, interpretability, and model complexity. In our study, only 20 feature representations (i.e., the 2D feature maps of the last convolutional layer) are sufficient to derive accurate predictions from histology images and correspond for example, to 4% of feature representations of ResNet18 [49] (i.e., 512 features), an off-the-shelf model commonly used in medical imaging research. Reducing the feature representation is associated with a better computation cost for downstream analysis and im proved human interpretation of these features. Moreover, our approach helps visualize and analyze three-dimensional representative features that preserve topological information at reasonable computation costs (e.g., analysis of 8,000,000 data points required ~30 min using parallel computing). In contrast, comparable studies that utilized off-the-shelf models are limited by extremely reduced feature granularity (1D) with loss of topological information for downstream analysis, given the high computation cost to analyze a large number of 3D representative features that these models have. Accordingly, comparable studies excluded the most information from the feature representation to conduct downstream analysis In contrast, our approach preserves the high granularity of the feature representation for downstream analysis and consequently improves the interpretability of our AI model.

Despite the strengths of our study, it is essential to acknowledge certain limitations Firstly, using slide images introduces potential variability in image quality due to factors such as diverse scanning technologies, staining variations, and image artifacts. These variations can introduce inconsistencies that may impact the accuracy and reliability of image analysis and interpretation. Nevertheless, we took measures to mitigate this concern by using PlexusNET to address the domain shift [16], conducting a comprehensive manual review involving domain experts and validating our findings on multicentric datasets. Additionally, we employed feature visualization techniques to identify the potential impact of artifacts and reviewed for the staining variations on the selected histological patterns Secondly, it is crucial to recognize that TCGA slide images offer a glimpse of a specific tumor region or patient sample, which may not fully capture the complex intra- and inter-tumor heterogeneity. Tumors can exhibit spatial and molecular heterogeneity, resulting in significant variations between different regions within the same tumor or among tumors of the same type. Analyzing only a subset of slide images may provide an incomplete representation of tumor characteristics. Nonetheless, it is noteworthy that the TCGA and PLCO study followed good research practices, aiming to select the most representative samples from each patient according to the existing technical feasibility. Moreover, the quality of survival data of TCGA was validated for overall survival analyses [18]. The good research practices and the data quality help mitigate this limitation to some extent. It is important to emphasize that TCGA slide images, obtained through the TCGA project, do not directly correspond to the specific sampling areas used for molecular examination. These images are prepared using Hematoxylin and Eosin (HE) staining, a common technique for histological analysis. In contrast, molecular examinations and profiling involve separate samples or portions of the tumor that undergo different processing steps. TCGA employs distinct protocols for various analyses, including genomic, transcriptomic, and proteomic profiling, which are not directly applied to the same tissue sections used for generating slide images. These protocols often utilize specialized techniques, such as DNA sequencing or protein expression analysis, requiring separate tissue preparation and processing. Hence, it is crucial to note that TCGA slide images, while they provide valuable histological information, do not directly correlate with the specific regions of the tumor that underwent molecular examination. Rather, they serve as representative snapshots of the tumor's morphology and architecture, offering valuable context for researchers analyzing the genomic and molecular data obtained from the TCGA project. We preferred slide images with formalin-fixed paraffin-embedded (FFPE) tissues as this approach offers standardized staining and more reliable histology images. In contrast, the process of preparing and staining frozen tissue slides are demanding and often result in associated artifacts; freezing can cause structural changes and cellular damage, while its staining consistency can be challenging due to variations in tissue quality and protocols [50–52]. Finally, histology images from frozen sections are also snapshots, contrary to a common misconception that assumes these images are direct complements to the entire TCGA samples.

The current study introduces a novel AI-based risk grouping system for survival derived from bladder cancer H&E slides. We show the linkage between our risk groups and multi-omics profiles for muscle-invasive bladder cancers. We highlight the concerns with predicting single molecular signatures (e.g., FGFR3) from histology images. While our approach has been rigorously tested and validated in the context of bladder cancer, its applicability extends beyond this specific disease.

Challenges and Future Directions

The present work underscores the significance of associating feature space distributions with prediction scores for the purpose of developing an interpretable scoring system for the mortality prediction. One of the prevailing challenges within the medical domain pertains to the divergence between the development dataset and unseen cohorts, which poses a persistent issue for existing algorithms. In response to this challenge, we have introduced a normalization strategy tailored for out-of-distribution cohorts, which seeks to mitigate skewness, following the principles of the central limit theorem. Our proposed normalization technique necessitates the utilization of a representative cohort to ensure the reliability of outcomes. Furthermore, we have put forth a continuous normalization approach with instantaneous threshold adjustments; this, however, requires either a latency period or initial representative data for accurate normalization. Another challenge that need to be addressed is the application boundary of our approach. The application boundary is generally determined by the image quality as well as the cohort characterization of the development set. One of the foremost challenges lies in harmonizing and integrating multi-omics data, including transcriptomics, genomics, and epigenomics. Future research should focus on developing robust methodologies and computational tools to streamline such a process, including all available data types. Integrating multi-omics analysis into the clinical workflow is a significant challenge.

Future efforts will focus on validating our approach for clinical utility to optimize the treatment management for bladder cancer. Digital biomarkers, such as histomics, have the potential to serve as companion variables for disease staging and patient selection. Future research should also explore integration with Electronic Health Records (EHRs) and decision support systems, ensuring clinicians can access and utilize the integrated data efficiently. Integrating multi-omics data can further our understanding of disease mechanisms, potentially leading to breakthroughs in treatment and prevention. Yet, it is not clear whether omics strategies provide superior clinical benefits compared to a single data modality. Finally, possessing a scoring system that captures the omics features of the underlying disease from a single image modality (in our case, FFPE histology images) may help justify customizing the molecular profiling in the clinical setting.

5. Conclusions

Our scoring system has the potential to facilitate shared decision making by offering comprehensive and precise risk stratification, treatment planning, and cost-effective preselection for expensive molecular characterization.

Supplementary Materials: The following supporting information can be downloaded at https://www.mdpi.com/article/10.3390/cancers15204998/s1. File S1: Supplementary Tables for "Artificial Intelligence Reveals Distinct Prognostic Subgroups of Muscle-Invasive Bladder Cancer on Histology Images".

Author Contributions: O.E.: Study design, data collection, analysis and writing original draft. S.-R.L.-B., S.F.S., L.-M.K., H.L., L.X. and M.A.: Critical review of data analysis and interpretation, review, and manuscript editing. M.A.: Supervision. All authors have read and agreed to the published version of the manuscript.

Funding: This research received no external funding.

Institutional Review Board Statement: The study was conducted in accordance with the Declaration of Helsinki. Since these data are publicly available, no specific ethics approval is required.

Informed Consent Statement: Informed consent was obtained from all subjects involved in the study and managed by the respective organizers (TCGA and PLCO). The current study was conducted in accordance with the Declaration of Helsinki, and the respective study organizers were responsible for obtaining the ethical approval.

Data Availability Statement: TCGA data and PCLO data are publicly available. The python jupyter notebook with python codes will be available at https://github.com/oeminaga/AI_BladderCancerMortality.git

Acknowledgments: We are thankful to the organization of PLCO and TCGA for data sharing to realize this work.

Conflicts of Interest: The authors declare no conflict of interest.

References

1. Lobo, N.; Afferi, L.; Moschini, M.; Mostafid, H.; Porten, S.; Psutka, S.P.; Gupta, S.; Smith, A.B.; Williams, S.B.; Lotan, Y. Epidemiology, Screening, and Prevention of Bladder Cancer. *Eur. Urol. Oncol.* **2022**, *5*, 628–639. [CrossRef] [PubMed]
2. Ferlay, J. GLOBOCAN 2008 v1. 2, Cancer Incidence and Mortality World-Wide: IARC Cancer Base No. 10. Available online: https://gco.iarc.fr (accessed on 10 May 2023).
3. The Cancer Genome Atlas Research Network. Comprehensive molecular characterization of urothelial bladder carcinoma. *Nature* **2014**, *507*, 315–322. [CrossRef]
4. Echle, A.; Rindtorff, N.T.; Brinker, T.J.; Luedde, T.; Pearson, A.T.; Kather, J.N. Deep learning in cancer pathology: A new generation of clinical biomarkers. *Br. J. Cancer* **2021**, *124*, 686–696. [PubMed]
5. Lafarge, M.W.; Koelzer, V.H. Towards computationally efficient prediction of molecular signatures from routine histology images. *Lancet Digit. Health* **2021**, *3*, e752–e753. [CrossRef]
6. Nayak, T.; Chadaga, K.; Sampathila, N.; Mayrose, H.; Gokulkrishnan, N.; Bairy, G.M.; Prabhu, S.; Swathi, K.S.; Umakanth, S. Deep learning based detection of monkeypox virus using skin lesion images. *Med. Nov. Technol. Devices* **2023**, *18*, 100243. [CrossRef]

7. Woerl, A.C.; Eckstein, M.; Geiger, J.; Wagner, D.C.; Daher, T.; Stenzel, P.; Fernandez, A.; Hartmann, A.; Wand, M.; Roth, W.; et al. Deep Learning Predicts Molecular Subtype of Muscle-invasive Bladder Cancer from Conventional Histopathological Slides. *Eur. Urol.* **2020**, *78*, 256–264. [CrossRef]

8. Mundhada, A.; Sundaram, S.; Swaminathan, R.; D'Cruze, L.; Govindarajan, S.; Makaram, N. Differentiation of urothelial carcinoma in histopathology images using deep learning and visualization. *J. Pathol. Inform.* **2023**, *14*, 100155. [CrossRef]

9. Zheng, Q.; Yang, R.; Ni, X.; Yang, S.; Xiong, L.; Yan, D.; Xia, L.; Yuan, J.; Wang, J.; Jiao, P.; et al. Accurate Diagnosis and Survival Prediction of Bladder Cancer Using Deep Learning on Histological Slides. *Cancers* **2022**, *14*, 5807. [CrossRef]

10. Nguyen, A.; Yosinski, J.; Clune, J. Deep neural networks are easily fooled: High confidence predictions for unrecognizable images. In Proceedings of the IEEE Conference on Computer Vision and Pattern Recognition, Boston, MA, USA, 7–12 June 2015; pp. 427–436.

11. Guo, C.; Pleiss, G.; Sun, Y.; Weinberger, K.Q. On calibration of modern neural networks. In Proceedings of the International Conference on Machine Learning, Sydney, Australia, 6–11 August 2017; pp. 1321–1330.

12. Team, P.P.; Gohagan, J.K.; Prorok, P.C.; Hayes, R.B.; Kramer, B.-S. The prostate, lung, colorectal and ovarian (PLCO) cancer screening trial of the National Cancer Institute: History, organization, and status. *Control. Clin. Trials* **2000**, *21*, 251S–272S.

13. Andriole, G.L.; Crawford, E.D.; Grubb III, R.L.; Buys, S.S.; Chia, D.; Church, T.R.; Fouad, M.N.; Gelmann, E.P.; Kvale, P.A.; Reding, D.J. Mortality results from a randomized prostate-cancer screening trial. *N. Engl. J. Med.* **2009**, *360*, 1310–1319. [CrossRef]

14. Hasson, M.A.; Fagerstrom, R.M.; Kahane, D.C.; Walsh, J.H.; Myers, M.H.; Caughman, C.; Wenzel, B.; Haralson, J.C.; Flickinger, L.M.; Turner, L.M.; et al. Design and evolution of the data management systems in the Prostate, Lung, Colorectal and Ovarian (PLCO) Cancer Screening Trial. *Control. Clin. Trials* **2000**, *21*, 329S–348S. [CrossRef] [PubMed]

15. Pinsky, P.F.; Prorok, P.C.; Yu, K.; Kramer, B.S.; Black, A.; Gohagan, J.K.; Crawford, E.D.; Grubb, R.L.; Andriole, G.L. Extended mortality results for prostate cancer screening in the PLCO trial with median follow-up of 15 years. *Cancer* **2017**, *123*, 592–599. [CrossRef] [PubMed]

16. Eminaga, O.; Abbas, M.; Shen, J.; Laurie, M.; Brooks, J.D.; Liao, J.C.; Rubin, D.L. PlexusNet: A neural network architectural concept for medical image classification. *Comput. Biol. Med.* **2023**, *154*, 106594. [CrossRef]

17. Robertson, A.G.; Kim, J.; Al-Ahmadie, H.; Bellmunt, J.; Guo, G.; Cherniack, A.D.; Hinoue, T.; Laird, P.W.; Hoadley, K.A.; Akbani, R.; et al. Comprehensive Molecular Characterization of Muscle-Invasive Bladder Cancer. *Cell* **2018**, *174*, 1033. [CrossRef]

18. Liu, J.; Lichtenberg, T.; Hoadley, K.A.; Poisson, L.M.; Lazar, A.J.; Cherniack, A.D.; Kovatich, A.J.; Benz, C.C.; Levine, D.A.; Lee, A.V.; et al. An Integrated TCGA Pan-Cancer Clinical Data Resource to Drive High-Quality Survival Outcome Analytics. *Cell* **2018**, *173*, 400–416.e411. [CrossRef] [PubMed]

19. Heagerty, P.J.; Lumley, T.; Pepe, M.S. Time-dependent ROC curves for censored survival data and a diagnostic marker. *Biometrics* **2000**, *56*, 337–344. [CrossRef]

20. Heller, G.; Mo, Q. Estimating the concordance probability in a survival analysis with a discrete number of risk groups. *Lifetime Data Anal.* **2016**, *22*, 263–279. [CrossRef]

21. Uno, H.; Cai, T.; Pencina, M.J.; D'Agostino, R.B.; Wei, L.J. On the C-statistics for evaluating overall adequacy of risk prediction procedures with censored survival data. *Stat. Med.* **2011**, *30*, 1105–1117. [CrossRef]

22. Sakamoto, Y.; Ishiguro, M.; Kitagawa, G. Akaike information criterion statistics. *J. Am. Stat. Assoc.* **1986**, *81*, 26853.

23. Vrieze, S.I. Model selection and psychological theory: A discussion of the differences between the Akaike information criterion (AIC) and the Bayesian information criterion (BIC). *Psychol. Methods* **2012**, *17*, 228. [CrossRef]

24. Neath, A.A.; Cavanaugh, J.E. The Bayesian information criterion: Background, derivation, and applications. *Wiley Interdiscip. Rev. Comput. Stat.* **2012**, *4*, 199–203. [CrossRef]

25. Benjamini, Y.; Hochberg, Y. Controlling the false discovery rate: A practical and powerful approach to multiple testing. *J. R. Stat. Soc. Ser. B* **1995**, *57*, 289–300. [CrossRef]

26. Schober, P.; Boer, C.; Schwarte, L.A. Correlation coefficients: Appropriate use and interpretation. *Anesth. Analg.* **2018**, *126*, 1763–1768. [CrossRef]

27. Mukaka, M.M. Statistics corner: A guide to appropriate use of correlation coefficient in medical research. *Malawi. Med. J.* **2012**, *24*, 69–71. [PubMed]

28. Craney, T.A.; Surles, J.G. Model-dependent variance inflation factor cutoff values. *Qual. Eng.* **2002**, *14*, 391–403.

29. Gulli, A.; Pal, S. *Deep Learning with Keras*; Packt Publishing Ltd.: Birmingham, UK, 2017.

30. Abadi, M.; Barham, P.; Chen, J.; Chen, Z.; Davis, A.; Dean, J.; Devin, M.; Ghemawat, S.; Irving, G.; Isard, M. Tensorflow: A system for large-scale machine learning. In Proceedings of the 12th USENIX Symposium on Operating Systems Design and Implementation ({OSDI} 16), Savannah, GA, USA, 2–4 November 2016; pp. 265–283.

31. Fuster, S.; Khoraminia, F.; Kiraz, U.; Kanwal, N.; Kvikstad, V.; Eftestøl, T.; Zuiverloon, T.C.; Janssen, E.A.; Engan, K. Invasive cancerous area detection in Non-Muscle invasive bladder cancer whole slide images. In Proceedings of the 2022 IEEE 14th Image, Video, and Multidimensional Signal Processing Workshop (IVMSP), Nafplio, Greece, 26–29 June 2022; pp. 1–5.

32. Wenger, K.; Tirdad, K.; Cruz, A.D.; Mari, A.; Basheer, M.; Kuk, C.; van Rhijn, B.W.; Zlotta, A.R.; van der Kwast, T.H.; Sadeghian, A. A semi-supervised learning approach for bladder cancer grading. *Mach. Learn. Appl.* **2022**, *9*, 100347.

33. Zhang, Z.; Chen, P.; McGough, M.; Xing, F.; Wang, C.; Bui, M.; Xie, Y.; Sapkota, M.; Cui, L.; Dhillon, J. Pathologist-level interpretable whole-slide cancer diagnosis with deep learning. *Nat. Mach. Intell.* **2019**, *1*, 236–245.

34. Lucas, M.; Jansen, I.; van Leeuwen, T.G.; Oddens, J.R.; de Bruin, D.M.; Marquering, H.A. Deep Learning-based Recurrence Prediction in Patients with Non-muscle-invasive Bladder Cancer. *Eur. Urol. Focus* **2022**, *8*, 165–172. [CrossRef]
35. Loeffler, C.M.L.; Ortiz Bruechle, N.; Jung, M.; Seillier, L.; Rose, M.; Laleh, N.G.; Knuechel, R.; Brinker, T.J.; Trautwein, C.; Gaisa, N.T.; et al. Artificial Intelligence-based Detection of FGFR3 Mutational Status Directly from Routine Histology in Bladder Cancer: A Possible Preselection for Molecular Testing? *Eur. Urol. Focus* **2022**, *8*, 472–479. [CrossRef]
36. Yu, K.-H.; Berry, G.J.; Rubin, D.L.; Re, C.; Altman, R.B.; Snyder, M. Association of omics features with histopathology patterns in lung adenocarcinoma. *Cell Syst.* **2017**, *5*, 620–627.e3.
37. Kather, J.N.; Heij, L.R.; Grabsch, H.I.; Loeffler, C.; Echle, A.; Muti, H.S.; Krause, J.; Niehues, J.M.; Sommer, K.A.; Bankhead, P. Pan-cancer image-based detection of clinically actionable genetic alterations. *Nat. Cancer* **2020**, *1*, 789–799. [PubMed]
38. Sirinukunwattana, K.; Domingo, E.; Richman, S.D.; Redmond, K.L.; Blake, A.; Verrill, C.; Leedham, S.J.; Chatzipli, A.; Hardy, C.; Whalley, C.M. Image-based consensus molecular subtype (imCMS) classification of colorectal cancer using deep learning. *Gut* **2021**, *70*, 544–554. [PubMed]
39. Coudray, N.; Tsirigos, A. Deep learning links histology, molecular signatures and prognosis in cancer. *Nat. Cancer* **2020**, *1*, 755–757. [CrossRef] [PubMed]
40. Diao, J.A.; Wang, J.K.; Chui, W.F.; Mountain, V.; Gullapally, S.C.; Srinivasan, R.; Mitchell, R.N.; Glass, B.; Hoffman, S.; Rao, S.K. Human-interpretable image features derived from densely mapped cancer pathology slides predict diverse molecular phenotypes. *Nat. Commun.* **2021**, *12*, 1–15.
41. Hong, R.; Liu, W.; DeLair, D.; Razavian, N.; Fenyö, D. Predicting endometrial cancer subtypes and molecular features from histopathology images using multi-resolution deep learning models. *Cell Rep. Med.* **2021**, *2*, 100400. [CrossRef]
42. Loriot, Y.; Necchi, A.; Park, S.H.; Garcia-Donas, J.; Huddart, R.; Burgess, E.; Fleming, M.; Rezazadeh, A.; Mellado, B.; Varlamov, S.; et al. Erdafitinib in Locally Advanced or Metastatic Urothelial Carcinoma. *N. Engl. J. Med.* **2019**, *381*, 338–348. [CrossRef]
43. Choudhury, N.J.; Campanile, A.; Antic, T.; Yap, K.L.; Fitzpatrick, C.A.; Wade, J.L., 3rd; Karrison, T.; Stadler, W.M.; Nakamura, Y.; O'Donnell, P.H. Afatinib Activity in Platinum-Refractory Metastatic Urothelial Carcinoma in Patients With ERBB Alterations. *J. Clin. Oncol.* **2016**, *34*, 2165–2171. [CrossRef]
44. Vanhaesebroeck, B.; Perry, M.W.D.; Brown, J.R.; Andre, F.; Okkenhaug, K. PI3K inhibitors are finally coming of age. *Nat. Rev. Drug Discov.* **2021**, *20*, 741–769. [CrossRef]
45. Ching, C.B.; Hansel, D.E. Expanding therapeutic targets in bladder cancer: The PI3K/Akt/mTOR pathway. *Lab. Investig.* **2010**, *90*, 1406–1414. [CrossRef]
46. Hurst, C.D.; Alder, O.; Platt, F.M.; Droop, A.; Stead, L.F.; Burns, J.E.; Burghel, G.J.; Jain, S.; Klimczak, L.J.; Lindsay, H.; et al. Genomic Subtypes of Non-invasive Bladder Cancer with Distinct Metabolic Profile and Female Gender Bias in KDM6A Mutation Frequency. *Cancer Cell* **2017**, *32*, 701–715.e707. [CrossRef]
47. Saporta, A.; Gui, X.; Agrawal, A.; Pareek, A.; Truong, S.Q.; Nguyen, C.D.; Ngo, V.-D.; Seekins, J.; Blankenberg, F.G.; Ng, A.Y. Benchmarking saliency methods for chest X-ray interpretation. *Nat. Mach. Intell.* **2022**, *4*, 867–878.
48. Bokadia, H.; Yang, S.C.H.; Li, Z.; Folke, T.; Shafto, P. Evaluating perceptual and semantic interpretability of saliency methods: A case study of melanoma. *Appl. AI Lett.* **2022**, *3*, e77.
49. He, K.; Zhang, X.; Ren, S.; Sun, J. Deep residual learning for image recognition. In Proceedings of the IEEE Conference on Computer Vision and Pattern Recognition, Las Vegas, NV, USA, 27–30 June 2016; pp. 770–778.
50. Desciak, E.B.; Maloney, M.E. Artifacts in frozen section preparation. *Dermatol. Surg.* **2000**, *26*, 500–504. [CrossRef] [PubMed]
51. Pech, P.; Bergström, K.; Rauschning, W.; Haughton, V.M. Attenuation values, volume changes and artifacts in tissue due to freezing. *Acta Radiol.* **1987**, *28*, 779–782. [CrossRef]
52. Rolls, G.O.; Farmer, N.J.; Hall, J.B. *Artifacts in Histological and Cytological Preparations*; Leica Microsystems: Wetzlar, Germany, 2008.

Article

Image-Based Deep Learning Detection of High-Grade B-Cell Lymphomas Directly from Hematoxylin and Eosin Images

Chava Perry [1,2], Orli Greenberg [2,3], Shira Haberman [1,2], Neta Herskovitz [1], Inbal Gazy [4], Assaf Avinoam [4], Nurit Paz-Yaacov [4], Dov Hershkovitz [2,3] and Irit Avivi [1,2,*]

1   Hematology Division, Tel Aviv Sourasky Medical Center, Tel Aviv 6423906, Israel
2   Sackler Faculty of Medicine, Tel Aviv University, Tel Aviv 6997801, Israel
3   Pathology Department, Tel Aviv Sourasky Medical Center, Tel Aviv 6492601, Israel
4   Imagene AI Ltd., Tel Aviv 6721409, Israel; nurit@imagene-ai.com (N.P.-Y.)
*   Correspondence: iritavi@tlvmc.gov.il

Simple Summary: Double/triple-hit lymphomas (DHLs/THLs) are an aggressive type of high-grade B-cell lymphomas (HGBLs), characterized by translocations in *MYC* and *BCL2/BCL6*. DHL patients respond poorly to standard chemoimmunotherapy regimens; thus, timely and accurate diagnosis is of paramount importance for their proper clinical management. The standard technique used for the identification of these translocations is fluorescence in situ hybridization (FISH), which is not routinely performed at every medical center to all potential patients. In the current study, we employed an image-based, artificial intelligence, deep learning algorithmic tool for the identification of DHL/THL cases by analyzing scanned histopathological H&E-stained tissue slide images. Our preliminary results demonstrate high performances, suggesting the potential use of such a solution in the clinical workflow to support the management of HGBL patients.

Abstract: Deep learning applications are emerging as promising new tools that can support the diagnosis and classification of different cancer types. While such solutions hold great potential for hematological malignancies, there have been limited studies describing the use of such applications in this field. The rapid diagnosis of double/triple-hit lymphomas (DHLs/THLs) involving *MYC*, *BCL2* and/or *BCL6* rearrangements is obligatory for optimal patient care. Here, we present a novel deep learning tool for diagnosing DHLs/THLs directly from scanned images of biopsy slides. A total of 57 biopsies, including 32 in a training set (including five DH lymphoma cases) and 25 in a validation set (including 10 DH/TH cases), were included. The DHL-classifier demonstrated a sensitivity of 100%, a specificity of 87% and an AUC of 0.95, with only two false positive cases, compared to FISH. The DHL-classifier showed a 92% predictive value as a screening tool for performing conventional FISH analysis, over-performing currently used criteria. The work presented here provides the proof of concept for the potential use of an AI tool for the identification of DH/TH events. However, more extensive follow-up studies are required to assess the robustness of this tool and achieve high performances in a diverse population.

Keywords: diffuse large B-cell lymphoma (DLBCL); high-grade B-cell lymphoma (HGBL); double hit; *MYC* rearrangement; *BCL2* rearrangement; artificial intelligence; deep learning

Citation: Perry, C.; Greenberg, O.; Haberman, S.; Herskovitz, N.; Gazy, I.; Avinoam, A.; Paz-Yaacov, N.; Hershkovitz, D.; Avivi, I. Image-Based Deep Learning Detection of High-Grade B-Cell Lymphomas Directly from Hematoxylin and Eosin Images. *Cancers* 2023, 15, 5205. https://doi.org/10.3390/cancers15215205

Academic Editors: Hamid Khayyam, Rahele Kafieh and Ali Hekmatnia

Received: 2 August 2023
Revised: 4 October 2023
Accepted: 17 October 2023
Published: 29 October 2023

1. Introduction

Diffuse large B-cell lymphoma (DLBCL) is an aggressive lymphoma. It is the most common type of non-Hodgkin's lymphoma (NHL), accounting for approximately 30–40% of NHL cases. More than 20 years ago, gene expression profiling stratified DLBCL into three main subgroups, according to the cell of origin (COO); germinal center B-cell-like (GCB), activated B-cell-like (ABC) and unclassified ("type 3") DLBCL [1,2]. These groups are characterized by different gene expression patterns, with a more favorable prognosis for GCB-type lymphoma compared to ABC.

An additional layer of B-cell lymphoma classification includes the presence of chromosomal rearrangements. MYC, a master regulator of multiple cellular processes such as cell proliferation, apoptosis and differentiation, is one of the most commonly rearranged oncogenes in B-cell lymphoma [3]. MYC deregulation can be found in many cancers, and translocations involving this oncogene occur in 5–15% of patients with DLBCL [4,5].

High-grade B-cell lymphoma (HGBL) was established as a distinct category of B-cell lymphomas in the 2016 revision of the World Health Organization (WHO) classification of lymphoid neoplasms [6]. A unique group among high-grade lymphomas, characterized by specific gene rearrangements, these lymphomas carry translocations in *MYC* together with one, or both, of the anti-apoptotic proto-oncogenes, *BCL2* and *BCL6*. HGBL, reported in less than 10% of cases of diffuse large B-cell lymphoma (DLBCL), have been referred to as double-hit (DH) or triple-hit (TH) lymphomas, if two or all three rearrangements are demonstrated, respectively.

Patients with HGBL-DH/TH poorly respond to R-CHOP (rituximab, cyclophosphamide, doxorubicin, vincristine and prednisone) chemoimmunotherapy and are at increased risk for central nervous system (CNS) involvement; therefore, other therapeutic regimens, aiming to overcome R-CHOP resistance and ensure CNS penetration, are often considered [7]. These include dose-adjusted R-EPOCH (rituximab, etoposide, prednisone, vincristine, cyclophosphamide, doxorubicin) [8], R-CODOX-M/IVAC (rituximab cyclophosphamide, vincristine, doxorubicin and high-dose methotrexate alternating with ifosfamide, etoposide and high-dose cytarabine) [9] and hyper-CVAD-R (cyclophosphamide, vincristine, adriamycin and dexamethasone with rituximab, alternating with methotrexate and cytosine arabinoside) [10,11].

The current diagnosis of DHL relies on fluorescence in situ hybridization (FISH) analysis, demonstrating *MYC* rearrangements together with *BCL2//BCL6* translocations [12]. However, using FISH is costly and requires elaborate laboratory protocols, which are not routinely performed at every medical center. Moreover, FISH results are usually not instantly attained, so the "DH molecular status" is often unknown by the time the first course of chemoimmunotherapy is administered. Thus, in most patients with DHL, the standard R-CHOP regimen is used at least in the first cycle of treatment and not a more intense protocol. Therefore, efforts are made to define restrictive criteria in order to identify the cases where FISH for DHL needs to be employed. The Ki-67 proliferation index (often high in patients with DHL), immunohistochemical (IHC) expression of c-MYC or diagnosis of the germinal GCB COO subtype have all been suggested [13], yet were found to be insufficiently specific and were associated with unacceptable rates of false negative cases [5].

The current diagnosis practices range from the performance of FISH analysis in all large B-cell lymphoma cases to a highly selective approach, restricting the analysis only to very suspicious clinical–pathological cases. While the first approach ensures accurate diagnosis, it is associated with considerable resources and high costs [14]. Conversely, the latter approach reduces efforts and costs, but increases the risk of missing DHL cases.

Deep learning (DL) applications are being extensively explored in digital pathology as novel solutions for cancer diagnosis [15,16]. The use of DL for both histopathology and molecular image-based analysis using the hematoxylin and eosin (H&E)-stained tissue slides can provide an immediate, objective and scalable solution that is exceedingly needed. This necessity is specifically prominent in hematopathology, where the microscopic diagnosis of hematological malignancies can be extremely challenging and molecular screening is not always available [17–19]. Indeed, there have been several publications in the last few years describing the potential use of DL for the identification and diagnosis of different subtypes of lymphoma, including DLBCL, using H&E-stained tissue slide images [20–25]. This approach of applying machine learning for the diagnosis of hematological neoplasms and DLBCL specifically is showing promising results and could significantly enhance the diagnostic process. However, the diagnosis and classification of hematological malignancies often rely not only on histopathological features but also on the characterization of genetic alterations [26].

Deep learning, which is adept in extracting relevant features from complex, variable data, is emerging as a powerful tool for identifying morphological patterns associated with molecular alterations from digitized histological slides [27]. Although the use of DL for the detection of a variety of molecular alterations directly from H&E slides is being widely investigated in diverse cancer types [28–31], its application for detecting genetic changes in lymphoma remains notably limited. A recent study, investigating the use of DL for the inference of *MYC* rearrangements from biopsies of patients with aggressive B-cell lymphoma, confirmed the potential value of this technology, demonstrating a sensitivity of 0.93 [32]. However, the specificity was only 0.52, attributing to high false positive rates of more than 30% [32]. Thus, despite their potential, DL solutions for the prediction of molecular alterations has not been broadly adopted in clinical settings yet, as they do not display high enough accuracies in robust large-scale studies [31].

Given the promising capabilities of DL and the importance of the identification of rearrangements for the proper management of DLBLC and HGBL patients, we chose to explore the abilities of AI tools in detecting DH/TH events in lymphoma specimen slides. In this study, we present a digital image-based approach, employing DL algorithms to differentiate between DHL/THL and non-DHL/THL cases by analyzing scanned images of H&E-stained tissue slides. While our work still requires further validation in a broader study, our DL classifier demonstrated high accuracies, suggesting that such an approach has the potential to be beneficial within hematological clinical settings.

2. Materials and Methods

2.1. Study Population

Patients diagnosed with aggressive B-cell lymphomas (DLBCL and HGL) between January 2017 and January 2022 at the Tel Aviv Sourasky Medical Center (TASMC), who were analyzed through FISH as part of their pathological workup, were included in this study. Patients with non-informative FISH results, attributed to technical issues, were excluded. The scanning of whole slide images (WSIs) of patients' H&E-stained diagnostic slides was performed at $40\times$ magnification, using a Philips Ultra-Fast Scanner (Philips Digital Pathology Solutions, Philips, Best, The Netherlands). Thirty-two biopsies from 30 patients were included in the training set, including 27 non-DH biopsies and 5 DH biopsies. The validation set included 25 cases, 15 non-DH and 10 DH/TH cases.

This study was approved by the Ethics Committee at the Tel Aviv Sourasky Medical Center (IRB 0308-22-TLV).

2.2. Data Collection

Data including patient demographics; clinical and laboratory characteristics at presentation (ECOG performance status, disease stage at diagnosis, lactic dehydrogenase (LDH) levels); the final results of histological diagnosis (DLBCL or HGBL); the results of immunohistochemistry staining focusing on BCL2, BCL6, c-MYC, CD10, MUM-1 and Ki-67; and the results of the FISH analysis for *MYC*, *BCL2* and *BCL6* were all collected from the patients' electronic medical records.

2.3. Histopathological Analysis and Immunohistochemical (IHC) Staining

Immunohistochemical staining was performed using the following antibodies: anti-CD10 (clone 56C6, Master Diagnostica, Sevilla, Spain), anti-BCL6 (clone GI191E/A8, Cell Marque, Rocklin, CA, USA), anti-MUM-1 (clone MRQ-8, Cell Marque, Rocklin, CA, USA), anti-c-Myc (clone EP121, Cell Marque, Rocklin, CA, USA), anti-BCL2 (clone E17, Cell Marque, Rocklin, CA, USA) and anti-Ki-67 (clone SP6, Cell Marque, Rocklin, CA, USA). Staining was performed on the Ventana Ultra Benchmark (Ventana Medical Systems, Tucson, AZ, USA) automatic slide stainer. Positivity for the expression of CD10, BCL6 and

MUM-1 was defined as 30% or more positive cells, 50% or more for BCL2 and 40% or more for c-MYC. The COO was determined using the HANS algorithm [33].

2.4. Fluorescence In Situ Hybridization (FISH) Analysis

FISH analysis was performed to assess *MYC* rearrangements using the Vysis MYC Break Apart FISH Probe Kit; *BCL2* rearrangements were assessed using the Vysis LSI IGH/BCL2 Dual Color, Dual Fusion Translocation Probe; and *BCL6* rearrangements were assessed using the Vysis LSI BCL6 (ABR) Dual Color Break Apart Rearrangement Probe (Abbott Molecular, Des Plains, IL, USA) and an automated fluorescence microscope scanning system (BioView Duet workstation; BioView Ltd., Rehovot, Israel). One hundred tumor cells at a minimum were evaluated per sample (except for rare cases where a minimum of 50 cells were evaluated). A cutoff of 10% was used to determine the positivity for each rearrangement.

2.5. Algorithm Development and Application

2.5.1. Model Training

For the training of the DHL model, a self-supervised scheme with dynamic data augmentation, combined with multiple instance learning (MIL) algorithms, were applied.

In the self-supervised step, the model is pre-trained on large numbers of unlabeled histopathology slides. This initial step establishes a foundation model that can be adapted to various downstream tasks using limited numbers of training samples. As such, this approach is particularly useful in the field of histopathology and is increasingly being adopted, given the limited availability of labeled samples [34,35].

In the following fine-tuning step, in addition to the foundation model, multiple instance learning (MIL) is utilized and training is performed on labeled data. Each WSI is subdivided into multiple smaller patches which are used as input for modeling. Given that the labeling is assigned on a slide level and the model receives multiple patches that collectively represent the entire slide, the MIL approach, which allows one to make use of such weakly labeled data and provides a single classification for the entire slide, is a powerful technique for classifying WSIs [35,36].

The self-supervised step was performed using untagged pan-cancer (not including lymphoma biopsies) WSIs of FFPE H&E-stained tissues scanned at 40× or 20× magnification from Imagene's internal database (including slides from the TCGA research network) All 40× images were transformed to 20× for analysis. Data augmentation was performed using over 20 techniques, including color jitter and channel shuffle. For model fine-tuning and the generation of the final DHL, the foundation model, together with MIL, were applied on the lymphoma WSIs training set described above, using patches of 384 × 384 pixels Training was performed for 20 epochs using a categorical cross-entropy loss, the Adam optimizer and a learning rate of 0.0001 (Figure 1).

2.5.2. Model Performance Evaluation

For the validation step, a categorical prediction (positive/negative) was made using the DHL-classifier model (comprising both the foundation model and the MIL algorithm) and the results were compared to the FISH results for the *MYC* and *BCL2/6* rearrangements (Figure 1).

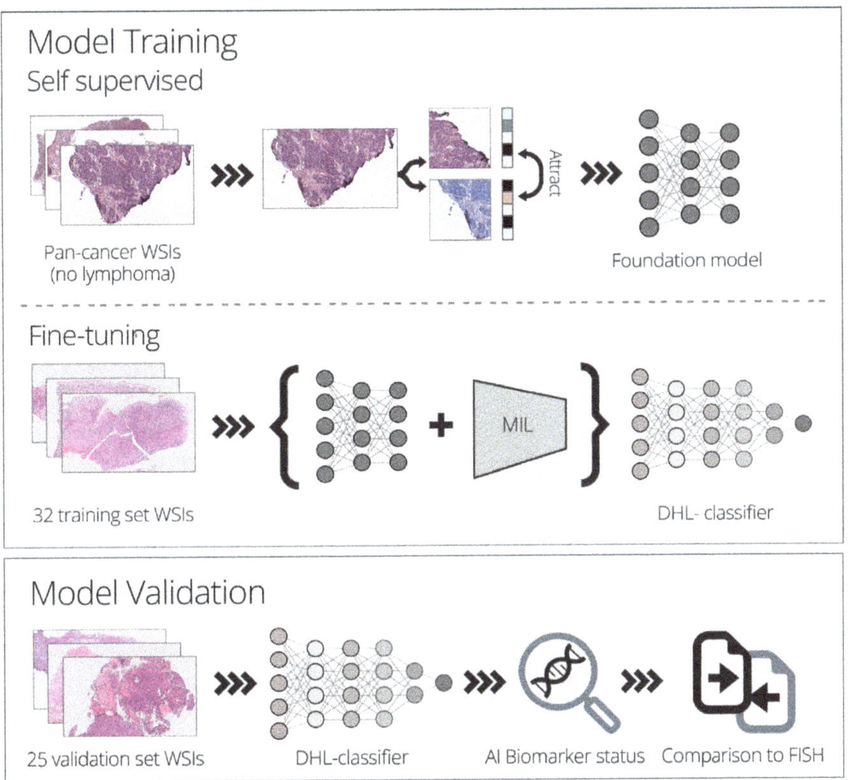

Figure 1. Study schematic representation. A self-supervised step was performed on a pan-cancer cohort (including cases of solid cancers with no lymphoma biopsies), establishing a foundation model, followed by a fine-tuning step using the training set's WSIs, generating the final DHL-classifier. For the DHL-classifier performance evaluation, the DH/TH status of 25 cases included in the validation set was evaluated, and the results were compared to the official results reported in the FISH analysis. WSI—whole slide image, MIL—multiple instance learning, DHL—double-hit lymphoma, DH—double-hit, TH—triple-hit.

3. Results

3.1. Patient Characteristics

This study included 57 biopsies from 55 patients, divided into a training set, which included 32 biopsies, and a validation set, which included 25 unique patients. The characteristics of the patients are presented in Table 1. Biopsies from lymph nodes represented approximately a third of the samples (39%, $n = 22$), and the rest were biopsies from extranodal tissues (61%, $n = 35$). The DHL/THL patients ($n = 15$) were mostly diagnosed with the germinal center B-cell (GCB) COO subtype determined based on IHC (73.3%, $n = 11$), with a median Ki-67 of 88% (range 40–100%) and c-MYC and BCL2 expression in 85.7% ($n = 12/14$, one with no available data) and 64.3% ($n = 9/14$, one with no available data) of the cases, respectively.

Table 1. Patient characteristics.

	Entire Cohort (*n* = 57)	Training Set (*n* = 32)	Validation Set (*n* = 25)
Male, *n* (%)	32 (56.1)	18 (56.3)	14 (56)
Age (years), median (range)	62 (8–84)	66.5 (8–84)	60 (17–77)
Tested tissue, *n* (%)			
Lymph node	22 (38.6)	16 (50.0)	6 (24.0)
Extra nodal	35 (61.4)	16 (50.0)	19 (76.0)
Procedure, *n* (%)			
Needle biopsy	43 (75.4)	24 (75.0)	19 (76.0)
Excisional	14 (24.6)	8 (25.0)	6 (24.0)
ECOG PS, *n* (%) *			
0/1	38 (82.6)	24 (82.8)	14 (82.4)
≥2	8 (17.4)	5 (17.2)	3 (17.6)
Disease stage *			
I/II	13 (24.5)	7 (23.3)	6 (26.1)
III/IV	40 (75.5)	23 (76.7)	17 (73.9)
LDH level *			
Normal	14 (28.0)	10 (32.3)	4 (21.1)
Increased	36 (72.0)	21 (67.7)	15 (78.9)
COO, *n* (%) #*			
GCB	27 (52.9)	13 (46.4)	14 (60.9)
IHC			
Ki67 *			
Median % (range)	80 (10–100)	80 (10–100)	85 (40–100)
Ki67 ≥ 90%	23 (41.8)	12 (37.5)	11 (47.8)
c-MYC expression *			
Positive/borderline positive	33 (61.1)	20 (64.5)	13 (56.5)
Negative	21 (38.9)	11 (35.5)	10 (43.5)
BCL2 expression *			
Positive/borderline positive	33 (62.3)	21 (67.7)	12 (54.5)
Negative	20 (37.7)	10 (32.3)	10 (45.5)
BCL6 expression *			
Positive/borderline positive	49 (89.1)	29 (90.6)	20 (87.0)
Negative	6 (10.9)	3 (9.4)	3 (13.0)
FISH			
DHL/THL, *n* (%)	15 (26.3)	5 (15.6)	10 (40)

ECOG PS—Eastern Cooperative Oncology Group performance status; COO—cell of origin; LDH—lactic dehydrogenase; GCB—germinal center B-cell; IHC—immunohistochemistry; FISH—fluorescence in situ hybridization; DHL/THL — double/triple-hit lymphoma; % are depicted from the total *n* with information per criteria. # GCB-based on HANS immunohistochemical criteria; * missing data: ECOG PS (*n* = 11), disease stage (*n* = 4), LDH (*n* = 7) COO (not determined/equivocal; *n* = 6), Ki-67% (*n* = 2), c-MYC (*n* = 3), BCL2 (*n* = 4) and BCL6 (*n* = 2).

3.2. Digital Imaging Analysis

The DHL-classifier was developed using H&E-stained slide images of 32 biopsies (as described in the methods section) (training set) that included five cases of DHL. The DHL-classifier was then blindly validated on an independent validation set, containing

25 DLBCL/HGL cases. The algorithm results were compared to the official clinical diagnosis that was based on the FISH analysis for the *MYC*, *BCL2* and *BCL6* rearrangements.

The validation set comprised nine DHL cases, one THL case and fifteen non-DH DLBCL cases. Altogether, the model correctly identified all 10 DHL/THL cases, demonstrating 100% sensitivity. The specificity was 86.7% due to two false positive cases (negative for FISH analysis). The accuracy rate was 92% and the area under the curve (AUC) was 0.95 (Figure 2A).

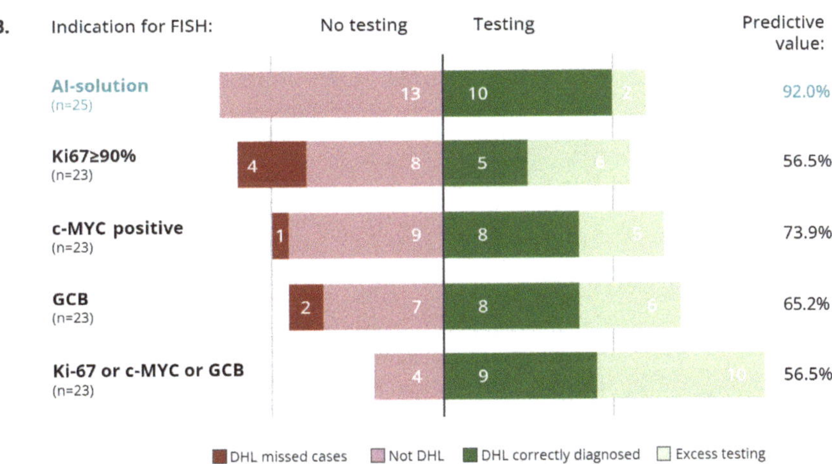

A.

	Official results		AI-classifier results				AI-classifier performance			
Total	# N	# P	# TN	# TP	# FN	# FP	Sensitivity	Specificity	Accuracy	AUC
25	15	10	13	10	0	2	100%	87%	92%	0.95

Figure 2. Performance of the DHL-classifier. (**A**) DHL-classifier results and performance in the validation cohort. N—negative, P—positive, TN—true negative, TP—true positive, FN—false negative, FP—false positive, AUC—area under the curve. (**B**) Predictive values of conventional methods vs. the DHL-classifier as a screening tool for FISH analysis. Presented are the number of samples in the relevant bars and predictive values for each screening method used. The number on the bars represents the number of cases in the relevant group.

3.3. AI Classifier as a Screening Tool for Selecting Cases for FISH Analysis

Current criteria for selecting cases for FISH analysis often rely on the presence of high Ki-67 ($\geq$90%), the IHC expression of c-MYC and the classification of the GCB subtype. Therefore, we assessed the performance of these criteria, as well as the AI DHL-classifier, as a screening tool for referring cases for FISH analysis. The DHL classifier was found to provide a predictive value of 92%, compared with 57–74% for any of the three IHC evaluable criteria (Figure 2B and Table S1). The AI DHL-classifier displayed the highest sensitivity (100%) and specificity (87%) rates, with only two excess unnecessary tests and no missed DHL cases. The conventional screening criteria showed variable sensitivities, ranging from 56% to 89%, and variable specificities ranging from 54% to 64% (Table S1). When all three IHC parameters were used together as screening criteria (Ki-67 $\geq$ 90% or increased c-MYC expression or GCB subtype), the accuracy remained low (57%), with an excess of non-valuable tests; only four cases out of the entire evaluated cohort (n = 23) did not meet the FISH screening criteria.

4. Discussion

The outcome of DHL patients treated with R-CHOP is generally poor. Intensive treatment regimens such as dose-adjusted R-EPOCH, R-CODOX-M/IVAC and hyper-

CVAD-R are often implemented in clinics to treat HGBL-DH/TH. Thus, the rapid and accurate diagnosis of DH/TH lymphoma is highly necessary. Unfortunately, diagnoses of DHL/THL are often delayed, or even missed, due to the limited availability of FISH tests which are essential for establishing the diagnosis.

Here, we describe a deep learning-based algorithmic tool trained to detect DHL/THL using H&E-stained biopsy slide images obtained from aggressive B-cell lymphoma patients. To evaluate the performance of our DHL-classifier, we used a cohort comprising images of samples that had been subjected to FISH during their diagnostic assessment and were not part of the training set. The cohort included, in total, 10 DHL/THL and 15 non-DH/TH samples. Our DHL-classifier identified all 10 DHL/THL cases, demonstrating 100% sensitivity. The specificity was 86.7%, with an AUC of 0.95 due to two false positive (FP) cases, where DH translocations were not identified through the FISH. Of note, cryptic rearrangements, undetected with FISH, may exist in up to 20% of DHL cases [37,38] raising the possibility that our FP cases might be due to "cryptic DHL changes". However, this speculation requires further evaluation.

The rapid identification of patients with DHL, enabling the early upfront administration of more intensified and compatible therapies, is imperative. FISH analysis is the currently used method for the identification of patients that are positive for DH/TH-associated gene rearrangements. Testing all high-grade lymphoma cases, although ensuring the detection of most DHL/THL patients, is associated with increased diagnostic costs and a high testing burden. Given that several studies have shown that more than 20% of DHL/THL cases are of a non-GCB origin, 15–30% of *MYC*-rearranged cases fail to overexpress c-MYC, and at least a third of DHL cases exhibit Ki-67 lower than 90% [39,40]; using these specific parameters for selecting cases for FISH testing seems to be inappropriate. These findings were also reflected in the current study, with less than 75% of DH cases being of GCB origin, ~50% with high proliferative index and ~15% not displaying high c-MYC expression. Therefore, new screening methods are direly needed.

In light of this unmet clinical need and considering that AI-based solutions offer accessible tools that can provide a biomarker status within minutes, we assessed whether the AI DHL-classifier could provide an alternative screening tool for aggressive DH B-cell lymphoma cases. Although our cohort size was small, our preliminary data demonstrated that the classifier effectively detected DHL cases with a predictive value of 92%, capturing all FISH-positive cases. Confirmation in a larger cohort, representing a more diverse group of patients, is required. However, our results, if confirmed, suggest that the DHL AI-based classifier may serve as a useful screening tool in places where FISH analysis is limited.

Our study has several limitations, mainly attributed to its retrospective nature and the small cohort size. During the study period, FISH was not routinely performed in our institution, but was reserved for patients with highly aggressive disease and/or a high Ki-67 proliferative index, introducing a selection bias of cases that were more likely to be positive. Moreover, the number of patients included in our study was small and all the samples were attained from a single center. All these factors, together with the known impact of scanning devices, specimen processing and staining protocols on histological slide images [41], emphasize the need for validating our results in a large prospective cohort, representing a more diverse group of patients, in order to assess the robustness and generalizability of our DHL-classifier. Additionally, NGS was not performed in any of the cases; thus, it is impossible to conclusively determine whether our false positive cases represent DHL cases with cryptic translocations. Therefore, evaluating samples with NGS in addition to the traditional FISH in follow-up studies can be of value for a more detailed investigation and characterization of the cohort, particularly in the case of false positives.

Moreover, the definition of HGBL-DH has been recently changed, referring now to patients with *MYC/BCL2* rearrangement only [42]. Therefore, a new algorithm, differentiating between *MYC/BCL2* and *MYC/BCL6*-rearraged cases, is currently warranted.

Lastly, while our algorithms' lack of interpretability poses challenges in model refinement and obtaining insights that can drive advances in the molecular pathology realm, we are optimistic that future advancements in explainable AI algorithms will facilitate substantial improvements in this field.

5. Conclusions

We presented here a proof of concept for the potential use of a DL algorithmic tool for the identification of DH/TH events, promoting the performance of FISH in "DH/TH-suspected" cases only. While further investigation, development and implementation of the proposed tool are required, it would be of great interest to further investigate the impact of integrating such a tool within the clinical workflow. Moreover, it would be of great interest to explore if such a tool could also be used for other lymphomas, for example, the identification of *MYC* translocations in Burkitt lymphoma, and to establish whether such AI solutions can improve hematological cancer patients' care.

Supplementary Materials: The following supporting information can be downloaded at: https://www.mdpi.com/article/10.3390/cancers15215205/s1, Table S1: Predictive values of conventional criteria *vs.* the AI DHL-classifier for correctly deciding whether to perform FISH testing.

Author Contributions: Conceptualization, I.A., D.H. and N.P.-Y. Investigation I.A., C.P., I.G. and N.P.-Y. Formal analysis, I.A., I.G. and N.P.-Y. Software A.A. Methodology, I.A., I.G., A.A. and N.P.-Y. Data curation, C.P., O.G., S.H., N.H., D.H. and I.A. Writing—original draft preparation, I.A., I.G. and N.P.-Y. All authors have read and agreed to the published version of the manuscript.

Funding: This work was funded by IMAGENE AI LTD and Imagene AI Inc.

Institutional Review Board Statement: This study was approved by the Ethics Committee at the Tel Aviv Sourasky Medical Center (IRB 0308-22-TLV 26 July 2022).

Data Availability Statement: The datasets generated during the current study are not publicly available and are available upon reasonable request.

Acknowledgments: The results published here are, in part, based upon images generated by the TCGA Research Network: https://www.cancer.gov/tcga (accessed on 1 October 2023).

Conflicts of Interest: C.P., O.G., S.H., N.H., D.H. and I.A. declare no conflict. I.G., A.A. and N.P-Y. are employees of IMAGENE AI LTD. and have stock options.

References

Alizadeh, A.A.; Eisen, M.B.; Davis, R.E.; Ma, C.; Lossos, I.S.; Rosenwald, A.; Boldrick, J.C.; Sabet, H.; Tran, T.; Yu, X.; et al. Distinct types of diffuse large B-cell lymphoma identified by gene expression profiling. *Nature* **2000**, *403*, 503–511. [CrossRef] [PubMed]

Rosenwald, A.; Wright, G.; Chan, W.C.; Connors, J.M.; Campo, E.; Fisher, R.I.; Gascoyne, R.D.; Muller-Hermelink, H.K.; Smeland, E.B.; Giltnane, J.M.; et al. The use of molecular profiling to predict survival after chemotherapy for diffuse large-B-cell lymphoma. *N. Engl. J. Med.* **2002**, *346*, 1937–1947. [CrossRef] [PubMed]

Nguyen, L.; Papenhausen, P.; Shao, H. The Role of c-MYC in B-Cell Lymphomas: Diagnostic and Molecular Aspects. *Genes* **2017**, *8*, 116. [CrossRef] [PubMed]

Barrans, S.; Crouch, S.; Smith, A.; Turner, K.; Owen, R.; Patmore, R.; Roman, E.; Jack, A. Rearrangement of MYC is associated with poor prognosis in patients with diffuse large B-cell lymphoma treated in the era of rituximab. *J. Clin. Oncol.* **2010**, *28*, 3360–3365. [CrossRef] [PubMed]

Savage, K.J.; Johnson, N.A.; Ben-Neriah, S.; Connors, J.M.; Sehn, L.H.; Farinha, P.; Horsman, D.E.; Gascoyne, R.D. MYC gene rearrangements are associated with a poor prognosis in diffuse large B-cell lymphoma patients treated with R-CHOP chemotherapy. *Blood* **2009**, *114*, 3533–3537. [CrossRef] [PubMed]

Swerdlow, S.H.; Campo, E.; Pileri, S.A.; Harris, N.L.; Stein, H.; Siebert, R.; Advani, R.; Ghielmini, M.; Salles, G.A.; Zelenetz, A.D.; et al. The 2016 revision of the World Health Organization classification of lymphoid neoplasms. *Blood* **2016**, *127*, 2375–2390. [CrossRef] [PubMed]

Phuoc, V.; Sandoval-Sus, J.; Chavez, J.C. Drug therapy for double-hit lymphoma. *Drugs Context* **2019**, *8*, 1–13. [CrossRef] [PubMed]

Dunleavy, K.; Fanale, M.A.; Abramson, J.S.; Noy, A.; Caimi, P.F.; Pittaluga, S.; Parekh, S.; Lacasce, A.; Hayslip, J.W.; Jagadeesh, D.; et al. Dose-adjusted EPOCH-R (etoposide, prednisone, vincristine, cyclophosphamide, doxorubicin, and rituximab) in untreated aggressive diffuse large B-cell lymphoma with MYC rearrangement: A prospective, multicentre, single-arm phase 2 study. *Lancet Haematol.* **2018**, *5*, e609–e617. [CrossRef]

9.   McMillan, A.K.; Phillips, E.H.; Kirkwood, A.A.; Barrans, S.; Burton, C.; Rule, S.; Patmore, R.; Pettengell, R.; Ardeshna, K.M.; Lawrie, A.; et al. Favourable outcomes for high-risk diffuse large B-cell lymphoma (IPI 3-5) treated with front-line R-CODOX-M/R-IVAC chemotherapy: Results of a phase 2 UK NCRI trial. *Ann. Oncol.* **2020**, *31*, 1251–1259. [CrossRef]

10.   Landsburg, D.J.; Falkiewicz, M.K.; Maly, J.; Blum, K.A.; Howlett, C.; Feldman, T.; Mato, A.R.; Hill, B.T.; Li, S.; Medeiros, L.J.; et al. Outcomes of Patients With Double-Hit Lymphoma Who Achieve First Complete Remission. *J. Clin. Oncol.* **2017**, *35*, 2260–2266. [CrossRef]

11.   Zhuang, Y.; Che, J.; Wu, M.; Guo, Y.; Xu, Y.; Dong, X.; Yang, H. Altered pathways and targeted therapy in double hit lymphoma. *Hematol. Oncol.* **2022**, *15*, 26. [CrossRef] [PubMed]

12.   Swerdlow, S.H. Diagnosis of 'double hit' diffuse large B-cell lymphoma and B-cell lymphoma, unclassifiable, with features intermediate between DLBCL and Burkitt lymphoma: When and how, FISH versus IHC. *Hematol. Am. Soc. Hematol. Educ. Program.* **2014**, *2014*, 90–99. [CrossRef] [PubMed]

13.   Thirunavukkarasu, B.; Bal, A.; Prakash, G.; Malhotra, P.; Singh, H.; Das, A. Screening Strategy for Detecting Double-Hit Lymphoma in a Resource-Limited Setting. *Appl. Immunohistochem. Mol. Morphol.* **2022**, *30*, 49–55. [CrossRef]

14.   Stephens, D.M.; Smith, S.M. Diffuse large B-cell lymphoma—Who should we FISH? *Ann. Lymphoma* **2018**, *2*, 8. [CrossRef]

15.   Shmatko, A.; Ghaffari Laleh, N.; Gerstung, M.; Kather, J.N. Artificial intelligence in histopathology: Enhancing cancer research and clinical oncology. *Nat. Cancer* **2022**, *3*, 1026–1038. [CrossRef]

16.   Baxi, V.; Edwards, R.; Montalto, M.; Saha, S. Digital pathology and artificial intelligence in translational medicine and clinical practice. *Mod. Pathol.* **2022**, *35*, 23–32. [CrossRef]

17.   Laurent, C.; Baron, M.; Amara, N.; Haioun, C.; Dandoit, M.; Maynadie, M.; Parrens, M.; Vergier, B.; Copie-Bergman, C.; Fabiani, B.; et al. Impact of Expert Pathologic Review of Lymphoma Diagnosis: Study of Patients From the French Lymphopath Network. *J. Clin. Oncol.* **2017**, *35*, 2008–2017. [CrossRef] [PubMed]

18.   Bowen, J.M.; Perry, A.M.; Laurini, J.A.; Smith, L.M.; Klinetobe, K.; Bast, M.; Vose, J.M.; Aoun, P.; Fu, K.; Greiner, T.C.; et al. Lymphoma diagnosis at an academic centre: Rate of revision and impact on patient care. *Br. J. Haematol.* **2014**, *166*, 202–208. [CrossRef]

19.   Matasar, M.J.; Shi, W.; Silberstien, J.; Lin, O.; Busam, K.J.; Teruya-Feldstein, J.; Filippa, D.A.; Zelenetz, A.D.; Noy, A. Expert second-opinion pathology review of lymphoma in the era of the World Health Organization classification. *Ann. Oncol.* **2012**, *23*, 159–166. [CrossRef]

20.   Syrykh, C.; Abreu, A.; Amara, N.; Siegfried, A.; Maisongrosse, V.; Frenois, F.X.; Martin, L.; Rossi, C.; Laurent, C.; Brousset, P. Accurate diagnosis of lymphoma on whole-slide histopathology images using deep learning. *NPJ Digit. Med.* **2020**, *3*, 63. [CrossRef]

21.   Li, D.; Bledsoe, J.R.; Zeng, Y.; Liu, W.; Hu, Y.; Bi, K.; Liang, A.; Li, S. A deep learning diagnostic platform for diffuse large B-cell lymphoma with high accuracy across multiple hospitals. *Nat. Commun.* **2020**, *11*, 6004. [CrossRef]

22.   Steinbuss, G.; Kriegsmann, M.; Zgorzelski, C.; Brobeil, A.; Goeppert, B.; Dietrich, S.; Mechtersheimer, G.; Kriegsmann, K. Deep Learning for the Classification of Non-Hodgkin Lymphoma on Histopathological Images. *Cancers* **2021**, *13*, 2419. [CrossRef] [PubMed]

23.   Achi, H.E.; Belousova, T.; Chen, L.; Wahed, A.; Wang, I.; Hu, Z.; Kanaan, Z.; Rios, A.; Nguyen, A.N.D. Automated Diagnosis of Lymphoma with Digital Pathology Images Using Deep Learning. *Ann. Clin. Lab. Sci.* **2019**, *49*, 153–160. [PubMed]

24.   Miyoshi, H.; Sato, K.; Kabeya, Y.; Yonezawa, S.; Nakano, H.; Takeuchi, Y.; Ozawa, I.; Higo, S.; Yanagida, E.; Yamada, K.; et al. Deep learning shows the capability of high-level computer-aided diagnosis in malignant lymphoma. *Lab. Investig.* **2020**, *100*, 1300–1310. [CrossRef] [PubMed]

25.   Mohlman, J.S.; Leventhal, S.D.; Hansen, T.; Kohan, J.; Pascucci, V.; Salama, M.E. Improving Augmented Human Intelligence to Distinguish Burkitt Lymphoma From Diffuse Large B-Cell Lymphoma Cases. *Am. J. Clin. Pathol.* **2020**, *153*, 743–759. [CrossRef] [PubMed]

26.   Taylor, J.; Xiao, W.; Abdel-Wahab, O. Diagnosis and classification of hematologic malignancies on the basis of genetics. *Blood* **2017**, *130*, 410–423. [CrossRef] [PubMed]

27.   Echle, A.; Rindtorff, N.T.; Brinker, T.J.; Luedde, T.; Pearson, A.T.; Kather, J.N. Deep learning in cancer pathology: A new generation of clinical biomarkers. *Br. J. Cancer* **2021**, *124*, 686–696. [CrossRef]

28.   Mayer, C.; Ofek, E.; Fridrich, D.E.; Molchanov, Y.; Yacobi, R.; Gazy, I.; Hayun, I.; Zalach, J.; Paz-Yaacov, N.; Barshack, I. Direct identification of ALK and ROS1 fusions in non-small cell lung cancer from hematoxylin and eosin-stained slides using deep learning algorithms. *Mod. Pathol.* **2022**, *35*, 1882–1887. [CrossRef] [PubMed]

29.   Rawat, R.R.; Ortega, I.; Roy, P.; Sha, F.; Shibata, D.; Ruderman, D.; Agus, D.B. Deep learned tissue "fingerprints" classify breast cancers by ER/PR/Her2 status from H&E images. *Sci. Rep.* **2020**, *10*, 7275. [CrossRef]

30.   Kather, J.N.; Heij, L.R.; Grabsch, H.I.; Loeffler, C.; Echle, A.; Muti, H.S.; Krause, J.; Niehues, J.M.; Sommer, K.A.J.; Bankhead, P.; et al. Pan-cancer image-based detection of clinically actionable genetic alterations. *Nat. Cancer* **2020**, *1*, 789–799. [CrossRef]

31.   Cifci, D.; Foersch, S.; Kather, J.N. Artificial intelligence to identify genetic alterations in conventional histopathology. *J. Pathol.* **2022**, *257*, 430–444. [CrossRef] [PubMed]

32.   Swiderska-Chadaj, Z.; Hebeda, K.M.; van den Brand, M.; Litjens, G. Artificial intelligence to detect MYC translocation in slides of diffuse large B-cell lymphoma. *Virchows Arch.* **2021**, *479*, 617–621. [CrossRef] [PubMed]

33. Hans, C.P.; Weisenburger, D.D.; Greiner, T.C.; Gascoyne, R.D.; Delabie, J.; Ott, G.; Muller-Hermelink, H.K.; Campo, E.; Braziel, R.M.; Jaffe, E.S.; et al. Confirmation of the molecular classification of diffuse large B-cell lymphoma by immunohistochemistry using a tissue microarray. *Blood* **2004**, *103*, 275–282. [CrossRef]

34. Chen, R.J.; Ding, T.; Lu, M.Y.; Williamson, D.F.K.; Jaume, G.; Chen, B.; Zhang, A.; Shao, D.; Song, A.H.; Shaban, M.; et al. A General-Purpose Self-Supervised Model for Computational Pathology. *arXiv* **2023**, arXiv:2308.15474. [CrossRef]

35. Saldanha, O.L.; Loeffler, C.M.L.; Niehues, J.M.; van Treeck, M.; Seraphin, T.P.; Hewitt, K.J.; Cifci, D.; Veldhuizen, G.P.; Ramesh, S.; Pearson, A.T.; et al. Self-supervised attention-based deep learning for pan-cancer mutation prediction from histopathology. *NPJ Precis. Oncol.* **2023**, *7*, 35. [CrossRef]

36. Ilse, M.; Tomczak, J.M.; Welling, M. Attention-based Deep Multiple Instance Learning. *arXiv* **2018**, arXiv:1802.04712. [CrossRef]

37. Hilton, L.K.; Tang, J.; Ben-Neriah, S.; Alcaide, M.; Jiang, A.; Grande, B.M.; Rushton, C.K.; Boyle, M.; Meissner, B.; Scott, D.W.; et al. The double-hit signature identifies double-hit diffuse large B-cell lymphoma with genetic events cryptic to FISH. *Blood* **2019**, *134*, 1528–1532. [CrossRef]

38. King, R.L.; McPhail, E.D.; Meyer, R.G.; Vasmatzis, G.; Pearce, K.; Smadbeck, J.B.; Ketterling, R.P.; Smoley, S.A.; Greipp, P.T.; Hoppman, N.L.; et al. False-negative rates for MYC fluorescence in situ hybridization probes in B-cell neoplasms. *Haematologica* **2019**, *104*, e248–e251. [CrossRef] [PubMed]

39. Landsburg, D.J.; Petrich, A.M.; Abramson, J.S.; Sohani, A.R.; Press, O.; Cassaday, R.; Chavez, J.C.; Song, K.; Zelenetz, A.D.; Gandhi, M.; et al. Impact of oncogene rearrangement patterns on outcomes in patients with double-hit non-Hodgkin lymphoma. *Cancer* **2016**, *122*, 559–564. [CrossRef]

40. Laude, M.C.; Lebras, L.; Sesques, P.; Ghesquieres, H.; Favre, S.; Bouabdallah, K.; Croizier, C.; Guieze, R.; Drieu La Rochelle, L.; Gyan, E.; et al. First-line treatment of double-hit and triple-hit lymphomas: Survival and tolerance data from a retrospective multicenter French study. *Am. J. Hematol.* **2021**, *96*, 302–311. [CrossRef]

41. Howard, F.M.; Dolezal, J.; Kochanny, S.; Schulte, J.; Chen, H.; Heij, L.; Huo, D.; Nanda, R.; Olopade, O.I.; Kather, J.N.; et al. The impact of site-specific digital histology signatures on deep learning model accuracy and bias. *Nat. Commun.* **2021**, *12*, 4423. [CrossRef]

42. Alaggio, R.; Amador, C.; Anagnostopoulos, I.; Attygalle, A.D.; Araujo, I.B.O.; Berti, E.; Bhagat, G.; Borges, A.M.; Boyer, D.; Calaminici, M.; et al. The 5th edition of the World Health Organization Classification of Haematolymphoid Tumours: Lymphoid Neoplasms. *Leukemia* **2022**, *36*, 1720–1748. [CrossRef]

Current Oncology

Article

Photon Absorption Remote Sensing Imaging of Breast Needle Core Biopsies Is Diagnostically Equivalent to Gold Standard H&E Histologic Assessment

James E. D. Tweel [1,2], Benjamin R. Ecclestone [1,2], Hager Gaouda [1,2], Deepak Dinakaran [2], Michael P. Wallace [3], Gilbert Bigras [4], John R. Mackey [2] and Parsin Haji Reza [1,*]

[1] PhotoMedicine Labs, University of Waterloo, Waterloo, ON N2L 3G1, Canada; jtweel@uwaterloo.ca (J.E.D.T.) benjamin.ecclestone@uwaterloo.ca (B.R.E.); hgaouda@uwaterloo.ca (H.G.)
[2] Illumisonics Inc., 22 King Street South, Suite 300, Waterloo, ON N2J 1N8, Canada; deepak@illumisonics.com (D.D.); john.mackey@illumisonics.com (J.R.M.)
[3] Department of Statistics and Actuarial Science, University of Waterloo, Waterloo, ON N2L 3G1, Canada; michael.wallace@uwaterloo.ca
[4] Department of Laboratory Medicine and Pathology, University of Alberta, Edmonton, AB T6G 2R3, Canada; gilbert.bigras@albertaprecisionlabs.ca
* Correspondence: phajirez@uwaterloo.ca

Abstract: Photon absorption remote sensing (PARS) is a new laser-based microscope technique that permits cellular-level resolution of unstained fresh, frozen, and fixed tissues. Our objective was to determine whether PARS could provide an image quality sufficient for the diagnostic assessment of breast cancer needle core biopsies (NCB). We PARS imaged and virtually H&E stained seven independent unstained formalin-fixed paraffin-embedded breast NCB sections. These identical tissue sections were subsequently stained with standard H&E and digitally scanned. Both the $40\times$ PARS and H&E whole-slide images were assessed by seven breast cancer pathologists, masked to the origin of the images. A concordance analysis was performed to quantify the diagnostic performances of standard H&E and PARS virtual H&E. The PARS images were deemed to be of diagnostic quality, and pathologists were unable to distinguish the image origin, above that expected by chance. The diagnostic concordance on cancer vs. benign was high between PARS and conventional H&E (98% agreement) and there was complete agreement for within-PARS images. Similarly, agreement was substantial (kappa > 0.6) for specific cancer subtypes. PARS virtual H&E inter-rater reliability was broadly consistent with the published literature on diagnostic performance of conventional histology NCBs across all tested histologic features. PARS was able to image unstained tissues slides that were diagnostically equivalent to conventional H&E. Due to its ability to non-destructively image fixed and fresh tissues, and the suitability of the PARS output for artificial intelligence assistance in diagnosis, this technology has the potential to improve the speed and accuracy of breast cancer diagnosis.

Keywords: photon absorption remote sensing (PARS); breast core biopsy; breast cancer; concordance analysis

Citation: Tweel, J.E.D.; Ecclestone, B.R.; Gaouda, H.; Dinakaran, D.; Wallace, M.P.; Bigras, G.; Mackey, J.R.; Reza, P.H. Photon Absorption Remote Sensing Imaging of Breast Needle Core Biopsies Is Diagnostically Equivalent to Gold Standard H&E Histologic Assessment. *Curr. Oncol.* **2023**, *30*, 9760–9771. https://doi.org/10.3390/curroncol30110708

Received: 11 September 2023
Revised: 27 October 2023
Accepted: 2 November 2023
Published: 6 November 2023

1. Introduction

A breast needle core biopsy (NCB) is a medical procedure in which a small, cylindrical piece of breast tissue is removed for examination and diagnosis, typically with the aid of imaging guidance (e.g., ultrasound) [1]. The procedure is performed when an abnormality is found in the breast, such as a palpable mass, or an area of suspicious tissue seen on a mammogram or other imaging tests. It is an established standard of care for obtaining accurate preoperative histological diagnosis of suspicious breast lesions [2–7]. In addition, it offers numerous advantages, including reduced cost and complication rates, over surgical biopsies primarily due its minimally invasive approach [8–11].

Following the NCB procedure, the samples undergo standard tissue processing and staining procedures to enable histological analysis. Samples are formalin-fixed and subsequently embedded in paraffin wax where they are thinly sectioned (~5 μm) and placed on glass slides for staining with hematoxylin and eosin (H&E) [12,13]. Hematoxylin stains anionic regions like the nuclei of cells blue-purple, while eosin stains cationic regions like the cytoplasm and extracellular matrix pink [13]. This creates a contrast between the different components of the tissue, allowing pathologists to identify different structures and cells within the sample. H&E is the gold standard staining method used in pathology to visualize the tissue structure of biopsy samples. It is the primary means by which pathologists assess breast NCB samples to distinguish between malignant and benign breast tissue, as well as to determine the type and grade of cancer [14].

The conventional tissue processing and staining procedures, despite being an essential part of histological analysis, are burdensome due to the significant costs, time, and expertise required [15]. However, advancements in label-free imaging technologies may have the potential to eliminate the need for these procedures, while preserving valuable biopsy samples for use in redundant or auxiliary screening procedures. Among the most promising ways of imaging tissue is an emerging technology called photon absorption remote sensing (PARS) microscopy. PARS enables simultaneous capture of contrast from both radiative and non-radiative relaxation processes following optical absorption, along with scattering contrasts in a tissue specimen [16]. The technique uses a picosecond-scale pulsed excitation laser to generate perturbations in the sample following absorption. The optical emissions from the radiative relaxation are broadly captured, while the non-radiative contrast is measured as a percentage modulation in the backward or forward scattering intensity of a secondary probe beam [16]. Depending on the excitation wavelength, PARS can provide sensitivity to a variety of chromophores including hemoglobin [17,18], DNA [16,19], collagen, elastin, cytochromes, and lipids [16,20,21]. Furthermore, by simultaneously capturing both absorption fractions, PARS provides additional contrast such as the quantum efficiency ratio (QER). QER is defined as the ratio of the non-radiative and radiative absorption portions (QER = P_{nr}/P_r) and is expected to yield additional biomolecular information [16].

Recent works have employed an ultraviolet-based (UV, 266 nm) PARS imaging system for label-free virtual histology [22,23]. Using the UV excitation source, PARS captures detailed nuclear contrast through the non-radiative relaxation of DNA absorption [24,25], as well as connective tissue contrast from the radiative relaxation of primarily collagen and elastin [26]. These contrasts are highly analogous to H&E staining and can be intelligently combined to virtually stain the sample. A deep learning-based image-to-image translation model is employed for H&E emulation and is trained on loosely registered PARS and H&E whole-slide images pairs [22]. The resulting virtual H&E images demonstrate a high degree of structural and colour similarity; however, its diagnostic efficacy has not been thoroughly measured. To assess diagnostic equivalence, a concordance analysis can be used to quantify the level of agreement between the PARS virtual H&E images and the gold standard H&E-stained samples.

Concordance rates refer to the degree of agreement between two or more pathologists who independently review the same tissue sample. Variability in the interpretation of breast core biopsies among pathologists can arise due to several contributing factors. These factors include the quality of the tissue sample obtained during the biopsy procedure and the level of experience of the pathologist. In addition, the amount of tissue available for histologic examination plays a role, wherein higher numbers of cores and longer cores tend to improve concordance among pathologists [27]. Therefore, in the context of breast NCBs, concordance rates are important to define because they reflect the diagnostic accuracy of the procedure, which typically informs subsequent surgical treatment decisions. In comparing PARS virtual H&E and true H&E, a high concordance rate would indicate a strong agreement, suggesting that the virtual histology method is successfully replicating the diagnostic information present in the traditional H&E staining. Evaluating concordance rates is crucial for validating PARS virtual H&E as a viable alternative to traditional H&E staining techniques in diagnostic applications.

We conducted a prospective study of seven independent breast tissue core biopsies representing a spectrum of known histologic findings spanning normal breast, ductal carcinoma in situ, invasive ductal carcinoma, and invasive lobular carcinoma. Unstained tissue was scanned via PARS microscopy to generate virtual H&E images, and then standard H&E staining of these same seven core biopsies was performed. The diagnostic characteristics of these images were assessed by multiple breast cancer expert pathologists, masked as to the origin of the images. We performed a concordance analysis to define the diagnostic performance of the two imaging modalities, standard H&E and PARS.

2. Materials and Methods

This study was approved by the Research Ethics Board of Alberta (Protocol ID: HREBA.CC 18-0277) and the University of Waterloo Health Research Ethics Committee (Protocol ID: 4027. Photoacoustic Remote Sensing Microscopy of Surgical Resection, Needle Biopsy, and Pathology Specimens). The ethics committees waived the need for patient consent as these archival tissues were no longer necessary for patient diagnostics. Researchers were not provided with any information pertaining to the identity of the patients. All human tissue experiments were conducted in accordance with the government of Canada guidelines and regulations, including "Ethical Conduct for Research Involving Humans (TCPS2)".

2.1. Patient Materials

Tissues were acquired from the Cross-Cancer Institute (Edmonton, AB, Canada through collaboration with clinical partners. The samples were obtained from anonymous patient donors, with all patient identifiers removed to ensure anonymity. The seven independent breast tissue core biopsies used in this study represented a spectrum of known histologic findings. Specifically, three of the breast core biopsy samples had invasive ductal carcinoma only, two samples had both invasive ductal carcinoma and ductal carcinoma in situ, one sample had invasive lobular carcinoma, and one was normal glandular tissue.

2.2. Sample Preparation Prior to PARS Imaging and Gold Standard H&E Staining

The breast core biopsy samples were obtained from patients using a hollow core needle and processed in a dedicated core facility. Immediately after excision, the collected tissue samples were placed in a formalin solution for fixation and preservation of the fresh tissue. The samples were stored in the formalin solution for a period of 24 to 28 h to ensure proper fixation. Following fixation, a skilled laboratory histotechnician performed a series of preparation steps. First, the samples were dehydrated using ethanol and then treated with xylene to remove any residual ethanol and fats. The samples were then subsequently embedded in paraffin wax, creating standard formalin-fixed paraffin-embedded (FFPE blocks. A microtome was then used to cut thin tissue sections ($\sim$4–5 μm) from the FFPE blocks. Tissue sections were placed on glass microscope slides and briefly baked at 60 °C to evaporate excess paraffin.

2.3. PARS Microscope Imaging

Whole-slide label-free PARS images were acquired from the unstained tissue section using a custom-built PARS microscope system. A more detailed recount of the PARS optical design, system schematic, and imaging process is reported in [28]. In brief, the sample is precisely targeted with focused excitation pulses from a 50 kHz 400 ps 266 nm UV laser (Wedge XF 266, RPMC; Bright Solutions, Pavia, Italy). To achieve 40× imaging magnification, these excitation events are spaced 250 nm apart, while three-axis mechanical stages move the sample across the objective lens in an "s"-shaped scanning pattern. At each excitation event, time-resolved radiative, non-radiative relaxation, and scattering signals are measured and compressed into single pixel intensity values. These intensity values collectively form the three co-registered label-free images.

To measure the radiative signal intensity, the spectrum of emitted photons is broadly collected with an avalanche photodiode (APD130A2; Thorlabs, Newton, NJ, USA) and the peak amplitude value is recorded. To measure the non-radiative relaxation effect, time-domain photothermal and photoacoustic signals are recorded. This is performed using a 405 nm continuous-wave probe beam (OBIS-LS405; Coherent, Santa Clara, CA, USA) which is coaxially aligned with the excitation spot. From this, a single non-radiative intensity value is extracted as the percentage modulation of the transmitted probe beam intensity before and after excitation. The scattering intensity of the sample is determined by calculating the average probe transmission intensity prior to excitation. Both the excitation and detection beams are focused onto the sample using a 0.42 numerical aperture (NA) UV objective lens (NPAL-50-UV-YSTF; OptoSigma, Santa Ana, CA, USA). The transmitted probe light and radiative photons are collected using a 0.7 NA objective lens (278-806-3; Mitutoyo, Aurora, IL, USA). The radiative spectrum (>266 nm) and 405 nm detection wavelength are then spectrally separated prior to measurement.

The entire sample is scanned in 500×500 µm parts which are later stitched back together into a single whole-slide image. The 405 nm scattering contrast is primarily used to find and maintain sharp focus across the sample while the radiative and non-radiative images are primarily used for virtual staining.

2.4. Gold Standard H&E Staining and Digital Image Acquisition

After all samples were imaged with the PARS microscope, standard H&E staining was performed on each of the seven core biopsies. Digital whole-slide H&E images were then acquired at $40\times$ resolution using a standard brightfield microscope (Morpholens 1; Morphle Digital Pathology, New York, NY, USA).

2.5. PARS Virtual H&E Colourization

A cycle-consistent generative adversarial network (CycleGAN), first developed by Zhu et al. [29], was employed to convert the PARS label-free data to virtual H&E images. While fixed-colour relationships have previously been applied to PARS data for emulating H&E staining [16], there are several advantages to using a deep learning-based virtual colouring process. For instance, a virtual staining algorithm can adaptively suppress data that is not directly necessary for generating a virtual H&E, as demonstrated in [22]. Additionally, the virtual colouring process has the capability to consider structural information when combining the raw PARS contrasts into a colourized H&E. The CycleGAN deep-learning based image-to-image translation model has previously been used for virtual H&E staining of PARS label-free contrast [22]. Here, with the exception of the Noise2Void denoising algorithm, the same training workflow and data preparation process was used, and a virtual H&E model was trained using a distinct set of whole-slide image pairs. These additional training samples underwent the same tissue processing, imaging, and staining procedures as the core needle biopsies.

In brief, the PARS label-free whole-slide images are first loosely registered to their corresponding ground truth H&E pairs using a simple affine transform with three registration points. Images are then cut into 512×512 px (128×128 µm) tile pairs for use in model training.

Prior to slicing, the PARS label-free contrasts are combined into a single total absorption (TA) coloured image where the radiative channel is blue, and the non-radiative channel is red. An example TA whole-slide image and corresponding training pairs can be seen in Figure 1 alongside its corresponding ground truth H&E image. The virtual staining model in this study was trained on roughly 1000 training pairs. Once the virtual staining model is trained, the same model was then applied to all seven breast core needle biopsies, forming the virtual and real H&E pairs.

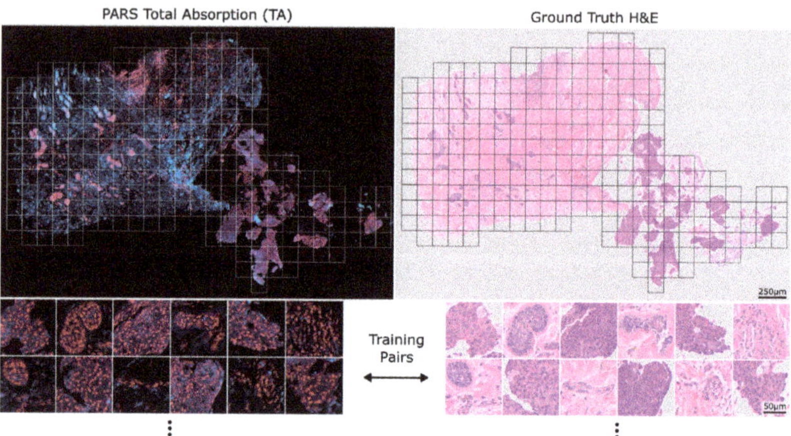

Figure 1. An example PARS total absorption (TA) image side-by-side with its corresponding ground truth H&E whole-slide image. Example 512 × 512 px pairs that can be used to train the model can be seen below each image.

2.6. Evaluation by Expert Pathologists

The PARS virtual H&E and true H&E images were randomly oriented and displayed in a pre-specified, custom random order generation algorithm designed to maximize the distance between the two image pairs (PARS and true H&E) for each individual sample. Each of the 14 images were placed on a customized web-based histology visualizing software platform without any identification except core biopsy #1 through core biopsy #14. The order of sample display was P2, P5, T1, T4, T3, T6, T2, P1, P7, P4, T5, T7, P3, and P6, where 'P' corresponds to PARS virtual H&E and 'T' corresponds to true H&E. Each of 14 images were provided independently to breast cancer focused board-certified anatomic pathologists, and 7 surveys were completed. The pathologists were masked to the clinicopathologic details of the cases and the origin of the digital images (either true H&E or PARS virtual H&E). Each pathologist was asked to score each image on the parameters shown in Table 1, including histologic diagnosis, grade of in situ disease, grade of invasive disease, and the origin of the digital image (Table 1).

Table 1. Survey questionnaire given to pathologists for each of the 14 total images.

1. The primary tissue diagnosis is:				
Invasive ductal carcinoma	Invasive lobular carcinoma	DCIS	Normal glandular tissue	Image inadequate for diagnosis
○	○	○	○	○

2. DCIS necrotic score:				
No in situ disease present	Grade 1	Grade 2	Grade 3	Not assessable
○	○	○	○	○

3. DCIS nuclear grade:				
No in situ disease present	Grade 1	Grade 2	Grade 3	Not assessable
○	○	○	○	○

4. Evaluation for invasive disease:	No invasive disease	Score 1	Score 2	Score 3	Not assessable
Tubule formation	○	○	○	○	○
Nuclear pleomorphism	○	○	○	○	○
Mitotic rate	○	○	○	○	○

5. Type of image: Is this image from FFPE H&E-stained tissue?		
Yes, this is H&E	No, this is not H&E	Uncertain
○	○	○

○ Symbol indicates the options provided to pathologists for each question.

2.7. Statistical Analysis

Concordance analysis is a statistical method used to measure the agreement between two or more raters or observers in their interpretation or classification of a set of data. Concordance may be measured through Cohen and Fleiss kappa coefficients. Cohen's kappa coefficient is a measure of inter-rater reliability that takes into account the possibility of agreement occurring by chance [30]. It is used to determine whether two raters agree beyond what would be expected by chance alone. Fleiss' kappa is an extension of Cohen's kappa to compare more than two raters [31].

Kappa values range from −1 to 1, with a value of 1 indicating perfect agreement and a value of 0 indicating agreement no better than chance. Negative values indicate agreement worse than chance. Interpretation of kappa values vary, but a value in excess of 0.6 is considered "substantial" [32] or "good" [33].

This method has several advantages over other measures of agreement, including its ability to account for chance agreement and its robustness to variations in the prevalence of different categories of data [30]. All calculations were performed in R statistical software (version 4.2.0) [34].

3. Results

3.1. Example Whole-Slide Image Pairs

Figures 2 and 3 show two exemplary sets of PARS virtual H&E and real H&E images employed in this study. At the top of each figure is the raw total absorption (TA) PARS image serving as the input to the virtual staining algorithm. Both of these figures show examples of invasive ductal carcinomas, with higher magnification regions showcasing irregular malignant glandular structures infiltrating a fibrofatty stroma, characteristic of invasive breast carcinoma. One benefit to the virtual H&E stains is that they all share consistent stain colouring, which matches the colouring of the training dataset. In contrast, staining colours for true H&E images can vary depending on specifics of the preparation, digitization and storage of the tissue samples [35]. As such, the virtual H&E images in Figures 2 and 3 share similar staining colours, whereas their true H&E counterparts exhibit a slight difference in colours. Nonetheless, both the virtual and H&E images achieve excellent epithelial and stromal contrast and highlight the same tissue structures.

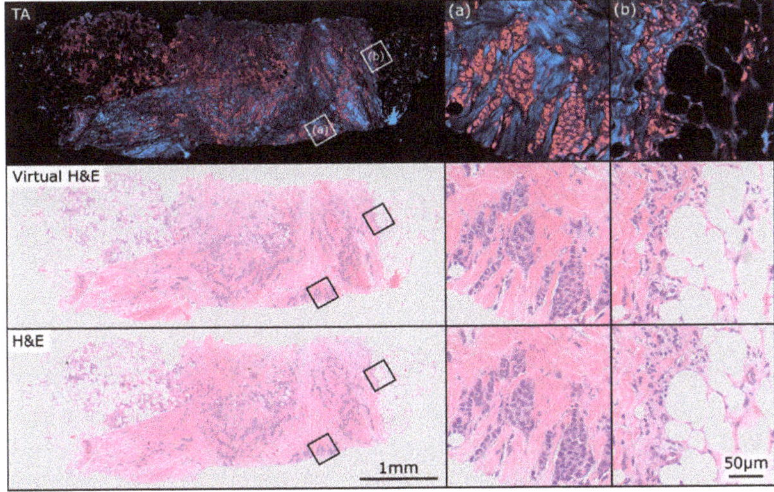

Figure 2. Example PARS total absorption (TA), virtual H&E, and true H&E pair used in this study. The sample exhibits closely matched staining colours between the virtual and real representations. (**a,b**) depict two regions of higher magnification on the sample.

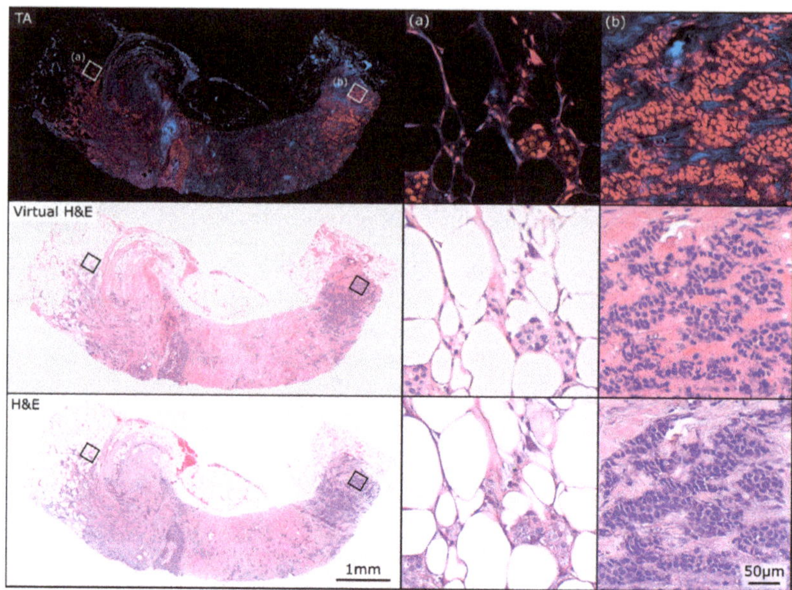

Figure 3. Example PARS total absorption (TA), virtual H&E, and true H&E pair used in this stud. Both the virtual and real H&E images exhibit excellent epithelial and stromal contrast and highligh. the same tissue structures. (**a,b**) depict two regions of higher magnification on the sample.

3.2. Image Origin

For each image, either true H&E or PARS virtual H&E, respondents were asked to identif. the origin of each image, whether it was a true FFPE H&E-stained slide (yes), not a true FFP. H&E-stained slide (no), or uncertain. Three raters responded 'Yes' when asked if a PARS imag. was a true H&E image for all seven images. The fourth respondent reported 'Uncertain' fo. all seven PARS and H&E images. For the remaining three pathologists, PARS images wer. misidentified as true H&E images 0/7, 1/7, 3/6 times and true H&E images were misidentifie. as virtual H&E images 1/7, 3/7, 3/6 times (the final pathologist only responded to this questio. for six image pairs). These results show that masked pathologists were unable to reliabl. distinguish between conventional H&E and PARS virtual H&E.

3.3. Primary Diagnosis

Respondents were asked to make a primary diagnosis for both the PARS virtual H&. images as well as the true H&E images. All respondents were able to make a primar. diagnosis for each whole-slide image with the exception of the fourth respondent, wh. selected 'Image inadequate for diagnosis' precisely once.

If all primary diagnosis responses are combined into either a high-level 'cance. or 'benign' category, out of the 48 image-pair assessments (excluding one diagnosis o. 'image inadequate'), there was only one disagreement between an H&E and PARS pai. (kappa = 0.921). In total, 40 image pairs were both assessed as cancerous, while the re. maining 7 image pairs were both assessed as benign. This indicates there was reliabl. discrimination between cancerous and benign cases. Here, 'cancer' comprises the diagnosi. of invasive lobular carcinoma, invasive ductal carcinoma, and DCIS.

For the specific cancer subtypes, a concordance analysis for the primary diagnosis amon. the seven pathologists was performed for both true H&E only and PARS virtual H&E only. Th. Fleiss' kappa value for agreement between rater for true H&E images was 0.639. The Fleis. kappa value for agreement between rater for PARS virtual H&E images was 0.620. Next, . pairwise comparison was conducted, calculating Cohen's kappa, to assess the concordanc. of the primary diagnosis between the PARS virtual H&E and true H&E images. Four of th.

seven pathologists (raters 3, 5, 6, 7) agreed on the primary diagnosis for all seven image pairs (kappa = 1). The first respondent disagreed on the primary diagnosis for a single image pair (kappa = 0.611). The second and fourth respondents disagreed on the primary diagnosis for two image pairs (rater 2, kappa = 0.364; rater 4, kappa = 0.417). Table 2 shows a comparison between the primary diagnosis given to the H&E and PARS image pairs.

Table 2. Summary of pathologist responses to question "the primary tissue diagnosis is".

PARS Diagnosis	H&E Diagnosis					Total
	IDC	ILC	DCIS	Benign	Image Inadequate	
IDC	36	0	1	1	0	38
ILC	1	1	0	0	0	2
DCIS	1	0	0	0	0	1
Benign	0	0	0	7	0	7
Image Inadequate	1	0	0	0	0	1
Total	39	1	1	8	0	49

Note: IDC, invasive ductal carcinoma; ILC, invasive lobular carcinoma; DCIS, ductal carcinoma in situ.

3.4. Evaluation of Tissue Gradings

A similar analysis to the primary diagnosis response was performed to assess the concordance for the evaluation of invasive tissue components: tubule formation score, nuclear pleomorphism score, and mitotic rate score. This analysis was also performed on the final Nottingham histological grade. In Table 3, a Fleiss' kappa coefficient was computed for the within-H&E-only invasiveness gradings and the within-PARS-only invasiveness gradings. This was performed to first observe the concordance among pathologists for H&E only and for PARS only and contrast it with the concordance result for a pairwise comparison of concordance between H&E and PARS. For the pairwise comparison, a Cohen's kappa coefficient was computed for each rater and the average coefficient is reported in Table 3. The kappa coefficients were computed from responses of all image pairs excluding image pair six. Image pair six was not involved because it was given the primary diagnosis of 'normal glandular tissue' from all raters except one (rater 2, PARS image).

Table 3. Inter-rater reliability of invasive cancer scores.

Evaluation Component	Comparison	Kappa Coefficient *
Invasive Tubule Formation Score	Within H&E	0.553
	Within PARS	0.300
	H&E–PARS	0.420
Invasive Nuclear Pleomorphism Score	Within H&E	0.051
	Within PARS	0.058
	H&E–PARS	0.188
Invasive Mitotic Rate Score	Within H&E	0.125
	Within PARS	0.148
	H&E–PARS	0.032
Nottingham Histological Grade	Within H&E	0.161
	Within PARS	0.126
	H&E–PARS	0.073

* The pairwise H&E–PARS comparisons are the mean Cohen's kappa coefficient values among raters.

In some cases, for both true H&E and PARS images, pathologists were unable to assess or submit a grading between 1 and 3. Table 4 summarizes the total number of 'not assessable' responses for each image type across all 49 image pair assessments. 'Both' is the number where the rater categorized both H&E and PARS images in the same pair as not assessable for that evaluation.

Table 4. Distribution of "not assessable" responses.

	DCIS Necrotic	DCIS Nuclear	Invasive Tubule Formation	Invasive Nuclear Pleomorphism	Invasive Mitotic Rate	Nottingham Histological Grade *
H&E	1	7	1	5	15	15
PARS	2	4	1	6	14	14
Both	1	3	0	3	8	8

* Nottingham grade "not assessable" if any of the invasive components are "not assessable".

For each rater, there was widespread agreement between which images were assessable or not. The accessibility was highest for the tubule formation score (48/49 H&E, 48/49 PARS), slightly lower for nuclear pleomorphism score (44/49 H&E, 43/49 PARS), and much lower for the mitotic rate score and the Nottingham histological grade (34/49 H&E, 35/49 PARS) Additionally, the number of image pairs where pathologists agreed that both PARS and H&E was either assessable or not assessable followed the same trend. Agreement was observed in 47/49 image pairs for the tubule formation score, 44/49 for nuclear pleomorphism score, and 36/49 for the mitotic rate score and Nottingham histological grade.

4. Discussion

4.1. Study Summary and Key Findings

In this pilot validation study, seven pairs of PARS virtual and conventional H&E images were assessed by seven pathologists masked with respect to the origin of the images Comparative analysis of PARS and H&E using the standardized synoptic reporting of the single core biopsy images demonstrated several key findings. Both PARS virtual and conventional H&E images were of diagnostic quality, and reliably allowed the discrimination of cancer (the aggregate diagnoses of invasive ductal carcinoma, invasive lobular carcinoma, and DCIS) from benign breast tissue. With respect to the primary diagnostic categories above, the raters showed almost identical agreement across H&E images as they did across PARS images. Furthermore, the diagnoses made by our pathologists viewing conventional H&E were comparable to the diagnoses of pathologists viewing PARS virtual H&E; with the granular categorization of each image into the five primary diagnostic results, overall concordance among pathologists was substantial (kappa > 0.6) and was not meaningfully higher with conventional H&E images rather than the PARS virtual H&E images. Four of the seven pathologists agreed on the primary diagnosis for all seven image pairs, one pathologist disagreed on the primary diagnosis for a single image pair, and two pathologists disagreed on the primary diagnosis for two image pairs.

4.2. Interpretation of Findings in Context

Diagnostic reproducibility in breast cancer histology remains suboptimal and underpins the difficult of evaluating new histologic techniques. Within the context of our study, PARS-based breast core biopsy imaging was equivalent to more traditional digital histopathology using H&E-stained slides. Inter-observer discordance appeared lower than previously reported intra-observer discordance. For example, pathologists given the same breast cancer biopsy material on which they had previously issued a diagnostic report, separated by a six month interval, exhibit surprisingly low intra-observer agreement rates of 92% for invasive breast cancer, 84% for DCIS, and 53% for benign with atypia [36]. Similarly, a pathology review of the original breast cancer needle core biopsy in a pre-operative quality assurance process identified 403 discordance cases out of 4950 (~8%) [37]. Furthermore, histologic interpretation and grading of core needle biopsies is dependent on the quantity of available tissue, which in our study was limited to a single core biopsy per case The authors could find no published literature on the inter-rater reliability of breast cancer core biopsy gradings, so it is difficult to put the grading data we generated in context, other than to state that grading concordance was poor (0.6 or less) with both conventional and PARS virtual H&E. As determination of mitotic rate requires the counting of mitoses in 10 high-power fields with invasive cancer, single core biopsies can have insufficient tissue for reliable assessment of mitotic rate. Consequently, in clinical practice, core biopsy pathology reports frequently omit overall grade [11,14], deferring definitive grading to the larger, surgical excision specimens.

4.3. Strengths, Limitations, and Future Research

Among the strengths of our study was the use of masked pathologists, the use of 40× digital scanning equivalent to standard digitized conventional H&E images, and efforts to reduce confounding observer recognition of serially presented images by re-orienting pairs of biopsies and maximizing their sequence separations. Furthermore, our study used the identical tissue for the two images, rather than adjacent slides, allowing direct cell-to-cell concordance of the images. Limitations of our study include its relatively small sample size, and the selection of tissues representing only the most common histologic findings on breast biopsy. Future research should aim to conduct larger studies with additional samples to substantiate and build upon our findings.

5. Conclusions

This prospective cohort study provides evidence supporting the effectiveness of PARS microscopy for the diagnostic interpretation of human breast tissue core biopsies. The images were deemed to be of diagnostic quality by expert breast cancer pathologists. The key consideration of cancer vs. benign tissue was reliably distinguished in both conventional and PARS virtual H&E histology images. Similarly, cancer subtypes were reliably distinguished with both techniques.

While the initial diagnosis of breast cancer is typically made via conventional H&E evaluation of core biopsies, the complexity of the tissue preparation and staining frequently requires one week or more before the pathology report is available. PARS is an imaging technique that can be applied not only to fixed, unstained tissues, as in this specific study, but also to freshly resected specimens and in vivo examination of tissue. As such, PARS has the potential to dramatically reduce diagnostic timelines.

Finally, PARS is a non-destructive process that generates a rich dataset suitable for analysis by artificial intelligence algorithms, which are being successfully applied to cancer diagnosis of digital histology [38,39]. The virtual colourization process already leverages in-house developed AI algorithms, and analysis via AI would be a natural extension of this process. The unstained tissue remains suitable for any additional subsequent analyses, which allows downstream standard-of-care processing of samples to be unaffected. Consequently, PARS virtual histology has the potential to both improve the speed and the accuracy of diagnostic interpretation of breast histology, reduces the consumption of limited biopsy tissue, and is, in principle, widely applicable to histologic evaluation of benign and malignant tissues of any origin. Moreover, this study was limited to human breast cancers but should be directly applicable to all other types of biopsied organs since tissue preservation and H&E staining are the same procedure for all types of tissues.

Author Contributions: Conceptualization, J.E.D.T., B.R.E., D.D., G.B., J.R.M. and P.H.R.; Data curation, J.E.D.T., H.G. and M.P.W.; Formal analysis, M.P.W.; Funding acquisition, P.H.R.; Methodology, J.E.D.T., B.R.E., H.G., D.D., M.P.W., G.B. and J.R.M.; Resources, H.G., D.D., G.B. and J.R.M.; Software, J.E.D.T. and B.R.E.; Supervision, J.R.M. and P.H.R.; Writing—original draft, J.E.D.T. and J.R.M.; Writing—review and editing, J.E.D.T., B.R.E., H.G., D.D., M.P.W., G.B., J.R.M. and P.H.R. All authors have read and agreed to the published version of the manuscript.

Funding: This research was funded by: Natural Sciences and Engineering Research Council of Canada (DGECR-2019-00143, RGPIN2019-06134, DH-2023-00371); Canada Foundation for Innovation (JELF #38000); Mitacs Accelerate (IT13594); University of Waterloo Startup funds; Centre for Bioengineering and Biotechnology (CBB Seed fund); illumiSonics Inc. (SRA #083181); New frontiers in research fund—exploration (NFRFE-2019-01012); The Canadian Institutes of Health Research (CIHR PJT 185984).

Institutional Review Board Statement: Tissues were acquired from the Cross-Cancer Institute (Edmonton, Alberta, Canada) through collaboration with clinical partners. Specimens were acquired under protocols (Protocol ID: HREBA.CC-18-0277) with the Research Ethics Board of Alberta and (Photoacoustic Remote Sensing Microscopy of Surgical Resection, Needle Biopsy, and Pathology Specimens; Protocol ID 40275) with the University of Waterloo Health Research Ethics Committee. All human tissue experiments were conducted in accordance with the government of Canada guidelines and regulations, including "Ethical Conduct for Research Involving Humans (TCPS2)". The seven independent breast tissue core biopsies used in this study represented a spectrum of known histologic findings, spanning normal breast, ductal carcinoma in situ cancer, invasive ductal carcinoma, and invasive lobular carcinoma.

Informed Consent Statement: The samples were obtained from anonymous patient donors, with all patient identifiers removed to ensure anonymity. The ethics committee waived the need for patient consent as these archival tissues were no longer necessary for patient diagnostics. Researchers were not provided with any information pertaining to the identity of the patients.

Data Availability Statement: Datasets used and analyzed during this study are available from the corresponding author on reasonable request.

Acknowledgments: The authors would like to thank the Cross-Cancer Institute in Edmonton, Alberta for providing human breast tissue samples. The authors would also like to thank Ellen Cai, Julinor Bacani, Judith Hugh, Murray Savard, Richard Berendt, Erene Farag, and Nadia Giannakopoulos for participating in the survey and questionnaire.

Conflicts of Interest: The authors JEDT, BRE, HG, DD, JRM, and PHR all have financial interests in IllumiSonics, which has provided funding to the PhotoMedicine Labs. The authors MPW and GB do not have any competing interests. The assessment data were generated by pathologists masked to the image origin, and these pathologists have no financial interest in the outcomes of the study. The image order was randomized by MPW, the statistician who designed and analyzed the study; MPW also has no financial interests in the outcomes. All the data gathered were included in the analysis. The raw dataset, generated using SurveyMonkey, was directly provided to the statistician. In addition, the authors also have other academic affiliations and are bound by their academic integrity requirements.

References

1. Bick, U.; Trimboli, R.M.; Athanasiou, A.; Balleyguier, C.; Baltzer, P.A.T.; Bernathova, M.; Borbély, K.; Brkljacic, B.; Carbonaro, L.A.; Clauser, P.; et al. Image-guided breast biopsy and localisation: Recommendations for information to women and referring physicians by the European Society of Breast Imaging. *Insights Imaging* **2020**, *11*, 12. [CrossRef] [PubMed]
2. Verkooijen, H.M.; Peeters, P.H.; Buskens, E.; Koot, V.C.; Borel Rinkes, I.H.; Mali, W.P.; van Vroonhoven, T.J. Diagnostic accuracy of large-core needle biopsy for nonpalpable breast disease: A meta-analysis. *Br. J. Cancer* **2000**, *82*, 1017–1021. [CrossRef] [PubMed]
3. Parker, S.H.; Burbank, F.; Jackman, R.J.; Aucreman, C.J.; Cardenosa, G.; Cink, T.M.; Coscia, J.L.; Eklund, G.W.; Evans, W.P.; Garver, P.R. Percutaneous large-core breast biopsy: A multi-institutional study. *Radiology* **1994**, *193*, 359–364. [CrossRef] [PubMed]
4. Crowe, J.P.; Rim, A.; Patrick, R.J.; Rybicki, L.A.; Grundfest-Broniatowski, S.F.; Kim, J.A.; Lee, K.B. Does core needle breast biopsy accurately reflect breast pathology? *Surgery* **2003**, *134*, 523–526; discussion 526–528. [CrossRef] [PubMed]
5. Verkooijen, H.M. Diagnostic accuracy of stereotactic large-core needle biopsy for nonpalpable breast disease: Results of a multicenter prospective study with 95% surgical confirmation. *Int. J. Cancer* **2002**, *99*, 853–859. [CrossRef]
6. Jackman, R.J.; Nowels, K.W.; Rodriguez-Soto, J.; Marzoni, F.A.; Finkelstein, S.I.; Shepard, M.J. Stereotactic, automated, large-core needle biopsy of nonpalpable breast lesions: False-negative and histologic underestimation rates after long-term follow-up. *Radiology* **1999**, *210*, 799–805. [CrossRef]
7. Youk, J.H.; Kim, E.-K.; Kim, M.J.; Oh, K.K. Sonographically Guided 14-Gauge Core Needle Biopsy of Breast Masses: A Review of 2,420 Cases with Long-Term Follow-Up. *Am. J. Roentgenol.* **2008**, *190*, 202–207. [CrossRef]
8. Hatmaker, A.R.; Donahue, R.M.J.; Tarpley, J.L.; Pearson, A.S. Cost-effective use of breast biopsy techniques in a veterans health care system. *Am. J. Surg.* **2006**, *192*, e37–e41. [CrossRef]
9. Bruening, W.; Fontanarosa, J.; Tipton, K.; Treadwell, J.R.; Launders, J.; Schoelles, K. Systematic Review: Comparative Effectiveness of Core-Needle and Open Surgical Biopsy to Diagnose Breast Lesions. *Ann. Intern. Med.* **2010**, *152*, 238–246. [CrossRef]
10. White, R.R.; Halperin, T.J.; Olson, J.A.; Soo, M.S.; Bentley, R.C.; Seigler, H.F. Impact of Core-Needle Breast Biopsy on the Surgical Management of Mammographic Abnormalities. *Ann. Surg.* **2001**, *233*, 769–777. [CrossRef]
11. Smith, D.N.; Christian, R.; Meyer, J.E. Large-core needle biopsy of nonpalpable breast cancers. The impact on subsequent surgical excisions. *Arch. Surg.* **1997**, *132*, 256–259; discussion 260. [CrossRef] [PubMed]
12. Day, C.E. (Ed.) *Histopathology: Methods and Protocols*; Methods in Molecular Biology; Springer: New York, NY, USA, 2014; Volume 1180, ISBN 978-1-4939-1049-6.
13. Gurina, T.S.; Simms, L. Histology, Staining. In *StatPearls*; StatPearls Publishing: Treasure Island, FL, USA, 2023.

14. Kwok, T.C.; Rakha, E.A.; Lee, A.H.S.; Grainge, M.; Green, A.R.; Ellis, I.O.; Powe, D.G. Histological grading of breast cancer on needle core biopsy: The role of immunohistochemical assessment of proliferation. *Histopathology* **2010**, *57*, 212. [CrossRef] [PubMed]

15. Brown, L. Improving histopathology turnaround time: A process management approach. *Curr. Diagn. Pathol.* **2004**, *10*, 444–452. [CrossRef]

16. Ecclestone, B.R.; Bell, K.; Sparkes, S.; Dinakaran, D.; Mackey, J.R.; Haji Reza, P. Label-free complete absorption microscopy using second generation photoacoustic remote sensing. *Sci. Rep.* **2022**, *12*, 8464. [CrossRef] [PubMed]

17. Hosseinaee, Z.; Abbasi, N.; Pellegrino, N.; Khalili, L.; Mukhangaliyeva, L.; Haji Reza, P. Functional and structural ophthalmic imaging using noncontact multimodal photoacoustic remote sensing microscopy and optical coherence tomography. *Sci. Rep.* **2021**, *11*, 11466. [CrossRef]

18. Restall, B.S.; Haven, N.J.M.; Kedarisetti, P.; Zemp, R.J. In vivo combined virtual histology and vascular imaging with dual-wavelength photoacoustic remote sensing microscopy. *OSA Contin.* **2020**, *3*, 2680–2689. [CrossRef]

19. Haven, N.J.M.; Bell, K.L.; Kedarisetti, P.; Lewis, J.D.; Zemp, R.J. Ultraviolet photoacoustic remote sensing microscopy. *Opt. Lett.* **2019**, *44*, 3586–3589. [CrossRef]

20. Bell, K.; Mukhangaliyeva, L.; Khalili, L.; Reza, P.H. Hyperspectral Absorption Microscopy Using Photoacoustic Remote Sensing. *Opt. Express* **2021**, *29*, 24338. [CrossRef]

21. Kedarisetti, P.; Haven, N.J.M.; Restall, B.S.; Martell, M.T.; Zemp, R.J. Label-free lipid contrast imaging using non-contact near-infrared photoacoustic remote sensing microscopy. *Opt. Lett.* **2020**, *45*, 4559–4562. [CrossRef]

22. Tweel, J.E.D.; Ecclestone, B.R.; Boktor, M.; Simmons, J.A.T.; Fieguth, P.; Reza, P.H. Virtual Histology with Photon Absorption Remote Sensing using a Cycle-Consistent Generative Adversarial Network with Weakly Registered Pairs. *arXiv* **2023**, arXiv:2306.08583.

23. Martell, M.T.; Haven, N.J.; Cikaluk, B.D.; Restall, B.S.; McAlister, E.A.; Mittal, R.; Adam, B.A.; Giannakopoulos, N.; Peiris, L.; Silverman, S.; et al. Deep learning-enabled realistic virtual histology with ultraviolet photoacoustic remote sensing microscopy. *Nat. Commun.* **2023**, *14*, 5967. [CrossRef]

24. Pecourt, J.-M.L.; Peon, J.; Kohler, B. DNA Excited-State Dynamics: Ultrafast Internal Conversion and Vibrational Cooling in a Series of Nucleosides. *J. Am. Chem. Soc.* **2001**, *123*, 10370–10378. [CrossRef] [PubMed]

25. Bricker, W.P.; Shenai, P.M.; Ghosh, A.; Liu, Z.; Enriquez, M.G.M.; Lambrev, P.H.; Tan, H.-S.; Lo, C.S.; Tretiak, S.; Fernandez-Alberti, S.; et al. Non-radiative relaxation of photoexcited chlorophylls: Theoretical and experimental study. *Sci. Rep.* **2015**, *5*, 13625. [CrossRef] [PubMed]

26. Soltani, S.; Ojaghi, A.; Robles, F.E. Deep UV dispersion and absorption spectroscopy of biomolecules. *Biomed. Opt. Express* **2019**, *10*, 487–499. [CrossRef] [PubMed]

27. Focke, C.M.; Decker, T.; van Diest, P.J. The reliability of histological grade in breast cancer core needle biopsies depends on biopsy size: A comparative study with subsequent surgical excisions. *Histopathology* **2016**, *69*, 1047–1054. [CrossRef] [PubMed]

28. Tweel, J.E.D.; Ecclestone, B.R.; Boktor, M.; Dinakaran, D.; Mackey, J.R.; Reza, P.H. Automated Whole Slide Imaging for Label-Free Histology using Photon Absorption Remote Sensing Microscopy. *arXiv* **2023**, arXiv:2304.13736.

29. Zhu, J.-Y.; Park, T.; Isola, P.; Efros, A.A. Unpaired Image-to-Image Translation using Cycle-Consistent Adversarial Networks. *arXiv* **2020**, arXiv:1703.10593.

30. McHugh, M.L. Interrater reliability: The kappa statistic. *Biochem. Med.* **2012**, *22*, 276–282. [CrossRef]

31. Fleiss, J.L. Measuring nominal scale agreement among many raters. *Psychol. Bull.* **1971**, *76*, 378–382. [CrossRef]

32. A Coefficient of Agreement for Nominal Scales—Jacob Cohen. 1960. Available online: https://journals.sagepub.com/doi/10.1177/001316446002000104 (accessed on 1 August 2023).

33. Practical Statistics for Medical Research. Available online: https://www.routledge.com/Practical-Statistics-for-Medical-Research/Altman/p/book/9780412276309 (accessed on 1 August 2023).

34. R: A Language and Environment for Statistical Computing. Available online: https://www.gbif.org/tool/81287/r-a-language-and-environment-for-statistical-computing (accessed on 1 August 2023).

35. Tosta, T.A.A.; de Faria, P.R.; Neves, L.A.; do Nascimento, M.Z. Color normalization of faded H&E-stained histological images using spectral matching. *Comput. Biol. Med.* **2019**, *111*, 103344. [CrossRef]

36. Jackson, S.L.; Frederick, P.D.; Pepe, M.S.; Nelson, H.D.; Weaver, D.L.; Allison, K.H.; Carney, P.A.; Geller, B.M.; Tosteson, A.N.A.; Onega, T.; et al. Diagnostic Reproducibility: What Happens When the Same Pathologist Interprets the Same Breast Biopsy Specimen at Two Points in Time? *Ann. Surg. Oncol.* **2017**, *24*, 1234–1241. [CrossRef] [PubMed]

37. Calle, C.; Zhong, E.; Hanna, M.G.; Ventura, K.; Friedlander, M.A.; Morrow, M.; Cody, H.; Brogi, E. Changes in the Diagnoses of Breast Core Needle Biopsies on Second Review at a Tertiary Care Center: Implications for Surgical Management. *Am. J. Surg. Pathol.* **2023**, *47*, 172–182. [CrossRef] [PubMed]

38. Polónia, A.; Campelos, S.; Ribeiro, A.; Aymore, I.; Pinto, D.; Biskup-Fruzynska, M.; Veiga, R.S.; Canas-Marques, R.; Aresta, G.; Araújo, T.; et al. Artificial Intelligence Improves the Accuracy in Histologic Classification of Breast Lesions. *Am. J. Clin. Pathol.* **2021**, *155*, 527–536. [CrossRef] [PubMed]

39. Liu, Y.; Han, D.; Parwani, A.V.; Li, Z. Applications of Artificial Intelligence in Breast Pathology. *Arch. Pathol. Lab. Med.* **2023**, *147*, 1003–1013. [CrossRef] [PubMed]

Article

Prediction of Neoadjuvant Chemoradiotherapy Response in Rectal Cancer Patients Using Harmonized Radiomics of Multcenter ^{18}F-FDG-PET Image

Hye-Min Ju [1,2], Jingyu Yang [2], Jung-Mi Park [3], Joon-Ho Choi [3], Hyejin Song [2], Byung-Il Kim [4], Ui-Sup Shin [5], Sun Mi Moon [5], Sangsik Cho [5] and Sang-Keun Woo [1,2,*]

[1] Radiological and Medico-Oncological Sciences, University of Science and Technology, Daejeon 34113, Republic of Korea; hmju@kirams.re.kr

[2] Division of Applied RI, Korea Institute of Radiological and Medical Sciences, Seoul 07812, Republic of Korea; jingue.yang@kirams.re.kr (J.Y.); songhj@kirams.re.kr (H.S.)

[3] Department of Nuclear Medicine, Soonchunhyang University Bucheon Hospital, Bucheon 14584, Republic of Korea; nm.jmipark@daum.net (J.-M.P.); 114780@schmc.ac.kr (J.-H.C.)

[4] Department of Nuclear Medicine, Korea Institute of Radiological and Medical Sciences, Seoul 07812, Republic of Korea; kimbi@kirams.re.kr

[5] Department of Surgery, Korea Institute of Radiological and Medical Sciences, Seoul 07812, Republic of Korea; uisupshin@kirams.re.kr (U.-S.S.); sms@kirams.re.kr (S.M.M.); whtkdtlr@kirams.re.kr (S.C.)

* Correspondence: skwoo@kirams.re.kr

Citation: Ju, H.-M.; Yang, J.; Park, J.-M.; Choi, J.-H.; Song, H.; Kim, B.-I.; Shin, U.-S.; Moon, S.M.; Cho, S.; Woo, S.-K. Prediction of Neoadjuvant Chemoradiotherapy Response in Rectal Cancer Patients Using Harmonized Radiomics of Multcenter ^{18}F-FDG-PET Image. *Cancers* 2023, 15, 5662. https://doi.org/10.3390/cancers15235662

Academic Editor: David Wong

Received: 5 October 2023
Revised: 7 November 2023
Accepted: 28 November 2023
Published: 30 November 2023

Simple Summary: Neoadjuvant chemotherapy is the standard treatment for locally advanced rectal cancer. Preoperative chemoradiotherapy yields clinically significant tumor regression; while some patients exhibit a minimal response, others exhibit a complete pathologic response. We developed deep learning and machine learning models to predict chemoradiotherapy response across external tests using multicenter data. The machine learning model, which used harmonized image features extracted from ^{18}F-FDG PET, showed higher performance and demonstrated reproducibility through external tests compared to the deep learning models using ^{18}F-FDG PET images. Our study highlights the feasibility of predicting the chemoradiotherapy response of individual patients using non-invasive and reliable image feature values.

Abstract: We developed machine and deep learning models to predict chemoradiotherapy in rectal cancer using ^{18}F-FDG PET images and harmonized image features extracted from ^{18}F-FDG PET/CT images. Patients diagnosed with pathologic T-stage III rectal cancer with a tumor size > 2 cm were treated with neoadjuvant chemoradiotherapy. Patients with rectal cancer were divided into an internal dataset (n = 116) and an external dataset obtained from a separate institution (n = 40), which were used in the model. AUC was calculated to select image features associated with radiochemotherapy response. In the external test, the machine-learning signature extracted from ^{18}F-FDG PET image features achieved the highest accuracy and AUC value of 0.875 and 0.896. The harmonized first-order radiomics model had a higher efficiency with accuracy and an AUC of 0.771 than the second-order model in the external test. The deep learning model using the balanced dataset showed an accuracy of 0.867 in the internal test but an accuracy of 0.557 in the external test. Deep-learning models using ^{18}F-FDG PET images must be harmonized to demonstrate reproducibility with external data. Harmonized ^{18}F-FDG PET image features as an element of machine learning could help predict chemoradiotherapy responses in external tests with reproducibility.

Keywords: harmonized radiomics; machine learning; deep learning; radiochemotherapy; ^{18}F-FDG PET

1. Introduction

More than 100,000 individuals worldwide are diagnosed with rectal cancer annually [1]. Rectal cancer is generally treated with neoadjuvant chemoradiotherapy, and tumor responses to therapy are diverse, with 54–75% of patients experiencing tumor downstaging [2]. The reasons for these changes in treatment response are poorly understood, and there is no exact method for predicting the treatment response [3]. Only 15–27% of patients show no residual viable tumors on pathological examination, pathological complete response (pCR) to chemoradiotherapy, and surgery [4]. An accurate imaging biomarker for predicting and evaluating chemotherapy could the early classification of patients into different prognostic groups and personalized treatment approaches. Early detection of patients who might respond poorly to chemoradiotherapy can provide them the opportunity to undergo surgery and receive enhanced treatments to maximize treatment response.

Medical imaging can be used to noninvasively evaluate therapeutic responses to chemotherapy. Jang et al. developed an MRI-based deep learning model for predicting chemotherapy response in rectal cancer and reported the area under receiver operating characteristic curve (AUC) of 0.76 and an accuracy of 0.85. ^{18}F-FDG PET/CT has also been widely used to monitor treatment response in many types of malignancies, stages, and diagnoses. ^{18}F-FDG PET can help detect glucose metabolism and reveal tumor characteristics. As the anatomical data obtained from CT in rectal cancer patients can help distinguish between physiological and pathological intestinal absorption [5], ^{18}F-FDG PET/CT is generally considered a standard tool for predicting the response to chemotherapy in rectal cancer. The radiomics features of ^{18}F-FDG PET/CT can also facilitate the prediction of chemoradiotherapy. Taking this into consideration, researchers are increasingly exploring the potential of incorporating radiomic features from ^{18}F-FDG PET/CT scans into predictive models to enhance the accuracy and reliability of forecasting responses to chemoradiotherapy.

Recently, the use of machine learning techniques for large and complex biological data analysis has increased. Deep learning techniques are considered among the most powerful tools and are frequently used in bioinformatics because they can allow the analysis of vast amounts of data. Many radiomics studies utilize features extracted by manual method, and these methods are significantly influenced by the knowledge and experience of individual researchers [6]. Consequently, deep learning techniques for computing task-adaptive feature representations by learning layers of complex features directly from medical images are considered suitable tools for predicting prognosis. Deep learning techniques that can automatically learn representative information from raw image data to decode the radiation expression type of tumors can assist in disease diagnosis, prognostic evaluation, and treatment sensitivity prediction [7]. The model performance of deeper hidden layers for pattern recognition has recently begun to surpass that of classical methods in different fields. One of the most popular deep neural networks is the Convolutional Neural Network (CNN). Random forest (RF) technology, which includes an ensemble of decision trees and naturally integrates feature selection and interaction during learning, is a popular choice in personalized medicine. It is nonparametric, efficient, and has a high predictive accuracy for many types of data. RF model is increasingly being adopted because of its advantages in dealing with small sample sizes, high-dimensional feature spaces, and complex data structures [8].

In oncology research, particularly when assessing rectal cancer responses to therapy, the role of SUVmax and SUVmean values derived from 18F-FDG PET/CT scans has been under critical evaluation, as illustrated by several independent studies. Two independent studies showed that the SUVmax predicted chemotherapy with a specificity and overall accuracy of only 35% and 44%, respectively [9,10]. SUVmean, dissimilarity, and contrast from the neighborhood intensity-difference matrix (NGTDM contrast) were significantly and independently associated with OS [11]. A decrease in metabolic tumor volume (MTV) and total lesion glycolysis (TLG) values was suggested to be an indicator of a positive response to chemotherapy [12]. Chemotherapy response predictions using ^{18}F-FDG PET/CT are not sufficiently accurate to distinguish patients showing treatment response from those who respond poorly to the treatment [13]. Several studies have reported that radiation features were scanner or protocol-sensitive, highlighting the importance of harmonizing image features to reduce multicenter variability before pooling data from multiple sites [14,15].

In the present study, we evaluated the use of machine learning to predict chemoradio therapy responses using radiomics harmonization and demonstrated the reproducibility and repeatability of the findings through rigorous external testing. Our effort is not only to address the limitations of the current methodologies but also to contribute to the development of a more robust and universally applicable predictive model for chemoradiotherapy responses in cancer treatment.

2. Materials and Methods

2.1. Patient Cohort

All patients were diagnosed with pathologic T-stage III rectal cancer, with tumor growth into the outer lining of the bowel wall without breaching its integrity. Patients with a tumor size > 2 cm were treated with neoadjuvant chemoradiotherapy before surgery. The internal and external cohorts comprised 116 patients from internal institutions (Korea Institute of Radiological and Medical Sciences) and 40 patients from independent institutions (Soonchunhyang University Bucheon Hospital). The internal cohort comprised 21 patients diagnosed with pCR and 95 patients diagnosed with non-pCR. The external cohort consisted of six patients diagnosed with pCR and 31 patients diagnosed with non-pCR. The rectal cancer region was cropped from an ^{18}F-FDG PET image (Figure 1).

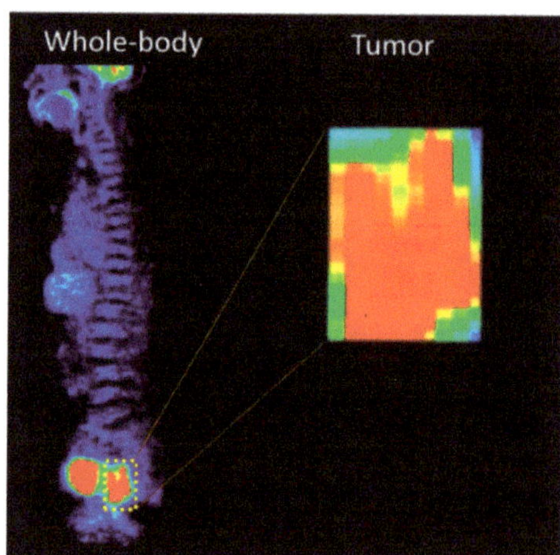

Figure 1. The corp process of rectal cancer region from ^{18}F-FDG PET image.

2.2. Image Feature Extraction

We utilized LIFEx (Local Image Features Extraction, version 4.90) software to calculate image features from 18F-FDG PET/CT images of rectal cancer patients. In total, 55 image features were extracted. The region of interest (ROI) was marked manually with an SUV threshold of 2.0 (Figure 2). Tumor lesions were identified in the area of ^{18}F-FDG uptake, which was pathologically increased and was in contrast to the CT images. To predict chemotherapy response in rectal cancer, first- and second-order images were used separately to compare intensity-based and GLCM-based image characteristics. The AUC was calculated to select the image features from the first- and second-order features using R (version 4.2.2) software (R Foundation for Statistical Computing, Vienna, Austria).

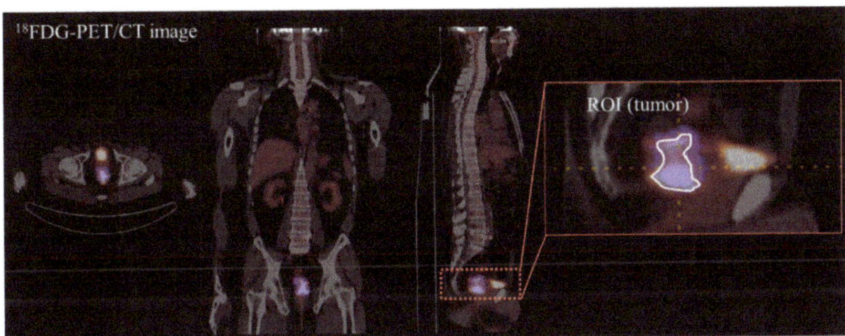

Figure 2. Radiomics extracted from ^{18}F-FDG PET/CT.

2.3. Harmonization Methodology

Harmonization of the image features from the internal and external ^{18}F-FDG PET/CT datasets was performed. Both of training set and test set were harmonized in separate manner. The harmonization (ComBat) method was used with an online application (https://forlhac.shinyapps.io/Shiny_ComBat/, accessed on 28 November 2023). ComBat is a batch-matching technology initially proposed for gene expression microarrays [16] and has been widely used in the field of imaging. The ComBat model is given by

$$y^{ij} = \alpha + \gamma_i + \delta_i \varepsilon_{ij}$$

where j indicates the specific measurement of image feature y, i indicates the setting of the scanner, protocol effect, or even observer effect (called the site effect), α represents the average value of the image features denoted as y, γ_i signifies additive batch effect influence on measurement, δ_i represents multiplicative batch effect, and ε_{ij} is an error term. Batch i represents the experimental settings employed for y measurement, including the possible scanner effect. Site effects γ_i and δ_i can be estimated using conditional posterior means and subsequently corrected using

$$y_{ij}{}^{ComBat} = \frac{y_{ij} - \hat{\alpha} - \hat{\gamma}_i}{\hat{\delta}_i} + \hat{\alpha}$$

where $\hat{\alpha}$, $\hat{\gamma}_i$ and $\hat{\delta}_i$ are estimators of α, γ_i and δ_i. $y_{ij}{}^{ComBat}$ is the converted y_{ij} measured value devoid of the site i effect.

2.4. Deep Learning and Machine Learning

The CNN structure consisted of input, convolution, batch normalization, ReLU, max pooling, linear, dropout, and output layers. The CNN parameters comprised the optimizer, learning rate, and epoch; the values were set to Adam, 0.0002, and 200, respectively. Two convolutional layers are used. The CNN structure was constructed using two-dimensional input slices taken from each patient. The chemotherapy prediction performance of the RF model was internally and externally evaluated using the scikit-learn library (version 1.2.0) in Python (version 3.10.11).

Augmentation techniques were employed to resolve the data imbalance between pCR and non-pCR. The "RandomRotation" function of PyTorch livery in Python were used to randomly rotate input images by a certain angle to increase the diversity of the training dataset. The "RandomResizedCrop" function of PyTorch livery in Python is employed to randomly select a portion of the input image and subsequently resize it, serving the purpose of augmenting the training dataset and enhancing its variety. The Synthetic minority oversampling technique was implemented on the training dataset for machine learning to mitigate data imbalance.

After splitting the internal dataset at a 7:3 ratio, internal test were performed for both models through evaluating AUC, accuracy, precision, and sensitivity. External test were proceed using independent institution dataset. Confusion matrix-based evaluation metrics including accuracy, sensitivity and precision were estimated and the threshold probability was adjusted to the value that maximizes Youden's index.

3. Results

3.1. Patients Cohort

^{18}F-FDG PET/CT images from 116 internal and 40 external datasets were used for model estimation. The average ages of the internal and external datasets were 61.85 years and 59.88, respectively. The internal cohort comprised 75 males (64.66%) and 41 females (35.34%). The external cohort comprised 27 males (67.5%) and 13 females (32.5%). A summary of the demographic characteristics and pathological TNM stages is presented in Table 1. The patient cohort included patients who developed lymph node- or distant organ-metastases.

Table 1. Characteristics of the study cohort.

Characteristics	Internal Dataset (n = 116)	External Dataset (n = 40)
Chemoradiotherapy response (%)		
pCR	21 (18.1)	6 (15)
non-pCR	95 (81.9)	34 (85)
Age (%)		
<65	69 (59.48)	23 (57.5)
≥65	47 (40.52)	17 (42.5)
Mean age (y)	61.85	59.88
Sex (%)		
Male	75 (64.66)	27 (67.5)
Female	41 (35.34)	13 (32.5)
Clinical T-stage, n (%)		
T3	116 (100)	40
Clinical N stage (%)		
N0	19 (16.38)	5 (12.5)
N1	31 (26.72)	8 (20)
N1a	2 (1.72)	
N1b	13 (11.21)	1 (2.5)
N2	37 (31.9)	6 (15)
N2a	13 (11.21)	12 (30)
N2b	1 (0.86)	8 (20)
Clinical M stage (%)		
M0	106 (91.38)	32 (80)
M1	6 (5.17)	
M1a	3 (2.59)	8 (50)
M1b	1 (0.86)	
Clinical stage (%)		
IIA		5 (12.5)
IIB	18 (15.52)	
IIC		
IIIA	42 (36.21)	21 (52.5)
IIIB	46 (39.66)	6 (15)
IIIC		8 (20)
IVA	10 (8.62)	

pCR: pathological complete response.

3.2. Evaluation of Deep Learning Model

The CNN model for rectal cancer chemoradiotherapy prediction was developed using ^{18}F-FDG PET images. The number of pCR data points from the internal and external data increased through augmentation to 84 and 24, respectively. To equalize the amount of pCR and non-pCR data, the pCR data from the internal and external cohorts were decreased by random sampling. The deep learning model showed a performance, with an accuracy of 0.867 and 0.789 in the internal test (Table 2). However, in the external test, the deep learning signature showed an accuracy of 0.557 and 0.355 (Table 3). The deep learning models showed higher performance in internal test then external test.

Table 2. Internal test of CNN model using ^{18}F-FDG PET images.

	Number of Data		Efficiency Evaluation			
Data Set	**pCR**	**Non-pCR**	**Accuracy**	**Precision**	**Sensitivity**	**AUC (95% CI)**
Imbalanced	21	21	0.867	0.871	0.871	0.903 (0.856–0.949)
Balanced	84	95	0.789	0.843	0.677	0.835 (0.804–0.866)

pCR: pathological complete response; AUC: area under receiver operating characteristic curve; CI: Confidence interval.

Table 3. External test of CNN model using ^{18}F-FDG PET images.

	Number of Data		Efficiency Evaluation			
Data Set	**pCR**	**Non-pCR**	**Accuracy**	**Precision**	**Sensitivity**	**AUC (95% CI)**
Imbalanced	6	6	0.557	0.542	0.495	0.498 (0.412–0.583)
Balanced	24	25	0.355	0.241	0.475	0.443 (0.378–0.509)

pCR: pathological complete response; AUC: area under receiver operating characteristic curve; CI: Confidence interval.

3.3. Image Feature Extraction and Harmonization

A total of 55 image featuers were quantitatively calculated from ^{18}F-FDG PET and CT images. The image features were separated into first-order features, including conventional indices, shapes, and histogram-based intensity values (n = 23). The image texture features were assigned as second-order features, including a Gray-level co-occurrence matrix (GLCM), neighborhood gray-level difference matrix (NGLDM), Gray-level run-length matrix (GLRLM), and Gray-level zone length matrix (GLZLM) (n = 22) (Figure 2). AUC was calculated to determine image features capable of distinguishing between chemotherapy and non-PCR cases. Subsequently, image features from the internal dataset were selected and used for machine learning. First-order features extracted from ^{18}F-FDG PET and CT with AUC over 0.65 and 0.55 were used for machine learning, respectively (Table 4). Second-order features extracted from ^{18}F-FDG PET and CT with AUC over 0.7 and 0.6 were used for machine learning, respectively (Table 5). Image feature values from internal and external institutions were harmonized to reduce multicenter variations. GLZLM GLNU, which had the largest change in the distribution of values before and after harmonization, was visualized (Figure 3).

Table 4. Extraction of first-order image features by AUC cut-off value.

First-Order Image Feature			
¹⁸F-FDG PET	**AUC**	**CT**	**AUC**
SHAPE Sphericity	0.715	Uniformity	0.663
SUVQ1	0.707	Entropy log10	0.659
SUVmean	0.694	Entropy log2	0.659
SUVQ3	0.692	SHAPE Compacity	0.618
SUVQ2	0.69	SHAPE Volume	0.604
Uniformity	0.681	SUVstd	0.6
Entropy log10	0.677	SUVmax	0.593
Entropy log2	0.677	SUVQ3	0.589
SUVstd	0.667	Kurtosis	0.582
SUVmin	0.65	ExcessKurtosis	0.582
		Volume	0.663
		Sphericity	0.579
		Skewness	0.578
		TLG	0.563

Abbreviations: SUVQ, Standardized Uptake Value Quotient; SUV, Standardized Uptake Value; SUVstd, Standardized Uptake Value Standard Deviation; SUVmin, Standardized Uptake Value Minimum; SHAPE, Sphericity Histogram Analysis, and Parametric Evaluation; SUVmax, Standardized Uptake Value Maximum; TLG, Total Lesion Glycolysis.

Table 5. Extraction of second-order image features by AUC cut-off value.

Second-Order Image Feature			
¹⁸F-FDG PET	**AUC**	**CT**	**AUC**
GLZLM LZLGE	0.766	NGLDM Contrast	0.704
GLZLM LZE	0.765	GLZLM ZP	0.698
GLRLM GLNU	0.763	GLRLM LRE	0.69
GLRLM SRE	0.756	GLRLM RP	0.69
GLRLM RP	0.755	GLRLM SRE	0.689
GLRLM LRE	0.753	GLZLM LZLGE	0.689
NGLDM Contrast	0.74	GLCM Homogeneity	0.689
GLZLM ZP	0.74	GLZLM LZE	0.685
GLZLM LZHGE	0.74	GLZLM LZHGE	0.683
GLCM Homogeneity	0.734	GLCM Energy	0.683
NGLDM Busyness	0.732	GLCM Entropy log10	0.667
GLRLM LRLGE	0.731	GLCM Entropy log2	0.667
GLCM Dissimilarity	0.71	GLCM Dissimilarity	0.661
GLCM Contrast	0.702	GLRLM GLNU	0.647
GLRLM LGRE	0.701	GLRLM LRHGE	0.633
		NGLDM Busyness	0.628
		GLRLM SRHGE	0.617
		GLCM Contrast	0.613
		GLRLM LRLGE	0.613

Abbreviations: GLZLM, Gray-Level Zone Length Matrix; LZLGE, Long Zone Low Gray-level Emphasis; LZE, Low Gray-level Zone Emphasis; GLRLM, Gray-Level Run Length Matrix; SRE, Short Run Emphasis; RP, Run Percentage; LRE, Gray-Level Run Length Matrix; NGLDM, Neighborhood Gray-Level Dependence Matrix; ZP, Zone Percentage; LZHGE, Long-Zone High-Grey level Emphasis; GLCM, Gray-Level Co-occurrence Matrix; LRLGE, Long Run Low Gray-level Emphasis; LGRE, Low Gray-level Run Emphasis.

3.4. Evaluation of Machine Learning Model

The extracted primary and secondary features were used as variables for the RF model, and each model was evaluated using internal and external tests. The RF model using harmonized first-order features showed an accuracy and AUC of 0.771, which is higher than before harmonization in the external test. The RF model using secondary features exhibited an accuracy and AUC of 0.675 and 0.603 in the external test after harmonization, lower than those without harmonization. The first-order features showed higher accuracy and AUC for the external datasets than the second-order features. In the external test set, the ^{18}F-FDG PET image feature as a machine learning signature achieved the highest accuracy with an AUC value of 0.875 and 0.896 (95% confidence interval 0.562–1) (Table 6).

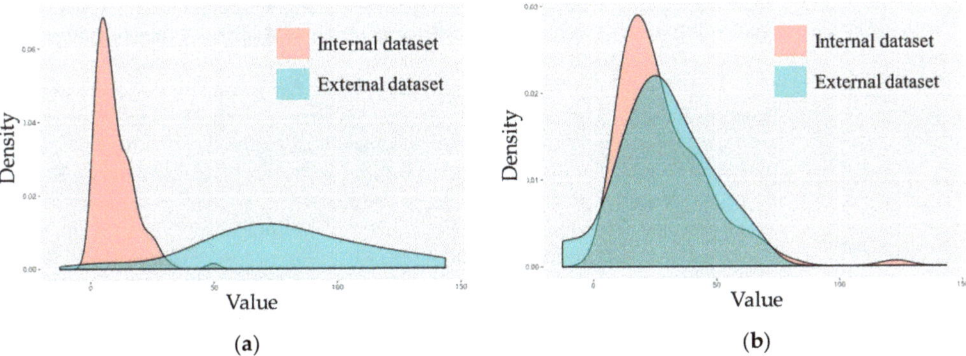

Figure 3. Distribution of GLZLM GLNU value before and after harmonization: (**a**) Distribution of GLZLM GLNU extracted from all T-stage patients before harmonization; (**b**) Distribution of GLZLM GLNU max extracted from all T-stage patients after harmonization.

Table 6. Internal and external test of RF model.

Image Feature	Value	Without Harmonization			Without Harmonization			With Harmonization		
		Internal Test			External Test			External Test		
		CT	PET	PET/CT	CT	PET	PET/CT	CT	PET	PET/CT
First order	Accuracy	0.54	0.62	0.56	0.55	0.7	0.525	0.6	0.646	0.771
	Precision	0.524	0.575	0.615	0.227	0.2	0.19	0.222	0.769	0.882
	Sensitivity	0.88	0.92	0.32	0.833	0.333	0.667	0.667	0.417	0.625
	AUC	0.54	0.62	0.56	0.667	0.549	0.583	0.627	0.646	0.771
	95% CI for AUC	-	-	-	0.412–0.921	0.291–0.807	0.325–0.842	0.37–0.885	0.469–0.962	0.429–0.934
Second order	Accuracy	0.52	0.64	0.7	0.425	0.525	0.7	0.65	0.583	0.675
	Precision	0.516	0.63	0.727	0.185	0.19	0.25	0.25	0.7	0.632
	Sensitivity	0.64	0.68	0.64	0.833	0.667	0.5	0.667	0.292	0.5
	AUC	0.52	0.64	0.7	0.593	0.583	0.618	0.657	0.583	0.603
	95% CI for AUC	-	-	-	0.334–0.852	0.325–0.842	0.36–0.876	0.402–0.912	0.562–1	0.344–0.862
All	Accuracy	0.68	0.76	0.7	0.65	0.675	0.775	0.425	0.875	0.725
	Precision	0.765	0.81	0.639	0.214	0.267	0.333	0.185	0.952	0.333
	Sensitivity	0.52	0.68	0.92	0.5	0.667	0.5	0.833	0.833	0.833
	AUC	0.68	0.76	0.7	0.588	0.672	0.662	0.593	0.896	0.77
	95% CI for AUC	-	-	-	0.329–0.847	0.418–0.925	0.556–1	0.334–0.852	0.562–1	0.536–1

AUC: area under receiver operating characteristic curve; CI: Confidence interval.

4. Discussion

The performance of the machine learning models in predicting chemoradiotherapy response using imaging features extracted from ^{18}F-FDG PET images was estimated using an external test. Conducting multicenter studies is one of the main objectives of clinical applications. However, medical images acquired from different institutions may introduce biases due to variations in imaging devices, data acquisition methods, and protocols [17,18]. Because radiomics is sensitive, variations in feature values may occur even in cases where the same feature is extracted from multiple organs. Large-scale radiomic data analysis is required to verify the reproducibility of radiomics, and radiomic features extracted from images acquired from different centers must be integrated. In this study, radiomics harmonization was performed to reduce batch effects. Our results indicated that the harmonization of image features extracted from multiple datasets is essential as a predictor

In several studies related to cancers, the RF model has shown a high potential in predicting clinical outcomes [19–22]. The RF model demonstrated reproducibility and repeatability in external tests when utilizing the features extracted from ^{18}F-FDG PET images. Because the RF model generates predictions by randomly selecting a decision tree, it mitigates the risk of overfitting. As it traverses the decision tree, it learns the image features that best encapsulate the discriminatory factors for distinguishing tumor characteristics. Moreover, it is expected to yield superior outcomes because it employs an optimal cut-off value for discriminating between pCR and non-pCR patients based on image features. These attributes of the RF model appear to have further enhanced its predictive accuracy and AUC in the context of chemoradiotherapy prognosis.

Medical imaging offers vital insights into the progress of patients with rectal cancer, and AI holds promise for developing quantitative treatment decision support tools. Some studies have shown that tumor metabolic changes on ^{18}F-FDG PET were more predictive than tumor morphological modifications on CT [23–25]. In our study, image features extracted from ^{18}F-FDG PET images showed higher machine learning performance than those extracted from CT images. The imaging features of CT in the external tests showed an accuracy and AUC of 0.425 and 0.593, whereas those extracted from ^{18}F-FDG PET showed an accuracy and AUC of 0.875 and 0.896. Our study indicate that the radiomics of ^{18}F-FDG PET have a more complementary effect then CT in predicting the pCR of rectal cancer. ^{18}F-FDG PET imaging is crucial for monitoring alterations in tumor metabolic activity, playing a vital role in prognostic predictions for patients undergoing concurrent chemoradiotherapy. Although CT imaging provides comprehensive details pertaining to the tumor's size and shape, excelling in anatomical delineation, it falls short in effectively predicting tumor responses to chemoradiotherapy. This discrepancy highlights a potential limitation in its prognostic utility for this specific therapeutic context. It has been observed that the integration of radiomic features extracted from both ^{18}F-FDG PET and CT into predictive models can lead to a decrement in performance, suggesting a paradoxical reduction in the model's efficacy despite the amalgamation of data from both imaging techniques. This underscores the need for careful consideration when combining features from different modalities to enhance the accuracy of treatment response predictions.

The first and second selected features for AUC values encompassed those previously identified as having prognostic significance in other investigations. The significance of SUVmax, SUVmean, and Uniformity, which are image feature values, has been demonstrated in previous studies. The secondary features based on GLRLM, NGLDM, and GLRM were incorporated as important variables in the radiochemotherapy prediction model These feature values have demonstrated their predictive utility in various cancers. When the chemoradiotherapy response was predicted using harmonized first-order features, it showed a higher performance than second-order features. The first-order features were derived from histograms, whereas the second-order features were based on the GLCM. As the first-order values exhibited significant alterations following harmonization, the impact of harmonization is noteworthy. Conversely, the second-order values displayed negligible changes after harmonization. Consequently, the model utilizing first-order features exhibited superior performance in predicting rectal cancer chemotherapy outcomes.

There are several ^{18}F-FDG PET/CT predictive radiomics for pCR to chemotherapy, including visual response, maximum standardized uptake value (SUVmax), percentage SUVmax reduction, TLG, and MTV [26–29]. Lovinfosse et al. revealed that SUVmean, dissimilarity, and contrast from contrast NGTDM were significantly and independently associated with OS in patients with rectal cancer. Jean-Emmanuel et al. predicted a complete response using a deep neural network after rectal chemoradiotherapy with 80% accuracy in a multicenter cohort using radiomics extracted from CT. Xiaolu M et al. The RF model for the degree of differentiation, T-stage, and N-stage were obtained using radiomics from MRI (AUC, 0.746; 95% CI, 0.622–0.872; sensitivity, 79.3%; and specificity, 72.2%). Giannini et al.

evaluated a logistic regression model using six texture features (five from PET and one from T2w MRI) to determine the chemotherapy outcomes (AUC = 0.86; sensitivity = 86%, and specificity = 83%).

We estimated the performance of the deep learning model in predicting the outcomes of neoadjuvant responses using multicenter ^{18}F-FDG PET images. However, the model performance proved insignificant in external tests conducted with datasets from independent institutions. Deep learning demonstrated subpar performance in external tests owing to the omission of dataset harmonization, which failed to account for potential biases between the internal and external datasets. In the case of machine learning, the difference between the internal and external datasets was drastically reduced through the harmonization of the image feature values shown in the ROI; thus, reproducibility as a predictor of machine learning was confirmed. Batch effects can be mitigated by preprocessing the images employed in deep learning, involving techniques such as slope distortion correction, bias slope distortion correction, bias field correction, and intensity normalization, which help standardize the data [30,31]. Reducing batch effects through harmonization at the image level is expected to show high performance in sufficiently predicting chemotherapy, even in external tests.

Our study has some limitations. Deep learning exhibited a lower performance in external tests than in internal tests. This outcome may be attributed to the absence of harmonization between internal and external datasets. Because the CNN model makes predictions using the image itself, it is necessary to harmonize the image. The number of patients within the presently registered external data may be relatively limited, leading to suboptimal performance in external tests. Deep learning techniques in the realm of medical image analysis are challenged by their black-box characteristics, which pose issues for interpretability. Additionally, given the extensive discussion in this article about how chemotherapy and radiotherapy can significantly increase the risk of infertility for women wishing to conceive in the future, we propose a more proactive approach. Women should be given greater autonomy over their reproductive timelines, particularly through the strategic use of oocyte vitrification prior to undergoing such medical interventions [32].

5. Conclusions

Our research underscores the critical significance of image harmonization in multicenter studies for accurate chemotherapy response prediction in pancreatic cancer while also highlighting the potential of noninvasive radiomics-based machine learning models in predicting neoadjuvant chemoradiotherapy response in rectal cancer. A machine learning model predicting radiochemotherapy outcomes for pancreatic cancer using harmonized ^{18}F-FDG PET imaging features was confirmed to be reproducible and repeatable in external testing using multicenter data. A deep model using ^{18}F-FDG PET images without the harmonization process performed poorly in predicting neoadjuvant chemoradiotherapy response, demonstrating the importance of image harmonization in multicenter studies. We confirmed the possibility of using a machine learning model to predict the chemoradiotherapy response of rectal cancer before treatment using radiomics, which can be obtained noninvasively.

Author Contributions: Conceptualization, H.-M.J., J.Y., U.-S.S., S.C. and S.-K.W.; Methodology, H.-M.J., J.Y. and S.-K.W.; Software, H.-M.J., J.Y. and S.-K.W.; validation, J.Y.; formal analysis, J.Y.; investigation, J.Y.; resources, S.-K.W.; data curation, J.-M.P., J.-H.C., H.S., B.-I.K., U.-S.S., S.M.M. and S.-K.W.; writing—original draft preparation, H.-M.J.; writing—review and editing, H.-M.J., U.-S.S. and S.-K.W.; visualization, J.Y.; supervision, S.-K.W.; project administration, S.-K.W. All authors have read and agreed to the published version of the manuscript.

Funding: This research was funded by the National Research Foundation of Korea (NRF) grant funded by the Ministry of Science and ICT (No. 2020M2D9A1094070).

Institutional Review Board Statement: This study was approved by the Institutional Review Board of KIRAMS (IRB No.: KIRAMS 2021-03-009).

Informed Consent Statement: Informed consent was obtained from all participants involved i
the study.

Data Availability Statement: The data presented in this study are available on request from th
corresponding author. The data are not publicly available due to privacy considerations.

Conflicts of Interest: The authors declare no conflict of interest.

References

1. Liu, Z.; Zhang, X.-Y.; Shi, Y.-J.; Wang, L.; Zhu, H.-T.; Tang, Z.; Wang, S.; Li, X.-T.; Tian, J.; Sun, Y.-S.; et al. Radiomics analysis fc
   evaluation of pathological complete response to neoadjuvant chemoradiotherapy in locally advanced rectal cancer. *Clin. Cance*
   Res. **2017**, *23*, 7253–7262. [CrossRef] [PubMed]
2. Valentini, V.; Coco, C.; Cellini, N.; Picciocchi, A.; Genovesi, D.; Mantini, G.; Barbaro, B.; Cogliandolo, S.; Mattana, C.; Ambesiimp
   ombato, F.; et al. Preoperative chemoradiation for extraperitoneal T3 rectal cancer: Acute toxicity, tumor response, and sphincte
   preservation. *Int. J. Radiat. Oncol. Biol. Phys.* **1998**, *40*, 1067–1075. [CrossRef] [PubMed]
3. Pham, T.T.; Liney, G.P.; Wong, K.; Barton, M.B. Functional MRI for quantitative treatment response prediction in locally advance
   rectal cancer. *Br. J. Radiol.* **2017**, *90*, 20151078. [CrossRef] [PubMed]
4. Maas, M.; Nelemans, P.J.; Valentini, V.; Das, P.; Rödel, C.; Kuo, L.J.; Beets, G.L. Long-term outcome in patients with a pathologica
   complete response after chemoradiation for rectal cancer: A pooled analysis of individual patient data. *Lancet Oncol.* **2010**, *1:*
   835–844. [CrossRef]
5. Maffione, A.; Chondrogiannis, S.; Capirci, C.; Galeotti, F.; Fornasiero, A.; Crepaldi, G.; Grassetto, G.; Rampin, L.; Marzola, M
   Rubello, D. Early prediction of response by 18F-FDG PET/CT during preoperative therapy in locally advanced rectal cancer: *
   systematic review. *Eur. J. Surg. Oncol. (EJSO)* **2014**, *40*, 1186–1194. [CrossRef] [PubMed]
6. Liu, X.; Li, K.W.; Yang, R.; Geng, L.S. Review of deep learning based automatic segmentation for lung cancer radiotherapy. *Fron*
   Oncol. **2021**, *11*, 717039. [CrossRef]
7. Zhong, Y.; She, Y.; Deng, J.; Chen, S.; Wang, T.; Yang, M.; Ma, M.; Song, Y.; Qi, H.; Wang, Y.; et al. Multi-omics Classifier fo
   Pulmonary Nodules (MISSION) Collaborative Group. Deep learning for prediction of N_2 metastasis and survival for clinica
   stage I non–small cell lung cancer. *Radiology* **2022**, *302*, 200–211. [CrossRef]
8. Kim, J.; Oh, J.E.; Lee, J.; Kim, M.J.; Hur, B.Y.; Sohn, D.K.; Lee, B. Rectal cancer: Toward fully automatic discrimination of T2 and T
   rectal cancers using deep convolutional neural network. *Int. J. Imaging Syst. Technol.* **2019**, *29*, 247–259. [CrossRef]
9. Palma, P.; Conde-Muíño, R.; Rodríguez-Fernández, A.; Segura-Jiménez, I.; Sánchez-Sánchez, R.; Martín-Cano, J.; Gómez-Ríc
   M.; Ferrón, J.A.; Llamas-Elvira, J.M. The value of metabolic imaging to predict tumour response after chemoradiation in locall
   advanced rectal cancer. *Radiat. Oncol.* **2010**, *5*, 1–8. [CrossRef]
10. Martoni, A.A.; Di Fabio, F.; Pinto, C.; Castellucci, P.; Pini, S.; Ceccarelli, C.; Cuicchi, D.; Iacopino, B.; Di Tullio, P.; Giaquinta, S.; et a
    Prospective study on the FDG–PET/CT predictive and prognostic values in patients treated with neoadjuvant chemoradiatio
    therapy and radical surgery for locally advanced rectal cancer. *Ann. Oncol.* **2011**, *22*, 650–656. [CrossRef]
11. Lovinfosse, P.; Polus, M.; Van Daele, D.; Martinive, P.; Daenen, F.; Hatt, M.; Hustinx, R. FDG PET/CT radiomics for predicting th
    outcome of locally advanced rectal cancer. *Eur. J. Nucl. Med. Mol. Imaging* **2018**, *45*, 365–375. [CrossRef]
12. Sun, W.; Xu, J.; Hu, W.; Zhang, Z.; Shen, W. The role of sequential 18F-FDG PET/CT in predicting tumour response afte
    preoperative chemoradiation for rectal cancer. *Color. Dis.* **2013**, *15*, e231–e238. [CrossRef] [PubMed]
13. Joye, I.; Deroose, C.M.; Vandecaveye, V.; Haustermans, K. The role of diffusion-weighted MRI and 18F-FDG PET/CT in th
    prediction of pathologic complete response after radiochemotherapy for rectal cancer: A systematic review. *Radiother. Oncol.* **2014**
    113, 158–165. [CrossRef] [PubMed]
14. Orlhac, F.; Eertink, J.J.; Cottereau, A.S.; Zijlstra, J.M.; Thieblemont, C.; Meignan, M.; Boellaard, R.; Buvat, I. A guide to ComBa
    harmonization of imaging biomarkers in multicenter studies. *J. Nucl. Med.* **2022**, *63*, 172–179. [CrossRef] [PubMed]
15. Sampaio, I.W.; Tassi, E.; Bellani, M.; Benedetti, F.; Poletti, S.; Spalletta, G.; Piras, F.; Bianchi, A.M.; Brambilla, P.; Maggioni, I
    Comparison of Multi-Site Neuroimaging Data Harmonization Techniques for Machine Learning Applications. In Proceedings o
    the IEEE EUROCON 2023-20th International Conference on Smart Technologies, Torino, Italy, 6–8 July 2023.
16. Johnson, W.E.; Li, C.; Rabinovic, A. Adjusting batch effects in microarray expression data using empirical Bayes method
    Biostatistics **2017**, *8*, 118–127. [CrossRef] [PubMed]
17. Jovicich, J.; Czanner, S.; Greve, D.; Haley, E.; van Der Kouwe, A.; Gollub, R.; Kennedy, D.; Schmitt, F.; Brown, G.; MacFall, J
    et al. Reliability in multi-site structural MRI studies: Effects of gradient non-linearity correction on phantom and human dat
    Neuroimage **2006**, *30*, 436–443. [CrossRef] [PubMed]
18. Shinohara, R.T.; Oh, J.; Nair, G.; Calabresi, P.A.; Davatzikos, C.; Doshi, J.; Henry, R.G.; Kim, G.; Linn, K.A.; Papinutto, N.; et a
    Volumetric analysis from a harmonized multisite brain MRI study of a single subject with multiple sclerosis. *Am. J. Neuroradio*
    2017, *38*, 1501–1509. [CrossRef]
19. Macaulay, B.O.; Aribisala, B.S.; Akande, S.A.; Akinnuwesi, B.A.; Olabanjo, O.A. Breast cancer risk prediction in African wome
    using random forest classifier. *Cancer Treat. Res. Commun.* **2021**, *28*, 100396. [CrossRef]

20. Kesler, S.R.; Rao, A.; Blayney, D.W.; Oakley-Girvan, I.A.; Karuturi, M.; Palesh, O. Predicting long-term cognitive outcome following breast cancer with pre-treatment resting state fMRI and random forest machine learning. *Front. Hum. Neurosci.* **2017**, *11*, 555. [CrossRef]

21. Li, N.; Luo, P.; Li, C.; Hong, Y.; Zhang, M.; Chen, Z. Analysis of related factors of radiation pneumonia caused by precise radiotherapy of esophageal cancer based on random forest algorithm. *Math. Biosci. Eng.* **2021**, *18*, 4477–4490. [CrossRef]

22. Bi, L.; Guo, Y. Development and Validation of the Random Forest Model via Combining CT-PET Image Features and Demographic Data for Distant Metastases among Lung Cancer Patients. *J. Healthc. Eng.* **2022**, *2022*, 7793533. [CrossRef]

23. Zhang, J.; Zhao, X.; Zhao, Y.; Zhang, J.; Zhang, Z.; Wang, J.; Wang, Y.; Dai, M.; Han, J. Value of pre-therapy 18F-FDG PET/CT radiomics in predicting EGFR mutation status in patients with non-small cell lung cancer. *Eur. J. Nucl. Med. Mol. Imaging* **2020**, *47*, 1137–1146. [CrossRef] [PubMed]

24. Kaira, K.; Higuchi, T.; Naruse, I.; Arisaka, Y.; Tokue, A.; Altan, B.; Suda, S.; Mogi, A.; Shimizu, K.; Sunaga, N.; et al. Metabolic activity by 18F-FDG-PET/CT is predictive of early response after nivolumab in previously treated NSCLC. *Eur. J. Nucl. Med. Mol. Imaging* **2018**, *45*, 56–66. [CrossRef]

25. Mu, W.; Tunali, I.; Gray, J.E.; Qi, J.; Schabath, M.B.; Gillies, R.J. Radiomics of 18F-FDG PET/CT images predicts clinical benefit of advanced NSCLC patients to checkpoint blockade immunotherapy. *Eur. J. Nucl. Med. Mol. Imaging* **2020**, *47*, 1168–1182. [CrossRef]

26. Huh, J.W.; Min, J.J.; Lee, J.H.; Kim, H.R.; Kim, Y.J. The predictive role of sequential FDG-PET/CT in response of locally advanced rectal cancer to neoadjuvant chemoradiation. *Am. J. Clin. Oncol.* **2012**, *35*, 340–344. [CrossRef]

27. Capirci, C.; Rubello, D.; Pasini, F.; Galeotti, F.; Bianchini, E.; Del Favero, G.; Panzavolta, R.; Crepaldi, G.; Rampin, L.; Facci, E.; et al. The role of dual-time combined 18-fluorodeoxyglucose positron emission tomography and computed tomography in the staging and restaging workup of locally advanced rectal cancer, treated with preoperative chemoradiation therapy and radical surgery. *Int. J. Radiat. Oncol. Biol. Phys.* **2009**, *74*, 1461–1469. [CrossRef]

28. Melton, G.B.; Lavely, W.C.; Jacene, H.A.; Schulick, R.D.; Choti, M.A.; Wahl, R.L.; Gearhart, S.L. Efficacy of preoperative combined 18-fluorodeoxyglucose positron emission tomography and computed tomography for assessing primary rectal cancer response to neoadjuvant therapy. *J. Gastrointest. Surg.* **2007**, *11*, 961–969. [CrossRef]

29. Guillem, J.G.; Puig-La Calle, J.; Akhurst, T.; Tickoo, S.; Ruo, L.; Minsky, B.D.; Gollub, M.J.; Klimstra, D.S.; Mazumdar, M.; Paty, P.B.; et al. Prospective assessment of primary rectal cancer response to preoperative radiation and chemotherapy using 18-fluorodeoxyglucose positron emission tomography. *Dis. Colon Rectum* **2000**, *43*, 18–24. [CrossRef]

30. Nyúl, L.G.; Udupa, J.K. On standardizing the MR image intensity scale. *Magn. Reson. Med. Off. J. Int. Soc. Magn. Reson. Med.* **1999**, *42*, 1072–1081. [CrossRef]

31. Shinohara, R.T.; Sweeney, E.M.; Goldsmith, J.; Shiee, N.; Mateen, F.J.; Calabresi, P.A.; Jarso, S.; Pham, D.L.; Reich, D.S.; Crainiceanu, C.M. Statistical normalization techniques for magnetic resonance imaging. *NeuroImage Clin.* **2014**, *6*, 9–19. [CrossRef]

32. Gullo, G.; Petousis, S.; Papatheodorou, A.; Panagiotidis, Y.; Margioula-Siarkou, C.; Prapas, N.; D'Anna, R.; Perino, A.; Cucinella, G.; Prapas, Y. Closed vs. Open oocyte vitrification methods are equally effective for blastocyst embryo transfers: Prospective study from a sibling oocyte donation program. *Gynecol. Obstet. Investig.* **2020**, *85*, 206–212. [CrossRef]

 *Current Oncology*

Article

Real-World Outcomes of Patients with Advanced Epidermal Growth Factor Receptor-Mutated Non-Small Cell Lung Cancer in Canada Using Data Extracted by Large Language Model-Based Artificial Intelligence

Ruth Moulson [1], Jennifer Law [2], Adrian Sacher [2], Geoffrey Liu [2], Frances A. Shepherd [2], Penelope Bradbury [2], Lawson Eng [2], Sandra Iczkovitz [3], Erica Abbie [3], Julia Elia-Pacitti [3], Emmanuel M. Ewara [3], Viktoriia Mokriak [1], Jessica Weiss [1], Christopher Pettengell [1] and Natasha B. Leighl [2,*]

[1] Pentavere, 460 College Street, Toronto, ON M6G 1A1, Canada; rmoulson@pentavere.com (R.M.)
[2] Department of Medical Oncology, Princess Margaret Cancer Centre, University Health Network, Toronto, ON M5G 2C1, Canada
[3] Janssen Inc., Toronto, ON M3C 1L9, Canada
* Correspondence: natasha.leighl@uhn.ca

Abstract: Real-world evidence for patients with advanced *EGFR*-mutated non-small cell lung cancer (NSCLC) in Canada is limited. This study's objective was to use previously validated DARWEN™ artificial intelligence (AI) to extract data from electronic heath records of patients with non-squamous NSCLC at University Health Network (UHN) to describe *EGFR* mutation prevalence, treatment patterns, and outcomes. Of 2154 patients with NSCLC, 613 had advanced disease. Of these, 136 (22%) had common sensitizing *EGFR* mutations (c*EGFR*m; ex19del, L858R), 8 (1%) had exon 20 insertions (ex20ins), and 338 (55%) had *EGFR* wild type. One-year overall survival (OS) (95% CI) for patients with c*EGFR*m, ex20ins, and *EGFR* wild type tumours was 88% (83, 94), 100% (100, 100), and 59% (53, 65), respectively. In total, 38% patients with ex20ins received experimental ex20ins targeting treatment as their first-line therapy. A total of 57 patients (36%) with c*EGFR*m received osimertinib as their first-line treatment, and 61 (39%) received it as their second-line treatment. One-year OS (95% CI) following the discontinuation of osimertinib was 35% (17, 75) post-first-line and 20% (9, 44) post-second-line. In this real-world AI-generated dataset, survival post-osimertinib was poor in patients with c*EGFR* mutations. Patients with ex20ins in this cohort had improved outcomes, possibly due to ex20ins targeting treatment, highlighting the need for more effective treatments for patients with advanced *EGFR*m NSCLC.

Keywords: real-world evidence; artificial intelligence; non-small cell lung cancer

Citation: Moulson, R.; Law, J.; Sacher, A.; Liu, G.; Shepherd, F.A.; Bradbury, P.; Eng, L.; Iczkovitz, S.; Abbie, E.; Elia-Pacitti, J.; et al. Real-World Outcomes of Patients with Advanced Epidermal Growth Factor Receptor-Mutated Non-Small Cell Lung Cancer in Canada Using Data Extracted by Large Language Model-Based Artificial Intelligence. *Curr. Oncol.* **2024**, *31*, 1947–1960. https://doi.org/10.3390/curroncol31040146

Received: 15 January 2024
Revised: 9 February 2024
Accepted: 25 March 2024
Published: 2 April 2024

1. Introduction

Lung cancer is the most common cancer diagnosis in Canada, with an estimated 1 in 15 Canadians receiving a diagnosis in their lifetime [1]. While the prognosis and outcomes of lung cancer have improved in recent decades, largely as a result of novel, innovative therapies and increased awareness of the risk factors, this disease remains the deadliest cancer in Canada [1,2]. Approximately 85% of patients with lung cancer present with NSCLC, with up to two-thirds harbouring actionable driver mutations, most commonly occurring in the *epidermal growth factor receptor* (*EGFR*) [3–5]. *EGFR* mutations can be categorized based on the type of mutation and the exon in which they occur. Exon 19 deletions (ex19del) and exon 21 L858R point mutations account for up to 90% of all *EGFR* mutations and are often referred to as common sensitizing *EGFR* mutations (c*EGFR*m) [6]. The third most frequently occurring mutations are exon 20 insertion mutations (ex20ins) and represent approximately 1–12% of all *EGFR* mutations, and 0.1–4% of all NSCLC mutations [7]. However, uncertainty in the real-world estimates of these mutations exist, partly due to the evolution of testing methods, with recent guidelines recommending next-generation sequencing (NGS) for identifying actionable driver alterations, such as *EGFR* [8,9]. This technique has improved sensitivity, can detect mutations using a smaller amount of DNA, and sequences a greater part of the gene compared with the historical standard, polymerase chain reaction (PCR), which is limited to specific loci and can miss up to 50% of ex20ins mutations, but it requires a smaller tissue sample than NGS [10–12].

The treatment of patients with *EGFR* mutations has been revolutionized by tyrosine kinase inhibitor (TKI) targeted therapy. The recommended first-line therapy for advanced-stage patients with cEGFRm in Canada is the third-generation kinase inhibitor, osimertinib [13,14]. However, the long-term benefit of this therapy is limited by the development of acquired resistance via multiple mechanisms [15]. Recently, multiple new options for overcoming osimertinib resistance have emerged, including amivantamab + lazertinib, chemotherapy, local therapy (surgery or radiation), chemotherapy + amivantamab/lazertinib, antibody-drug conjugates (ADCs), including patritumab deruxtecan and datopotamab deruxtecan, and combined targeted therapies against emergent targetable alterations (e.g., for MET amplification: osimertinib + savolitinib and tepotinib + osimertinib) [16]. These emerging treatment options are particularly important as many patients with cEGFRm who are treated with a first-line TKI die before receiving a second-line one [17]; thus, there remains a high unmet need for effective and safe therapies early in patients' treatment journeys, and there is currently a lack of real-world evidence (RWE), specifically in the Canadian setting, on patients with cEGFRm who may benefit from these therapies.

Independent of acquired resistance, ex20ins are associated with limited response to TKIs [18]. Compared with other *EGFR* mutations, patients with ex20ins have especially poor prognosis, with markedly reduced sensitivity to approved *EGFR* kinase inhibitors [18–20]. Until recently, there have been limited treatment options for patients with ex20ins, with the recommended first-line treatment being either platinum-based chemotherapy or clinical trial [13]. However, the Canadian treatment landscape is evolving, as the results from the phase III PAPILLON study have established amivantamab + chemotherapy as a new first-line standard for this patient population [21]. As the treatment landscape changes, there is a need to gain a better understanding of the patients who may benefit from these newer therapies.

Over the past two decades, the generation of RWE from electronic health record (EHR) systems has contributed new insights into the prevalence of lung cancer subtypes and the disease characteristics and clinical outcomes for these patients. Through the routine collection of clinical evidence, real-world data (RWD) from EHRs can be harnessed to study disease progression, treatment patterns, and measure survival outcomes over time. Recent advances in artificial intelligence (AI) and Natural Language Processing (NLP) have enabled the extraction and analysis of RWD from clinical documentation and unstructured text (such as clinical notes and lab results) housed within EHR systems, with higher accuracy and at a significantly greater scale than manual abstraction, the current standard practice for extracting RWD from EHRs [22,23]. It is increasingly being recognized that these technologies play an important role in clinical medicine by allowing clinician's and researchers access to previously inaccessible data, which can be used to inform clinical decision making and enhance clinical care [24].

The aim of this study was to leverage the previously validated, commercially available AI technology, Pentavere's DARWEN^{TM}, to identify patients and extract RWD from EHRs at the University Health Network Princess Margaret Cancer Centre (UHN-PMCC), the largest cancer-treating centre in Canada, to understand the prevalence, treatment patterns, and clinical outcomes of patients diagnosed with advanced cEGFRm (ex19del and exon 21 L858R) and ex20ins mutations.

2. Materials and Methods

2.1. Study Design

This was a retrospective cohort study of data elements from EHRs stored at the UHN-PMCC using AI technology. The AI engine combines large language models and an ensemble of other techniques that have previously been evaluated and validated against manual abstraction across multiple disease domains, including lung cancer [22,25], breast cancer [26], dermatology [27], and infectious diseases [28] at multiple Canadian institutions including the UHN-PMCC.

The study period extended from 1 January 2017 to 1 March 2022 and used the institutional Cancer Registry. All adult patients who were $\geq$18 years of age with non-squamous NSCLC and seen at the UHN-PMCC during the study period were included in the study. Follow-up data from EHRs were included up to the extent that they were available within the study period. The initial list of patients was provided from the UHN-PMCC's Molecular Testing Database.

2.2. Data Extraction

Clinical features extracted included mutation status, clinical and demographic characteristics, treatment information, and clinical outcomes. Data were extracted directly from the EHRs of all patients with non-squamous NSCLC seen at the UHN-PMCC between 1 January 2017 and 1 March 2022. The AI engine was installed on the UHN-PMCC's infrastructure and used to extract relevant data variables directly from the source systems where available. Clinical outcomes were derived using the extracted data, including time to treatment discontinuation (TTD) and overall survival (OS). All features were extracted following a set of pre-defined rules and definitions developed by the UHN-PMCC Principal Investigator. DARWEN™ AI has previously been validated against the manual chart review for the same clinical features at the UHN-PMCC, the process for which has previously been described [22].

2.3. Outcomes

The primary outcome of interest was mutation prevalence. Other outcomes of interest included the frequency of patients receiving each type of therapy by line of therapy (LoT), time from diagnosis to treatment initiation per LoT, and clinical outcomes, including TTD, OS, and OS post-osimertinib. TTD was measured from the date of the treatment initiation of one line of therapy to the last known date of the treatment of the same line of therapy. TTD was derived for first-line, second-line, and third-line therapies. OS was measured from date of diagnosis to date of death, and from date of treatment initiation to date of death for first-line and second-line therapies. Patients who did not experience the event before the study's end period were censored at their date of last follow-up or the study's end date, whichever came first. Overall survival, specifically for patients who had discontinued osimertinib, was explored and measured from the stop date of osimertinib to date of death. Patients who did not experience the event before the study's end period were censored at their date of last follow-up or the study's end date, whichever came first. OS was derived from the end of first-line osimertinib and the end of second-line osimertinib.

2.4. Statistical Analyses

Descriptive analyses were performed to summarize the patients' demographics, disease characteristics, treatment patterns, and outcomes of interest across the study cohort. Continuous variables were described using mean and standard deviation (SD) and the median and range. Categorical variables were described by frequencies and related percentages. The number of missing observations was reported for all variables. Time to event(s) was described using Kaplan–Meier curves that visually estimated the distribution of times to some events (e.g., OS) and accounted for patients for whom the event had not yet occurred, i.e., following standard censoring rules. Numbers at risk and the cumulative number of events were reported for each curve.

3. Results

3.1. Patients

Between 1 January 2017 and 1 March 2022, 2154 patients were identified with non-squamous NSCLC and were seen at the UHN-PMCC. Of these patients, 613 patients had advanced-stage disease, of which 136 (22%) patients had cEGFRm at diagnosis, 8 (1%) had ex20ins at diagnosis, 338 (55%) had *EGFR* wild type tumours at diagnosis, and 131 (21%) did not have mutation testing at diagnosis conducted at the UHN-PMCC. A flow diagram of the included patients is presented in Supplementary Figure S1.

Across all 613 patients with advanced-stage disease, median (range) age at advanced diagnosis was 67 years (27–96); 51% of patients were male, 84% had adenocarcinoma, and 38% had never been smokers. At advanced diagnosis, 30% of patients presented with bone metastases, and 14% had brain metastases (Table 1). The majority of patients (81%) were diagnosed at the UHN-PMCC. Of the 131 patients who did not have mutation testing at the UHN-PMCC, 56% were also not diagnosed at the UHN-PMCC, and all 131 were not included in the clinical outcome analyses. The median (range) duration of the follow-up from diagnosis for all patients was 12.3 months (0.0–61.8) (Table 1). AI validation metrics for the AI-extracted clinical features are presented in Supplementary Table S1.

Table 1. Clinical, demographic, and disease characteristics of advanced-stage NSCLC patients stratified by EGFR mutation status at diagnosis.

	Common Sensitizing *EGFR* (N = 136)	*EGFR* Wild Type [b] (N = 338)	Exon 20 Insertion (N = 8)	*EGFR* Test Not Conducted at UHN (N = 131)	Total (N = 613)
Age at diagnosis					
Mean (SD)	65.1 (11.6)	67.6 (11.6)	59.9 (19.3)	65.3 (10.6)	66.5 (11.6)
Median (range)	65.0 (34.0, 91.0)	68.0 (27.0, 96.0)	59.0 (38.0, 88.0)	66.0 (32.0, 88.0)	67.0 (27.0, 96.0)
Sex					
Female	89 (65.4%)	146 (43.2%)	4 (50.0%)	64 (48.9%)	303 (49.4%)
Male	47 (34.6%)	192 (56.8%)	4 (50.0%)	67 (51.1%)	310 (50.6%)
Histology					
Adenocarcinoma	129 (94.9%)	276 (81.7%)	7 (87.5%)	102 (77.9%)	514 (83.8%)
Adenosquamous	0 (0.0%)	2 (0.6%)	1 (12.5%)	0 (0.0%)	3 (0.5%)
Large cell	2 (1.5%)	21 (6.2%)	0 (0.0%)	16 (12.2%)	39 (6.4%)
Sarcomatoid	0 (0.0%)	5 (1.5%)	0 (0.0%)	6 (4.6%)	11 (1.8%)
Non-small cell (unspecified)	5 (3.7%)	34 (10.1%)	0 (0.0%)	7 (5.3%)	46 (7.5%)
Smoking status					
Smoker	8 (5.9%)	101 (29.9%)	0 (0.0%)	30 (22.9%)	139 (22.7%)
Former smoker	29 (21.3%)	152 (45.0%)	2 (25.0%)	50 (38.2%)	233 (38.0%)
Never smoked	98 (72.1%)	83 (24.6%)	6 (75.0%)	48 (36.6%)	235 (38.3%)
Missing	1 (0.7%)	2 (0.6%)	0 (0.0%)	3 (2.3%)	6 (1.0%)
Weight Category					
<80 kg	105 (77.2%)	241 (71.3%)	6 (75.0%)	98 (74.8%)	450 (73.4%)
≥80 kg	17 (12.5%)	59 (17.5%)	0 (0.0%)	21 (16.0%)	97 (15.8%)
Missing	14 (10.3%)	38 (11.2%)	2 (25.0%)	12 (9.2%)	66 (10.8%)

Table 1. *Cont.*

	Common Sensitizing *EGFR* (N = 136)	*EGFR* Wild Type [b] (N = 338)	Exon 20 Insertion (N = 8)	*EGFR* Test Not Conducted at UHN (N = 131)	Total (N = 613)
ECOG at diagnosis					
0	23 (16.9%)	34 (10.1%)	1 (12.5%)	17 (13.0%)	75 (12.2%)
1	84 (61.8%)	180 (53.3%)	5 (62.5%)	64 (48.9%)	333 (54.3%)
2	15 (11.0%)	59 (17.5%)	1 (12.5%)	20 (15.3%)	95 (15.5%)
3	9 (6.6%)	28 (8.3%)	1 (12.5%)	13 (9.9%)	51 (8.3%)
4	1 (0.7%)	5 (1.5%)	0 (0.0%)	2 (1.5%)	8 (1.3%)
Missing	4 (2.9%)	32 (9.5%)	0 (0.0%)	15 (11.5%)	51 (8.3%)
Organ level metastatic sites at diagnosis [a]					
Bone	46 (33.8%)	119 (35.2%)	3 (37.5%)	18 (13.7%)	186 (30.3%)
Brain	21 (15.4%)	46 (13.6%)	1 (12.5%)	16 (12.2%)	84 (13.7%)
Lung	22 (16.2%)	55 (16.3%)	0 (0.0%)	14 (10.7%)	91 (14.8%)
Liver	19 (14.0%)	42 (12.4%)	2 (25.0%)	17 (13.0%)	80 (13.1%)
Diagnosed at UHN					
True	120 (88.2%)	314 (92.9%)	5 (62.5%)	58 (44.3%)	497 (81.1%)
False	16 (11.8%)	24 (7.1%)	3 (37.5%)	73 (55.7%)	116 (18.9%)
Follow-up time since diagnosis (months)					
Mean (SD)	21.4 (14.1)	13.6 (14.0)	24.0 (19.8)	19.0 (16.0)	16.6 (14.9)
Median (range)	19.2 (0.4, 58.9)	8.0 (0.3, 59.4)	20.2 (0.4, 61.0)	14.7 (0.0, 61.8)	12.3 (0.0, 61.8)

[a] Patients could have had multiple metastatic sites at diagnosis, and therefore percentages may not add up to 100%. Further, patients may have had metastases to body parts other than the bone, brain, lung, and liver, which also explains why percentages may not add up to 100%. [b] Includes patients with a negative *EGFR* test within 3 months of NSCLC diagnosis but does not exclude the possibility of other mutations. ECOG: Eastern Cooperative Oncology Group; NSCLC: non-small-cell lung cancer; SD: standard deviation; UHN: University Health Network.

3.2. Treatment Patterns

Treatment patterns were assessed from the date of diagnosis until date of death, date of last follow-up, or the end of the study period, whichever came first. For advanced-stage patients with cEGFRm at diagnosis, 129/136 (95%) received first-line therapy, of which 124/129 (96%) received an *EGFR* TKI in their first-line treatment (Figure 1A; Supplementary Table S2). Of patients with cEGFRm, 62/136 (46%) did not go on to receive second-line treatment during the study period (Figure 1A) (34 of which received osimertinib in their first-line therapy and 19 of which received gefitinib in their first-line therapy), and 21/62 (34%) of these patients died. Of patients who did go on to receive second-line (74/136 [54%]) and third-line therapies (27/136 [20%]), the most common treatment type was also *EGFR* TKIs in those lines (Figure 1A). Between 2017 and 2019, gefitinib was the most common first-line *EGFR* TKI administered for patients with cEGFRm, with 81% of patients who initiated an *EGFR* TKI in 2017–2019 receiving gefitinib (Table 2). Coincident with provincial funding as of January 2020, osimertinib was the most frequently used first-line *EGFR* TKI from 2020 to 2022, with 93% of patients who initiated an *EGFR* TKI in this period receiving osimertinib (Table 2).

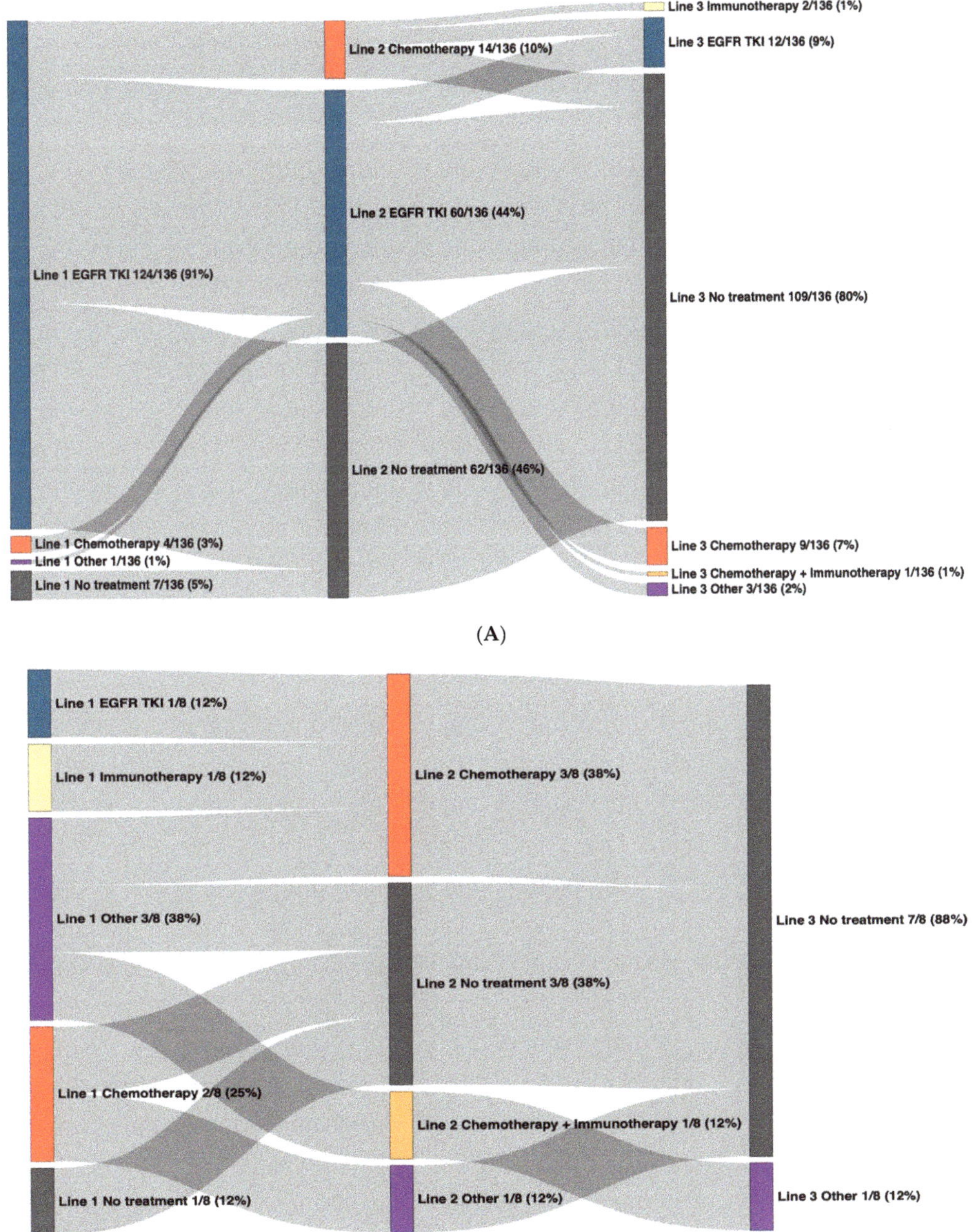

Figure 1. Overall treatment patterns in advanced-stage NSCLC patients by mutation status at diagnosis. Line of therapy is denoted by the number followed by the treatment regimen, with first-line on the left and subsequent lines to the right. "Other" includes capmatinib, savolitinib, poziotinib, mobocertinib, lazertinib, and telisotuzumab. *EGFR: epidermal growth factor receptor*; NSCLC: non-small-cell lung cancer; TKI: tyrosine kinase inhibitor. (**A**) Common sensitizing *EGFR* mutations. (**B**) Exon 20 insertion mutations.

For advanced-stage patients with ex20ins at diagnosis, 7/8 (88%) received first-line therapy (Figure 1B), and 3/8 (38%) received the experimental ex20ins targeting TKI, pozotinib (Supplementary Table S1). Second-line therapy was received by 5/8 (63%) patients (4/8 received chemotherapy), and 1/8 (13%) went on to receive third-line therapy (Figure 1B). For advanced-stage patients with *EGFR* wild type tumours at diagnosis, treatment patterns were generally heterogeneous across all lines of therapy (Supplementary Table S1).

Table 2. First-line EGFR TKI treatment patterns in advanced-stage NSCLC patients stratified by year of initiating treatment and mutation status at diagnosis.

	Common Sensitizing *EGFR*						***EGFR* Wild Type**			**Exon 20 Insertion**
	2017 (N = 17)	**2018** (N = 36)	**2019** (N = 28)	**2020** (N = 17)	**2021** (N = 24)	**2022** (N = 2)	**2017** (N = 1)	**2019** (N = 3)	**2021** (N = 5)	**2018** (N = 1)
Afatinib	1 (5.9%)	6 (16.7%)	2 (7.1%)	1 (5.9%)	0 (0.0%)	0 (0.0%)	0 (0.0%)	1 (33.3%)	4 (80.0%)	1 (100.0%)
Erlotinib	1 (5.9%)	0 (0.0%)	0 (0.0%)	1 (5.9%)	0 (0.0%)	0 (0.0%)	0 (0.0%)	0 (0.0%)	0 (0.0%)	0 (0.0%)
Gefitinib	15 (88.2%)	30 (83.3%)	21 (75.0%)	1 (5.9%)	0 (0.0%)	0 (0.0%)	1 (100.0%)	2 (66.7%)	0 (0.0%)	0 (0.0%)
Osimertinib	0 (0.0%)	0 (0.0%)	5 (17.9%)	14 (82.4%)	24 (100.0%)	2 (100.0%)	0 (0.0%)	0 (0.0%)	1 (20.0%)	0 (0.0%)

Bolded N includes patients who initiated a first-line *EGFR* TKI in the specified year. *EGFR: epidermal growth factor receptor*; NSCLC: non-small-cell lung cancer; TKI: tyrosine kinase inhibitor.

The median time from advanced diagnosis to first-line treatment initiation for patients with c*EGFR*m, ex20ins, and *EGFR* wild type tumours was 0.8 months, 2.5 months, and 1.5 months, respectively (Supplementary Table S1). Longer time from advanced diagnosis to first-line treatment initiation was observed for patients with ex20ins, likely due to a lack of clear treatment options for these patients, and time required for clinical trial enrolment.

3.3. Clinical Outcomes

For patients with c*EGFR*m, median TTD1 (95% CI), TTD2, and TTD3 was 9.0 months (7.0, 10.3), 6.7 months (4.8, 10.7), and 2.9 months (1.6, 6.8), respectively (Table 3, Figure 2). For patients with ex20ins median, TTD1 (95% CI) and TTD2 were 5.0 months (3.5, NA) and 7.9 months (5.7, NA), respectively (Table 3, Figure 2). For patients with *EGFR* wild type tumours, treatment duration was generally shorter, with median TTD1 (95% CI), TTD2, and TTD3 of 4.0 months (3.3, 4.6), 2.8 months (1.9, 4.8), and 2.1 months (1.3, 5.8), respectively (Table 3, Figure 2).

Table 3. Time to event analyses for patients stratified by mutation status at diagnosis.

Clinical Outcome	12 Months (95% CI)	24 Months (95% CI)	Median (95% CI)
		TTD1	
Exon 20 insertion	14% (2, 88)	14% (2, 88)	5 months (3.5, NA)
Common sensitizing *EGFR*	34% (27, 43)	12% (7, 19)	9 months (7, 10.3)
EGFR wild type	20% (15, 26)	7% (4, 11)	4 months (3.3, 4.6)
		TTD2	
Exon 20 insertion	NA	NA	7.9 months (5.7, NA)
Common sensitizing *EGFR*	34% (25, 46)	8% (4, 17)	6.7 months (4.8, 10.7)
EGFR wild type	15% (10, 24)	4% (1, 10)	2.8 months (1.9, 4.8)

Table 3. *Cont.*

Clinical Outcome	12 Months (95% CI)	24 Months (95% CI)	Median (95% CI)
TTD3			
Exon 20 insertion	NA	NA	2.8 months (NA, NA)
Common sensitizing *EGFR*	11% (4, 32)	4% (1, 25)	2.9 months (1.6, 6.8)
EGFR wild type	11% (4, 27)	3% (0, 19)	2.1 months (1.3, 5.8)
OS from diagnosis			
Exon 20 insertion	100% (100, 100)	80% (52, 100)	NA months (32.1, NA)
Common sensitizing *EGFR*	88% (83, 94)	63% (54, 73)	30.1 months (25.2, 38.9)
EGFR wild type	59% (53, 65)	38% (32, 44)	16.2 months (13.2, 20.5)
OS from first-line			
Exon 20 insertion	100% (100, 100)	60% (29, 100)	NA months (18.4, NA)
Common sensitizing *EGFR*	85% (78, 92)	57% (48, 69)	26.4 months (23.2, 36.8)
EGFR wild type	62% (56, 69)	38% (32, 46)	19.3 months (14.2, 22.6)
OS from second-line			
Exon 20 insertion	75% (43, 100)	NA	13.1 months (11, NA)
Common sensitizing *EGFR*	56% (45, 69)	42% (31, 58)	20.3 months (11, 40.2)
EGFR wild type	48% (39, 59)	26% (17, 38)	10.6 months (7.6, 15.3)
OS from end of first-line osimertinib			
Common sensitizing *EGFR*	35% (17, 75)	-	5.6 months (3.2, NA)
OS from end of second-line osimertinib			
Common sensitizing *EGFR*	20% (9, 44)	-	3.3 months (2, 10.4)

CI: confidence interval; *EGFR: epidermal growth factor receptor*; NA: Not applicable either due to small sample size or confidence interval not reached; OS: overall survival; TTD: time to treatment discontinuation.

The 1-year OS from diagnosis for patients with cEGFRm, ex20ins, and *EGFR* wild type was 88% (83, 94), 100% (100, 100), and 59% (53, 65), respectively (Table 3). OS from first-line and second-line therapies can be found in Table 3.

Of advanced-stage patients with cEGFRm, 57 (36%) received first-line osimertinib, and 61 (39%) received second-line osimertinib. After discontinuing osimertinib treatment, OS was low: 1-year OS (95% CI) was 35% (17, 75) post-first-line osimertinib and 20% (9, 44) post-second-line. Median OS was 5.6 months (3.2, NA) post-first-line osimertinib and 3.3 months (2.0, 10.4) post-second-line (Table 3, Figure 3).

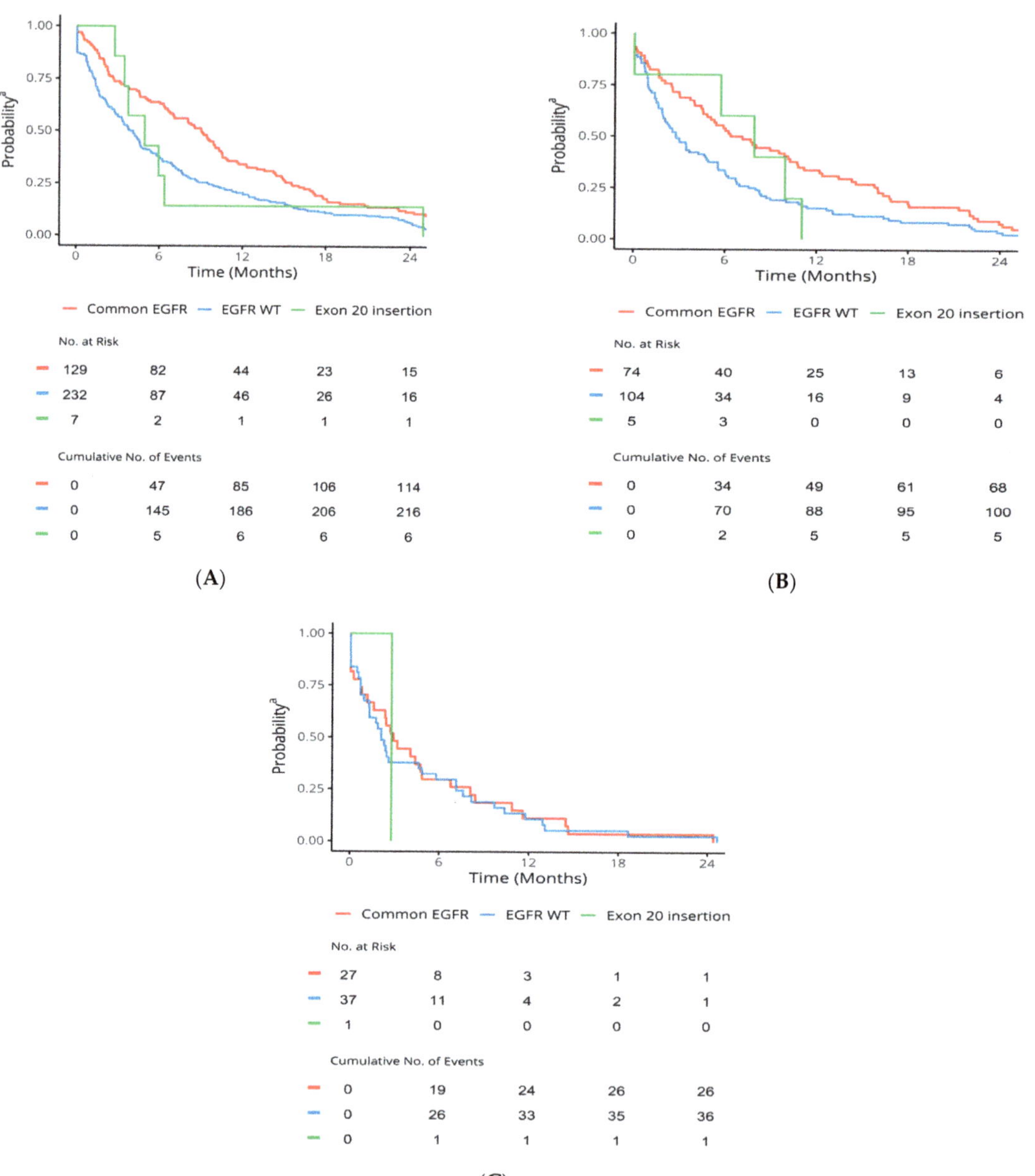

Figure 2. TTD in advanced-stage NSCLC patients stratified by mutation status. [a] Probability of staying on the line treatment. *EGFR: epidermal growth factor receptor*; NSCLC: non-small-cell lung cancer; TTD: time to treatment discontinuation. (**A**) TTD1. (**B**) TTD2. (**C**) TTD3.

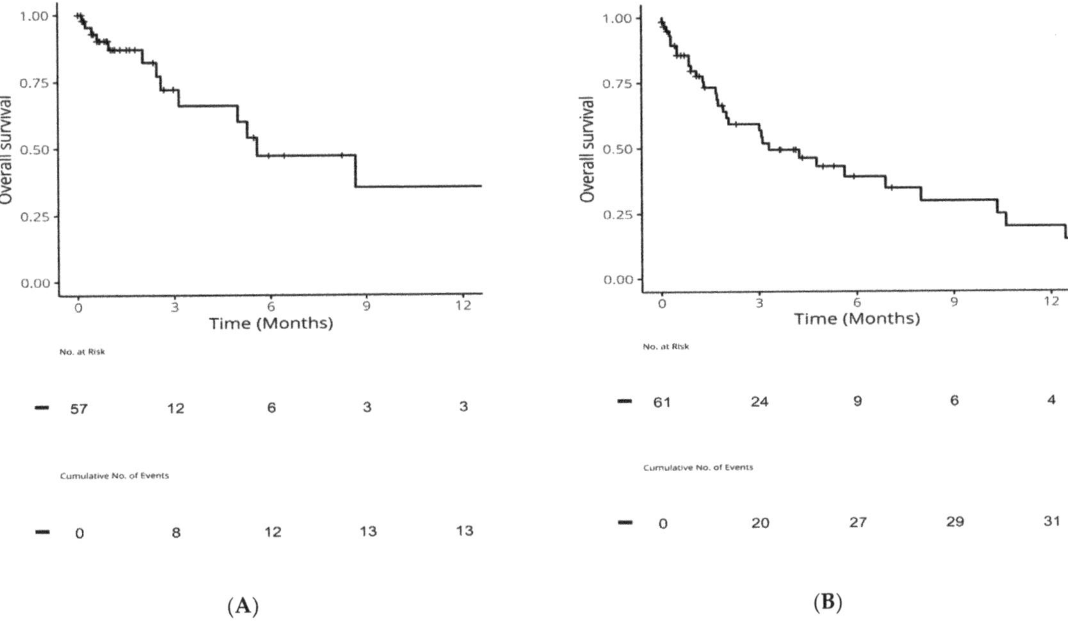

(A)

(B)

Figure 3. OS from end of first-line or second-line in patients with common sensitizing *EGFR* mutations who received osimertinib. (**A**): OS from end of first-line osimertinib; (**B**): OS from end of second-line osimertinib. *EGFR: epidermal growth factor receptor*; OS: overall survival.

4. Discussion

This study identified Canadian patients with non-squamous NSCLC at the largest cancer treatment centre in Canada and described the real-world characteristics, treatment patterns, and clinical outcomes for patients with advanced ex19del, exon 21 L858R, and ex20ins *EGFR* mutations using AI-extracted data. It was found that, as expected, patients with c*EGFR*m were primarily treated with *EGFR* TKIs. TKI treatment use changed over time with the approval of novel therapies. From 2020, osimertinib emerged as the most frequently administered *EGFR* TKI, in line with the treatment guidelines. Importantly, it was found that patients with c*EGFR*m treated with osimertinib progressed on therapy and exhibited poor survival rates after discontinuing treatment, emphasizing the need for more efficacious therapies earlier in patients' treatment journeys. It was also found that several patients with ex20ins were treated with the experimental ex20ins TKI, poziotinib, and may have had better survival as a result.

Among 2154 patients with non-squamous NSCLC and seen at the UHN-PMCC during the study period, 613 had advanced disease, of which 1% had ex20ins at diagnosis, consistent with other real-world estimates in Canada, and at the UHN-PMCC [29–31], median time from advanced diagnosis to initiating first-line therapy was longer for patients with ex20ins in comparison to patients with c*EGFR*m (2.5 months versus 0.8 months, respectively), likely due to the absence of a clear first-line targeted treatment option for these patients, coupled with the time required for clinical trial enrolment.

A recent European RWE registry study investigated the use of different treatment types and their impact on survival rates among patients with EGFR ex20ins mutations. Novel targeted agents, including amivantamab, mobocertinib, and poziotinib, were associated with improved survival rates in the first-line setting. As well, in the multivariate analysis, type of treatment (novel targeted therapy versus chemotherapy) had a significant effect on OS ($p = 0.03$) [32]. In this study, of patients with ex20ins, 38% received the experimental exon 20 targeting TKI, poziotinib, in their first-line therapy and achieved better survival than patients with c*EGFR*m or *EGFR* wild type, emphasizing the benefit of novel, targeted therapies; although, it is important to acknowledge the limitation of the survival analyses for the ex20ins patient group in this study due to the small sample size associated with this rare mutation. However, in the phase II trial of poziotinib, serious adverse events were observed, including grade ≥ 3 diarrhoea and rash, leading to treatment interruptions, which could explain the shorter TTD1 for patients with ex20ins in this study compared with c*EGFR*m. Further, the recent phase III trial of mobocertinib in first-line therapy for ex20ins patients was terminated early due to futility. These results highlight the need for efficacious and safe exon 20 targeting therapies to improve survival outcomes for these patients, in alignment with the evolving treatment landscape.

Over the study period, treatment patterns for patients with c*EGFR*m evolved with the introduction of novel third-generation *EGFR* TKIs. From 2017 to 2019, gefitinib was the predominant first-line *EGFR* TKI, followed by osimertinib in 2020–2022. However, it is noteworthy that 62/136 (46%) of patients with c*EGFR*m (34 of which received osimertinib in their first-line treatment) did not go on to receive second-line therapy during the study period, and of these patients, 21/62 (34%) died. For patients with c*EGFR*m who received osimertinib either in their first-line or second-line therapies, OS following the discontinuation of osimertinib was poor (1-year OS [95% CI] was 35% (17, 75) post-first-line osimertinib), aligning with findings observed in the RWE study of US databases conducted by Girard et al. (2023) [33]. These observations highlight the importance of effective novel treatment options early in patients' treatment journeys. Further studies may wish to investigate the specific risk factors associated with the mortality of patients prior to receiving second-line therapy.

As this study was a retrospective study of data extracted from EHRs, limitations due to the availability and accuracy of data captured in the EHR were observed. For example, many patient deaths occurred in the community setting rather than the hospital, and dates of death are only collected when hospitals are notified of a patient's death, which may have resulted in missing mortality data. This could have led to higher levels of data censoring in Kaplan–Meier curves and survival analyses. Additionally, at the UHN-PMCC, oral therapy prescription data are only dictated into the clinical notes and, therefore, these records are susceptible to incompleteness and human dictation error. Further, as this study was conducted at one urban treatment site in Toronto, Ontario, the cohort may not accurately represent the wider provincial or national population and may not be directly reproducible; however, the prevalence rates observed in this study do align with previous studies in Canada and at the UHN-PMCC [29–31].

5. Conclusions

This study identified patients with non-squamous NSCLC at one of Canada's largest cancer treatment centres using previously validated AI technology. Using these types of technologies allows for the extraction of previously unavailable data in a more consistent, efficient, and scalable way compared to manual chart review [22]. The results from this study highlight the importance of effective novel targeted therapies for improving survival outcomes in patients with ex20ins *EGFR* mutations, in alignment with the evolving treatment landscape for first-line therapy. The findings also emphasize the need for optimal therapies early in the treatment of patients with c*EGFR*m.

Supplementary Materials: The following supporting information can be downloaded at: https://www.mdpi.com/article/10.3390/curroncol31040146/s1, Table S1: Evaluation of DARWEN™ AI. Table S2: First-line treatments in advanced-stage NSCLC patients stratified by mutation status at diagnosis. Figure S1: Summary of included patients.

Author Contributions: R.M.: Conceptualization; Data Curation; Formal Analysis; Funding Acquisition; Investigation; Methodology; Project Administration; Resources; Software; Supervision; Validation; Visualization; Roles/Writing—Original Draft; Writing—Review and Editing. J.L.: Conceptualization; Data Curation; Formal Analysis; Funding Acquisition; Investigation; Methodology;

Project Administration; Resources; Supervision; Validation; Visualization; Roles/Writing—Original Draft; Writing—Review and Editing. A.S.: Conceptualization; Data Curation; Funding Acquisition; Investigation; Methodology; Resources; Validation; Roles/Writing—Original Draft; Writing—Review and Editing. G.L.: Conceptualization; Data Curation; Funding Acquisition; Investigation; Methodology; Resources; Validation; Roles/Writing—Original Draft; Writing—Review and Editing. F.A.S.: Conceptualization; Data Curation; Funding Acquisition; Investigation; Methodology; Resources; Validation; Roles/Writing—Original Draft; Writing—Review and Editing. P.B.: Conceptualization; Data Curation; Funding Acquisition; Investigation; Methodology; Resources; Validation; Roles/Writing—Original Draft; Writing—Review and Editing. L.E.: Conceptualization; Data Curation; Funding Acquisition; Investigation; Methodology; Resources; Validation; Roles/Writing—Original Draft; Writing—Review and Editing. S.I.: Conceptualization; Formal Analysis; Funding Acquisition; Investigation; Methodology; Project Administration; Resources; Supervision; Validation; Visualization; Roles/Writing—Original Draft; Writing—Review and Editing. E.A.: Conceptualization; Formal Analysis; Funding Acquisition; Investigation; Methodology; Project Administration; Resources; Supervision; Validation; Visualization; Roles/Writing—Original Draft; Writing—Review and Editing. J.E.-P.: Conceptualization; Formal Analysis; Funding Acquisition; Investigation; Methodology; Project Administration; Resources; Supervision; Validation; Visualization; Roles/Writing—Original Draft; Writing—Review and Editing. E.M.E.: Conceptualization; Formal Analysis; Funding Acquisition; Investigation; Methodology; Project Administration; Resources; Supervision; Validation; Visualization; Roles/Writing—Original Draft; Writing—Review and Editing. V.M.: Conceptualization; Data Curation; Formal Analysis; Methodology; Software; Validation; Visualization; Roles/Writing—Original Draft; Writing—Review and Editing. J.W.: Conceptualization; Data Curation; Formal Analysis; Methodology; Validation; Visualization; Roles/Writing—Original Draft; Writing—Review and Editing. C.P.: Conceptualization; Data Curation; Formal Analysis; Funding Acquisition; Investigation; Methodology; Project Administration; Resources; Software; Supervision; Validation; Visualization; Roles/Writing—Original Draft; Writing—Review and Editing. N.B.L.: Conceptualization; Data Curation; Formal Analysis; Funding Acquisition; Investigation; Methodology; Project Administration; Resources; Software; Supervision; Validation; Visualization; Roles/Writing—Original Draft; Writing—Review and Editing. All authors have read and agreed to the published version of the manuscript.

Funding: This work was funded by Janssen Canada and supported by the Princess Margaret Cancer Foundation.

Institutional Review Board Statement: The study was conducted according to the guidelines of the Declaration of Helsinki and approved by the Institutional Review Board of the UHN-PMCC (protocol code: 22-5330 and date of approval: 9 August 2022).

Informed Consent Statement: Patient consent was waived by the UHN-PMCC IRB as this study was retrospective; information abstracted for the study was already collected as part of routine clinical care and contained within the patient's health record. No additional contact with participants for information was needed. All information was kept strictly confidential.

Data Availability Statement: The data presented in this study are not publicly available due to the privacy of individuals. The data presented in this study may be available on reasonable request from the corresponding author.

Conflicts of Interest: R.M., J.W., V.M. and C.P. were employees of Pentavere Research Group Inc., Toronto, ON, Canada, at the time of this study. J.L., P.B. and L.E. have no conflicts of interest to declare. A.S. has received institutional funding from Amgen, AstraZeneca, Bristol-Myers Squibb, CRISPR Therapeutics AG, Eli Lilly, Genentech, GlaxoSmithKline, Lovance, Merck, Pfizer, and Spectrum Pharmaceuticals and has consulted on the advisory boards for AstraZeneca and Genentech. G.L. has received funding for advisory boards from Takeda, AstraZeneca, Pfizer, Amgen, Bayer, Eli Lily, Merck, Novartis, Jazz, Roche, AbbVie, J&J, and BMS; has received funding for grants from Takeda, AstraZeneca, Amgen, Bayer, and Boehringer Ingelheim; and has received funding for education sessions from Takeda, AstraZeneca, Pfizer, and Bayer. F.A.S. has received institutional funding from Lilly, Pfizer, Bristol-Myers Squibb, AstraZeneca/MedImmune, and Roche Canada; honoraria from AstraZeneca, Merck Serono, Takeda, and Daiichi Sankyo; consulted for AstraZeneca and Merck Sorono; and has stocks/ownership interest in Lilly and AstraZeneca. S.I., E.A., J.E.-P. and E.M.E. are employees of Janssen Inc. and are shareholders in Johnson & Johnson. N.B.L. has received institutional funding from Roche Canada, Guardant Health, MSD, EMD Serono, Lilly, AstraZeneca Canada, Takeda, Amgen, Bayer, MSD Oncology, and Merck Sharp and Dohme.

References

1. Canadian Cancer Society. Lung and Bronchus cancer Statistics. Available online: https://cancer.ca/en/cancer-information/cancer-types/lung/statistics (accessed on 26 August 2023).
2. Howlader, N.; Forjaz, G.; Mooradian, M.J.; Meza, R.; Kong, C.Y.; Cronin, K.A.; Mariotto, A.B.; Lowy, D.R.; Feuer, E.J. The Effect of Advances in Lung-Cancer Treatment on Population Mortality. *N. Engl. J. Med.* **2020**, *383*, 640–649. [CrossRef] [PubMed]
3. Black, R.C.; Khurshid, H. NSCLC: An Update of Driver Mutations, Their Role in Pathogenesis and Clinical Significance. *R. I. Med J.* **2015**, *98*, 25–28.
4. Sharma, S.V.; Bell, D.W.; Settleman, J.; Haber, D.A. Epidermal Growth Factor Receptor Mutations in Lung Cancer. *Nat. Rev. Cancer* **2007**, *7*, 169–181. [CrossRef] [PubMed]
5. Chevallier, M.; Borgeaud, M.; Addeo, A.; Friedlaender, A. Oncogenic Driver Mutations in Non-Small Cell Lung Cancer: Past, Present and Future. *World J. Clin. Oncol.* **2021**, *12*, 217–237. [CrossRef] [PubMed]
6. Melosky, B.; Blais, N.; Cheema, P.; Couture, C.; Juergens, R.; Kamel-Reid, S.; Tsao, M.S.; Wheatley-Price, P.; Xu, Z.; Ionescu, D.N. Standardizing Biomarker Testing for Canadian Patients with Advanced Lung Cancer. *Curr. Oncol.* **2018**, *25*, 73–82. [CrossRef] [PubMed]
7. Burnett, H.; Emich, H.; Carroll, C.; Stapleton, N.; Mahadevia, P.; Li, T. Epidemiological and Clinical Burden of EGFR Exon 20 Insertion in Advanced Non-Small Cell Lung Cancer: A Systematic Literature Review. *PLoS ONE* **2021**, *16*, e0247620. [CrossRef]
8. Planchard, D.; Popat, S.; Kerr, K.; Novello, S.; Smit, E.F.; Faivre-Finn, C.; Mok, T.S.; Reck, M.; Van Schil, P.E.; Hellmann, M.D.; et al. Metastatic Non-Small Cell Lung Cancer: ESMO Clinical Practice Guidelines for Diagnosis, Treatment and Follow-up. *Ann. Oncol.* **2018**, *29*, iv192–iv237. [CrossRef] [PubMed]
9. Lindeman, N.I.; Cagle, P.T.; Aisner, D.L.; Arcila, M.E.; Beasley, M.B.; Bernicker, E.H.; Colasacco, C.; Dacic, S.; Hirsch, F.R.; Kerr, K.; et al. Updated Molecular Testing Guideline for the Selection of Lung Cancer Patients for Treatment With Targeted Tyrosine Kinase Inhibitors: Guideline From the College of American Pathologists, the International Association for the Study of Lung Cancer, and the Association for Molecular Pathology. *Arch. Pathol. Lab. Med.* **2018**, *142*, 321–346. [CrossRef] [PubMed]
10. Ellison, G.; Zhu, G.; Moulis, A.; Dearden, S.; Speake, G.; McCormack, R. *EGFR* Mutation Testing in Lung Cancer: A Review of Available Methods and Their Use for Analysis of Tumour Tissue and Cytology Samples. *J. Clin. Pathol.* **2013**, *66*, 79. [CrossRef] [PubMed]
11. Bauml, J.M.; Viteri, S.; Minchom, A.; Bazhenova, L.; Ou, S.; Schaffer, M.; Le Croy, N.; Riley, R.; Mahadevia, P.; Girard, N. FP07.12 Underdiagnosis of EGFR Exon 20 Insertion Mutation Variants: Estimates from NGS-Based Real-World Datasets. *J. Thorac. Oncol.* **2021**, *16* (Suppl. S3), S208–S209. [CrossRef]
12. Yip, S.; Christofides, A.; Banerji, S.; Downes, M.R.; Izevbaye, I.; Lo, B.; MacMillan, A.; McCuaig, J.; Stockley, T.; Yousef, G.M.; et al. A Canadian Guideline on the Use of Next-Generation Sequencing in Oncology. *Curr. Oncol.* **2019**, *26*, 241–254. [CrossRef] [PubMed]
13. Melosky, B.; Banerji, S.; Blais, N.; Chu, Q.; Juergens, R.; Leighl, N.B.; Liu, G.; Cheema, P. Canadian Consensus: A New Systemic Treatment Algorithm for Advanced Egfr-Mutated Non-Small-Cell Lung Cancer. *Curr. Oncol.* **2020**, *27*, e146–e155. [CrossRef] [PubMed]
14. Ettinger, D.S.; Wood, D.E.; Aisner, D.L.; Akerley, W.; Bauman, J.R.; Bharat, A.; Bruno, D.S.; Chang, J.Y.; Chirieac, L.R.; D'Amico, T.A.; et al. Non–Small Cell Lung Cancer, Version 3.2022, NCCN Clinical Practice Guidelines in Oncology. *J. Natl. Compr. Cancer Netw.* **2022**, *20*, 497–530. [CrossRef] [PubMed]
15. Leonetti, A.; Sharma, S.; Minari, R.; Perego, P.; Giovannetti, E.; Tiseo, M. Resistance Mechanisms to Osimertinib in EGFR-Mutated Non-Small Cell Lung Cancer. *Br. J. Cancer* **2019**, *121*, 725–737. [CrossRef] [PubMed]
16. Fu, K.; Xie, F.; Wang, F.; Fu, L. Therapeutic Strategies for EGFR-Mutated Non-Small Cell Lung Cancer Patients with Osimertinib Resistance. *J. Hematol. Oncol.* **2022**, *15*, 173. [CrossRef] [PubMed]
17. Nieva, J.; Reckamp, K.L.; Potter, D.; Taylor, A.; Sun, P. Retrospective Analysis of Real-World Management of EGFR-Mutated Advanced NSCLC, After First-Line EGFR-TKI Treatment: US Treatment Patterns, Attrition, and Survival Data. *Drugs Real World Outcomes* **2022**, *9*, 333–345. [CrossRef]
18. Remon, J.; Hendriks, L.E.L.; Cardona, A.F.; Besse, B. *EGFR* Exon 20 Insertions in Advanced Non-Small Cell Lung Cancer: A New History Begins. *Cancer Treat Rev.* **2020**, *90*, 102105. [CrossRef] [PubMed]
19. Yang, M.; Xu, X.; Cai, J.; Ning, J.; Wery, J.P.; Li, Q.-X. NSCLC Harboring EGFR Exon-20 Insertions after the Regulatory C-Helix of Kinase Domain Responds Poorly to Known EGFR Inhibitors. *Int. J. Cancer* **2016**, *139*, 171–176. [CrossRef] [PubMed]
20. Wu, J.-Y.; Yu, C.-J.; Shih, J.-Y. Effectiveness of Treatments for Advanced Non–Small-Cell Lung Cancer With Exon 20 Insertion Epidermal Growth Factor Receptor Mutations. *Clin. Lung Cancer* **2019**, *20*, e620–e630. [CrossRef]
21. Zhou, C.; Tang, K.-J.; Cho, B.C.; Liu, B.; Paz-Ares, L.; Cheng, S.; Kitazono, S.; Thiagarajan, M.; Goldman, J.W.; Sabari, J.K.; et al. Amivantamab plus Chemotherapy in NSCLC with EGFR Exon 20 Insertions. *N. Engl. J. Med.* **2023**, *389*, 2039–2051. [CrossRef] [PubMed]
22. Gauthier, M.P.; Law, J.H.; Le, L.W.; Li, J.J.N.; Zahir, S.; Nirmalakumar, S.; Sung, M.; Pettengell, C.; Aviv, S.; Chu, R.; et al. Automating Access to Real-World Evidence. *JTO Clin. Res. Rep.* **2022**, *3*, 100340. [CrossRef] [PubMed]

3. Yang, X.; Mu, D.; Peng, H.; Li, H.; Wang, Y.; Wang, P.; Wang, Y.; Han, S. Research and Application of Artificial Intelligence Based on Electronic Health Records of Patients With Cancer: Systematic Review. *JMIR Med. Inform.* **2022**, *10*, e33799. [CrossRef] [PubMed]

4. Maddox, T.M.; Rumsfeld, J.S.; Payne, P.R.O. Questions for Artificial Intelligence in Health Care. *JAMA* **2019**, *321*, 31–32. [CrossRef] [PubMed]

5. Law, J.H.; Pettengell, C.; Le, L.W.; Aviv, S.; DeMarco, P.; Merritt, D.C.; Lau, S.C.M.; Sacher, A.G.; Leighl, N.B. Generating Real-World Evidence: Using Automated Data Extraction to Replace Manual Chart Review. *J. Clin. Oncol.* **2019**, *37* (Suppl. S15), e18096. [CrossRef]

6. Petch, J.; Kempainnen, J.; Pettengell, C.; Aviv, S.; Butler, B.; Pond, G.; Saha, A.; Bogach, J.; Allard-Coutu, A.; Sztur, P.; et al. Developing a Data and Analytics Platform to Enable a Breast Cancer Learning Health System at a Regional Cancer Center. *JCO Clin. Cancer Inform.* **2023**, *7*, e2200182. [CrossRef] [PubMed]

7. Vender, R.; Lynde, C. AI-Powered Patient Identification to Optimize Care. Canadian Dermatology Association. 2023. Available online: https://event.fourwaves.com/cda2023/abstracts (accessed on 28 August 2023).

8. Petch, J.; Batt, J.; Murray, J.; Mamdani, M. Extracting Clinical Features From Dictated Ambulatory Consult Notes Using a Commercially Available Natural Language Processing Tool: Pilot, Retrospective, Cross-Sectional Validation Study. *JMIR Med. Inform.* **2019**, *7*, e12575. [CrossRef] [PubMed]

9. Mittal, A.; Hueniken, K.; Cheng, S.; Zhan, L.J.; Brown, C.; Mai, V.; Lee, J.; Adewole, O.O.; Herman, J.; Sabouhanian, A.; et al. EP04.01-001 Prevalence And Outcomes of EGFR Exon 20 Insertion Mutation In NSCLC: Princess Margaret Cancer Center Experience. *J. Thorac. Oncol.* **2022**, *17* (Suppl. S9), S245. [CrossRef]

10. O'Sullivan, D.E.; Jarada, T.N.; Yusuf, A.; Hu, L.; Xun, Y.; Gogna, P.; Brenner, D.R.; Abbie, E.; Rose, J.B.; Eaton, K.; et al. Prevalence, Treatment Patterns, and Outcomes of Individuals with EGFR Positive Metastatic Non-Small Cell Lung Cancer in a Canadian Real-World Setting: A Comparison of Exon 19 Deletion, L858R, and Exon 20 Insertion EGFR Mutation Carriers. *Curr. Oncol.* **2022**, *29*, 7198–7208. [CrossRef] [PubMed]

11. Kuang, S.; Fung, A.S.; Perdrizet, K.A.; Chen, K.; Li, J.J.N.; Le, L.W.; Cabanero, M.; Karsaneh, O.A.A.; Tsao, M.S.; Morganstein, J.; et al. Upfront Next Generation Sequencing in Non-Small Cell Lung Cancer. *Curr. Oncol.* **2022**, *29*, 4428–4437. [CrossRef] [PubMed]

12. Mountzios, G.; Planchard, D.; Metro, G.; Tsiouda, D.; Prelaj, A.; Lampaki, S.; Shalata, W.; Riudavets, M.; Christopoulos, P.; Girard, N.; et al. Molecular Epidemiology and Treatment Patterns of Patients With EGFR Exon 20-Mutant NSCLC in the Precision Oncology Era: The European EXOTIC Registry. *JTO Clin. Res. Rep.* **2023**, *4*, 100433. [CrossRef] [PubMed]

13. Girard, N.; Leighl, N.; Ohe, Y.; Min Kim, T.; Demirdjian, L.; Bourla, A.; Sultan, A.A.; Mahadevia, P.; Bauml, J.; Sabari, J. 19P—Mortality Among EGFR-Mutated Advanced NSCLC Patients After Frontline Osimertinib Treatment: A Real-World, US Attrition Analysis. *J. Thorac. Oncol.* **2023**, *18*, S51–S52. [CrossRef]

Article

Real-World Treatment Patterns and Clinical Outcomes among Patients Receiving CDK4/6 Inhibitors for Metastatic Breast Cancer in a Canadian Setting Using AI-Extracted Data

Ruth Moulson [1,*], Guillaume Feugère [2], Tracy S. Moreira-Lucas [2], Florence Dequen [2], Jessica Weiss [1], Janet Smith [3] and Christine Brezden-Masley [3]

1   Pentavere, Toronto, ON M6G 1A1, Canada
2   Pfizer Canada ULC, Kirkland, QC H9J 2M5, Canada
3   Mount Sinai Hospital, Toronto, ON M5G 1X5, Canada; christine.brezden@sinaihealth.ca (C.B.-M.)
*   Correspondence: rmoulson@pentavere.com

Abstract: Cyclin-dependent kinase 4/6 inhibitors (CDK4/6i) are widely used in patients with hormone receptor-positive (HR+)/human epidermal growth factor receptor 2 negative (HER2−) advanced/metastatic breast cancer (ABC/MBC) in first line (1L), but little is known about their real-world use and clinical outcomes long-term, in Canada. This study used Pentavere's previously validated artificial intelligence (AI) to extract real-world data on the treatment patterns and outcomes of patients receiving CDK4/6i+endocrine therapy (ET) for HR+/HER2− ABC/MBC at Sinai Health in Toronto, Canada. Between 1 January 2016 and 1 July 2021, 48 patients were diagnosed with HR+/HER2− ABC/MBC and received CDK4/6i + ET. A total of 38 out of 48 patients received CDK4/6i + ET in 1L, of which 34 of the 38 (89.5%) received palbociclib + ET. In 2L, 12 of the 21 (57.1%) patients received CDK4/6i + ET, of which 58.3% received abemaciclib. In 3L, most patients received chemotherapy (10/12, 83.3%). For the patients receiving CDK4/6i in 1L, the median (95% CI) time to the next treatment was 42.3 (41.2, NA) months. The median (95% CI) time to chemotherapy was 46.5 (41.4, NA) months. The two-year overall survival (95% CI) was 97.4% (92.4, 100.0), and the median (range) follow-up was 28.7 (3.4–67.6) months. Despite the limitations inherent in real-world studies and a limited number of patients, these AI-extracted data complement previous studies demonstrating the effectiveness of CDK4/6i + ET in the Canadian real-world 1L, with most patients receiving palbociclib as CDK4/6i in 1L.

Keywords: real-world evidence; CDK4/6 inhibitors; AI; HR+; HER2−; metastatic breast cancer

Citation: Moulson, R.; Feugère, G.; Moreira-Lucas, T.S.; Dequen, F.; Weiss, J.; Smith, J.; Brezden-Masley, C. Real-World Treatment Patterns and Clinical Outcomes among Patients Receiving CDK4/6 Inhibitors for Metastatic Breast Cancer in a Canadian Setting Using AI-Extracted Data. *Curr. Oncol.* **2024**, *31*, 2172–2184. https://doi.org/10.3390/curroncol31040161

Received: 20 February 2024
Revised: 20 March 2024
Accepted: 7 April 2024
Published: 9 April 2024

1. Introduction

Breast cancer is the most common global cancer diagnosis and accounts for one out of four cancer cases and one out of six cancer deaths in females [1]. In Canada, the age standardized mortality rate for breast cancer has declined by 48% since the 1980s due to improved screening and more effective targeted systemic therapies [2]. However, despite this trend, 5-year survival differs between stage 0–I (100%), stage II (93%), stage III (72%), and stage IV advanced/metastatic breast cancer (ABC/MBC) (22%) [3]. Following the introduction of cyclin-dependent kinase 4/6 inhibitors (CDK4/6i), palbociclib, ribociclib and abemaciclib, over the last 8 years, CDK4/6i with endocrine therapy (ET) have become the standard of care for patients with hormone receptor-positive (HR+)/human epidermal growth factor receptor 2 negative (HER2-) ABC/MBC in first line (1L). The combination is recommended by all treatment guidelines, including the National Comprehensive Cancer Network (NCCN), the Canadian Cancer Society, and Canadian oncologists, and is supported by several phase III trials and RWE studies in the US [4–14]. However, there remains a lack of evidence on longer-term treatment patterns and clinical outcomes in patients with HR+/HER2− ABC/MBC in the Canadian real-world setting.

Real-world evidence (RWE) is increasingly being used to understand treatment use and outcomes in clinical practice and can complement the findings from randomized clinical trials (RCTs) [15–17]. For example, in the multicenter, heterogenous US cohort study by Rugo et al., (2022), palbociclib plus the aromatase inhibitor demonstrated greater median real-world progression-free survival (rwPFS) versus the aromatase inhibitor alone (19.3 [17.5–20.7] versus 13.9 [12.5–15.2] months; hazard ratio, 0.70 [95% CI, 0.62–0.78]; $p < 0.0001$), complementing PFS from the phase III PALOMA-2 study of palbociclib and letrozole versus letrozole and placebo (24.8 months [95% CI, 22.1–NA] versus 14.5 [95% CI, 12.9–17.1] months; hazard ratio, 0.58; [95% CI, 0.46 to 0.72]; $p < 0.001$) [7,12].

Recently, electronic health records (EHRs) have been leveraged as a rich source of real-world data (RWD), as they can provide a comprehensive overview of patients' disease in a centralized location, allowing researchers to study disease progression, treatment patterns, and clinical outcomes over time. Still, complexities exist in harnessing data from the EHR. Basic patient information, such as demographics, is typically easier to collect, as it is held within structured fields of the EHR, but it may be incomplete or incorrect. Other valuable features, such as evidence of metastases, are often found within the unstructured fields, which are less easy to collect. A manual chart review is commonly used for extracting RWD from the EHR [18]. However, due to the complexities of the EHR, this is time-consuming, prone to human error, lacks scalability, and can result in inconsistent data. These challenges have contributed to the limited translation of EHR adoption into enhanced clinical care [19–21].

To overcome these limitations, artificial intelligence (AI) has proven its ability to extract data from structured and unstructured fields of the EHR to produce reliable, structured clinical data in a more consistent, efficient, and scalable manner compared to manual abstraction [18,22,23]. This technology allows clinicians and researchers to access previously unavailable RWD and is being used for patient and disease identification, pharmacovigilance, and the development of learning health systems [23–28].

Complexities also exist for the AI extraction of RWD from the EHR as a result of inconsistencies in the sections of the EHR where information is stored, variations and complexity in the narrative used within clinical text, and the need to coordinate multiple pieces of evidence temporally. This can result in uncertainty regarding the validity and transferability of such technologies [29]. The commercially available AI engine, DARWEN™ (Darwen, UK), has been evaluated against manual abstraction for the same clinical features in multiple disease areas, including breast cancer [25], lung cancer [18,30–34], ambulatory care diseases [23], and dermatology [28] at multiple Canadian institutions, validating its use to extract RWD more accurately and efficiently than a manual chart review. Sinai Health is a leading Canadian cancer center and has been using EHR systems since 2006 with the goal of leveraging technology to harness data from the EHRs to inform clinician decision-making.

In this study, we describe how the AI extraction of RWD was used to describe and better understand the treatment patterns and clinical outcomes of Canadian patients receiving CDK4/6i + ET for HR+/HER2− ABC/MBC in a real-world setting, with a longer follow-up. RWE is necessary to understand these trends to inform targeted sequencing and future treatment decisions in this population.

2. Materials and Methods

2.1. Study Design

This was a retrospective chart review of the data from the EHRs of patients diagnosed with HR+/HER2− ABC/MBC between 1 January 2016 and 1 July 2021, receiving CDK4/6i treatment at Sinai Health, Toronto. Included patients were as follows: women aged ≥ 18 years old, diagnosed with HR+/HER2− ABC/MBC between 1 January 2016 and 1 July 2021, and treated with CDK4/6i. The study period encompassed 1 January 2016 to 1 October 2021 to capture all patients treated with CDK4/6i since their approval and allowed for a minimum three-month follow-up period.

2.2. Clinical Feature Extraction

The clinical features extracted from Sinai Health's EHRs included patient demographics, clinical characteristics, treatment information, and clinical outcomes. Data were extracted from the patient EHRs using DARWEN™ AI technology, or for three specific features (radiation treatment, date of ABC/MBC diagnosis, and treatment start/stop date), the data were extracted manually. Finally, some features were derived using the extracted data, such as age at ABC/MBC diagnosis and clinical outcomes, including the time to the next treatment (TTNT), time to chemotherapy (TTC), and overall survival (OS).

DARWEN™—which has been previously described and validated in detail—combines multiple state-of-the-art approaches to extract relevant data from structured and unstructured EHR fields [18,23]. DARWEN™ uses a "twin-engine design", which allows model training to begin on one task while learnings and adjustments can be made quickly and easily for adjacent tasks. This provides knowledge transfer between tasks, flexibility, and adaptability, reducing the overall number of models required and hence the compound error, thus achieving high accuracy with the results that are aligned with clinician expertise.

All features were extracted following pre-defined rules and definitions developed by the Sinai Health Principal Investigator (PI). Based on the reality of the available data at Sinai Health, the definitions and rules were updated in an iterative process until a finalized set of rules was agreed upon with the PI. A full list of the features extracted, as well as the feature definitions and data sources, can be found in Supplementary Table S1.

For the features extracted from the unstructured EHR field, DARWEN™ algorithms were pre-trained on general medical and other ABC/MBC datasets and then fine-tuned and validated on the Sinai Health data, as detailed below.

Using the initial data provisioned by Sinai Health (which included all patients at Sinai Health who received a CDK4/6i and were aged $\geq$ 18 years), one subset of patient data was used for the fine-tuning and testing of the algorithms based on the finalized feature definitions and extraction rules, until accuracy, precision (positive predictive value), recall (sensitivity), and F1 (the harmonic mean of precision and recall) score targets were achieved. The AI training and tuning methods have previously been reported [18]. Subsequently, the models were applied to a second subset of data (distinct from the first one) to generate validation metrics against data unseen by the model. The steps were repeated if necessary until the results on both subsets met the target scores and were sufficiently stable. Finally, the models were run on all the remaining data, which had not been part of either the first or second subset, to produce the final dataset. See Figure 1 for the workflow and methodology used throughout this study. All extracted data (irrespective of the extraction method) was reviewed by the PI to confirm that the findings aligned with their clinical expectations.

2.3. Outcomes

The primary outcome was to characterize real-world treatment patterns among patients with HR+/HER2− ABC/MBC receiving CDK4/6i. Other outcomes of interest included clinical outcomes: TTNT for 1L, TTC from diagnosis, and OS. TTNT was measured from the date of the initiation (first dose) of treatment to the date of the initiation of the subsequent line of therapy. Patients who did not progress on to a subsequent line of therapy were censored at their last known date of treatment. TTC was measured from the date of ABC/MBC diagnosis to the date of chemotherapy initiation. Patients still on treatment and who did not start chemotherapy were censored at the date of their last follow-up or death, whichever came first. Patients who died before starting their next line of therapy were also censored. OS was measured from the date of ABC/MBC diagnosis to the date of death. For patients where no death event was found, the date of the last follow-up was used, and these patients were censored in the survival analyses.

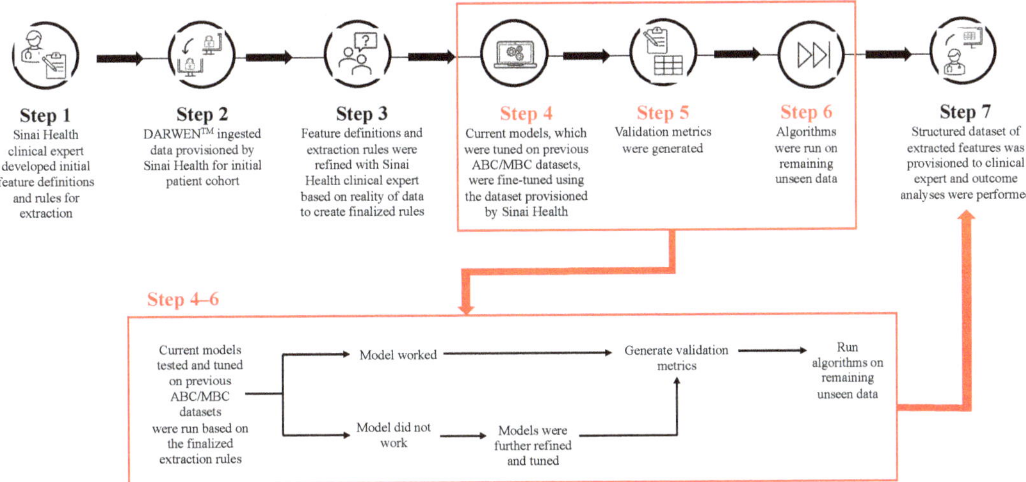

Figure 1. Workflow and methodology used to refine, test, and validate models. ABC/MBC: Advanced/metastatic breast cancer.

2.4. Statistical Analyses

Descriptive analyses summarized patients' demographics, clinical characteristics, and outcomes of interest across the study cohort. Continuous variables were described using mean and standard deviation (SD), median, and the first and third quartiles. Categorical variables were described by frequencies and percentages. Kaplan–Meier (KM) curves were used to describe the time to event(s) and followed standard censoring rules.

3. Results

DARWEN^{TM} was used to extract nine features found within the unstructured fields of the EHR. AI performance for the extracted features is shown in Supplementary Table S2. An F1 score (the harmonic mean of precision and recall) of 1.00 was achieved for three features: histology, ER receptor status, and PR receptor status, and an overall accuracy (the number of correctly identified predictions) of above 90% was achieved for all AI-extracted features. These results are consistent with the previous validations of DARWEN^{TM} [18,23]. Radiation treatment, the date of ABC/MBC diagnosis, and treatment (start/stop date) were extracted manually due to the limitations imposed by the data captured in the EHR. Radiation treatment is administered at sites outside of Sinai Health; therefore, information on a patient's radiation therapy was not consistently captured in the Sinai Health patient EHR. The date of the ABC/MBC diagnosis is also often inconsistently reported in the patient's EHR, with the ABC/MBC diagnosis often being reported as suspicious but not confirmed. Additionally, patients were often diagnosed with ABC/MBC at other sites and referred to Sinai Health. Prescription information is not stored electronically in the EHR system at Sinai Health but rather in paper format, dictated into clinical notes. Before data extraction using either method, the pre-defined rules and definitions for each clinical feature were finalized with the Sinai Health PI (Supplementary Table S1).

3.1. Patients

In total, DARWEN^{TM} ingested a total of 5052 patient reports, including clinical, pathology, and radiology reports for 87 patients at Sinai Health who received a CDK4/6i and were aged $\geq$ 18 years. A total of 48 patients were identified as having HR+/HER2− ABC/MBC diagnosed between 1 January 2016 and 1 July 2021 and were treated with a CDK4/6i during the study period.

The baseline characteristics for the 48 included patients can be found in Table 1. In this cohort, the median age was 60.5 years. The majority of patients (70.8%) had recurrent

ABC/MBC, and 29.2% had de novo disease; 66.7% of the patients had ductal carcinoma and 18.8% were pre-menopausal. A total of 31.2% of patients presented with bone-only metastases at their ABC/MBC diagnosis. A total of 39.6% of patients had lung metastases during the study period, and 37.5% had liver metastases during the study period. A total of 45.8% of patients had one metastatic site during the study period. Of the patients with reported Eastern Cooperative Oncology Group (ECOG) performance scores at diagnosis (22/48), the majority had an ECOG score of 0/1 (18/22 [81.8%]). At ABC/MBC diagnosis, the most common comorbidity was hypertension (37.5%), followed by diabetes (14.6%). The tumor grade at ABC/MBC diagnosis was not consistently reported across the patients, with 29 out of 48 (60.4%) missing tumor grades at the time of their ABC/MBC diagnosis. Of the 48 patients, 38 received a CDK4/6i in the 1L setting. Baseline demographics for the 38 patients who received a CDK4/6i in 1L were similar to the full patient cohort (Table 1). Of the full cohort, 21 out of 48 (43.8%) patients went on to receive a second line (2L) therapy during the study, and 12 of the 48 (25.0%) went on to receive a third line (3L) therapy during the study.

Table 1. Demographics and baseline characteristics of all patients and patients who received a CDK4/6i in 1L.

	All Patients (N = 48)	Patients Receiving CDK4/6i in 1L (N = 38)
Age at ABC/MBC diagnosis		
Mean (SD)	57.9 (14.0)	58.4 (13.0)
Median	60.5	61.0
Q1, Q3	48.8, 67.0	50.0, 65.5
Range	23.0–89.0	23.0–84.0
Year of ABC/MBC diagnosis [a]		
2016–2018	20 (41.7%)	16 (42.1%)
2019–2021	28 (58.3%)	22 (57.9%)
Sex		
Female	48 (100.0%)	38 (100.0%)
Tumor histology		
Ductal	32 (66.7%)	25 (65.8%)
Lobular	7 (14.6%)	≤5 (NR)
Mixed	≤5 (NR)	≤5 (NR)
Other	8 (16.7%)	8 (21.1%)
De novo/recurrent at initial BC diagnosis		
De novo	14 (29.2%)	10 (26.3%)
Recurrent	34 (70.8%)	28 (73.7%)
HER2 status at ABC/MBC diagnosis		
Negative	48 (100.0%)	38 (100.0%)
ER status at ABC/MBC diagnosis		
Positive	48 (100.0%)	38 (100.0%)
PR status at ABC/MBC diagnosis		
Negative	13 (27.1%)	11 (28.9%)
Positive	30 (62.5%)	23 (60.5%)
Unknown	≤5 (NR)	≤5 (NR)

Table 1. *Cont.*

	All Patients (N = 48)	Patients Receiving CDK4/6i in 1L (N = 38)
ECOG at ABC/MBC diagnosis		
0	6 (12.5%)	≤5 (NR)
1	12 (25.0%)	10 (26.3%)
2	≤5 (NR)	≤5 (NR)
3	≤5 (NR)	≤5 (NR)
Unknown	26 (54.2%)	21 (55.3%)
Tumor grade at ABC/MBC diagnosis		
1	≤5 (NR)	≤5 (NR)
2	11 (22.9%)	8 (21.1%)
3	≤5 (NR)	≤5 (NR)
Unknown	29 (60.4%)	24 (63.2%)
Organ-level metastatic sites [b]		
Bone	35 (72.9%)	26 (68.4%)
Bone-only metastases	15 (31.2%)	12 (31.6%)
Brain	≤5 (NR)	≤5 (NR)
Lung	19 (39.6%)	16 (42.1%)
Liver	18 (37.5%)	12 (31.6%)
Number of metastatic sites during study period		
0	≤5 (NR)	≤5 (NR)
1	22 (45.8%)	19 (50.0%)
2	14 (29.2%)	9 (23.7%)
3	≤5 (NR)	≤5 (NR)
4	≤5 (NR)	≤5 (NR)
Comorbidities at ABC/MBC diagnosis [b]		
Atrial Fibrillation	≤5 (NR)	≤5 (NR)
Hypertension	18 (37.5%)	12 (31.6%)
Diabetes	7 (14.6%)	≤5 (NR)
Coronary Artery Disease	≤5 (NR)	≤5 (NR)
Radiotherapy for ABC/MBC [b]		
Any radiotherapy	18 (37.5%)	13 (34.2%)
Follow-up since diagnosis (months)		
Mean (SD)	28.8 (16.7)	28.7 (16.9)
Median	29.3	28.7
Q1, Q3	17.5, 37.2	17.8, 39.3
Range	3.4–67.6	3.4–67.6

[a] 2016–2018 represents the first half of the study period, and 2019–2021 represents the second half of the study period. [b] Denominator for the table is the patient population number. Percentages will not add up to 100%, as some patients may have multiple values. Pre-menopausal was defined as patients who are 50 years old or younger and are on an LHRH antagonist at any point. ABC/MBC: advanced/metastatic breast cancer; CDK4/6i: cyclin-dependent kinase 4 and 6 inhibitors; ECOG: Eastern Cooperative Oncology Group; LHRH: luteinizing hormone-releasing hormone; NR: not reported (data are suppressed to protect privacy, as per site's requirement); SD: standard deviation.

3.2. Treatment Patterns

Treatment patterns were assessed from the date of ABC/MBC until the date of death, date of last follow-up, or the end of the study period, whichever came first (the median duration of follow-up for all patients was 28.7 months). Throughout the study period, across all patients, CDK4/6i included abemaciclib, palbociclib, and ribociclib. ET included tamoxifen, anastrozole, letrozole, exemestane, and fulvestrant. Chemotherapy included the following agents (either a single agent or in combination): capecitabine, cisplatin, cyclophosphamide, paclitaxel, docetaxel, doxorubicin, eribulin, and/or gemcitabine.

Of 38 out of 48 patients who received a CDK4/6i in 1L, 34 of 38 (89.5%) received palbociclib + ET (Table 2; Supplementary Figure S1). Letrozole was the most common ET given with CDK4/6i in 1L (30/38 [78.9%]). A total of 27 out of 48 (56%) patients did not go on to receive a 2L during the study period (for reasons including that the patient remained on 1L, the patient died, or the patient was lost to follow-up). Of the 21 patients who went on to 2L treatment during the study period, 12 out of 21 (58.3%) of these patients received a CDK4/6i, of which 7 out of 12 received abemaciclib + ET (Table 2; Supplementary Figure S1). Fulvestrant was the most commonly prescribed ET with CDK4/6i in 2L (9/12 [75.0%]). The majority of patients who progressed to a 3L therapy received chemotherapy (10/12 [83.3%]) (Table 2; Supplementary Figure S1).

Table 2. Treatment patterns for all patients across 1L, 2L, and 3L of treatment.

1L Treatment Regimen	All Patients (N = 48)
Palbociclib + ET	30 (62.5%)
Palbociclib + ET + goserelin	≤5 (NR)
Other CDK4/6i + ET	≤5 (NR)
Other CDK4/6i + ET + goserelin	≤5 (NR)
ET	≤5 (NR)
Chemotherapy	7 (14.6%)
2L treatment regimen	**All patients (N = 21)**
Palbociclib + ET	≤5 (NR)
Palbociclib + ET + goserelin	≤5 (NR)
Other CDK4/6i + ET	7 (33.3%)
Other CDK4/6i + ET + goserelin	≤5 (NR)
Alpelisib + ET	≤5 (NR)
ET	≤5 (NR)
Chemotherapy	6 (28.6%)
3L treatment regimen	**All patients (N = 12)**
Palbociclib + ET + goserelin	≤5 (NR)
Other CDK4/6i	≤5 (NR)
Chemotherapy	10 (83.3%)

CDK4/6i: cyclin-dependent kinase 4 and 6 inhibitors; ET: endocrine therapy; NR: not reported (data are suppressed to protect privacy, as per site's requirement).

3.3. Clinical Outcomes

For the patients who received a CDK4/6i in 1L, the median (95% confidence interval [CI]) time to the next treatment for 1L (TTNT1) was 42.3 (41.2, NA) months (Figure 2). The median (95% CI) TTC for these patients was 45.1 months (41.2, NA; Figure 3). A median (95%) OS was not reached, and the 2-year OS rate (95% CI) was 97% (92%, 100%; Figure 4)

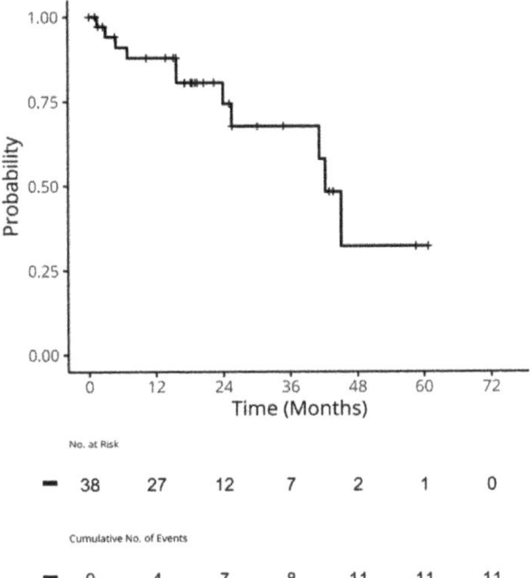

Figure 2. TTNT1 in patients who received a CDK4/6i in 1L. Median TTNT (95% CI): 42.3 (41.2, NA) months. CDK4/6i: cyclin-dependent kinase 4/6 inhibitors; CI: confidence interval; NA: not applicable; TTNT1: the time to next treatment was calculated by subtracting the start date of 1L from the start date of 2L. Patients who did not go on to receive 2L were censored at their last known date of treatment.

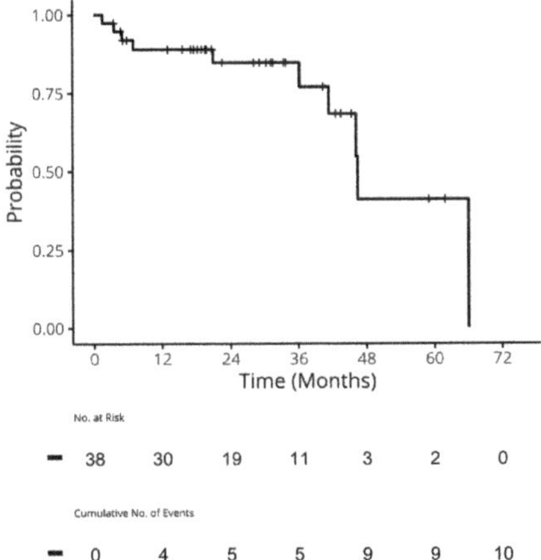

Figure 3. TTC in patients who received a CDK4/6i in 1L. Median (95% CI) TTC: 46.5 (41.4, NA) months. CDK4/6i: cyclin-dependent kinase 4/6 inhibitors; CI: confidence interval; NA: not applicable; TTC: the time to chemotherapy was calculated by subtracting the start date of chemotherapy from the date of ABC/MBC diagnosis. Patients who did not receive chemotherapy were censored at their date of last follow-up or death. Patients who experienced death before starting their next treatment were also censored.

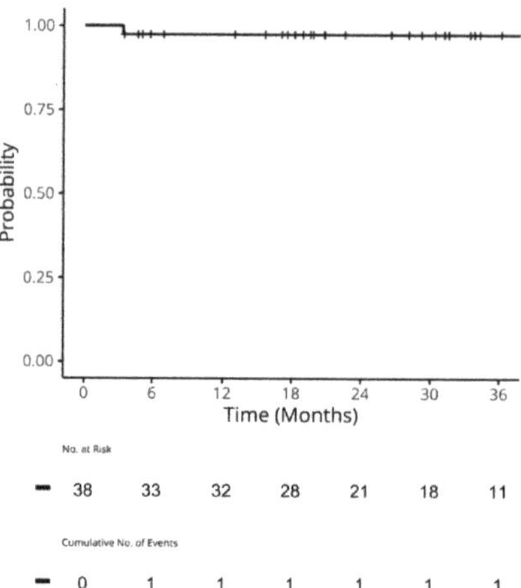

Figure 4. OS in patients who received a CDK4/6i in 1L. Median OS (95% CI): NA (NA, NA) months. Two-year OS (95% CI) was 97.4% (92.4, 100.0). CDK4/6i: cyclin-dependent kinase 4/6 inhibitors; CI: confidence interval; NA: not applicable; OS: the overall survival was calculated by subtracting the date of starting 1L from the date of death. Patients who did not die were censored at their last date of follow-up or the study's end date.

4. Discussion

This study illustrates the validity of using AI technologies for identifying patients with HR+/HER2− ABC/MBC and generating RWE, including the treatment patterns and clinical outcomes for patients. AI was used to extract nine crucial features from the patient EHR, which were validated and reviewed by a breast cancer expert. The results from this study complement the findings from previous RWE studies and demonstrate the effectiveness of CDK4/6i in the Canadian real-world 1L setting (particularly palbociclib, as most patients in this study received 1L palbociclib) over a longer follow-up period than previous real-world Canadian studies (up to 69 months versus 62 and 24) [35,36].

Recently, much progress has been made in the implementation of AI tools in healthcare including assisting radiologists in detecting abnormalities and disease from X-rays, MRIs and CT scans, personalized medicine and predicting which treatments are likely to benefit a patient, clinical decision support systems and AI-remote monitoring and telemedicine platforms [37–39]. Additionally, AI tools used for the extraction of clinical text can make sense of and analyze vast amounts of unstructured clinical text from pathology reports, clinical notes, and radiology reports. These tools, such as DARWEN™, are being used for patient and disease identification, pharmacovigilance, and the development of learning health systems [23–28]. However, many tools, such as ClinicalBERT, rely on open-source datasets, such as the MIMIC-III dataset of de-identified hospital records from intensive care units [40,41]. These datasets have limited insight into the entirety of the patient journey and may not be appropriate for investigating diseases such as breast cancer, for which care is provided in many different settings outside of the intensive care unit and over long periods of time. Further, many of these tools only focus on a single clinical feature, e.g., a diagnosis of a certain condition or the development of metastases, with few investigating multiple distinct medical features [42–44]. In comparison, this study investigated multiple complex features throughout the patient's journey, which are critical for determining knowledge gaps and unmet needs for patients with breast cancer.

While AI holds immense promise in improving cancer diagnosis, treatment, and outcomes, it is important to recognize the challenges and limitations of the technology specifically related to accuracy and precision. AI algorithms are only as reliable as the data they are trained on, and biases in training data can lead to inaccurate outcomes, particularly in underrepresented populations. In the context of this study, limitations imposed by the data captured in the EHR were observed, which hindered the ability to extract certain features using a completely AI approach. For example, the administration of radiation treatment outside of Sinai Health resulted in inconsistent reporting of such treatment within the clinical notes. This inconsistency posed challenges for AI in capturing all instances of when the patients received radiation therapy. However, the incomplete documentation presented similar difficulties for manual abstractors. Additionally, prescription information is not stored electronically in the EHR system at Sinai Health, but rather in paper format, dictated into clinical notes. Consequently, these records are susceptible to missingness, incompleteness, and human error. Notably, it was found that clinicians tended to document the initiation of treatment more consistently than its termination. Limited use of imputation methods was used for missing treatment dates; however, if only the month and year were present for a date, then "15" was input as the date to create a complete observation. The date of death was also not consistently reported in the EHR, as these data are only collected when hospitals are notified of a patient's death and are provided with the exact date. These limitations are consistent with previous applications of AI tools for the extraction of oncology EHR data, but it is important to note that these limitations also impact the manual curation of data, highlighting a broader limitation in generating RWE from EHR systems [18,45].

Currently, at Sinai Health and other Canadian institutions, more sophisticated and universal EHR systems are being implemented (e.g., EPIC and Cerner Solutions), which will likely improve the ability of AI to extract data efficiently and accurately for generating RWE. AI data extraction from EHRs could allow institutions such as Sinai Health to more quickly and easily understand how they are performing compared to the currently published metrics, enabling them to perform meaningful QA projects and enhance patient care.

As this study was conducted at a single institution, there was a limited number of eligible patients, and in accordance with the hospital data privacy regulations in place at Sinai Health, observations that included less than or equal to five patients were suppressed. Future studies may hope to include additional sites to increase the number of patients included and potentially increase the diversity of patient cases represented in the results. Future work at Sinai Health may also hope to expand the use of AI technologies to harness data from the EHR system across breast cancer cohorts and disease domains and for further applications, such as patient and disease identification and the development of learning health systems, for ongoing prospective data collection.

Despite the limitations inherent to RWE studies using EHR data and the limited sample size, the real-world clinical outcomes observed in this study complemented those previously reported in the US and Canada. For Canadian patients receiving a C DK4/6i in 1L (97.8% of whom received palbociclib + ET), the 1-year OS was 97% (92%, 100%). This is similar to the 1-year survival rate reported in the Ibrance® Real World Insights Study (IRIS) (the 1-year survival rate was 95.6% for palbociclib + AI and 100% for palbociclib and fulvestrant) [35]. Further, the median (95% CI) for TTNT1 was 42.3 (41.2, NA) months for patients receiving a CDK4/6i in 1L, which is longer than the median rwPFS for palbociclib combination treatment from the US DeMichele et al. (2022) study (20.0 months [95% CI, 17.5–21.9]) for 1L [11]. Additionally, the validation metrics for AI-extracted data are consistent with the previous validations of DARWEN^{TM}, which has been evaluated against a manual abstraction for the same clinical features in breast cancer [25], lung cancer [18,30–34], ambulatory care diseases [23], and dermatology [28] at multiple Canadian institutions.

5. Conclusions

This study highlights the validity of AI technology in identifying patients with HR+/HER2− ABC/MBC and generating RWE, including treatment patterns and clinical outcomes for patients. This type of technology allows for a more efficient, consistent, and scalable extraction of data from EHR systems. AI was used to extract nine crucial features from the patient EHR, which were validated and reviewed by a breast cancer expert, and the accuracy metrics were consistent with the previous validations of the AI technology. The results from this study demonstrate the effectiveness of CDK4/6i + ET in the Canadian real-world 1L, with most patients receiving palbociclib as the CDK4/6i in 1L over a longer follow-up period than in previous real-world Canadian studies.

Supplementary Materials: The following supporting information can be downloaded at: https //www.mdpi.com/article/10.3390/curroncol31040161/s1, Table S1: Clinical features extracted, definitions, and data sources; Table S2: Evaluation of AI algorithms compared with manual chart review; Figure S1: Sankey diagram of treatment patterns in all patients.

Author Contributions: R.M.: Conceptualization; Data curation; Formal analysis; Funding acquisition, Investigation; Methodology; Project administration; Resources; Software; Supervision; Validation; Visualization; Roles/Writing—original draft; Writing—review and editing; G.F.: Conceptualization; Formal analysis; Funding acquisition; Investigation; Methodology; Project administration; Resources; Supervision; Validation; Visualization; Roles/Writing—original draft; Writing—review and editing.; T.S.M.-L.: Conceptualization; Formal analysis; Funding acquisition; Investigation; Methodology; Project administration; Resources; Supervision; Validation; Visualization; Roles/Writing—original draft; Writing—review and editing; F.D.: Conceptualization; Formal analysis; Funding acquisition, Investigation; Methodology; Project administration; Resources; Supervision; Validation; Visualization; Roles/Writing—original draft; Writing—review and editing; J.W.: Conceptualization; Data curation, Formal analysis; Methodology; Validation; Visualization; Roles/Writing—original draft; Writing— review and editing; J.S.: Conceptualization; Data curation; Formal analysis; Funding acquisition; Investigation; Methodology; Project administration; Resources; Supervision; Validation; Visualization, Roles/Writing—original draft; Writing—review and editing.; C.B.-M.: Conceptualization; Data curation; Formal analysis; Funding acquisition; Investigation; Methodology; Project administration; Resources; Software; Supervision; Validation; Visualization; Roles/Writing—original draft; Writing— review and editing. All authors have read and agreed to the published version of the manuscript.

Funding: This study was sponsored by Pfizer Canada ULC. Pentavere Research Group Inc. received financial support from Pfizer Canada ULC in connection with the conduct of this study, AI technology, the study analyses, and the medical writing of this manuscript. Sinai Health and Dr. Brezden-Masley received an unrestricted grant from Pfizer Canada ULC for data access and for providing clinical expertise oversight on the study, respectively.

Institutional Review Board Statement: This study was conducted in accordance with the legal and regulatory requirements, as well as with scientific purpose, value, and rigor, and followed the accepted research practices described in *Guidelines for Good Pharmacoepidemiology Practices* (GPP) issued by the International Society for Pharmacoepidemiology (ISPE), as well as the ethical principles that have their origin in the Declaration of Helsinki. Research ethics approval was obtained from Mount Sinai Hospital REB for this study (protocol code: 21-0263-C and date of approval: 5 April 2022).

Informed Consent Statement: Patient consent was waived by Mount Sinai Hospital REB as this study was retrospective; information abstracted for the study was already collected as part of the patient's routine clinical care and was contained within the patient's health record. No additional contact with participants for information was needed. All information was kept strictly confidential

Data Availability Statement: The data presented in this study are not publicly available due to the privacy of individuals. The data presented in this study may be available upon reasonable request from the senior author.

Conflicts of Interest: Ruth Moulson and Jessica Weiss were employees of Pentavere Research Group Inc., Toronto, ON, Canada, at the time of this study. Guillaume Feugère, Tracy S Moreira-Lucas, and Florence Dequen were employees of Pfizer Canada ULC, Kirkland, QC, Canada, at the time of this study. Janet Smith has no conflicts of interest to disclose. Christine Brezden-Masley has received research funding from Novartis Pharmaceuticals Canada Inc., Eli Lilly Canada Inc., and Pfizer Canada ULC; honoraria and has been involved as a consultant for Astellas, AstraZeneca, Bristol-Myers Squibb, Eisai, Eli Lilly Canada Inc., Gilead, Knight, Merck Sharp and Dohme LLC, Novartis Pharmaceuticals Canada Inc., Pfizer Canada ULC, Seagen, and Taiho Pharma Canada Inc.; speaker fees from Novartis Pharmaceuticals Canada Inc., Eli Lilly Canada Inc., Astellas, AstraZeneca, Knight, and Seagen.

References

1. Sung, H.; Ferlay, J.; Siegel, R.L.; Laversanne, M.; Soerjomataram, I.; Jemal, A.; Bray, F. Global Cancer Statistics 2020: GLOBOCAN Estimates of Incidence and Mortality Worldwide for 36 Cancers in 185 Countries. *CA Cancer J. Clin.* **2021**, *71*, 209–249. [CrossRef] [PubMed]
2. Release Notice—Canadian Cancer Statistics 2021. *Health Promot. Chronic Dis. Prev. Can.* **2021**, *41*, 399. [CrossRef]
3. Survival Statistics for Breast Cancer | Canadian Cancer Society. Available online: https://cancer.ca/en/cancer-information/cancer-types/breast/prognosis-and-survival/survival-statistics (accessed on 7 June 2023).
4. Migliaccio, I.; Bonechi, M.; McCartney, A.; Guarducci, C.; Benelli, M.; Biganzoli, L.; Di Leo, A.; Malorni, L. CDK4/6 inhibitors: A focus on biomarkers of response and post-treatment therapeutic strategies in hormone receptor-positive HER2− negative breast cancer. *Cancer Treat. Rev.* **2021**, *93*, 102136. [CrossRef] [PubMed]
5. Gradishar, W.J.; Moran, M.S.; Abraham, J.; Aft, R.; Agnese, D.; Allison, K.H.; Anderson, B.; Burstein, H.J.; Chew, H.; Dang, C.; et al. Breast Cancer, Version 3.2022, NCCN Clinical Practice Guidelines in Oncology. *J. Natl. Compr. Cancer Netw.* **2022**, *20*, 691–722. [CrossRef]
6. Jerzak, K.J.; Bouganim, N.; Brezden-Masley, C.; Edwards, S.; Gelmon, K.; Henning, J.-W.; Hilton, J.F.; Sehdev, S. HR+/HER2− Advanced Breast Cancer Treatment in the First-Line Setting: Expert Review. *Curr. Oncol.* **2023**, *30*, 5425–5447. [CrossRef]
7. Finn, R.S.; Martin, M.; Rugo, H.S.; Jones, S.; Im, S.-A.; Gelmon, K.; Harbeck, N.; Lipatov, O.N.; Walshe, J.M.; Moulder, S.; et al. Palbociclib and Letrozole in Advanced Breast Cancer. *N. Engl. J. Med.* **2016**, *375*, 1925–1936. [CrossRef]
8. Rugo, H.S.; Finn, R.S.; Diéras, V.; Ettl, J.; Lipatov, O.; Joy, A.A.; Harbeck, N.; Castrellon, A.; Iyer, S.; Lu, D.R.; et al. Palbociclib plus letrozole as first-line therapy in estrogen receptor-positive/human epidermal growth factor receptor 2-negative advanced breast cancer with extended follow-up. *Breast Cancer Res. Treat.* **2019**, *174*, 719–729. [CrossRef]
9. Hortobagyi, G.N.; Stemmer, S.M.; Burris, H.A.; Yap, Y.-S.; Sonke, G.S.; Paluch-Shimon, S.; Campone, M.; Blackwell, K.L.; André, F.; Winer, E.P.; et al. Ribociclib as First-Line Therapy for HR-Positive, Advanced Breast Cancer. *N. Engl. J. Med.* **2016**, *375*, 1738–1748. [CrossRef]
10. Goetz, M.P.; Toi, M.; Campone, M.; Sohn, J.; Paluch-Shimon, S.; Huober, J.; Park, I.H.; Trédan, O.; Chen, S.-C.; Manso, L.; et al. MONARCH 3: Abemaciclib As Initial Therapy for Advanced Breast Cancer. *J. Clin. Oncol.* **2017**, *35*, 3638–3646. [CrossRef]
11. DeMichele, A.; Cristofanilli, M.; Brufsky, A.; Liu, X.; Mardekian, J.; McRoy, L.; Layman, R.M.; Emir, B.; Torres, M.A.; Rugo, H.S.; et al. Comparative effectiveness of first-line palbociclib plus letrozole versus letrozole alone for HR+/HER2− metastatic breast cancer in US real-world clinical practice. *Breast Cancer Res.* **2021**, *23*, 37. [CrossRef]
12. Rugo, H.S.; Brufsky, A.; Liu, X.; Li, B.; McRoy, L.; Chen, C.; Layman, R.M.; Cristofanilli, M.; Torres, M.A.; Curigliano, G.; et al. Real-world study of overall survival with palbociclib plus aromatase inhibitor in HR+/HER2− metastatic breast cancer. *npj Breast Cancer* **2022**, *8*, 114. [CrossRef] [PubMed]
13. Goyal, R.K.; Chen, H.; Abughosh, S.M.; Holmes, H.M.; Candrilli, S.D.; Johnson, M.L. Overall survival associated with CDK4/6 inhibitors in patients with HR+/HER2− metastatic breast cancer in the United States: A SEER-Medicare population-based study. *Cancer* **2023**, *129*, 1051–1063. [CrossRef] [PubMed]
14. Canadian Cancer Society. Treatments for Stage 4 Breast Cancer. Available online: https://cancer.ca/en/cancer-information/cancer-types/breast/treatment/stage-4 (accessed on 7 June 2023).
15. Maio, M.D.; Perrone, F.; Conte, P. Real-World Evidence in Oncology: Opportunities and Limitations. *Oncologist* **2020**, *25*, e746. [CrossRef] [PubMed]
16. Corrigan-Curay, J.; Sacks, L.; Woodcock, J. Real-World Evidence and Real-World Data for Evaluating Drug Safety and Effectiveness. *JAMA* **2018**, *320*, 867–868. [CrossRef] [PubMed]
17. Cejuela, M.; Gil-Torralvo, A.; Castilla, M.Á.; Domínguez-Cejudo, M.Á.; Falcón, A.; Benavent, M.; Molina-Pinelo, S.; Ruiz-Borrego, M.; Salvador Bofill, J. Abemaciclib, Palbociclib, and Ribociclib in Real-World Data: A Direct Comparison of First-Line Treatment for Endocrine-Receptor-Positive Metastatic Breast Cancer. *Int. J. Mol. Sci.* **2023**, *24*, 8488. [CrossRef]
18. Gauthier, M.P.; Law, J.H.; Le, L.W.; Li, J.J.N.; Zahir, S.; Nirmalakumar, S.; Sung, M.; Pettengell, C.; Aviv, S.; Chu, R.; et al. Automating Access to Real-World Evidence. *JTO Clin. Res. Rep.* **2022**, *3*, 100340. [CrossRef] [PubMed]
19. Yanamadala, S.; Morrison, D.; Curtin, C.; McDonald, K.; Hernandez-Boussard, T. Electronic Health Records and Quality of Care: An Observational Study Modeling Impact on Mortality, Readmissions, and Complications. *Medicine* **2016**, *95*, e3332. [CrossRef]
20. Evans, R.S. Electronic Health Records: Then, Now, and in the Future. *Yearb. Med. Inform.* **2016**, *25*, S48–S61. [CrossRef]
21. Zhou, L.; Soran, C.S.; Jenter, C.A.; Volk, L.A.; Orav, E.J.; Bates, D.W.; Simon, S.R. The relationship between electronic health record use and quality of care over time. *J. Am. Med. Inform. Assoc.* **2009**, *16*, 457–464. [CrossRef]
22. Pons, E.; Braun, L.M.M.; Hunink, M.G.M.; Kors, J.A. Natural Language Processing in Radiology: A Systematic Review. *Radiology* **2016**, *279*, 329–343. [CrossRef]
23. Petch, J.; Batt, J.; Murray, J.; Mamdani, M. Extracting Clinical Features From Dictated Ambulatory Consult Notes Using a Commercially Available Natural Language Processing Tool: Pilot, Retrospective, Cross-Sectional Validation Study. *JMIR Med. Inform.* **2019**, *7*, e12575. [CrossRef] [PubMed]
24. Luo, Y.; Thompson, W.K.; Herr, T.M.; Zeng, Z.; Berendsen, M.A.; Jonnalagadda, S.R.; Carson, M.B.; Starren, J. Natural Language Processing for EHR-Based Pharmacovigilance: A Structured Review. *Drug Saf.* **2017**, *40*, 1075–1089. [CrossRef] [PubMed]

25. Petch, J.; Kempainnen, J.; Pettengell, C.; Aviv, S.; Butler, B.; Pond, G.; Saha, A.; Bogach, J.; Allard-Coutu, A.; Sztur, P.; et al. Developing a Data and Analytics Platform to Enable a Breast Cancer Learning Health System at a Regional Cancer Center. *JCO Clin. Cancer Inform.* **2023**, *7*, e2200182. [CrossRef] [PubMed]

26. Carrell, D.S.; Cronkite, D.; Palmer, R.E.; Saunders, K.; Gross, D.E.; Masters, E.T.; Hylan, T.R.; Von Korff, M. Using natural language processing to identify problem usage of prescription opioids. *Int. J. Med. Inform.* **2015**, *84*, 1057–1064. [CrossRef] [PubMed]

27. Melton, G.B.; Hripcsak, G. Automated detection of adverse events using natural language processing of discharge summaries. *Am. Med. Inform. Assoc.* **2005**, *12*, 448–457. [CrossRef] [PubMed]

28. Vender, R.; Lynde, C. *AI-Powered Patient Identification to Optimize Care*; Canadian Dermatology Association: Ottawa, ON, Canada, 2023.

29. Birtwhistle, R.; Williamson, T. Primary care electronic medical records: A new data source for research in Canada. *CMAJ: Can. Med. Assoc. J.* **2015**, *187*, 239. [CrossRef] [PubMed]

30. Law, J.H.; Pettengell, C.; Le, L.W.; Aviv, S.; DeMarco, P.; Merritt, D.C.; Lau, S.C.M.; Sacher, A.G.; Leighl, N.B. Generating real-world evidence: Using automated data extraction to replace manual chart review. *J. Clin. Oncol.* **2019**, *37*, e18096. [CrossRef]

31. Law, J.; Pettengell, C.; Chen, L.; Le, L.; Sung, M.; Aviv, S.; Lau, S.; Sacher, A.; Merritt, D.; Demarco, P.; et al. EP1.16-05 Real World Outcomes of Advanced NSCLC Patients with Liver Metastases. *J. Thorac. Oncol.* **2019**, *14*, S1066. [CrossRef]

32. Pettengell, C.; Law, J.; Le, L.; Sung, M.; Lau, S.; Sacher, A.; Merritt, D.; Demarco, P.; Leighl, N. P1.16-07 Real World Evidence of the Impact of Immunotherapy in Patients with Advanced Lung Cancer. *J. Thorac. Oncol.* **2019**, *14*, S588. [CrossRef]

33. Cheung, W.Y.; Farrer, C.; Darwish, L.; Pettengell, C.; Stewart, E.L. 82P Exploring treatment patterns and outcomes of patients with advanced lung cancer (aLC) using artificial intelligence (AI)-extracted data. *Ann. Oncol.* **2021**, *32*, S1407. [CrossRef]

34. Grant, B.M.; Zarrin, A.; Zhan, L.; Ajaj, R.; Darwish, L.; Khan, K.; Patel, D.; Chiasson, K.; Balaratnam, K.; Chowdhury, M.T.; et al. Abstract 4087: Developing a standardized framework for curating oncology datasets generated by manual abstraction and artificial intelligence. *Cancer Res.* **2022**, *82*, 4087. [CrossRef]

35. Mycock, K.; Zhan, L.; Taylor-Stokes, G.; Milligan, G.; Mitra, D. Real-World Palbociclib Use in HR+/HER2− Advanced Breast Cancer in Canada: The IRIS Study. *Curr. Oncol.* **2021**, *28*, 678–688. [CrossRef] [PubMed]

36. Tripathy, D.; Rocque, G.; Blum, J.L.; Karuturi, M.S.; McCune, S.; Kurian, S.; Moezi, M.M.; Anderson, D.; Gauthier, E.; Zhang, Z.; et al. 251P Real-world clinical outcomes of palbociclib plus endocrine therapy (ET) in hormone receptor–positive advanced breast cancer: Results from the POLARIS trial. *Ann. Oncol.* **2022**, *33*, S651–S652. [CrossRef]

37. Siddique, S.; Chow, J. Artificial intelligence in radiotherapy. *Rep. Pract. Oncol. Radiother.* **2020**, *25*, 656–666. [CrossRef]

38. Schork, N.J. Artificial Intelligence and Personalized Medicine. In *Precision Medicine in Cancer Therapy*; Von Hoff, D.D., Han, H., Eds.; Springer International Publishing: Cham, Switzerland, 2019; pp. 265–283.

39. Davenport, T.; Kalakota, R. The potential for artificial intelligence in healthcare. *Future Healthc. J.* **2019**, *6*, 94. [CrossRef] [PubMed]

40. Johnson, A.E.W.; Pollard, T.J.; Shen, L.; Lehman, L.H.; Feng, M.; Ghassemi, M.; Moody, B.; Szolovits, P.; Anthony Celi, L.; Mark, R.G. MIMIC-III, a freely accessible critical care database. *Sci. Data* **2016**, *3*, 160035. [CrossRef]

41. Turchin, A.; Masharsky, S.; Zitnik, M. Comparison of BERT implementations for natural language processing of narrative medical documents. *Inform. Med. Unlocked* **2023**, *36*, 101139. [CrossRef]

42. Chase, H.S.; Mitrani, L.R.; Lu, G.G.; Fulgieri, D.J. Early recognition of multiple sclerosis using natural language processing of the electronic health record. *BMC Med. Inform. Decis. Mak.* **2017**, *17*, 24. [CrossRef] [PubMed]

43. Afzal, N.; Sohn, S.; Abram, S.; Liu, H.; Kullo, I.J.; Arruda-Olson, A.M. Identifying Peripheral Arterial Disease Cases Using Natural Language Processing of Clinical Notes. In Proceedings of the 2016 IEEE-EMBS International Conference on Biomedical and Health Informatics (BHI), Las Vegas, NV, USA, 24–27 February 2016.

44. Banerjee, I.; Bozkurt, S.; Caswell-Jin, J.L.; Kurian, A.W.; Rubin, D.L. Natural Language Processing Approaches to Detect the Timeline of Metastatic Recurrence of Breast Cancer. *JCO Clin. Cancer Inform.* **2019**, *3*, 1–12. [CrossRef]

45. Benedum, C.M.; Sondhi, A.; Fidyk, E.; Cohen, A.B.; Nemeth, S.; Adamson, B.; Estévez, M.; Bozkurt, S. Replication of Real-World Evidence in Oncology Using Electronic Health Record Data Extracted by Machine Learning. *Cancers* **2023**, *15*, 1853. [CrossRef]

MDPI
St. Alban-Anlage 66
4052 Basel
Switzerland
www.mdpi.com

MDPI Books Editorial Office
E-mail: books@mdpi.com
www.mdpi.com/books